Classics in Mathematics

Richard Courant • Fritz John

Introduction to Calculus and Analysis
Volume II/1

Springer
Berlin
Heidelberg
New York
Barcelona
Hong Kong
London
Milan
Paris
Singapore
Tokyo

Richard Courant was born in 1888 in a small town of what is now Poland, and died in New Rochelle, N.Y. in 1972. He received his doctorate from the legendary David Hilbert in Göttingen, where later he founded and directed its famed Mathematics Institute, a Mecca for mathematicians in the twenties. In 1933 the Nazi government dismissed Courant for being Jewish, and he emigrated to the United States. He found, in New York, what he called "a reservoir of talent" to be tapped. He built, at New York University, a new Mathematical Sciences Institute that shares the philosophy of its illustrious predecessor and rivals it in worldwide influence.

For Courant mathematics was an adventure, with applications forming a vital part. This spirit is reflected in his books, in particular in his influential calculus text, revised in collaboration with his brilliant younger colleague, Fritz John. *(P. D. Lax)*

Fritz John was born on June 14, 1910, in Berlin. After his school years in Danzig (now Gdansk, Poland), he studied in Göttingen and received his doctorate in 1933, just when the Nazi regime came to power. As he was half-Jewish and his bride Aryan, he had to flee Germany in 1934. After a year in Cambridge, UK, he accepted a position at the University of Kentucky, and in 1946 joined Courant, Friedrichs and Stoker in building up at New York University the institute that later became the Courant Institute of Mathematical Sciences. He remained there until his death in New Rochelle on February 10, 1994.

John's research and the books he wrote had a strong impact on the development of many fields of mathematics, foremost in partial differential equations. He also worked on Radon transforms, ill-posed problems, convex geometry, numerical analysis, elasticity theory. In connection with his work in the latter field, he and Nirenberg introduced the space of the BMO-functions (bounded mean oscillations). Fritz John's work exemplifies the unity of mathematics as well as its elegance and its beauty. *(J. Moser)*

Richard Courant • Fritz John

Introduction to Calculus and Analysis

Volume II/1
Chapters 1-4
Reprint of the 1989 Edition

Springer

Originally published in 1974 by Interscience Publishers, a division of John Wiley and Sons, Inc.
Reprinted in 1989 by Springer-Verlag New York, Inc.

Mathematics Subject Classification (1991): 26xx, 26-01

Cataloging-in-Publication Data applied for

Die Deutsche Bibliothek - CIP-Einheitsaufnahme
Courant, Richard:
Introduction to calculus and analysis / Richard Courant; Fritz John.- Reprint.- Berlin; Heidelberg; New York; Barcelona; Hong Kong; London; Milan; Paris; Singapore; Tokyo: Springer
(Classics in mathematics)
Vol.2. / With the assistance of Albert A. Blank and Alan Solomon 1. Chapter 1-4.- Reprint of the 1989 ed.- 2000
ISBN 3-540-66569-2

Photograph of Richard Courant from: C. Reid, *Courant in Göttingen and New York. The Story of an Improbable Mathematician*, Springer New York, 1976

Photograph of Fritz John by kind permission of The Courant Institute of Mathematical Sciences, New York

ISSN 1431-0821
ISBN 3-540-66569-2 Springer-Verlag Berlin Heidelberg New York

Springer-Verlag Berlin Heidelberg New York
a member of BertelsmannSpringer Science+Business Media GmbH

SPIN 10991329 41/3111CK - 5 4 3 2 - Printed on acid-free paper

Richard Courant Fritz John

Introduction to Calculus and Analysis

Volume II

With the assistance of
Albert A. Blank and Alan Solomon

With 120 Illustrations

Springer

Richard Courant (1888 - 1972) Fritz John
Courant Institute of Mathematical Sciences
New York University
New York, NY 10012

Originally published in 1974 by Interscience Publishers, a division of John Wiley and Sons, Inc.

Mathematical Subject Classification: 26xx, 26-01

Printed on acid-free paper.

Printed and bound by Edwards Brothers, Inc., Ann Arbor, MI.
Printed in the United States of America.

9 8 7 6 5 4 3

ISBN 0-387-97152-1 Springer-Verlag New York Berlin Heidelberg
ISBN 0-540-97152-1 Springer-Verlag Berlin Heidelberg New York SPIN 10691322

Preface

Richard Courant's Differential and Integral Calculus, Vols. I and II, has been tremendously successful in introducing several generations of mathematicians to higher mathematics. Throughout, those volumes presented the important lesson that meaningful mathematics is created from a union of intuitive imagination and deductive reasoning. In preparing this revision the authors have endeavored to maintain the healthy balance between these two modes of thinking which characterized the original work. Although Richard Courant did not live to see the publication of this revision of Volume II, all major changes had been agreed upon and drafted by the authors before Dr. Courant's death in January 1972.

From the outset, the authors realized that Volume II, which deals with functions of several variables, would have to be revised more drastically than Volume I. In particular, it seemed desirable to treat the fundamental theorems on integration in higher dimensions with the same degree of rigor and generality applied to integration in one dimension. In addition, there were a number of new concepts and topics of basic importance, which, in the opinion of the authors, belong to an introduction to analysis.

Only minor changes were made in the short chapters (6, 7, and 8) dealing, respectively, with Differential Equations, Calculus of Variations, and Functions of a Complex Variable. In the core of the book, Chapters 1–5, we retained as much as possible the original scheme of two roughly parallel developments of each subject at different levels: an informal introduction based on more intuitive arguments together with a discussion of applications laying the groundwork for the subsequent rigorous proofs.

The material from linear algebra contained in the original Chapter 1 seemed inadequate as a foundation for the expanded calculus structure. Thus, this chapter (now Chapter 2) was completely rewritten and now presents all the required properties of nth order determinants and matrices, multilinear forms, Gram determinants, and linear manifolds.

The new Chapter 1 contains all the fundamental properties of linear differential forms and their integrals. These prepare the reader for the introduction to higher-order exterior differential forms added to Chapter 3. Also found now in Chapter 3 are a new proof of the implicit function theorem by successive approximations and a discussion of numbers of critical points and of indices of vector fields in two dimensions.

Extensive additions were made to the fundamental properties of multiple integrals in Chapters 4 and 5. Here one is faced with a familiar difficulty: integrals over a manifold M, defined easily enough by subdividing M into convenient pieces, must be shown to be independent of the particular subdivision. This is resolved by the systematic use of the family of Jordan measurable sets with its finite intersection property and of partitions of unity. In order to minimize topological complications, only manifolds imbedded smoothly into Euclidean space are considered. The notion of "orientation" of a manifold is studied in the detail needed for the discussion of integrals of exterior differential forms and of their additivity properties. On this basis, proofs are given for the divergence theorem and for Stokes's theorem in n dimensions. To the section on Fourier integrals in Chapter 4 there has been added a discussion of Parseval's identity and of multiple Fourier integrals.

Invaluable in the preparation of this book was the continued generous help extended by two friends of the authors, Professors Albert A. Blank of Carnegie-Mellon University, and Alan Solomon of the University of the Negev. Almost every page bears the imprint of their criticisms, corrections, and suggestions. In addition, they prepared the problems and exercises for this volume.[1]

Thanks are due also to our colleagues, Professors K. O. Friedrichs and Donald Ludwig for constructive and valuable suggestions, and to John Wiley and Sons and their editorial staff for their continuing encouragement and assistance.

FRITZ JOHN
New York
September 1973

[1]In contrast to Volume I, these have been incorporated completely into the text; their solutions can be found at the end of the volume.

Contents

Chapter 1 Functions of Several Variables and Their Derivatives

1.1 Points and Points Sets in the Plane and in Space **1**
a. Sequences of points. Convergence, **1** **b.** Sets of points in the plane, **3** **c.** The boundary of a set. Closed and open sets, **6** **d.** Closure as set of limit points, **9** **e.** Points and sets of points in space, **9**

1.2 Functions of Several Independent Variables **11**
a. Functions and their domains, **11** **b.** The simplest types of functions, **12** **c.** Geometrical representation of functions, **13**

1.3 Continuity **17**
a. Definition, **17** **b.** The concept of limit of a function of several variables, **19** **c.** The order to which a function vanishes, **22**

1.4 The Partial Derivatives of a Function **26**
a. Definition. Geometrical representation, **26** **b.** Examples, **32** **c.** Continuity and the existence of partial derivatives, **34**

d. Change of the order of differentiation, **36**

1.5 **The Differential of a Function and Its Geometrical Meaning** **40**
a. The concept of differentiability, **40** b. Directional derivatives, **43** c. Geometric interpretation of differentiability, The tangent plane, **46** d. The total differential of a function, **49** e. Application to the calculus of errors, **52**

1.6 **Functions of Functions (Compound Functions) and the Introduction of New Independent Variables** **53**
a. Compound functions. The chain rule, **53** b. Examples, **59** c. Change of independent variables, **60**

1.7 **The Mean Value Theorem and Taylor's Theorem for Functions of Several Variables** **64**
a. Preliminary remarks about approximation by polynomials, **64**
b. The mean value theorem, **66**
c. Taylor's theorem for several independent variables, **68**

1.8 **Integrals of a Function Depending on a Parameter** **71**
a. Examples and definitions, **71**
b. Continuity and differentiability of an integral with respect to the parameter, **74** c. Interchange of integrations. Smoothing of functions, **80**

1.9 **Differentials and Line Integrals** **82**
a. Linear differential forms, **82**

b. Line integrals of linear differential forms, **85** **c.** Dependence of line integrals on endpoints, **92**

1.10 The Fundamental Theorem on Integrability of Linear Differential Forms **95**
a. Integration of total differentials, **95** **b.** Necessary conditions for line integrals to depend only on the end points, **96** **c.** Insufficiency of the integrability conditions, **98** **d.** Simply connected sets, **102** **e.** The fundamental theorem, **104**

APPENDIX

A.1. The Principle of the Point of Accumulation in Several Dimensions and Its Applications **107**
a. The principle of the point of accumulation, **107** **b.** Cauchy's convergence test. Compactness, **108** **c.** The Heine-Borel covering theorem, **109** **d.** An application of the Heine-Borel theorem to closed sets contains in open sets, **110.**

A.2. Basic Properties of Continuous Functions **112**

A.3. Basic Notions of the Theory of Point Sets **113**
a. Sets and sub-sets, **113** **b.** Union and intersection of sets, **115** **c.** Applications to sets of points in the plane, **117.**

A.4. Homogeneous functions. **119**

Chapter 2 Vectors, Matrices, Linear Transformations

2.1 **Operations with Vectors** **122**
a. Definition of vectors, **122** b. Geometric representation of vectors, **124** c. Length of vectors. Angles between directions, **127** d. Scalar products of vectors, **131** e. Equation of hyperplanes in vector form, **133** f. Linear dependence of vectors and systems of linear equations, **136**

2.2 **Matrices and Linear Transformations** **143**
a. Change of base. Linear spaces, **143** b. Matrices, **146** c. Operations with matrices, **150** d. Square matrices. The reciprocal of a matrix. Orthogonal matrices. **153**

2.3 **Determinants** **159**
a. Determinants of second and third order, **159** b. Linear and multilinear forms of vectors, **163** c. Alternating multilinear forms. Definition of determinants, **166** d. Principal properties of determinants, **171** e. Application of determinants to systems of linear equations. **175**

2.4 **Geometrical Interpretation of Determinants** **180**
a. Vector products and volumes of parallelepipeds in three-dimensional space, **180** b. Expansion of a determinant with respect to a column. Vector products in higher dimensions, **187** c. Areas of parallelograms and volumes of parallelepipeds in

higher dimensions, **190** **d.** Orientation of parallelepipeds in n-dimensional space, **195** **e.** Orientation of planes and hyperplanes, **200** **f.** Change of volume of parallelepipeds in linear transformations, **201**

2.5 Vector Notions in Analysis **204**
a. Vector fields, **204** **b.** Gradient of a scalar, **205** **c.** Divergence and curl of a vector field, **208** **d.** Families of vectors. Application to the theory of curves in space and to motion of particles, **211**

Chapter 3 Developments and Applications of the Differential Calculus

3.1 Implicit Functions **218**
a. General remarks, **218** **b.** Geometrical interpretation, **219** **c.** The implicit function theorem, **221** **d.** Proof of the implicit function theorem, **225** **e.** The implicit function theorem for more than two independent variables, **228**

3.2 Curves and Surfaces in Implicit Form **230**
a. Plane curves in implicit form, **230** **b.** Singular points of curves, **236** **c.** Implicit representation of surfaces, **238**

3.3 Systems of Functions, Transformations, and Mappings **241**
a. General remarks, **241** **b.** Curvilinear coordinates, **246** **c.** Extension to more than two independent variables, **249** **d.** Differentiation formulae for the inverse functions,

252 e. Symbolic product of mappings, **257** f. General theorem on the inversion of transformations and of systems of implicit functions. Decomposition into primitive mappings, **261** g. Alternate construction of the inverse mapping by the method of successive approximations, **266** h. Dependent functions, **268** i. Concluding remarks, **275**

3.4 **Applications** 278
a. Elements of the theory of surfaces, **278** b. Conformal transformation in general, **289**

3.5 **Families of Curves, Families of Surfaces, and Their Envelopes** 290
a. General remarks, **290** b. Envelopes of one-parameter families of curves, **292** c. Examples, **296** d. Endevelopes of families of surfaces, **303**

3.6 **Alternating Differential Forms** 307
a. Definition of alternating differential forms, **307** b. Sums and products of differential forms, **310** c. Exterior derivatives of differential forms, **312** d. Exterior differential forms in arbitrary coordinates, **316**

3.7 **Maxima and Minima** 325
a. Necessary conditions, **325** b. Examples, **327** c. Maxima and minima with subsidiary conditions, **330** d. Proof of the method of undetermined multipliers in the simplest case, **334** e. Generalization of the method of undetermined multipliers, **337** f. Examples, **340**

APPENDIX

A.1 Sufficient Conditions for Extreme Values **345**

A.2 Numbers of Critical Points Related to Indices of a Vector Field **352**

A.3 Singular Points of Plane Curves **360**

A.4 Singular Points of Surfaces **362**

A.5 Connection Between Euler's and Lagrange's Representation of the motion of a Fluid **363**

A.6 Tangential Representation of a Closed Curve and the Isoperimetric Inequality **365**

Chapter 4 Multiple Integrals

4.1 Areas in the Plane **367**
a. Definition of the Jordan measure of area, **367** **b.** A set that does not have an area, **370** **c.** Rules for operations with areas, **372**

4.2 Double Integrals **374**
a. The double integral as a volume, **374** **b.** The general analytic concept of the integral, **376** **c.** Examples, **379** **d.** Notation. Extensions. Fundamental rules, **381** **e.** Integral estimates and the mean value theorem, **383**

4.3 Integrals over Regions in three and more Dimensions **385**

4.4 Space Differentiation. Mass and Density **386**

4.5 Reduction of the Multiple Integral to Repeated Single Integrals **388**
a. Integrals over a rectangle, **388** **b.** Change of order of integration. Differentiation under the integral sign, **390** **c.** Reduction of double integrals to single integrals for more general regions, **392** **d.** Extension of the results to regions in several dimensions, **397**

4.6 Transformation of Multiple Integrals **398**
a. Transformation of integrals in the plane, **398** **b.** Regions of more than two dimensions, **403**

4.7 Improper Multiple Integrals **406**
a. Improper integrals of functions over bounded sets, **407** **b.** Proof of the general convergence theorem for improper integrals, **411**
c. Integrals over unbounded regions, **414**

4.8 Geometrical Applications **417**
a. Elementary calculation of volumes, **417** **b.** General remarks on the calculation of volumes. Solids of revolution. Volumes in spherical coordinates, **419** **c.** Area of a curved surface, **421**

4.9 Physical Applications **431**
a. Moments and center of mass, **431** **b.** Moments of inertia, **433**
c. The compound pendulum, **436**
d. Potential of attracting masses, **438**

4.10 Multiple Integrals in Curvilinear Coordinates **445**

a. Resolution of multiple integrals, **445** b. Application to areas swept out by moving curves and volumes swept out by moving surfaces. Guldin's formula. The polar planimeter, **448**

4.11 Volumes and Surface Areas in Any Number of Dimensions **453**

a. Surface areas and surface integrals in more than three dimensions, **453** b. Area and volume of the *n*-dimensional sphere, **455** c. Generalizations. Parametric Representations, **459**

4.12 Improper Single Integrals as Functions of a Parameter **462**

a. Uniform convergence. Continuous dependence on the parameter, **462** b. Integration and differentiation of improper integrals with respect to a parameter, **466** c. Examples, **469** d. Evaluation of Fresnel's integrals, **473**

4.13 The Fourier Integral **476**

a. Introduction, **476** b. Examples, **479** c. Proof of Fourier's integral theorem, **481** d. Rate of convergence in Fourier's integral theorem, **485** e. Parseval's identity for Fourier transforms, **488** f. The Fourier transformation for functions of several variables, **490**

4.14 The Eulerian Integrals (Gamma Function) **497**

a. Definition and functional equation, **497** **b.** Convex functions. Proof of Bohr and Mollerup's theorem, **499** **c.** The infinite products for the gamma function, **503** **d.** The nextensio theorem, **507** **e.** The beta function, **508** **f.** Differentiation and integration of fractional order. Abel's integral equation, **511**

APPENDIX: DETAILED ANALYSIS OF THE PROCESS OF INTEGRATION

A.1 Area **515**
a. Subdivisions of the plane and the corresponding inner and outer areas, **515** **b.** Jordan-measurable sets and their areas, **517** **c.** Basic properties of areas, **519**

A.2 Integrals of Functions of Several Variables **524**
a. Definition of the integral of a function $f(x, y)$, **524** **b.** Integrability of continuous functions and integrals over sets, **526** **c.** Basic rules for multiple integrals, **528** **d.** Reduction of multiple integrals to repeated single integrals, **531**

A.3 Transformation of Areas and Integrals **534**
a. Mappings of sets, **534** **b.** Trans formation of multiple integrals, **539**

A.4 Note on the Definition of the Area of a Curved Surface **540**

Chapter 5 Relations Between Surface and Volume Integrals

5.1 Connection Between Line Integrals and Double Integrals in the Plane (The Integral Theorems of Gauss, Stokes, and Green) **543**

5.2 Vector Form of the Divergence Theorem. Stokes's Theorem **551**

5.3 Formula for Integration by Parts in Two Dimensions. Green's Theorem **556**

5.4 The Divergence Theorem Applied to the Transformation of Double Integrals **558**
a. The case of 1–1 mappings, **558**
b. Transformation of integrals and degree of mapping, **561**

5.5 Area Differentiation. Transformation of Δu to Polar Coordinates **565**

5.6 Interpretation of the Formulae of Gauss and Stokes by Two-Dimensional Flows **569**

5.7 Orientation of Surfaces **575**
a. Orientation of two-dimensional surfaces in three-space, **575** **b.** Orientation of curves on oriented surfaces, **587**

5.8 Integrals of Differential Forms and of Scalars over Surfaces **589**
a. Double integrals over oriented plane regions, **589** **b.** Surface

integrals of second-order differential forms, **592** **c.** Relation between integrals of differential forms over oriented surfaces to integrals of scalars over unoriented surfaces, **594**

5.9 Gauss's and Green's Theorems in Space **597**
a. Gauss's theorem, **597** **b.** Application of Gauss's theorem to fluid flow, **602** **c.** Gauss's theorem applied to space forces and surface forces, **605** **d.** Integration by parts and Green's theorem in three dimensions, **607** **e.** Application of Green's theorem to the transformation of ΔU to spherical coordinates, **608**

5.10 Stokes's Theorem in Space **611**
a. Statement and proof of the theorem, **611** **b.** Interpretation of Stokes's theorem, **615**

5.11 Integral Identities in Higher Dimensions **622**

APPENDIX: GENERAL THEORY OF SURFACES AND OF SURFACE INTEGALS

A.1 Surfaces and Surface Integrals in Three dimensions **624**
a. Elementary surfaces, **624** **b.** Integral of a function over an elementary surface, **627** **c.**Oriented elementary surfaces, **629** **d.** Simple surfaces, **631** **e.** Partitions of unity and integrals over simple surfaces, **634**

A.2 The Divergence Theorem 637
a. Statement of the theorem and its invariance, **637** **b.** Proof of the theorem, **639**

A.3 Stokes's Theorem 642

A.4 Surfaces and Surface Integrals in Euclidean Spaces of Higher Dimensions 645
a. Elementary surfaces, **645**
b. Integral of a differential form over an oriented elementary surface, **647**
c. Simple m-dimensional surfaces, **648**

A.5 Integrals over Simple Surfaces, Gauss's Divergence Theorem, and the General Stokes Formula in Higher Dimensions 651

Chapter 6 Differential Equations

6.1 The Differential Equations for the Motion of a Particle in Three Dimensions 654
a. The equations of motion, **654**
b. The principle of conservation of energy, **656** **c.** Equilibrium. Stability, **659** **d.** Small oscillations about a position of equilibrium, **661**
e. Planetary motion, **665** **f.** Boundary value problems. The loaded cable and the loaded beam, **672**

6.2 The General Linear Differential Equation of the First Order 678
a. Separation of variables, **678**
b. The linear first-order equation, **680**

6.3 Linear Differential Equations of Higher Order **683**
a. Principle of superposition. General solutions, **683** **b.** Homogeneous differential equations of the second second order, **688** **c.** The non-homogeneous differential equations. Method of variation of parameters, **691**

6.4 General Differential Equations of the First Order **697**
a. Geometrical interpretation, **697** **b.** The differential equation of a family of curves. Singular solutions. Orthogonal trajectories, **699** **c.** Theorem of the existence and uniqueness of the solution, **702**

6.5 Systems of Differential Equations and Differential Equations of Higher Order **709**

6.6 Integration by the Method of Undermined Coefficients **711**

6.7 The Potential of Attracting Charges and Laplace's Equation **713**
a. Potentials of mass distributions, **713** **b.** The differential equation of the potential, **718** **c.** Uniform double layers, **719** **d.** The mean value theorem, **722** **e.** Boundary value problem for the circle. Poisson's integral, **724**

6.8 Further Examples of Partial Differential Equations from Mathematical Physics **727**
a. The wave equation in one dimension, **727** **b.** The wave equation

in three-dimensional space, **728**
c. Maxwell's equations in free space, **731**

Chapter 7 Calculus of Variations

7.1 Functions and Their Extrema 737

7.2 Necessary conditions for Extreme Values of a Functional 741
a. Vanishing of the first variation, **741** **b.** Deduction of Euler's differential equation, **743** **c.** Proofs of the fundamental lemmas, **747** **d.** Solution of Euler's differential equation in special cases. Examples, **748** **e.** Identical vanishing of Euler's expression, **752**

7.3 Generalizations 753
a. Integrals with more than one argument function, **753** **b.** Examples, **755** **c.** Hamilton's principle. Lagrange's equations, **757** **d.** Integrals involving higher derivatives, **759** **e.** Several independent variables, **760**

7.4 Problems Involving Subsidiary Conditions. Lagrange Multipliers 762
a. Ordinary subsidiary conditions, **762** **b.** Other types of subsidiary conditions, **765**

Chapter 8 Functions of a Complex Variable

8.1 Complex Functions Represented by Power Series 769
a. Limits and infinite series with complex terms, **769** **b.** Power

series, **772** **c.** Differentiation and integration of power series, **773** **d.** Examples of power series, **776**

8.2 Foundations of the General Theory of Functions of a Complex Variable **778**
a. The postulate of differentiability, **778** **b.** The simplest operations of the differential calculus, **782** **c.** Conformal transformation. Inverse functions, **785**

8.3 The Integration of Analytic Functions **787**
a. Definition of the integral, **787** **b.** Cauchy's theorem, **789** **c.** Applications. The logarithm, the exponential function, and the general power function, **792**

8.4 Cauchy's Formula and Its Applications **797**
a. Cauchy's formula, **797** **b.** Expansion of analytic functions in power series, **799** **c.** The theory of functions and potential theory, **802** **d.** The converse of Cauchy's theorem, **803** **e.** Zeros, poles, and residues of an analytic function, **803**

8.5 Applications to Complex Integration (Contour Integration) **807**
a. Proof of the formula (8.22), **807** **b.** Proof of the formula (8.22), **808** **c.** Application of the theorem of residues to the integration of rational functions, **809** **d.** The theorem of residues and linear differential equations with constant coefficients, **812**

8.6 Many-Valued Functions and Analytic Extension 814

List of Biographical Dates 941*

Index 943**

* page 543 of this edition
** page 545 of this edition

Introduction to Calculus and Analysis

Volume II

CHAPTER 1

Functions of Several Variables and Their Derivatives

The concepts of limit, continuity, derivative, and integral, as developed in Volume I, are also basic in two or more independent variables. However, in higher dimensions many new phenomena, which have no counterpart at all in the theory of functions of a single variable, must be dealt with. As a rule, a theorem that can be proved for functions of *two* variables may be extended easily to functions of more than two variables without any essential change in the proof. In what follows, therefore, we often confine ourselves to functions of two variables, where relations are much more easily visualized geometrically, and discuss functions of three or more variables only when some additional insight is gained thereby; this also permits simpler geometrical interpretations of our results.

1.1 Points and Point Sets in the Plane and in Space

a. Sequences of Points: Convergence

An ordered pair of values (x, y) can be represented geometrically by the point P having x and y as coordinates in some Cartesian coordinate system. The distance between two points $P = (x, y)$ and $P' = (x', y')$ is given by the formula

$$\overline{PP'} = \sqrt{(x' - x)^2 + (y' - y)^2},$$

which is basic for euclidean geometry. We use the notion of distance to define the neighborhoods of a point. The ε-*neighborhood* of a point

$C = (\alpha, \beta)$ consists of all the points $P = (x, y)$ whose distance from C is less than ε; geometrically this is the circular disk[1] of center C and radius ε that is described by the inequality

$$(x - \alpha)^2 + (y - \beta)^2 < \varepsilon^2.$$

We shall consider *infinite sequences* of points

$$P_1 = (x_1, y_1), \; P_2 = (x_2, y_2), \; \ldots, \qquad P_n = (x_n, y_n), \; \ldots$$

For example, $P_n = (n, n^2)$ defines a sequence all of whose points lie on the parabola $y = x^2$. The points in a sequence do not all have to be distinct. For example, the infinite sequence $P_n = (2, (-1)^n)$ has only two distinct elements.

The sequence $P_1, P_2, \ldots$ is *bounded* if a disk can be found containing all of the P_n, that is, if there is a point Q and a number M such that $\overline{P_nQ} < M$ for all n. Thus the sequence $P_n = (1/n, 1/n^2)$ is bounded, and the sequence (n, n^2), unbounded.

The most important concept associated with sequences is that of *convergence*. We say that a sequence of points $P_1, P_2, \ldots$ converges to a point Q, or that

$$\lim_{n\to\infty} P_n = Q,$$

if the distances $\overline{P_nQ}$ converge to 0. Thus, $\lim_{n\to\infty} P_n = Q$ means that for every $\varepsilon > 0$ there exists a number N such that P_n lies in the ε-neighborhood of Q for all $n > N$.[2]

For example, for the sequence of points defined by $P_n = (e^{-n/4} \cos n, e^{-n/4} \sin n)$, we have $\lim_{n\to\infty} P_n = (0, 0) = Q$, since here

$$\overline{P_nQ} = e^{-n/4} \longrightarrow 0 \qquad \text{for} \qquad n \longrightarrow \infty \cdot$$

We note that the P_n approach the origin Q along the logarithmic spiral with equation $r = e^{-\theta/4}$ in polar coordinates r, θ (see Fig. 1.1).

Convergence of the sequence of points $P_n = (x_n, y_n)$ to the point

[1]The word "circle," as used ordinarily, is ambiguous, referring either to a curve or to the region bounded by it. We shall follow the current practice of reserving the term "circle" for the curve only, and the term "circular region" or "disk" for the two-dimensional region. Similarly, in space we distinguish the "sphere" (i.e., the spherical surface) from the solid three-dimensional "ball" that it bounds.

[2]Equivalently, any disk with center Q contains all but a finite number of the P_n. The notation $P_n \to Q$ for $n \to \infty$ will also be used.

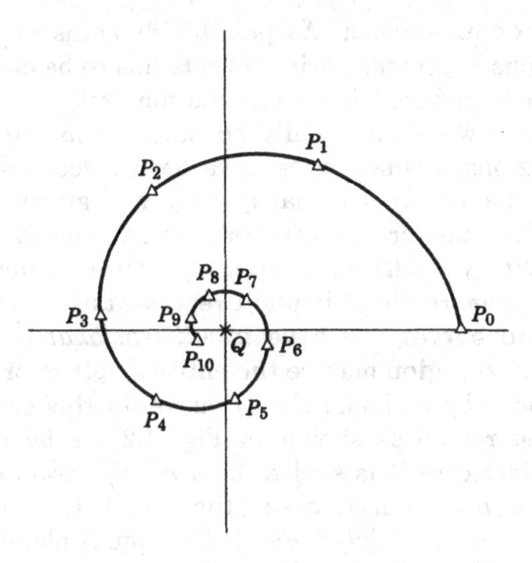

Figure 1.1 Converging sequence P_n.

$Q = (a, b)$ means that the two sequences of numbers x_n and y_n *converge separately* and that

$$\lim_{n\to\infty} x_n = a, \qquad \lim_{n\to\infty} y_n = b.$$

Indeed, smallness of $\overline{P_nQ}$ implies that both $x_n - a$ and $y_n - b$ are small, since $|x_n - a| \leqq \overline{P_nQ}$, $|y_n - b| \leqq \overline{P_nQ}$; conversely,

$$\overline{P_nQ} = \sqrt{(x_n - a)^2 + (y_n - b)^2} \leqq |x_n - a| + |y_n - b|,$$

so that $P_nQ \longrightarrow 0$ when both $x_n \longrightarrow a$ and $y_n \longrightarrow b$.

Just as in the case of sequences of numbers, we can prove that a sequence of points converges, without knowing the limit, using *Cauchy's intrinsic convergence test.* In two dimensions this asserts: For the convergence of a sequence of points $P_n = (x_n, y_n)$ it is necessary and sufficient that for every $\varepsilon > 0$ the inequality $\overline{P_nP_m} < \varepsilon$ holds for all n, m exceeding a suitable value $N = N(\varepsilon)$. The proof follows immediately by applying the Cauchy test for sequences of numbers to each of the sequences x_n and y_n.

b. Sets of Points in the Plane

In the study of functions of a single variable x we generally permitted x to vary over an "interval," which could be either closed or open, bounded or unbounded. As possible domains of functions in higher dimensions, a greater variety of sets has to be considered and terms have to be introduced describing the simplest properties of such sets. In the plane we shall usually consider either curves or two-dimensional regions. Plane curves have been discussed extensively in Volume I (Chapter 4). Ordinarily they are given either "non-parametrically" in the form $y = f(x)$ or "parametrically" by a pair of functions $x = \phi(t)$, $y = \psi(t)$, or "implicitly" by an equation $F(x, y) = 0$ (we shall say more about implicit representations in Chapter 3).

In addition to curves, we have *two-dimensional* sets of points, forming a *region*. A region may be the entire *xy*-plane or a portion of the plane bounded by a simple closed curve (in this case forming a *simply connected* region as shown in Fig. 1.2) or by several such curves. In the last case it is said to be a *multiply connected* region, the number of boundary curves giving the so-called *connectivity*; Fig. 1.3, for example, shows a *triply connected* region. A plane set may not be connected[1] at all, consisting of several separate portions (Fig. 1.4).

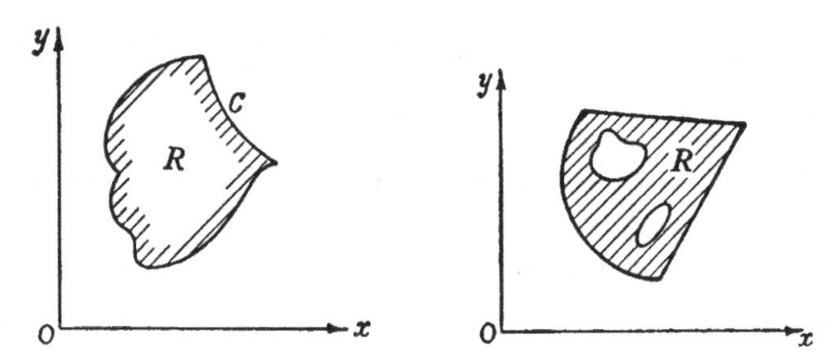

Figure 1.2 A simply connected region. **Figure 1.3** A triply connected region.

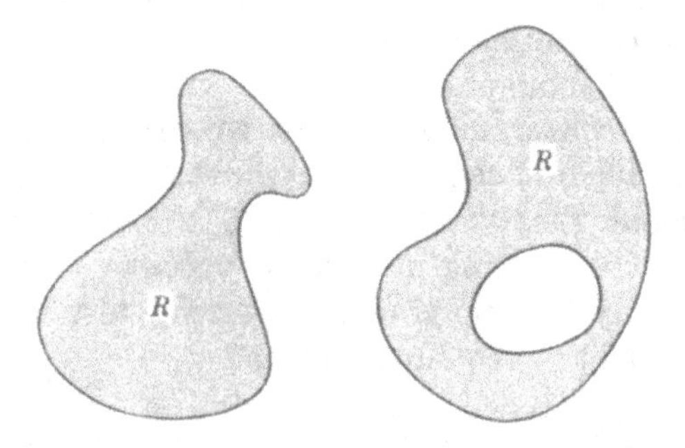

Figure 1.4 A nonconnected region *R*.

[1]For a precise definition of "connected," see p. 102.

Ordinarily the boundary curves of the regions to be considered are *sectionally smooth*. That is, every such curve consists of a finite number of arcs, each of which has a continuously turning tangent at all of its points, including the end points. Such curves, therefore, can have at most a finite number of corners.

In most cases we shall describe a region by one or more inequalities, the equal sign holding on some portion of the boundary. The two most important types of regions, which recur again and again, are the rectangular regions (with sides parallel to the coordinate axes) and the circular disks. A *rectangular region* (Fig. 1.5) consists of the points (x, y) whose coordinates satisfy inequalities of the form

$$a < x < b, \qquad c < y < d;$$

each coordinate is restricted to a definite interval, and the point (x, y) varies over the interior of a rectangle. As defined here, our rectangular region is *open;* that is, it does not contain its boundary.

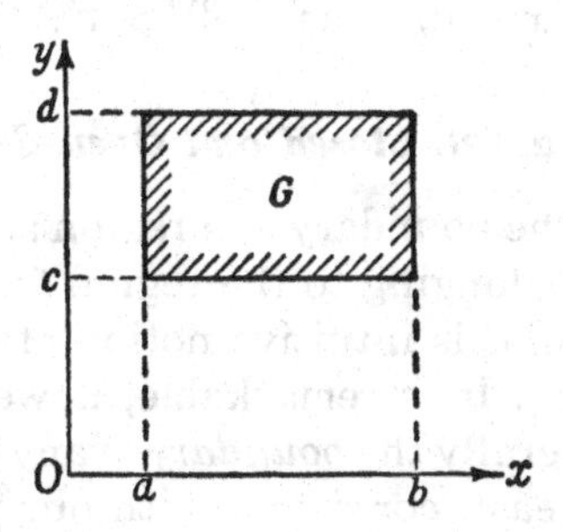

Figure 1.5 A rectangular region.

The boundary curves are obtained by replacing one or more of the inequalities defining the region by equality and permitting (but not requiring) the equal sign in the others. For example,

$$x = a, \qquad c \leqq y \leqq d$$

defines one of the sides of the rectangle. The *closed* rectangle obtained by adding all the boundary points to the set is described by the inequalities

$$a \leqq x \leqq b, \qquad c \leqq y \leqq d.$$

The *circular disk* with center (α, β) and radius r (Fig. 1.6) is, as seen before, given by the inequality

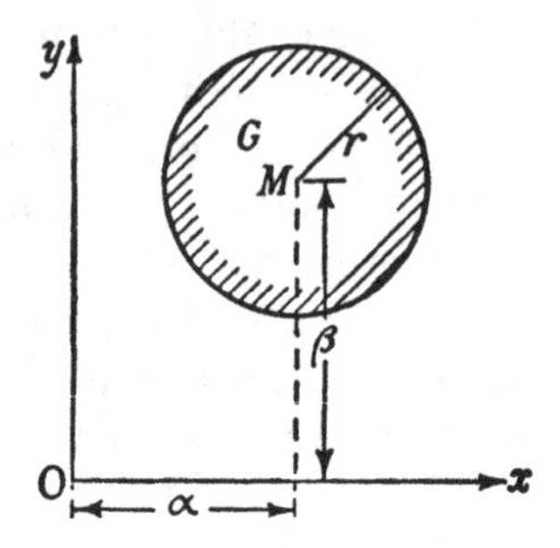

Figure 1.6 A circular disk.

$$(x-\alpha)^2+(y-\beta)^2<r^2.$$

Adding the boundary circle to this "open" disk, we obtain the "closed disk" described by

$$(x-\alpha)^2+(y-\beta)^2\leqq r^2.$$

c. *The Boundary of a Set. Closed and Open Sets*

One might think of the boundary of a region as a kind of membrane separating the points belonging to the region from those that do not belong. As we shall see, this intuitive notion of boundary would not always have a meaning. It is remarkable, however, that there is a way to define quite generally the *boundary* of any point set whatsoever in a way which is, at least, consistent with our intuitive notion. We say that *a point P is a boundary point of a set S of points if every neighborhood of P contains both points belonging to S and points not belonging to S.* Consequently, if P is not a boundary point, there exists a neighborhood of P that contains only one kind of point; that is, we either can find a neighborhood of P that consists entirely of points of S, in which case we call P an *interior point* of S, or we can find a neighborhood of P entirely free of points of S, in which case we call P an *exterior point* of S. Thus, *for a given set S of points, every point in the plane is either boundary point or interior point or exterior point of S and belongs to only one of these classes.* The set of boundary points of S forms the *boundary* of S, denoted by the symbol ∂S.

For example, let S be the rectangular region

$$a<x<b,\qquad c<y<d.$$

Obviously, we can find for any point P of S a small circular disk with center $P = (\alpha, \beta)$ that is entirely contained in S; we only have to take an ε-neighborhood of P in which ε is positive and so small that

$$a < \alpha - \varepsilon < \alpha + \varepsilon < b, \qquad c < \beta - \varepsilon < \beta + \varepsilon < d.$$

This shows that here every point of S is an interior point. The boundary points P of S are just the points lying either on one of the sides or at a corner of the rectangle; in the first case, one-half of every sufficiently small neighborhood of P will belong to S and one-half will not. In the second case, one-quarter of every neighborhood belongs to S and three-quarters do not (Fig. 1.7).

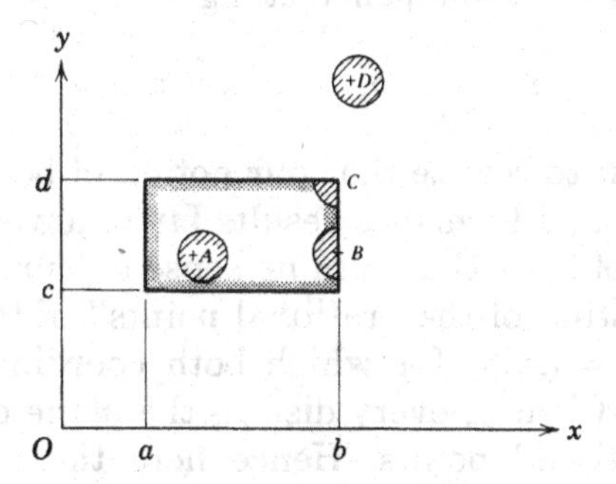

Figure 1.7 Interior point A, exterior point D, boundary points B, C of rectangular region.

By definition, every interior point P of set S is necessarily a point of S, for there is a neighborhood of P consisting entirely of points of S, and P belongs to that neighborhood. Similarly, any exterior point of S definitely does not belong to S. On the other hand, the boundary points of a set sometimes do, and sometimes do not belong to the set.[1] The open rectangle

$$a < x < b, \qquad c < y < d$$

does not contain its boundary points, while the closed rectangle

$$a \leqq x \leqq b, \qquad c \leqq y \leqq d$$

does.

[1]Observe the distinction between "not belonging to S" and "exterior to S." A boundary point of S never is exterior, even when it does not belong to S.

Generally we call a set S of points *open* if no boundary point of S belongs to S (i.e., if S consists entirely of interior points). S is called *closed* if it contains its boundary. From any set S we can always obtain a closed set by adding to S all its boundary points, insofar as they do not belong to S already. We then obtain a new set, the *closure* $\bar{S}$ of S. The reader can easily verify that the closure of S is a closed set. The exterior points are exactly those that do not belong to the closure of S. Similarly, we define the *interior* S^o of S as the set of interior points of S, that is, the set obtained by removing the boundary points from S. The interior of S is open.

It should be observed that sets do not have to be either open or closed. We can easily construct a set S containing only part of its boundary, such as the semiopen rectangle

$$a \leqq x < b, \qquad c \leqq y < d.$$

It is also important to realize that our notion of boundary applies to quite general sets and furnishes results far removed from intuition. A prime example of a set that is in no sense a "curve" or a "region" is the set S consisting of the "rational points" of the plane, that is, of those points $P = (x, y)$ for which both coordinates x and y are rational numbers. Clearly, every disk in the plane contains both rational and nonrational points. Hence here there is no boundary "curve"; the boundary ∂S consists of the whole plane. There exist neither interior nor exterior points.

Even in cases where the boundary is one-dimensional, not all of it serves to *separate* interior from exterior points. For example, the inequalities

$$(x - \alpha)^2 + (y - \beta)^2 < r^2, \qquad y \neq \beta$$

describe a disk with one diameter cut out; here the boundary con-

Figure 1.8 Disk with diameter removed.

sists of the circle $(x - \alpha)^2 + (y - \beta)^2 = r^2$, and of the diameter

$$y = \beta, \qquad |x - \alpha| < r.$$

Any sufficiently small neighborhood of a point of that diameter contains no exterior points at all (Fig. 1.8).

d. Closure as Set of Limit Points

The notions of "interior," "boundary," and "exterior" of a set S are of importance when we consider limits of sequences of points $P_1, P_2, \ldots$ all of which belong to the set S.[1] Clearly, a point Q exterior to S cannot be the limit of the sequence, since there is a neighborhood of Q free of points of S, which prevents the P_k from coming arbitrarily close to Q. Hence, the limit of a sequence of points in S must either be a boundary point or an interior point of S. Since the interior and boundary points of S form the closure of S it follows that *limits of sequences in S belong to the closure of S.*

Conversely, every point Q of the closure of S is actually the limit of some sequence $P_1, P_2, \ldots$ of points of S, for if Q is a point of the closure, then Q either belongs to S or to its boundary. In the first case we have trivially in $Q, Q, Q, \ldots$ a sequence of points of S converging to S. In the second case, for any $\varepsilon > 0$ the ε-neighborhood of Q contains at least one point of S. For every natural number n we may choose a point P_n of S belonging to the ε-neighborhood of Q with $\varepsilon = 1/n$. Clearly, the P_n converge to Q.

e. Points and Sets of Points in Space

An ordered triple of numbers (x, y, z) can be represented in the usual manner by a point P in space. Here the numbers x, y, z, the Cartesian coordinates of P, are the (signed) distances of P from three mutually perpendicular planes. The distance $\overline{PP'}$ between the two points $P = (x, y, z)$ and $P' = (x', y', z')$ is given by

$$\overline{PP'} = \sqrt{(x' - x)^2 + (y' - y)^2 + (z' - z)^2}.$$

The ε-neighborhood of the point $Q = (a, b, c)$ consists of the points $P = (x, y, z)$ for which $\overline{PQ} < \varepsilon$; these points form the *ball* given by the inequality

$$(x - a)^2 + (y - b)^2 + (z - c)^2 < \varepsilon^2.$$

[1]The points P_k do not have to be *distinct* from one another.

The analogues to the rectangular plane regions are the rectangular parallelepipeds[1] described by a system of inequalities of the form

$$a < x < b, \qquad c < y < d, \qquad e < z < f.$$

All the notions developed for plane sets—boundary, closure, and so on—carry over to sets in three dimensions in an obvious way.

When we are dealing with ordered quadruples like x, y, z, w, our visual intuition fails to provide a geometrical interpretation. Still, it is convenient to make use of geometrical terminology, attributing to (x, y, z, w) a "point in four-dimensional space." The quadruples (x, y, z, w) satisfying an inequality of the form

$$(x-a)^2 + (y-b)^2 + (z-c)^2 + (w-d)^2 < \varepsilon^2$$

constitute, by definition, the ε-neighborhood of the point (a, b, c, d). A rectangular region[2] is described by a system of inequalities of the form

$$a < x < b, \qquad c < y < d, \qquad e < z < f, \qquad g < w < h.$$

Of course, there is nothing mysterious in this idea of "points" in four dimensions; it is just a convenient terminology and implies nothing about the physical reality of four-dimensional space. Indeed, nothing prevents us from calling an "n-tuple" $(x_1, \ldots, x_n)$ a "point" in n-dimensional space, where n can be any natural number. For many applications it is quite useful and suggestive to represent a system described by n quantities in this way by a single point in some higher-dimensional space.[3] Often analogies with geometric interpretations in three-dimensional space provide guidance for operating in more than three dimensions.

Exercises 1.1

1. A point (x, y) of the plane may be represented by a complex number (Volume I, p. 103) in the form $z = x + iy$. Investigate the convergence

---

[1]Parallel *epipedon* (Greek for "plane").

[2]The terms "cell" and "interval" are also used to describe rectangular regions of this type in higher dimensions.

[3]Thus the system of molecules of a gas in a container can be described by the position of a single point in a "phase-space" with a very high number of dimensions. Going even further, it is customary in some parts of analysis to represent an infinite sequence of numbers $x_1, x_2, \ldots$ by a point $(x_1, x_2, \ldots)$ in a space with *infinitely many dimensions.*

for different values of z of the sequences

(a) z^n

(b) $z^{1/n}$ where $z^{1/n}$ is defined as the *primitive nth root of z*, that is, as the root with minimum positive amplitude.

2. Prove for $P_n = (x_n + \xi_n,\ y_n + \eta_n)$ that $\lim_{n\to\infty} P_n = (x + \xi,\ y + \eta)$ where the limits $x = \lim_{n\to\infty} x_n$, $\xi = \lim_{n\to\infty} \xi_n$, $y = \lim_{n\to\infty} y_n$, $\eta = \lim_{n\to\infty} \eta_n$ are presumed to exist.
3. Show that every point of the disk $x^2 + y^2 < 1$ is an interior point. Is this also true for $x^2 + y^2 \le 1$? Explain.
4. Show that the set S of points (x, y) with $y > x^2$ is open.
5. What is the boundary of a line segment considered as a subset of the x, y-plane?

Problems 1.1

1. Let P be a boundary point of the set S that does not belong to S. Prove that there exists a sequence of *distinct* points $P_1, P_2, \ldots$ in S having P as limit.
2. Prove that the closure of a set is closed.
3. Let P be any point of a set S, and let Q be any point outside the set. Prove that the line segment PQ contains a boundary point of S.
4. Let G be the set of points (x, y) for which $|x| < 1$, $|y| < 1/2$ and for which $y < 0$ if $x = 1/2$. Does G contain only interior points? Give evidence.

1.2 Functions of Several Independent Variables

a. Functions and Their Domains

Equations of the form

$$u = x + y, \qquad u = x^2y^2, \qquad \text{or} \qquad u = \log(1 - x^2 - y^2)$$

assign a *functional value* u to a pair of values (x, y). In the first two of these examples, a value of u is assigned to *every* pair of values (x, y), while in the third the correspondence has a meaning only for those pairs of values (x, y) for which the inequality $x^2 + y^2 < 1$ is true.

In general, we say that u is a *function* of the *independent variables* x and y whenever some law f assigns a unique value of u, the *dependent variable*, to each pair of values (x, y) belonging to a certain specified set, the *domain* of the function. A function $u = f(x, y)$ thus defines a *mapping* of a set of points in the x, y-plane, the domain of f, onto a certain set of points on the u-axis, the *range* of f. Similarly, we say that u is a function of the n variables $x_1, x_2, \ldots, x_n$ if for each set of values $(x_1, \ldots, x_n)$ belonging to a certain specified set there is assigned a corresponding unique value of u.[1]

Thus, for example, the volume $u = xyz$ of a rectangular parallelepiped is a function of the length of the three sides x, y, z; the magnetic declination is a function of the latitude, the longitude, and the time; the sum $x_1 + x_2 + \cdots + x_n$ is a function of the n terms $x_1, x_2, \ldots, x_n$.

It is to be noted that the domain of a function f is an indispensable part of its description. In cases where $u = f(x, y)$ is given by an explicit expression, it is natural to take as domain of f all (x, y) for which this expression makes sense. However, functions given by the same expression but having smaller domains can be defined by "restriction." Thus the formula $u = x^2 + y^2$ can be used to define a function with domain $x^2 + y^2 < 1/2$.

Just as in the case of functions of one variable, a functional correspondence $u = f(x, y)$ associates a *unique* value of u with the system of independent variables x, y. Thus, no functional value is assigned by an analytic expression that is multivalued, such as arc tan y/x, unless we specify, for example, that the "arc tangent" is to stand for the *principal branch* with values lying between $-\pi/2$ and $+\pi/2$ (see Volume I, p. 214); in addition we have to exclude the line $x = 0$.[2]

b. The Simplest Types of Functions

Just as in the case of one independent variable, the simplest functions of more than one variable are the *rational integral* functions or *polynomials*. The most general polynomial of the first degree, or *linear* function, has the form

$$u = ax + by + c,$$

where a, b, and c are constants. The general polynomial of the second degree has the form

[1]Often we think of functions f as assigning a value to a *point* P rather than to the pair (x, y) of coordinates describing P. We write then $f(P)$ for $f(x, y)$. This notation is particularly useful when the functional relation between points P and values $f(P)$ is defined geometrically without reference to a specific x, y-coordinate system.

[2]Taking the principal value, we see that $u = \arctan y/x$ for $x > 0$ is nothing but the polar angle of the point (x, y) counted from the positive x-axis. This polar angle can still be defined geometrically in an obvious way as a univalued function with values between $-\pi$ and π if we just exclude the origin and the points on the negative x-axis, but the polar angle is then no longer given by arc tan y/x in the extended region, if we understand the arc tangent to mean the principal branch.

$$u = ax^2 + bxy + cy^2 + dx + ey + f.$$

Its domain is the whole x, y-plane. The general polynomial of any degree is a sum of a finite number of terms $a_{mn}x^m y^n$ (called *monomials*), where m and n are nonnegative integers and the coefficients a_{mn} are arbitrary.

The *degree* of the monomial $a_{mn}x^m y^n$ is the sum $m + n$ of the exponents of x and y, provided the coefficient a_{mn} does not vanish. The degree of a polynomial is the highest degree of any monomial with nonvanishing coefficient (after combining terms with the same powers of x and y). A polynomial consisting of monomials all of which have the same degree N is called a *homogeneous polynomial* or a *form* of degree N. Thus $x^2 + 2xy$ or $3x^3 + (7/5)\, x^2y + 2y^3$ are forms.

By extracting roots of rational functions we obtain certain *algebraic* functions,[1] for example,

$$u = \sqrt{\frac{x-y}{x+y}} + \sqrt[3]{\frac{(x+y)^2}{x^3 + xy}}\ .$$

Most of the more complicated functions of several variables that we shall use here can be described in terms of the well-known functions of one variable, such as

$$u = \sin\,(x \operatorname{arc\,cos} y) \qquad \text{or} \qquad u = \log_x y.$$

c. *Geometrical Representation of Functions*

Just as we represent functions of one variable by curves, we may represent functions of two variables geometrically by surfaces. To this end, we consider a rectangular x,y,u-coordinate system in space, and mark off above each point (x, y) of the domain R of the function in the x, y-plane the point P with the third coordinate $u = f(x, y)$. As the point (x, y) ranges over the region R, the point P describes a surface in space. This surface we take as the geometrical representation of the function.

Conversely, in analytical geometry, surfaces in space are represented by functions of two variables, so that between such surfaces and functions of two variables there is a reciprocal relation. For example, to the function

$$u = \sqrt{1 - x^2 - y^2}$$

[1]For a general definition of the term "algebraic function," see p. 229.

there corresponds the hemisphere lying above the x, y-plane, with unit radius and center at the origin. To the function $u = x^2 + y^2$ there corresponds a so-called *paraboloid of revolution*, obtained by rotating the parabola $u = x^2$ about the u-axis (Fig. 1.9). To the functions $u = x^2 - y^2$ and $u = xy$, there correspond *hyperbolic paraboloids* (Fig. 1.10). The linear function $u = ax + by + c$ has for its "graph" a plane in space. If in the function $u = f(x, y)$ one of the independent variables, say y, does not occur, so that u depends on x only, say $u = g(x)$, the function is represented in x,y,u-space by a cylindrical surface generated by the perpendiculars to the u,x-plane at the points of the curve $u = g(x)$.

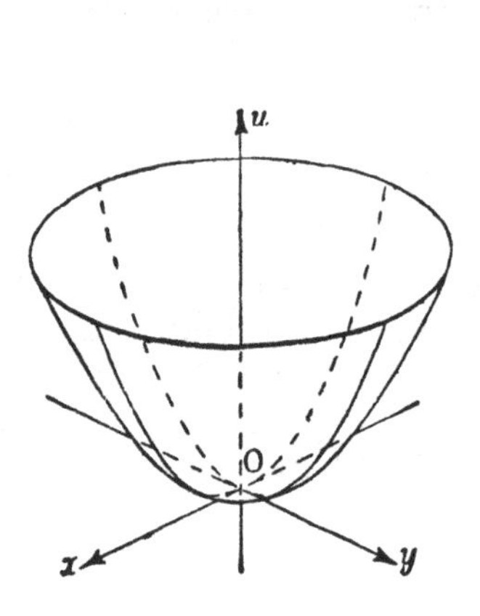

Figure 1.9 $u = x^2 + y^2$.

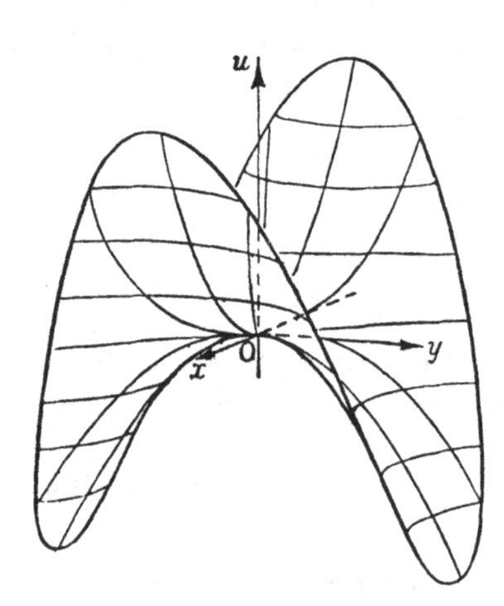

Figure 1.10 $u = x^2 - y^2$.

This representation by means of rectangular coordinates has, however, two disadvantages. First, geometric visualization fails us whenever we have to deal with three or more independent variables. Second, even for two independent variables it is often more convenient to confine the discussion to the x,y-plane alone, since in the plane we can sketch and can perform geometrical constructions without difficulty. From this point of view, another geometrical representation of a function of two variables, by means of contour lines, is sometimes preferable. In the x,y-plane we take all the points for which $u = f(x, y)$ has a constant value, say $u = k$. These points will usually lie on a curve or curves, the so-called *contour line*, or *level line*, for the given constant value k of the function. We can also obtain these curves by cutting the surface $u = f(x, y)$ by the plane $u = k$ parallel to the x, y-plane and projecting the curves of intersection perpendicularly onto the x, y-plane.

The system of these contour lines, marked with the corresponding values $k_1, k_2, \ldots$ of the height k, gives us a representation of the function. In practice, k is assigned values in arithmetic progression, say $k = \nu h$, where $\nu = 1, 2, \ldots$. The distance between the contour lines then gives us a measure of the steepness of the surface $u = f(x, y)$, for between every two neighboring lines the value of the function changes by the same amount. Where the contour lines are close together, the function rises or falls steeply; where the lines are far apart, the surface is flattish. This is the principle on which contour maps such as those of the U.S. Geological Survey are constructed.

In this method the linear function $u = ax + by + c$ is represented by a system of parallel straight lines $ax + by + c = k$. The function $u = x^2 + y^2$ is represented by a system of concentric circles (cf. Fig. 1.11). The function $u = x^2 - y^2$, whose surface is "saddle-shaped" (Fig. 1.10), is represented by the system of hyperbolas shown in Fig. 1.12.

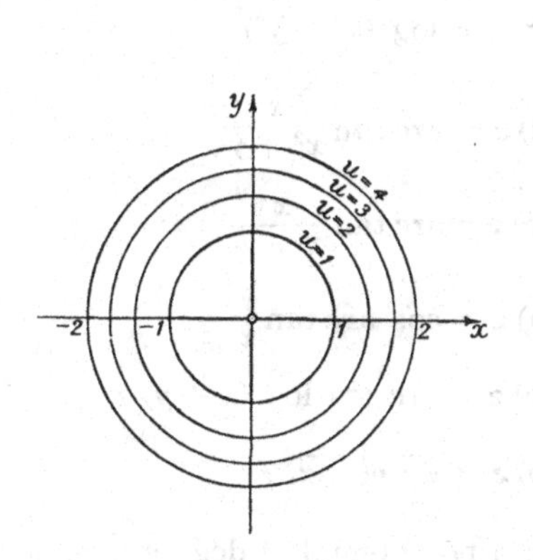

Figure 1.11 Contour lines of $u = x^2 + y^2$.

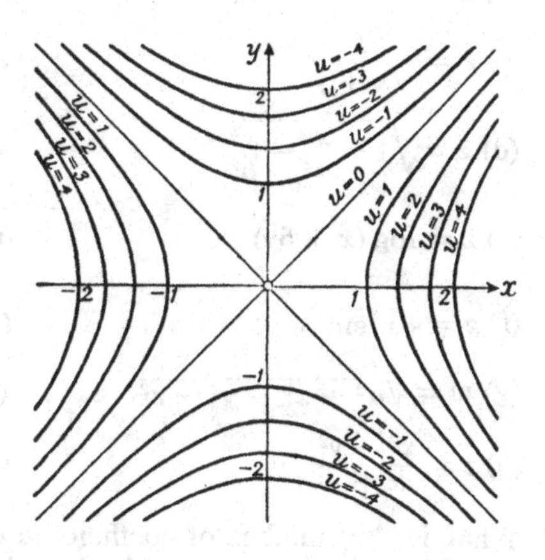

Figure 1.12 Contour lines of $u = x^2 - y^2$.

The method of representing the function $u = f(x, y)$ by contour lines has the advantage of being capable of extension to functions of three independent variables. Instead of the contour lines we then have the *level surfaces* $f(x, y, z) = k$, where k is a constant to which we can assign any suitable sequence of values. For example, the level surfaces for the function $u = x^2 + y^2 + z^2$ are spheres concentric about the origin of the x, y, z-coordinate system.

Exercises 1.2

1. Evaluate the following functions at the points indicated:

   (a) $z = \left(\dfrac{\text{arc cot } (x+y)}{\text{arc tan } (x-y)}\right)^3$ for $x = \dfrac{1+\sqrt{3}}{2}$, $y = \dfrac{1-\sqrt{3}}{2}$

   (b) $w = e^{\cos z(x+y)}$, for $x = y = \dfrac{\pi}{2}$, $z = -1$

   (c) $z = y^{x \cos xy}$, $x = e$, $y = \log \pi$

   (d) $z = \cosh (x+y)$, $x = \log \pi$, $y = \log \dfrac{1}{2}$

   (e) $z = \dfrac{x+y}{x-y}$, $x = \dfrac{1}{2}$, $y = \dfrac{1}{3}$.

2. As in Volume I, unless we make an explicit exception, we consider the domain of a function defined by a formal expression to be the set of all points for which the expression is meaningful. Give the domain and range of each of the following functions:

   (a) $z = \sqrt{x+y}$

   (b) $z = \sqrt{2x - y^2}$

   (c) $z = \dfrac{1}{\sqrt{x+y}}$

   (d) $z = \sqrt{1 - \dfrac{x^2}{a^2} - \dfrac{y^2}{b^2}}$

   (e) $z = \log (x + 5y)$

   (f) $z = \sqrt{x \sin y}$

   (g) $w = \sqrt{a^2 - x^2 - y^2 - z^2}$

   (h) $z = \dfrac{x^2 - y^2}{x+y}$

   (i) $z = \sqrt{3 - x^2 - 2y^2}$

   (j) $z = \sqrt{-x^2 - y^2}$

   (k) $z = \log (x^2 - y^2)$

   (l) $z = \text{arc tan} \dfrac{x^2}{x^2 + y^2}$

   (m) $z = \text{arc tan} \dfrac{x}{x+y}$

   (n) $z = \cos \text{ arc tan} \dfrac{y}{x}$

   (o) $z = \text{arc cos } \log (x+y)$

   (p) $z = \sqrt{y \cos x}$.

3. What is the number of coefficients of a polynomial of degree n in two variables? In three variables? In k variables?

4. For each of the following functions sketch the contour lines corresponding to $z = -2, -1, 0, 1, 2, 3$:

   (a) $z = x^2 y$

   (b) $z = x^2 + y^2 - 1$

   (c) $z = x^2 - y^2$

   (d) $z = y^2$

   (e) $z = y\left(1 - \dfrac{1}{x^2 + y^2}\right)$.

5. Draw the contour lines for $z = \cos(2x + y)$ corresponding to $z = 0, \pm 1, \pm 1/2$.
6. Sketch the surfaces defined by
   (a) $z = 2xy$
   (b) $z = x^2 + y^2$
   (c) $z = x - y$.
   (d) $z = x^2$
   (e) $z = \sin(x + y)$.
7. Find the level lines of the function
$$z = \log \frac{1 + \sqrt{x^2 + y^2}}{1 - \sqrt{x^2 + y^2}}.$$
8. Find the surfaces on which the function $u = 2(x^2 + y^2)/z$ is constant.

1.3 Continuity

a. Definition

As in the theory of functions of a single variable, the concept of continuity figures prominently when we consider functions of several variables. The statement that the function $u = f(x, y)$ is continuous at the point (ξ, η) should mean, roughly speaking, that for all points (x, y) near (ξ, η) the value of $f(x, y)$ differs but little from the value $f(\xi, \eta)$. We express this idea more precisely as follows: *If f has the domain R and $Q = (\xi, \eta)$ is a point of R, then f is continuous at Q if for every $\varepsilon > 0$ there exists a $\delta > 0$ such that*

$$|f(P) - f(Q)| = |f(x, y) - f(\xi, \eta)| < \varepsilon \tag{1}$$

for all $P = (x, y)$ in R for which[1]

$$\overline{PQ} = \sqrt{(x - \xi)^2 + (y - \eta)^2} < \delta. \tag{2}$$

If a function is continuous at every point of a set D of points, we say that it is *continuous in D*.

The following facts are almost obvious: The sum, difference, and

[1]Instead of confining (x, y) to a small disk with center (ξ, η) we could use a small square. Thus condition (2) in the definition of continuity can be replaced by

$$|x - \xi| < \delta \quad \text{and} \quad |y - \eta| < \delta. \tag{2'}$$

product of continuous functions are also continuous. The quotient of continuous functions defines a continuous function at points where the denominator does not vanish (for the proof see the next section, p. 00). In particular, all polynomials are continuous, and all rational functions are continuous at the points where the denominator does not vanish. Continuous functions of continuous functions are themselves continuous (cf. p. 22).

A function of several variables may have discontinuities of a much more complicated type than a function of a single variable. For example, discontinuities may occur along whole arcs of curves, not just at isolated points. This is the case for the function defined by

$$u = y/x \qquad \text{for} \qquad x \neq 0; \qquad u = 0 \qquad \text{for} \qquad x = 0,$$

which is discontinuous along the whole line $x = 0$. Moreover, a function $f(x, y)$ may be continuous in x for each fixed value of y and continuous in y for each fixed value of x, and yet be discontinuous as a function of the point (x, y). This is exemplified by

$$f(x, y) = \frac{2xy}{x^2 + y^2} \qquad \text{for} \qquad (x, y) \neq (0, 0), \quad f(0, 0) = 0.$$

For any fixed $y \neq 0$, this function is obviously continuous as a function of x, as the denominator cannot vanish. For $y = 0$ we have $f(x, 0) = 0$, which also is continuous as a function of x. Similarly, $f(x, y)$ is continuous as a function of y for any fixed x. But at every point of the line $y = x$ except at the point $x = y = 0$ we have $f(x, y) = 1$, and there are points of this line arbitrarily close to the origin. Hence, $f(x, y)$ is discontinuous at the point $(0, 0)$.

Just as in the case of functions of a single variable, a function $f(P) = f(x, y)$ is called *uniformly continuous* in the set R of the x, y-plane if f is defined at the points of R and if for every $\varepsilon > 0$ there exists a positive $\delta = \delta(\varepsilon)$ such that $|f(P) - f(Q)| < \varepsilon$ for any two points P, Q in R of distance $< \delta$.[1] The quantity $\delta = \delta(\varepsilon)$ is called a *modulus of continuity* for f. We have the basic theorem:

A function f that is defined and continuous in a closed and bounded set R is uniformly continuous in R. (For the proof see the Appendix to this chapter.)

Particularly important is the case in which we can find a modulus of continuity that is proportional to ε (see Volume I, p. 43). The

[1]The essential requirement making the continuity *uniform* is that δ depends on ε but not on P or Q.

function $f(P)$ defined in R is called *Lipschitz-continuous* if there exists a constant L such that

$$|f(P) - f(Q)| \leqq L\,\overline{PQ} \quad \text{for all points } P, Q \text{ in } R. \tag{3}$$

(L is called the "Lipschitz constant," relation (3) the "Lipschitz condition.") It is clear that a Lipschitz-continuous function f is uniformly continuous and has $\delta = \varepsilon/L$ as modulus of continuity.[1]

b. The Concept of Limit of a Function of Several Variables

The notion of limit of a function is closely related to the notion of continuity. Let us suppose that $f(x, y)$ is a function with domain R. Let $Q = (\xi, \eta)$ be a point of the closure of R. We say that f *has the limit* L *for* (x, y) *tending to* (ξ, η) *and write*

$$\lim_{(x,y)\to(\xi,\eta)} f(x, y) = L \quad \text{or} \quad \lim_{P\to Q} f(P) = L,^{2} \tag{4}$$

if for every $\varepsilon > 0$ we can find a neighborhood

$$\overline{PQ} = \sqrt{(x-\xi)^2 + (y-\eta)^2} < \delta \tag{5}$$

of (ξ, η) such that

$$|f(P) - L| = |f(x, y) - L| < \varepsilon$$

for all $P = (x, y)$ belonging to R in that neighborhood.[3]

In case the point (ξ, η) belongs to the domain of f we have in $(x, y) = (\xi, \eta)$ a point of R satisfying (5) for all $\delta > 0$. Then (4) implies in particular that

$$|f(\xi, \eta) - L| < \varepsilon$$

[1]The still wider class of "Hölder-continuous" functions f is obtained when we replace the Lipschitz condition (3) by the Hölder condition

$$|f(P) - f(Q)| \leqq L\,\overline{PQ}^{\alpha} \quad \text{for all } P, Q \text{ in } R.$$

L and α are constants and $0 < \alpha \leqq 1$ (see Volume I, p. 44). These functions also are uniformly continuous, and we can choose as modulus of continuity the quantity

$$\delta = (\varepsilon/L)^{1/\alpha}$$

[2]Or else $\lim f(x, y) = L$ for $(x, y) \to (\xi, \eta)$ or $\lim_{\substack{x\to\xi \\ y\to\eta}} f(x, y) = L$.

[3]The notion makes no sense for points (ξ, η) *exterior* to R since then there exist no points arbitrarily close to (ξ, η) in which f is defined, and every L could be considered as limit.

for all $\varepsilon > 0$ and hence that $L = f(\xi, \eta)$. But then, by definition, the relation

$$\lim_{(x, y)\to(\xi, \eta)} f(x, y) = f(\xi, \eta)$$

is identical with the condition for continuity of f at (ξ, η). Hence, *continuity of the function f at the point (ξ, η) is equivalent to the statement that f is defined at (ξ, η) and that $f(x, y)$ has the limit $f(\xi, \eta)$ for (x, y) tending to (ξ, η).*

If f is not defined at the boundary point (ξ, η) of its domain but has a limit L for $(x, y) \to (\xi, \eta)$, we can naturally extend the definition of f to the point (ξ, η) by putting $f(\xi, \eta) = L$; the function f extended in this way will then be continuous at (ξ, η). If $f(x, y)$ is continuous in its domain R, we can extend the definition of f as limit not just to a single boundary point (ξ, η) but simultaneously to all boundary points of R for which f has a limit. The resulting extended function is again continuous, as the reader may verify as an exercise. Take, for example, the function

$$f(x, y) = e^{-x^2/y}$$

defined for all (x, y) with $y > 0$. This function obviously is continuous at all points of its domain R, the upper half-plane. Consider a boundary point $(\xi, 0)$. For $\xi \neq 0$ we have clearly

$$\lim_{(x, y)\to(\xi, \eta)} f(x, y) = \lim_{s\to\infty} e^{-s} = 0$$

when y is restricted to positive values. If then we define the extended function $f^*(x, y)$ by

$$f^*(x, y) = f(x, y) = e^{-x^2/y}$$

for $y > 0$ and all x, and by

$$f^*(x, 0) = 0$$

for $x \neq 0$. the function f^* will be continuous in its domain R^* where R^* is the closed upper half-plane $y \geqq 0$ with the exception of the point $(0, 0)$. At the origin f^* does not have a limit, and hence it is not possible to define $f^*(0, 0)$ in such a way that the extension is continuous at the origin. Indeed, for (x, y) on the parabola $y = kx^2$, we have

$$f(x, y) = e^{-1/k}.$$

Approaching the origin along different parabolas leads to different limiting values, so that there exists no single limit of $f(x, y)$ for $(x, y) \to 0$.

We can also relate the concept of *limit of a function* $f(x, y)$ to that of *limit of a sequence* (cf. Volume I, p. 82). Suppose f has the domain R and

$$\lim_{(x, y)\to(\xi, \eta)} f(x, y) = L.$$

Let $P_n = (x_n, y_n)$ for $n = 1, 2, \ldots$, be any sequence of points in R for which $\lim_{n\to\infty} P_n = (\xi, \eta)$. Then the sequence of numbers $f(x_n, y_n)$ has the limit L. For $f(x, y)$ will differ arbitrarily little from L for all (x, y) in R sufficiently close to (ξ, η), and (x_n, y_n) will be sufficiently close to (ξ, η) if only n is sufficiently large. Conversely, $\lim_{n\to\infty} f(x, y)$ for $(x, y) \to (\xi, \eta)$ exists and has the value L if for every sequence of points (x_n, y_n) in R with limit (ξ, η) we have $\lim_{n\to\infty} f(x_n, y_n) = L$. The proof can easily be supplied by the reader. If we restrict ourselves to points (ξ, η) in the domain of f, we obtain the statement that *continuity of f in its domain R means just that*

$$\lim_{n\to\infty} f(x_n, y_n) = f(\xi, \eta) \tag{6}$$

whenever $\lim_{n\to\infty} (x_n, y_n) = (\xi, \eta)$ *or that*

$$\lim_{n\to\infty} f(x_n, y_n) = f(\lim_{n\to\infty} x_n, \lim_{n\to\infty} y_n),$$

where we only consider sequences (x_n, y_n) in R that converge and have their limits in R. Essentially, then, continuity of a function f allows the interchange of the symbol for f with that for limit.

It is clear that the notions of limit of a function and of continuity apply just as well when the domain of f is not a two-dimensional region but a curve or any other point set. For example, the function

$$f(x + y) = (x + y)!$$

is defined in the set R consisting of all the lines $x + y = \text{const.} = n$, where n is a positive integer. Obviously, f is continuous in its domain R.

It was mentioned earlier (p. 17) that when $f(x, y)$ and $g(x, y)$ are continuous at a point (ξ, η), then $f + g$, $f - g$, $f \cdot g$, and for $g(\xi, \eta) \neq 0$ also f/g are continuous at (ξ, η). These rules follow immediately from the formulation of continuity in terms of convergence of sequences. For any sequence (x_n, y_n) of points belonging to the domains of f and g and converging to (ξ, η), we have by (6)

$$\lim_{n\to\infty} f(x_n, y_n) = f(\xi, \eta), \qquad \lim_{n\to\infty} g(x_n, y_n) = g(\xi, \eta).$$

The convergence of $f(x_n, y_n) + g(x_n, y_n)$ and so on follows then from the rules for operating with sequences (Volume I, p. 72).

c. *The Order to Which a Function Vanishes*

If the function $f(x, y)$ is continuous at the point (ξ, η), the difference $f(x, y) - f(\xi, \eta)$ tends to 0 as x tends to ξ and y tends to η. By introducing the new variables $h = x - \xi$ and $k = y - \eta$, we can express this as follows: The function $\phi(h, k) = f(\xi + h, \eta + k) - f(\xi, \eta)$ of the variables h and k tends to 0 as h and k tend to 0.

We shall frequently meet with functions $\phi(h, k)$ which tend to 0 as h and k do. As in the case of one independent variable, for many purposes it is useful to describe the behavior of $\phi(h, k)$ for $h \to 0$ and $k \to 0$ more precisely by distinguishing between different "orders of vanishing" or "orders of magnitude" of $\phi(h, k)$. For this purpose we base our comparisons on the distance

$$\rho = \sqrt{h^2 + k^2} = \sqrt{(x - \xi)^2 + (y - \eta)^2}$$

of the point with coordinates $x = \xi + h$ and $y = \eta + k$ from the point with coordinates ξ and η and make use of the following definition:

A function $\phi(h, k)$ vanishes as $\rho \to 0$ to at least the same order as $\rho = \sqrt{h^2 + k^2}$, provided that there is a constant C independent of h and k such that the inequality

$$\left|\frac{\phi(h, k)}{\rho}\right| \leqq C$$

holds for all sufficiently small values of ρ; that is, provided there is a $\delta > 0$ such that the inequality holds for all values of h and k such that

[1]In order to avoid confusion, we expressly point out that a *higher* order of vanishing for $\rho \to 0$ implies *smaller* values in the neighborhood of $\rho = 0$; for example, ρ^2 vanishes to a higher order than ρ and ρ^2 is smaller than ρ when ρ is nearly 0.

$0 < \sqrt{h^2+k^2} < \delta$. We write, then, symbolically: $\phi(h,k) = O(\rho)$. Further, we say that $\phi(h, k)$ *vanishes to a higher order*[1] *than* ρ *if the quotient* $\phi(h, k)/\rho$ *tends to* 0 *as* $\rho \to 0$. This will be expressed by the symbolical notation $\phi(h, k) = o(\rho)$ for $(h, k) \to 0$ (see Volume I, p. 253, where the symbols "o" and "O" are explained for functions of a single variable).

Let us consider some examples. Since

$$\frac{|h|}{\sqrt{h^2+k^2}} \leqq 1 \quad \text{and} \quad \frac{|k|}{\sqrt{h^2+k^2}} \leqq 1,$$

the components h and k of the distance ρ in the direction of the x and y-axes vanish to at least the same order as the distance itself. The same is true for a linear homogeneous function $ah + bk$ with constants a and b or for the function $\rho \sin 1/\rho$. For fixed values of α greater than 1, the power ρ^α of the distance vanishes to a higher order than ρ; symbolically, $\rho^\alpha = o(\rho)$ for $\alpha > 1$. Similarly, a homogeneous quadratic polynomial $ah^2 + bhk + ck^2$ in the variables h and k vanishes to a higher order than ρ as $\rho \to 0$:

$$ah^2 + bhk + ck^2 = o(\rho).$$

More generally, the following definition is used. If the comparison function $\omega(h, k)$ is defined for all nonzero values of (h, k) in a sufficiently small circle about the origin and is not equal to 0, then $\phi(h, k)$ *vanishes to at least the same order as* $\omega(h, k)$ as $\rho \to 0$ if for some suitably chosen constant C the relation

$$\left|\frac{\phi(h, k)}{\omega(h, k)}\right| \leqq C$$

holds in a neighborhood of the point $(h, k) = (0, 0)$. We indicate this by the symbolic equation $\phi(h, k) = O(\omega(h, k))$. Similarly, $\phi(h, k)$ *vanishes to a higher order than* $\omega(h, k)$, or $\phi(h, k) = o(\omega(h, k))$, if $\dfrac{\phi(h, k)}{\omega(h, k)} \to 0$ when $\rho \to 0$.

For example, the homogeneous polynomial $ah^2 + bhk + ck^2$ is at least of the same order as ρ^2, since

$$|ah^2 + bhk + ck^2| \leqq \left(|a| + \frac{1}{2}|b| + |c|\right)(h^2 + k^2)$$

Also $\rho = o(1/|\log \rho|)$, since $\lim\limits_{\rho\to 0} (\rho \log \rho) = 0$ (Volume I, p. 252).

Exercises 1.3

1. The function $z = (x - y)/(x + y)$ is discontinuous along $y = -x$. Sketch the level lines of its surface for $z = 0, \pm 1, \pm 2$. What is the appearance of the level lines for $z = \pm m$, and m large?
2. Examine the continuity of the function $z = (x^2 + y) - \sqrt{x^2 + y^2}$, where $z = 0$ for $x = y = 0$. Sketch the level lines $z = k$ ($k = -4, -2, 0, 2, 4$). Exhibit (on one graph) the behavior of z as a function of x alone for $y = -2, -1, 0, 1, 2$. Similarly, exhibit the behavior of z as a function of y alone for $x = 0, \pm 1, \pm 2$. Finally, exhibit the behavior of z as a function of ρ alone when θ is constant (ρ, θ being polar coordinates).
3. Verify that the functions
   (a) $f(x, y) = x^3 - 3xy^2$
   (b) $g(x, y) = x^4 - 6x^2y^2 + y^4$
   are continuous at the origin by determining the modulus of continuity $\delta(\varepsilon)$. To what order does each function vanish at the origin?
4. Show that the following functions are continuous:

   (a) $\sin(x^2 + y)$

   (b) $\dfrac{\sin xy}{\sqrt{x^2 + y^2}}$

   (c) $\dfrac{x^3 + y^3}{x^2 + y^2}$

   (d) $x^2 \log(x^2 + y^2)$
   where in each case the function is defined at (0, 0) to be equal to the limit of the given expression.
5. Find a modulus of continuity, $\delta = \delta(\varepsilon, x, y)$, for the continuous functions

   (a) $f(x, y) = \sqrt{1 + x^2 + 2y^2}$

   (b) $f(x, y) = \sqrt{1 + e^{xy}}$.
6. Where is the function $z = 1/(x^2 - y^2)$ discontinuous?
7. Where is the function $z = \tan \pi y / \cos \pi x$ discontinuous?
8. For what set of values (x, y) is the function $z = \sqrt{y \cos x}$ continuous?
9. Show that the function $z = 1/(1 - x^2 - y^2)$ is continuous in the unit disk $x^2 + y^2 < 1$.
11. Find the condition that the polynomial

$$P = ax^2 + 2bxy + cy^2$$

   has exactly the same order as ρ^2 in the neighborhood of $x = 0$, $y = 0$ (i.e., that both P/ρ^2 and ρ^2/P are bounded).
12. Find whether or not the following functions are continuous, and if not, where they are discontinuous:

   (a) $\sin \dfrac{y}{x}$

(b) $\dfrac{x^3 + y^2}{x^2 + y^2}$

(c) $\dfrac{x^3 + y^2}{x^3 + y^3}$

(d) $\dfrac{x^3 + y^2}{x^2 + y}$.

13. Show that the functions

$$f(x, y) = \frac{x^4y^4}{(x^2 + y^4)^3}, \qquad g(x, y) = \frac{x^2}{x^2 + y^2 - x}$$

tend to 0 if (x, y) approaches the origin along any straight line but that f and g are discontinuous at the origin.

14. Determine whether the following functions have limits at $x = y = 0$ and give the limit when it exists.

(a) $\dfrac{x^2 - y^2}{x^2 + y^2}$

(b) $\dfrac{x^2 + 2xy + y^2}{x^2 + y^2}$

(c) $\dfrac{x^2 + 3xy + y^2}{x^2 + 4xy + y^2}$

(d) $-\dfrac{|x - y|}{x^2 - 2xy + y^2}$

(e) $\exp[-|x - y|/(x^2 - 2xy + y^2)]$

(f) $|x|^y$

(g) $|x|^{|1/y|}$

(h)* $\dfrac{|y|^{|x|}\sqrt{x^2 + y^2}}{\sqrt{x^2 + y^2} + |y/x|}$

15. Find a modulus of continuity $\delta(\varepsilon)$ for those functions of Exercise 14 that have limits at $x = y = 0$, where the functions are defined at the origin by their limiting values.

16. Show that $f(x, y, z) = (x^2 + y^2 - z^2)/(x^2 + y^2 + z^2)$ is not continuous at $(0, 0, 0)$.

17. Prove that if $P(x, y)$ and $Q(x, y)$ are each polynomials of degree $n > 0$, vanishing at the origin,

$$R(x, y) = \frac{P(x, y)}{Q(x, y)}$$

is not continuous at the origin.

18. Find the limits of the following expressions as (x, y) tends to $(0, 0)$ in an arbitrary manner:

(a) $\dfrac{\sin(x^2 + y^2)}{x^2 + y^2}$

(b) $\dfrac{\sin(x^4 + y^4)}{x^2 + y^2}$

(c) $\dfrac{e^{-1/(x^2+y^2)}}{x^4 + y^4}$.

19. Show that the function $z = 3(x - y)/(x + y)$ can tend to any limit as (x, y) tends to $(0, 0)$. Give examples of variations of (x, y) such that

 (a) $\lim_{\substack{x \to 0 \\ y \to 0}} z = 2$

 (b) $\lim_{\substack{x \to 0 \\ y \to 0}} z = -1$

 (c) $\lim_{\substack{x \to 0 \\ y \to 0}} z$ does not exist

20. If $f(x, y) \to 0$ as $(x, y) \to (0, 0)$ along all straight lines passing through the origin, does $f(x, y) \to 0$ as $(x, y) \to (0, 0)$ along any path?
21. Investigate the behavior of $z = y \log x$ in a neighborhood of the origin $(0, 0)$.
22. For $z = f(x, y) = (x^2 - y)/2x$, draw the graphs of

 (a) $z = f(x, x^2)$

 (b) $z = f(x, 0)$

 (c) $z = f(x, 1)$

 (d) $z = f(x, x)$

 Does the limit of $f(x, y)$ as $(x, y) \to (0, 0)$ exist?
23. Give a geometrical interpretation of the following statement: $\phi(h, k)$ vanishes to the same order as $\rho = \sqrt{h^2 + k^2}$.

Problems 1.3

1. Let the continuous function f be extended to the function f^* defined so that $f^* = f$ on the domain of f and $f^*(Q) = \lim_{P \to Q} f(P)$ for all points Q on the boundary of f where the limit exists. Prove that f^* is continuous.
2. Prove that $\lim f(x, y)$ for $(x, y) \to (\xi, \eta)$ exists and has the value L if and only if for every sequence of points (x_n, y_n) in the domain of f with limit (ξ, η) we have $\lim_{n \to \infty} f(x_n, y_n) = L$.

1.4 The Partial Derivatives of a Function

a. Definition. Geometrical Representation

If in a function of several variables we assign definite numerical values to all but one of the variables and allow only that variable, say x, to vary, the function becomes a function of a single variable. We consider a function $u = f(x, y)$ of the two variables x and y and assign to y a definite fixed value $y = y_0 = c$. The resulting function $u = f(x, y_0)$ of the single variable x may be represented geometrically by cutting the surface $u = f(x, y)$ by the plane $y = y_0$ (cf. Figs. 1.13

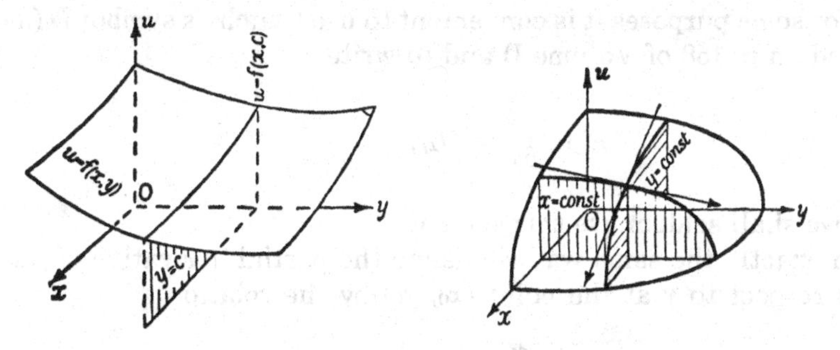

Figure 1.13 and **Figure 1.14** Sections of $u = f(x, y)$.

and 1.14). The curve of intersection thus formed in the plane is represented by the equation $u = f(x, y_0)$. If we differentiate this function in the usual way at the point $x = x_0$, assuming that f is defined in a neighborhood of (x_0, y_0) and that the derivative exists,[1] we obtain the *partial derivative* of $f(x, y)$ *with respect to* x at the point (x_0, y_0):

$$\lim_{h\to 0} \frac{f(x_0 + h, y_0) - f(x_0, y_0)}{h}.$$

Geometrically, this partial derivative denotes the tangent of the angle between a parallel to the x-axis and the tangent line to the curve $u = f(x, y_0)$. It is therefore the *slope of the surface* $u = f(x, y)$ *in the direction of the x-axis.*

To represent these partial derivatives several different notations are used, one of which is the following:

$$\lim_{h\to 0} \frac{f(x_0 + h, y_0) - f(x_0, y_0)}{h} = f_x(x_0, y_0) = u_x(x_0, y_0).$$

If we wish to emphasize that the partial derivative is the limit of a difference quotient, we denote it by

$$\frac{\partial f}{\partial x} \quad \text{or} \quad \frac{\partial}{\partial x} f.$$

Here we use the special round letter ∂ instead of the ordinary d used in the differentiation of functions of one variable in order to show that we are dealing with a function of several variables and differentiating with respect to one of them.

[1]We shall not try to define a derivative at boundary points of the domain (except, on occasion, as limit of the values of partial derivatives as the boundary point is approximated by interior points).

For some purposes it is convenient to use Cauchy's symbol D (mentioned on p. 158 of Volume I) and to write

$$\frac{\partial f}{\partial x} = D_x f,$$

but we shall seldom use this symbol.

In exactly the same way we define the partial derivative of $f(x, y)$ with respect to y at the point (x_0, y_0) by the relation

$$\lim_{k \to 0} \frac{f(x_0, y_0 + k) - f(x_0, y_0)}{k} = f_y(x_0, y_0) = D_y f(x_0, y_0).$$

This represents the slope of the curve of intersection of the surface $u = f(x, y)$ with the plane $x = x_0$ perpendicular to the x-axis (Fig. 1.14).

Let us now think of the point (x_0, y_0), hitherto considered fixed, as variable and accordingly omit the subscripts 0. In other words, we think of the differentiation as carried out at any point (x, y) of the region of definition of $f(x, y)$. Then the two derivatives are themselves functions of x and y,

$$u_x(x, y) = f_x(x, y) = \frac{\partial f(x, y)}{\partial x} \qquad \text{and} \qquad u_y(x, y) = f_y(x, y) = \frac{\partial f(x, y)}{\partial y}.$$

For example, the function $u = x^2 + y^2$ has the partial derivatives $u_x = 2x$ (in differentiation with respect to x the term y^2 is regarded as a constant and so has the derivative 0) and $u_y = 2y$. The partial derivatives of $u = x^3y$ are $u_x = 3x^2y$ and $u_y = x^3$.

Similarly, for a function of any number n of independent variables, we define partial derivatives by

$$\frac{\partial f(x_1, x_2, \ldots, x_n)}{\partial x_1} = \lim_{h \to 0} \frac{f(x_1 + h, x_2, \ldots, x_n) - f(x_1, x_2, \ldots, x_n)}{h}$$

$$= f_{x_1}(x_1, x_2, \ldots, x_n) = D_{x_1} f(x_1, x_2, \ldots, x_n),$$

it being assumed that the limit exists.

Of course, we can also form *higher partial derivatives* of $f(x, y)$ by again differentiating the partial derivatives of the "first order," $f_x(x, y)$ and $f_y(x, y)$, with respect to one of the variables and repeating this process. We indicate the order in which the differentiations are carried out by the order of the subscripts or by the order of the symbols ∂x and ∂y in the "denominator" from right to left[1] and use the following symbols for the second derivatives:

$$\frac{\partial}{\partial x}\left(\frac{\partial f}{\partial x}\right) = \frac{\partial^2 f}{\partial x^2} = f_{xx} = (D_x)^2 f,$$

$$\frac{\partial}{\partial x}\left(\frac{\partial f}{\partial y}\right) = \frac{\partial^2 f}{\partial x\,\partial y} = f_{xy} = D_x D_y f,$$

$$\frac{\partial}{\partial y}\left(\frac{\partial f}{\partial x}\right) = \frac{\partial^2 f}{\partial y\,\partial x} = f_{yx} = D_y D_x f,$$

$$\frac{\partial}{\partial y}\left(\frac{\partial f}{\partial y}\right) = \frac{\partial^2 f}{\partial y^2} = f_{yy} = (D_y)^2 f.$$

We likewise denote the third partial derivatives by

$$\frac{\partial}{\partial x}\left(\frac{\partial^2 f}{\partial x^2}\right) = \frac{\partial^3 f}{\partial x^3} = f_{xxx},$$

$$\frac{\partial}{\partial y}\left(\frac{\partial^2 f}{\partial x^2}\right) = \frac{\partial^3 f}{\partial y\,\partial x^2} = f_{yxx},$$

$$\frac{\partial}{\partial x}\left(\frac{\partial^2 f}{\partial x\,\partial y}\right) = \frac{\partial^3 f}{\partial x^2\,\partial y} = f_{xxy},$$

and so on, and in general the nth derivatives by

$$\frac{\partial}{\partial x}\left(\frac{\partial^{n-1} f}{\partial x^{n-1}}\right) = \frac{\partial^n f}{\partial x^n} = f_{x^n},$$

$$\frac{\partial}{\partial y}\left(\frac{\partial^{n-1} f}{\partial x^{n-1}}\right) = \frac{\partial^n f}{\partial y\,\partial x^{n-1}} = f_{yx^{n-1}},$$

and so on.

The different notations for partial derivatives have their respective advantages. Writing $\partial f(x, y)/\partial x$ or $D_x f(x, y)$ for the partial derivative of the function $f(x, y)$ with respect to its first argument emphasizes that differentiation has the character of an *operator* D_x or $\partial/\partial x$ acting on the function, written symbolically as a *factor* multiplying the function. The notation for higher derivatives is consistent with this idea of a product:

$$\frac{\partial}{\partial y}\left(\frac{\partial}{\partial x} f\right) = \frac{\partial^2}{\partial y\,\partial x} f = D_y D_x f.$$

[1]This is consistent with the general notation for symbolic products of operators (see Volume I, p. 53). Actually, the order in which differentiations are carried out turns out to be immaterial in most cases of interest (see p. 36).

A disadvantage of the operator notation is its clumsiness when it comes to indicating for what values of the independent variables the derivatives are taken. For example, if $f(x, y) = x^2 + 2xy + 4y^2$, then its x-derivative at the point $x = 1$, $y = 2$ can be written as

$$\left(\frac{\partial f(x, y)}{\partial x}\right)_{\substack{x=1\\y=2}} = f_x(1, 2) = (2x + 2y)_{\substack{x=1\\y=2}} = 6.$$

We should not write it simply as

$$\frac{\partial f(1, 2)}{\partial x}$$

since $f(1, 2)$ has the constant value 21 and hence has 0 as its x-derivative.

Just as in the case of one independent variable, the possession of derivatives is a special property of a function, not enjoyed even by all continuous functions.[1] All the same, this property is possessed by all functions of practical importance, except perhaps at isolated exceptional points or curves.

Exercises 1.4 a

1. Find $\partial z/\partial x$, $\partial z/\partial y$ for each of the following:
   (a) $z = ax^n + by^m$, a, b, m, n constants
   (b) $z = 2xe^{y^2} + 3y$
   (c) $z = 2\dfrac{x}{y} + 3\dfrac{y}{x}$
   (d) $z = \arctan \dfrac{y}{x^2}$
   (e) $z = x^2y^{3/2}$
   (f) $z = y^x$
   (g) $z = x^{1/2}\, y^{3/4}$
   (h) $z = 3^{x/y}$
   (i) $z = \log\left(x + \dfrac{y}{x^2}\right)$
   (j) $z = \cos\,(x^2 + y)$
   (k) $z = \tan\,(xy^3 + e^x)$
   (l) $z = \dfrac{\cos x}{\sin y}$
   (m) $z = xe^y + ye^x$
   (n) $z = x\sqrt{x^2 + y^2}$

2. Find the first partial derivatives of the following:
   (a) $\sqrt[3]{x^2 + y^2}$
   (b) $\sin\,(x^2 - y)$
   (d) $\dfrac{1}{\sqrt{1 + x + y^2 + z^2}}$
   (e) $y \sin xz$

[1]For an explanation of the term "differentiable", which implies more than that the partial derivatives with respect to x and y exist, see pp. 41–42.

(c) e^{x-y} (f) $\log \sqrt{1+x^2+y^2}$

3. Find all the first and second partial derivatives of the following:
   (a) xy
   (b) $\log xy$
   (c) $\tan(\text{arc tan } x + \text{arc tan } y)$
   (d) x^y
   (e) $e^{(x^y)}$
4. Let $w = f(x, y, z) = (\cos x/\sin y)e^z$. Find f_x, f_y, f_z, for $x = \pi$, $y = \pi/2$, $z = \log 3$.
5. For $f(x, y) = y \cosh x + x \sinh y$, find $f_x{}^2 + f_y{}^2$ at $x = 0$, $y = 0$.
6. Show that the functions $u = e^x \cos y$, $v = e^x \sin y$, satisfy the conditions $u_x = v_y$, $u_y = -v_x$.
7. Show that the functions of Exercise 6 satisfy the partial differential equation

$$f_{xx} + f_{yy} = 0.$$

   Do the same for the functions
   (a) $\log \sqrt{x^2+y^2}$
   (b) $\text{arc tan} \dfrac{y}{x}$
   (c) $\dfrac{y}{x^2+y^2}$
   (d) $3x^2y - y^3$
   (e) $\sqrt{x+\sqrt{x^2+y^2}}$
8. For $r = \sqrt{x^2+y^2+z^2}$, find $r_{xx} + r_{yy} + r_{zz}$.
9. Find a constant a for which if $z = y^3 + ayx^2$, then $z_{xx} + z_{yy} = 0$.
10. Prove that the function

$$f(x_1, x_2, \ldots, x_n) = \frac{1}{(x_1{}^2 + x_2{}^2 + \cdots + x_n{}^2)^{(n-2)/2}}$$

    satisfies the equation

$$f_{x_1x_1} + f_{x_2x_2} + \cdots + f_{x_nx_n} = 0.$$

Problems 1.4 a

1. How many nth derivatives has a function of three variables? of k variables?
2. Give an example of a function $f(x, y)$ for which f_x exists and f_y does not.
3. Find a function $f(x, y)$ that is a function of $(x^2 + y^2)$ and is also a product of the form $\psi(x)\psi(y)$; that is, solve the equation

$$f(x, y) = \phi(x^2 + y^2) = \psi(x)\psi(y)$$

   for the unknown functions.

4. Prove that any function of the form

$$u(x, y, z) = \frac{f(t+r)}{r} + \frac{g(t-r)}{r}$$

(where $r^2 = x^2 + y^2 + z^2$), satisfies the equation

$$u_{xx} + u_{yy} + u_{zz} = u_{tt}.$$

b. Examples

In practice, partial differentiation involves nothing that the student has not already met. For, according to the definition, all the independent variables are to be kept constant except the one with respect to which we are differentiating. Therefore, we have merely to regard the other variables as constants and carry out the differentiation according to the rules by which we differentiate functions of a single independent variable. We list some partial derivatives of several simple functions.

1. Function:

$$f(x, y) = xy$$

First derivatives:

$$f_x = y, \qquad f_y = x$$

Second derivatives:

$$f_{xx} = 0, \qquad f_{xy} = f_{yx} = 1, \qquad f_{yy} = 0$$

2. Function:

$$f(x, y) = \sqrt{x^2 + y^2}$$

First derivatives:

$$f_x = \frac{x}{\sqrt{x^2 + y^2}} \qquad f_y = \frac{y}{\sqrt{x^2 + y^2}}$$

[Thus, for the radius vector $r = \sqrt{x^2 + y^2}$ from the origin to the point (x, y), the partial derivatives with respect to x and to y are given by $\cos \phi = x/r$, and $\sin \phi = y/r$, where ϕ is the angle that the radius vector makes with the positive direction of the x-axis.]

Second derivatives:

$$f_{xx} = \frac{y^2}{\sqrt{x^2 + y^2)^3}} = \frac{\sin^2 \phi}{r},$$

$$f_{xy} = f_{yx} = -\frac{xy}{\sqrt{(x^2 + y^2)^3}} = -\frac{\sin \phi \cos \phi}{r},$$

$$f_{yy} = \frac{x^2}{\sqrt{(x^2 + y^2)^3}} = \frac{\cos^2 \phi}{r}.$$

3. Reciprocal of the radius vector in three dimensions:

$$f(x, y, z) = \frac{1}{\sqrt{x^2+y^2+z^2}} = \frac{1}{r}$$

First derivatives:

$$f_x = -\frac{x}{\sqrt{(x^2 + y^2 + z^2)^3}} = -\frac{x}{r^3},$$

$$f_y = -\frac{y}{\sqrt{(x^2 + y^2 + z^2)^3}} = -\frac{y}{r^3},$$

$$f_z = -\frac{z}{\sqrt{(x^2 + y^2 + z^2)^3}} = -\frac{z}{r^3};$$

Second derivatives:

$$f_{xx} = -\frac{1}{r^3} + \frac{3x^2}{r^5}, \quad f_{yy} = -\frac{1}{r^3} + \frac{3y^2}{r^5}, \quad f_{zz} = -\frac{1}{r^3} + \frac{3z^2}{r^5},$$

$$f_{xy} = f_{yx} = \frac{3xy}{r^5}, \quad f_{yz} = f_{zy} = \frac{3yz}{r^5}, \quad f_{zx} = f_{xz} = \frac{3zx}{r^5}.$$

From this we see that for the function $f = \dfrac{1}{\sqrt{x^2 + y^2 + z^2}}$ the equation

$$f_{xx} + f_{yy} + f_{zz} = -\frac{3}{r^3} + \frac{3(x^2 + y^2 + z^2)}{r^5} = 0$$

holds for all values of x, y, z except 0, 0, 0; we say, the function $f(x, y, z) = 1/r$ satisfies the *partial differential equation* ("Laplace equation")

$$f_{xx} + f_{yy} + f_{zz} = 0.$$

4. Function:

$$f(x, y) = \frac{1}{\sqrt{y}} e^{-(x-a)^2/4y}$$

First derivatives:

$$f_x = \frac{-(x-a)}{2y^{3/2}} e^{-(x-a)^2/4y},$$

$$f_y = \left(\frac{-1}{2y^{3/2}} + \frac{(x-a)^2}{4y^{5/2}}\right) e^{-(x-a)^2/4y}$$

Second derivatives:

$$f_{xx} = \left(\frac{-1}{2y^{3/2}} + \frac{(x-a)^2}{4y^{5/2}}\right) e^{-(x-a)^2/4y},$$

$$f_{xy} = f_{yx} = \left(\frac{3}{4}\frac{x-a}{y^{5/2}} - \frac{(x-a)^3}{8y^{7/2}}\right) e^{-(x-a)^2/4y},$$

$$f_{yy} = \left(\frac{3}{4}\frac{1}{y^{5/2}} - \frac{1}{2}\frac{(x-a)^2}{y^{7/2}} + \frac{(x-a)^4}{16y^{9/2}}\right) e^{-(x-a)^2/4y}.$$

The partial differential equation $f_{xx} - f_y = 0$ is therefore satisfied identically in x and y.

c. *Continuity and the Existence of Partial Derivatives*

For a function of a single variable, the existence of the derivative at a point implies the continuity of the function at that point (cf. Volume I, p. 166). In contrast to this, the possession of partial derivatives does *not* imply the continuity of a function of two variables: for example, the function $u(x, y) = 2xy/(x^2 + y^2)$, with $u(0, 0) = 0$, has partial derivatives everywhere, and yet we have already seen (p. 18) that it is discontinuous at the origin. Geometrically speaking, the existence of partial derivatives restricts the behavior of the function in the directions of the x- and y-axes only and not in other directions. Nevertheless, the possession of *bounded* partial derivatives does imply continuity, as is stated by the following theorem:

If a function $f(x, y)$ has partial derivatives f_x and f_y everywhere in an open set R, and these derivatives everywhere satisfy the inequalities

$$|f_x(x, y)| < M, \qquad |f_y(x, y)| < M,$$

where M is independent of x and y, then $f(x, y)$ is continuous everywhere in R.[1]

For the proof, we consider two points with coordinates (x, y) and $(x + h, y + k)$, respectively, both lying in the region R. We further assume that the two line segments joining these points to the point $(x + h, y)$ both lie entirely in R; this is certainly true if (x, y) is a point interior to R and the point $(x + h, y + k)$ lies sufficiently close to (x, y). We then have

$$f(x + h, y + k) - f(x, y) = \{f(x + h, y + k) - f(x + h, y)\} + \{f(x + h, y) - f(x, y)\}. \tag{7}$$

The two terms in the first bracket on the right differ only in y; those in the second bracket, only in x. We can therefore apply the ordinary mean value theorem of the differential calculus (Volume I, p. 174) to the first bracket as a function of y alone and to the second bracket as a function of x alone. We thus obtain the relation

$$f(x + h, y + k) - f(x, y) = kf_y(x + h, y + \theta_1 k) + hf_x(x + \theta_2 h, y), \tag{8}$$

where θ_1 and θ_2 are numbers between 0 and 1. In other words, the derivative with respect to y is to be formed for a point of the vertical line joining $(x + h, y)$ to $(x + h, y + k)$, and the derivative with respect to x is to be formed for a point of the horizontal line joining (x, y) and $(x + h, y)$. Since by hypothesis both derivatives are less than M in absolute value, it follows that

$$|f(x + h, y + k) - f(x, y)| \leq M(|h| + |k|). \tag{9}$$

For sufficiently small values of h and k the right-hand side is itself arbitrarily small, and the continuity of $f(x, y)$ is proved.[2]

[1]This applies even, as the proof shows, to *boundary points* of the domain, provided they can be joined to any neighboring points of the domain by a broken line consisting of two segments parallel to the axes and f is defined properly at the boundary point.

[2]If the domain of f is a rectangle with sides parallel to the axes, the inequality holds for any two points (x, y) and $(x + h, y + k)$ in the domain. It follows then that f is even Lipschitz-continuous (see p. 19).

Exercises 1.4c

1. State and prove for a function of three variables $f(x, y, z)$ that the existence and boundedness of the first partial derivatives are sufficient for the continuity of f.
2. Show that the following functions $f(x, y)$ are continuous:

(a) $f(x, y) = \begin{cases} e^{-1/(x^2 + y^2)}, & x, y \neq 0 \\ 0, & x = 0, y = 0 \end{cases}$

(b) $f(x, y) = \begin{cases} (x^4 + y^4) \log (x^2 + y^2), & x, y \neq 0 \\ 0, & x = y = 0. \end{cases}$

d. Change of the Order of Differentiation

In all examples of partial differentiation given on pp. 32–34 we find that $f_{yx} = f_{xy}$; in other words, it makes no difference whether we differentiate first with respect to x and then with respect to y or first with respect to y and then with respect to x. This is true generally under the conditions of the following theorem:

If the "mixed" partial derivatives f_{xy} and f_{yx} of a function $f(x, y)$ are continuous in an open set R, then the equation

$$f_{yx} = f_{xy} \tag{10}$$

holds throughout R; that is, the order of differentiation with respect to x and to y is immaterial.

The proof, like that of the previous subsection, is based on the mean value theorem of the differential calculus. We consider the four points (x, y), $(x + h, y)$, $(x, y + k)$, and $(x + h, y + k)$, where $h \neq 0$ and $k \neq 0$. If (x, y) is a point of the open set R and if h and k are small enough, all four of these points belong to R. We now form the expression

$$A = f(x + h, y + k) - f(x + h, y) - f(x, y + k) + f(x, y). \tag{11}$$

By introducing the function

$$\phi(x) = f(x, y + k) - f(x, y)$$

of the variable x and regarding the variable y merely as a "parameter," A assumes the form

$$A = \phi(x + h) - \phi(x).$$

Applying the mean value theorem of differential calculus yields

$$A = h\phi'(x + \theta h),$$

where θ lies between 0 and 1. From the definition of $\phi(x)$, however, we have

$$\phi'(x) = f_x(x, y + k) - f_x(x, y),$$

and since we have assumed that the "mixed" second partial derivative f_{yx} does exist, we can again apply the mean value theorem and find that

$$A = hkf_{yx}(x + \theta h, y + \theta' k), \tag{12}$$

where θ and θ' denote two unspecified numbers between 0 and 1.

In exactly the same way we may introduce the function

$$\psi(y) = f(x + h, y) - f(x, y)$$

and express A as

$$A = \psi(y + k) - \psi(y).$$

We thus arrive at the equation

$$A = hkf_{xy}(x + \theta_1 h, y + \theta_1' k),$$

where $0 < \theta_1 < 1$ and $0 < \theta_1' < 1$, and if we equate the two expressions for A, we obtain the equation

$$f_{yx}(x + \theta h, y + \theta' k) = f_{xy}(x + \theta_1 h, y + \theta_1' k).$$

If here we let h and k tend simultaneously to 0 and recall that the derivatives $f_{xy}(x, y)$ and $f_{yx}(x, y)$ are continuous at the point (x, y), we immediately obtain

$$f_{yx}(x, y) = f_{xy}(x, y),$$

which was to be proved.[1]

[1]For more refined investigations it is often useful to know that the theorem on the reversibility of the order of differentiation can be proved with weaker hypotheses. It is, in fact, sufficient to assume that in addition to the first partial derivatives f_x and f_y, only one mixed partial derivative, say f_{yx}, exists and that this derivative is continuous at the point in question. To prove this, we return to equation (11), divide by hk, and then let k alone tend to 0. Then the right-hand side has a limit, and therefore the left-hand side also has a limit, and

$$\lim_{k\to 0} \frac{A}{kh} = \frac{f_y(x + h, y) - f_y(x, y)}{h}.$$

Further, it was proved above with the sole assumption that f_{yx} exists that

$$\frac{A}{hk} = f_{yx}(x + \theta h, y + \theta' k).$$

By virtue of the assumed continuity of f_{yx}, we find that for arbitrary $\varepsilon > 0$ and for

The theorem on the reversibility of the order of differentiation (i.e., on the commutativity of the differentiation operators D_x and D_y) has far-reaching consequences. In particular, we see that the number of distinct derivatives of the second order and of higher orders of functions of several variables is decidedly smaller than we might at first have expected. If we assume that all the derivatives that we are about to form are continuous functions of the independent variables in the region under consideration and if we apply our theorem to the functions $f_x(x, y)$, $f_y(x, y)$, $f_{xy}(x, y)$, and so on, instead of to the function $f(x, y)$, we arrive at the equations

$$f_{xxy} = f_{xyx} = f_{yxx},$$
$$f_{xyy} = f_{yxy} = f_{yyx},$$
$$f_{xxyy} = f_{xyxy} = f_{xyyx} = f_{yxxy} = f_{yxyx} = f_{yyxx},$$

and in general we have the following result:

In the repeated differentiation of a function of two independent variables the order of the differentiations may be changed at will, provided only that the derivatives in question are continuous functions.[1]

---

all sufficiently small values of h and k

$$f_{yx}(x, y) - \varepsilon < f_{yx}(x + \theta h, y + \theta' k) < f_{yx}(x, y) + \varepsilon,$$

whence it follows that

$$f_{yx}(x, y) - \varepsilon \leqq \frac{f_y(x + h, y) - f_y(x, y)}{h} \leqq f_{yx}(x, y) + \varepsilon$$

or

$$\lim_{h \to 0} \frac{f_y(x + h, y) - f_y(x, y)}{h} = f_{yx}(x, y),$$

that is,

$$f_{xy}(x, y) = f_{yx}(x, y).$$

[1]It is of fundamental interest to show by means of an example that without the assumption of the continuity of the second derivative f_{xy} or f_{yx} the theorem need not be true and f_{xy} can differ from f_{yx}. This is exemplified by the function

$$f(x, y) = xy \frac{x^2 - y^2}{x^2 + y^2}, \quad f(0, 0) = 0,$$

for which all the partial derivatives of second order exist but are not continuous. We find that

$$f_x(0, y) = \lim_{x \to 0} \frac{f(x, y) - f(0, y)}{x} = \lim_{x \to 0} y \frac{x^2 - y^2}{x^2 + y^2} = -y,$$

$$f_y(x, 0) = \lim_{y \to 0} \frac{f(x, y) - f(x, 0)}{y} = \lim_{y \to 0} x \frac{x^2 - y^2}{x^2 + y^2} = x,$$

With our assumptions about continuity, a function of two variables has *three* partial derivatives of the second order,

$$f_{xx}, \quad f_{xy}, \quad f_{yy};$$

four partial derivatives of the third order,

$$f_{xxx}, \quad f_{xxy}, \quad f_{xyy}, \quad f_{yyy};$$

and in general $(n + 1)$ partial derivatives of the nth order,

$$f_{x^n}, f_{x^{n-1}y}, f_{x^{n-2}y^2}, \dots, f_{xy^{n-1}}, f_{y^n}.$$

It is obvious that similar statements also hold for functions of more than two independent variables. For we can apply our proof equally well to the interchange of differentiations with respect to x and z or with respect to y and z, and so on, for each interchange of two successive differentiations involves only two independent variables at a time.

Exercise 1.4d

1. Obtain $\partial^2 z/(\partial x\, \partial y)$ and $\partial^2 z/(\partial y\, \partial x)$ to confirm their equality.
   (a) $z = (ax + by)^2$
   (b) $z = \sqrt{ax + by}$
   (c) $z = f(ax + by)$
   (d) $z = y\, e^x$
   (e) $z = \log \dfrac{x + y}{x}$
   (f) $z = e^{\cos(y^2 + x)}$
2. Find all partial derivatives through the third order of the following functions:
   (a) $f(x, y) = x^y$
   (b) $f(x, y) = \cosh xy$
   (c) $f(x, y) = ax^2 + bxy + cy^2$
   (d) $f(x, y) = \dfrac{x}{y} + \dfrac{y}{x}$
   (e) $f(x, y) = 2 \cos x + 3 \sin (y - x)$.
3. Show for $f(x, y) = \log (e^x + e^y)$ that $f_x + f_y = 1$ and $f_{xx} f_{yy} - (f_{xy})^2 = 0$.

Problems 1.4d

1. (a) Show that a function of the form $u(x, y) = f(x)\, g(y)$ satisfies the partial differential equation

---

and consequently

$$f_{yx}(0, 0) = -1 \quad \text{and} \quad f_{xy}(0, 0) = +1.$$

These two expressions are different, which by the above theorem can only be caused by the discontinuity of f_{xy} at the origin.

$$u\,u_{xy} - u_x u_y = 0.$$

(b) Prove the converse statement.

2. Define $f(x, y)$ as:

$$f(x, y) = \begin{cases} x^2 \operatorname{arc\,tan} \dfrac{y}{x} - y^2 \operatorname{arc\,tan} \dfrac{x}{y}, & x, y \neq 0, \\ 0 & \text{for } x = 0 \text{ or } y = 0. \end{cases}$$

Show that $f_{xy}(0, 0) = -1$, $f_{yx} = 1$.

1.5 The Total Differential of a Function and Its Geometrical Meaning

a. The Concept of Differentiability

For functions $y = f(x)$ of one variable, the existence of a derivative is intimately connected with the possibility of approximating the function f in the neighborhood of a value x by a linear function; geometrically, this corresponds to approximating the graph of f by its tangent. By definition, the function f has a derivative at the point x if the limit

$$\lim_{h \to 0} \frac{f(x + h) - f(x)}{h} = A$$

exists; the value A of the limit is denoted by $f'(x)$. Thus, differentiability of f at the point x means that for fixed x the increment $\Delta f = f(x + h) - f(x)$ corresponding to the increment $h = \Delta x$ of the independent variable can be written in the form

$$\Delta f = f(x + h) - f(x) = Ah + \varepsilon h,$$

where A does not depend on h and $\lim_{h \to 0} \varepsilon = 0$. Letting $x + h = \xi$, we may say that $f(\xi)$ *is approximated by a linear function of* ξ, *namely* $\phi(\xi) = f(x) + A(\xi - x)$, *with an error that is of higher than the first order in* $\xi - x$:

$$f(\xi) - \phi(\xi) = \varepsilon \cdot (\xi - x) = o(\xi - x) \qquad \text{for} \qquad \xi \to x.$$

Of course, the graph of this linear function $\eta = \phi(\xi) = f(x) + f'(x)(\xi - x)$ in running coordinates ξ, η is just the tangent to the graph of f at the point $(x. y)$. Formulated differently, differentiability of f at x means that the increment Δf considered as a function of $h = \Delta x$ can be approximated by the linear function $df = f'(x)\,h = f'(x)\,dx$ within an error that is of higher than the first order in h.[1]

[1]For the independent variable x we have $dx = 1 \cdot h = h = \Delta x$.

These ideas can be extended in a perfectly natural way to functions of two and more variables.

We say that the function $u = f(x, y)$ is *differentiable* at the point (x, y) if it can be approximated in the neighborhood of this point by a linear function, that is, if it can be represented in the form

$$f(x + h, y + k) = Ah + Bk + C + \varepsilon\sqrt{h^2 + k^2} \tag{13}$$

where A, B, and C are independent of the variables h and k and where ε tends to 0 as h and k do. In other words, the difference between the function $f(x + h, y + k)$ at the point $(x + h, y + k)$ and the function $Ah + Bk + C$, which is linear in h and k, must be of order of magnitude $o(\rho)$, where $\rho = \sqrt{h^2 + k^2}$ denotes the distance of the point $(x + h, y + k)$ from the point (x, y).

If such an approximate representation is possible, it follows at once that the function $f(x, y)$ is continuous and has partial derivatives with respect to x and to y at the point (x, y) and that

$$A = f_x(x, y), \qquad B = f_y(x, y), \qquad C = f(x, y).$$

For first of all we find from (13) for $h = k = 0$ that $f(x, y) = C$. Moreover, $\lim\limits_{\substack{h \to 0 \\ k \to 0}} f(x + h, y + k) = C = f(x, y)$.

Thus f is continuous at the point (x, y). Setting $k = 0$ in (13) and dividing by h yields the relation

$$\frac{f(x + h, y) - f(x, y)}{h} = A + \varepsilon.$$

Since ε tends to 0 as h tends to 0, the left-hand side has a limit, and that limit is A. Similarly, we obtain the equation $f_y(x, y) = B$.

Conversely, we shall prove the fundamental fact:

A function $u = f(x, y)$ is differentiable in the sense just defined—that is, it can be approximated by a linear function with an error $o(\rho)$ as in (13)—if it possesses *continuous* derivatives of the first order at the point in question.

Indeed, we can write the increment

$$\Delta u = f(x + h, y + k) - f(x, y)$$

of the function in the form

$$\Delta u = f(x + h, y + k) - f(x, y + k) + f(x, y + k) - f(x, y).$$

As before (p. 31), the two parentheses can be expressed in the form

$$\Delta u = hf_x(x + \theta_1 h, y + k) + kf_y(x, y + \theta_2 k),$$

where $0 < \theta_1, \theta_2 < 1$, using the ordinary mean value theorem of differential calculus. Since by hypothesis the partial derivatives f_x and f_y are continuous at the point (x, y), we can write

$$f_x(x + \theta_1 h, y + k) = f_x(x, y) + \varepsilon_1$$

and

$$f_y(x, y + \theta_2 k) = f_y(x, y) + \varepsilon_2$$

where the numbers ε_1 and ε_2 tend to 0 as h and k do. We thus obtain

$$\begin{aligned}\Delta u &= hf_x(x, y) + kf_y(x, y) + \varepsilon_1 h + \varepsilon_2 k \\ &= hf_x(x, y) + kf_y(x, y) + o(\sqrt{h^2 + k^2}),\end{aligned}$$

and this equation expresses the differentiability of f.[1]

We shall occasionally refer to a function with continuous first partial derivatives as a *continuously differentiable* function or as a function *of class* C^1. We see that functions of class C^1 are differentiable. If in addition all the second-order partial derivatives are continuous, we say that the function is *twice continuously differentiable*, or *of class* C^2, and so on. The continuous functions are also referred to as the functions of class C^0.[2]

Exercises 1.5a

1. Show that each of the following functions is not differentiable at the origin:

   (a) $f(x, y) = \sqrt{x} \cos y$

   (b) $f(x, y) = \sqrt{|xy|}$

[1]If we assume merely the existence, and not the continuity, of the derivatives f_x and f_y, the function need not be differentiable (cf. p. 34).

[2]These definitions of class C^1, C^2, and so on apply only to functions f whose domain is an *open* set, since partial derivatives have been defined only for interior points of the domain. One can extend the notion of class to functions f with a nonopen domain R; it then means that the derivatives of f in question exist at all interior points of R and coincide at those points with functions that are defined and continuous throughout R.

(c) $f(x, y) = \begin{cases} \dfrac{2xy}{\sqrt{x^2 + y^2}}, & (x, y) \neq (0, 0) \\ 0, & (x, y) = (0, 0). \end{cases}$

2. For $g(x)$, $h(y)$ continuous functions of x, y in the intervals $[x_0, x_1]$, $[y_0, y_1]$, respectively, show that the function $f(x, y) = \left(\int_{x_0}^{x} g(s)\, ds\right) \times \left(\int_{y_0}^{y} h(t)\, dt\right)$ is differentiable at (x, y) for $x_0 \leq x \leq x_1$, $y_0 \leq y \leq y_1$.

Problems 1.5a

1. Suppose that in a neighborhood of the point (a, b), $f(x, y) = f(a, b) + h f_x(a, b) + k f_y(a, b) + o(\sqrt{h^2 + k^2})$, where $h = x - a$ and $k = y - b$. On the assumption that f_x and f_y exist at (a, b) but are not necessarily continuous there, prove that f is continuous at (a, b).

b. Directional Derivatives

A basic property of differentiable functions f is that they not only possess partial derivatives with respect to x and y—or, as we also say, in the x- and y-directions—but that they have derivatives in any direction and that these derivatives can all be expressed in terms of f_x and f_y. By the ***derivative in the direction*** α we mean the rate of change of f at the point (x, y) with respect to distance as we approach (x, y) along the ray that forms the angle α with the positive x-axis. The points $(x + h, y + k)$ of the ray are the ones for which h and k have the form

$$h = \rho \cos \alpha, \qquad k = \rho \sin \alpha,$$

where $\rho = \sqrt{h^2 + k^2}$ is the distance of $(x + h, y + k)$ from (x, y). Along the ray f becomes a function of ρ given by

$$f(x + \rho \cos \alpha, y + \rho \sin \alpha).$$

The derivative of f at the point (x, y) in the direction α is defined as the derivative of $f(x + \rho \cos \alpha, y + \rho \sin \alpha)$ with respect to ρ at $\rho = 0$ and denoted by $D_{(\alpha)} f(x, y)$. Thus,

$$\begin{aligned} D_{(\alpha)} f(x, y) &= \left(\frac{d}{d\rho} f(x + \rho \cos \alpha, y + \rho \sin \alpha)\right)_{\rho=0} \\ &= \lim_{\rho \to 0} \frac{f(x + \rho \cos \alpha, y + \rho \sin \alpha) - f(x, y)}{\rho}, \end{aligned}$$

provided the limit exists. In particular, we obtain for $\alpha = 0$ and $\alpha = \pi/2$ the partial derivatives of f:

$$D_{(0)}f(x, y) = \lim_{\rho\to 0}\frac{f(x+\rho, y) - f(x, y)}{\rho} = f_x(x, y)$$

$$D_{(\pi/2)}f(x, y) = \lim_{\rho\to 0}\frac{f(x, y+\rho) - f(x, y)}{\rho} = f_y(x, y).$$

If $f(x, y)$ is differentiable, we have

$$f(x+h, y+k) - f(x, y) = hf_x + kf_y + \varepsilon\rho \tag{14}$$
$$= \rho(f_x \cos\alpha + f_y \sin\alpha + \varepsilon)$$

Let ρ tend to 0; then, since ε tends to 0, we obtain for the derivative of f in the direction α the expression

$$D_{(\alpha)}f(x, y) = f_x \cos\alpha + f_y \sin\alpha. \tag{14a}$$

Thus the directional derivative $D_{(\alpha)}f$ is a linear combination of the derivatives f_x and f_y in the x- and y-directions with the coefficients $\cos\alpha$ and $\sin\alpha$. This result holds in particular whenever the derivatives f_x and f_y exist and are continuous at the point in question.

Taking, for example, for $f(x, y)$ the distance $r = \sqrt{x^2 + y^2}$ from the origin to the point (x, y), we have the partial derivatives

$$r_x = \frac{x}{\sqrt{x^2+y^2}} = \frac{x}{r} = \cos\theta \qquad \text{and} \qquad r_y = \frac{y}{\sqrt{x^2+y^2}} = \frac{y}{r} = \sin\theta,$$

where θ denotes the angle that the radius vector makes with the x-axis. Consequently, in the direction α the function r has the derivative

$$D_{(\alpha)}r = r_x \cos\alpha + r_y \sin\alpha = \cos\theta\cos\alpha + \sin\theta\sin\alpha = \cos(\theta - \alpha);$$

in particular, in the direction of the radius vector itself (i.e., in the direction away from the origin), this derivative has the value 1, while in the directions perpendicular to the radius vector, it has the value 0.

The function x has, in the direction of the radius vector, the derivative $D_\theta(x) = \cos\theta$, and the function y, the derivative $D_\theta(y) = \sin\theta$; in the direction perpendicular to the radius vector these functions have the derivatives $D_{(\theta+\pi/2)}\, x = -\sin\theta$ and $D_{(\theta+\pi/2)}\, y = \cos\theta$, respectively.

The derivative of a function $f(x, y)$ in the direction of the radius vector is in general denoted by $\partial f(x,y)/\partial r$. It is really the partial derivative with respect to r of $f(r \cos \theta, r \sin \theta)$ considered as a function of r and θ. Thus, we have the relation

$$\frac{\partial f}{\partial r} = \cos \theta \frac{\partial f}{\partial x} + \sin \theta \frac{\partial f}{\partial y},$$

which we write conveniently in symbolic form as the identity

$$\frac{\partial}{\partial r} = \cos \theta \frac{\partial}{\partial x} + \sin \theta \frac{\partial}{\partial y}$$

between the differentiation operators $\partial/\partial r$, $\partial/\partial x$, $\partial/\partial y$.

It is worth noting that we also obtain the derivative of the function $f(x, y)$ in the direction α if, instead of allowing the point Q with coordinates $(x + h, y + k)$ to approach the point P with coordinates (x, y) along a straight line with the direction α, we let Q approach P along an arbitrary curve whose tangent at P has the direction α. For then if the line PQ has the direction β, we can write $h = \rho \cos \beta$, $k = \rho \sin \beta$, and in the formulae (14) used in the proof above we have to replace α by β. But since by hypothesis β tends to α as $\rho \to 0$, we obtain the same expression as for $D_{(\alpha)} f(x, y)$.

In the same way, a differentiable function $f(x, y, z)$ of three independent variables can be differentiated in a given direction. We suppose that the direction is specified by the cosines of the three angles that it forms with the coordinate axes. If we call these three angles α, β, γ and if we consider two points (x, y, z) and $(x + h, y + k, z + l)$, where

$$h = \rho \cos \alpha, \qquad k = \rho \cos \beta, \qquad l = \rho \cos \gamma,$$

then just as in (14a), we obtain the expression

$$f_x \cos \alpha + f_y \cos \beta + f_z \cos \gamma \tag{14b}$$

for the derivative in the direction given by the angles (α, β, γ).

Exercises 1.5b

1. What is the geometrical interpretation of the derivative $D_{(\alpha)} f(x, y)$ of the function f in the direction defined by the angle of inclination α?

2. Find $D_{(\alpha)}f(x_0, y_0)$, $\alpha = 0, 30°, 60°, 90°$ for the following functions:
   (a) $f(x, y) = ax + by$, a, b constants, $x_0 = y_0 = 0$
   (b) $f(x, y) = ax^2 + y^2b$, $x_0 = y_0 = 1$, (a, b constants)
   (c) $f(x, y) = x^2 - y^2$, $x_0 = 1$, $y_0 = 2$
   (d) $f(x, y) = \sin x + \cos y$, $x_0 = y_0 = 0$
   (e) $f(x, y) = e^x \cos y$, $x_0 = 0$, $y_0 = \pi$
   (f) $f(x, y) = \sqrt{2x^2 + y^2}$, $x_0 = 1$, $y_0 = 1$
   (g) $f(x, y) = \cos(x + y)$, $x_0 = 0$, $y_0 = 0$.
3. Find the directional derivatives of each of the following functions as indicated:
   (a) $z^2 - x^2 - y^2$ at $(1, 0, 1)$ in the direction of $(4, 3, 0)$.
   (b) $xyz - xy - yz - zx + x + y + z$ at $(2, 2, 1)$
   in the direction of $(2, 2, 0)$.
   (c) $xz^2 + y^2 + z^3$ at $(1, 0, -1)$ in the direction of $(2, 1, 0)$.
4. Give an example of a function that has derivatives in every direction at a point yet is not differentiable at that point.
5. Show for $f(x, y) = \sqrt[3]{xy}$ that f is continuous and that the partial derivatives $\partial z/\partial x$ and $\partial z/\partial y$ exist at the origin but that the directional derivatives in all other directions do not exist.
6. Let $f(x,y) = xy + \sqrt{2x^2 + y^2}$, $r = \sqrt{x^2 + y^2}$, $y/x = \tan\theta$. Find $\partial^2 f/\partial r^2$ for $\theta = 0°, 30°, 60°, 90°$, and $x, y = 1$.

c. *Geometrical Interpretation of Differentiability. The Tangent Plane*

For a function $z = f(x, y)$ all these concepts can easily be illustrated geometrically. We recall that the partial derivative with respect to x is the slope of the tangent to the curve in which the surface representing the relation $z = f(x, y)$ is intersected by a plane perpendicular to the x,y-plane and parallel to the x-axis. In the same way, the derivative in the direction α gives the slope of the tangent to the curve in which the surface is intersected by a plane through (x, y, z) that is perpendicular to the x, y-plane and makes the angle α with the x-axis. The formula $D_{(\alpha)}f(x, y) = f_x \cos\alpha + f_y \sin\alpha$ now enables us to calculate the slopes of the tangents to all such curves, that is, of all tangents to the surface at a given point, from the slopes of two such tangents.[1]

[1]For points (ξ, η, ζ) in that plane we have $\xi = x + \rho\cos\alpha$, $\eta = y + \rho\sin\alpha$, and thus for points on the curve of intersection,

We have approximated the differentiable function $\zeta = f(\xi, \eta)$ in the neighborhood of the point (x, y) by the linear function

$$\phi(\xi, \eta) = f(x, y) + (\xi - x)f_x + (\eta - y)f_y,$$

where ξ and η are the current coordinates. Geometrically, this linear function is represented by a plane, which by analogy with the tangent line to a curve we shall call the *tangent plane* to the surface. The difference between this linear function and the function $f(\xi, \eta)$ vanishes to a higher order than $\sqrt{h^2 + k^2}$ as $\xi - x = h$ and $\eta - y = k$ tend to 0. Recalling the definition of the tangent to a plane curve, however, this means that the line of intersection of the tangent plane with any plane perpendicular to the x, y-plane is the tangent to the corresponding curve of intersection. *We thus see that all these tangent lines to the surface at the point (x, y, z) lie in one plane, the tangent plane.*

This property is the geometrical expression of the differentiability of the function at the point (x, y, z) where $z = f(x, y)$. In running coordinates (ξ, η, ζ), the equation of the tangent plane at the point (x, y, z) is

$$\zeta - z = (\xi - x)f_x + (\eta - y)f_y.$$

As has already been shown on p. 41, the function is differentiable at a given point provided that the partial derivatives are continuous there. In contrast with the case of functions of one independent variable, the mere *existence* of the partial derivatives f_x and f_y is *not* sufficient to ensure the differentiability of the function. If the derivatives are not continuous at the point in question, the tangent plane to the surface at this point may fail to exist; or, analytically speaking, the difference between $f(x + h, y + k)$ and the function $f(x, y) + hf_x(x, y) + kf_y(x, y)$, which is linear in h and k, may fail to vanish to a higher order than $\sqrt{h^2 + k^2}$. This is clearly shown by a simple example:

$$\zeta = f(x + \rho \cos \alpha, y + \rho \sin \alpha).$$

Using ρ and ζ as coordinates, the slope of the tangent to the curve at $\zeta = z$, $\rho = 0$ is given by

$$\left(\frac{d\zeta}{d\rho}\right)_{\rho=0} = D_{(\alpha)}f(x, y).$$

Hence, the tangent has the equation

$$\zeta = z + \rho D_{(\alpha)}f(x,y) = f(x, y) + \rho \cos \alpha \, f_x(x, y) + \rho \sin \alpha \, f_y(x, y).$$

$$u = f(x, y) = \frac{xy}{\sqrt{x^2 + y^2}} \quad \text{if} \quad x^2 + y^2 \neq 0,$$

$$u = 0 \quad \text{if} \quad x = 0, y = 0.$$

If we introduce polar coordinates this becomes

$$u = \frac{r}{2} \sin 2\theta.$$

The first derivatives with respect to x and to y exist everywhere in the neighborhood of the origin and have the value 0 at the origin itself. These derivatives, however, are not continuous at the origin, for

$$u_x = y\left(\frac{1}{\sqrt{x^2 + y^2}} - \frac{x^2}{\sqrt{(x^2 + y^2)^3}}\right) = \frac{y^3}{\sqrt{(x^2 + y^2)^3}}.$$

If we approach the origin along the x-axis, u_x tends to 0, while if we approach along the y-axis, u_x tends to 1. This function is not differentiable at the origin; at that point no tangent plane to the surface $z = f(x, y)$ exists. For the equations $f_x(0, 0) = f_y(0, 0) = 0$ show that the tangent plane would have to coincide with the plane $z = 0$. But at the points of the line $\theta = \pi/4$, we have $\sin 2\theta = 1$ and $z = f(x, y) = r/2$; thus, the distance z of the point of the surface from the point of the plane does not, as must be the case with a tangent plane, vanish to a higher order than r. The surface is a cone with vertex at the origin, whose generators do not all lie in one plane.

Exercises 1.5c

1. Find the equation of the tangent plane to the surface defined by $z = f(x, y)$ at the point $P = (x_0, y_0)$ in each of the following cases:

   (a) $f(x, y) = 3x^2 + 4y^2$, $P = (0, 1)$

   (b) $f(x, y) = 2\cos(x - y) + 3\sin x$, $P = \left(\pi, \frac{\pi}{2}\right)$

   (c) $f(x, y) = \cosh(x + y)$, $P = (0, \log 2)$

   (d) $f(x, y) = \sqrt{x^2 + y^2}$, $P = (1, 2)$

   (e) $f(x, y) = e^{x \cos y}$, $P = \left(1, \frac{\pi}{4}\right)$

   (f) $f(x, y) = \cos \pi\, e^{xy}$, $P = (\log 2, 1)$

   (g) $f(x, y) = \int_0^{x^2+y^2} e^{-t^2}\, dt$, $P = (1, 1)$

   (h) $f(x, y) = ax^3 + bx^2y + cxy^2 + dy^3$, $P = (1, 1)$, (a, b, c, d constants)

2. Show that all tangent planes to a surface $z = y\,f(x/y)$ meet in a common point where f is any differentiable function of one variable.
3. Show that the tangent plane to the surface S: $z = f(x, y)$ at the point $P_0 = (x_0, y_0)$ is the limiting position of the plane passing through the three points (x_i, y_i, z_i), $i = 0, 1, 2$, of S where $P_1 = (x_1, y_1)$ and $P_2 = (x_2, y_2)$ approach P_0 from distinct directions, making an angle not equal to 0° or 180°.
4. Prove that the tangent plane to the quadric surface
$$ax^2 + by^2 + cz^2 = 1$$
at the point (x_0, y_0, z_0) is
$$ax_0x + by_0y + cz_0z = 1.$$

d. The Differential of a Function

As for functions of one variable, it is often convenient to have a special name and symbol for the linear part of the increment of a differentiable function $u = f(x, y)$ which occurs in formula (14),

$$\Delta u = f(x + h, y + k) - f(x, y) = hf_x(x, y) + kf_y(x, y) + \varepsilon\sqrt{h^2 + k^2}.$$

We call this linear part the *differential* of the function, and write

$$du = df(x, y) = \frac{\partial f}{\partial x}h + \frac{\partial f}{\partial y}k = \frac{\partial f}{\partial x}\Delta x + \frac{\partial f}{\partial y}\Delta y. \tag{15a}$$

The differential, sometimes called the *total differential*, is a function of *four* independent variables, namely, the coordinates x and y of the point under consideration and the increments h and k of the independent variables. We emphasize again that this has nothing to do with the vague concept of "infinitely small quantities." It simply means that du approximates to the increment $\Delta u = f(x + h, y + k) - f(x, y)$ of the function, with an error that is an arbitrarily small fraction ε of $\sqrt{h^2 + k^2}$, provided that h and k are sufficiently small quantities. For the independent variables x and y we find from (15a) that

$$dx = \frac{\partial x}{\partial x}\Delta x + \frac{\partial x}{\partial y}\Delta y = \Delta x \quad \text{and} \quad dy = \frac{\partial y}{\partial x}\Delta x + \frac{\partial y}{\partial x}\Delta y = \Delta y.$$

Hence, the differential $df(x, y)$ is written more commonly

$$df(x, y) = \frac{\partial f}{\partial x}dx + \frac{\partial f}{\partial y}dy = f_x(x, y)\,dx + f_y(x, y)\,dy. \tag{15b}$$

Incidentally, the differential completely determines the first partial derivatives of f. For example, we obtain the partial derivative $\partial f/\partial x$ from df, by putting $dy = 0$ and $dx = 1$.

We emphasize that the total differential of a function $f(x, y)$ as the linear approximation to Δf has no meaning unless the function is differentiable in the sense defined above (for which the continuity, but not the mere existence, of the two partial derivatives suffices).

If the function $f(x, y)$ also has continuous partial derivatives of higher order, we can form the differential of the differential $df(x, y)$; that is, we can multiply its partial derivatives with respect to x and y by $h = dx$ and $k = dy$, respectively, and then add these products. In this differentiation, we regard h and k as constants, corresponding to the fact that the differential $df = hf_x(x, y) + kf_y(x, y)$ is a function of the four independent variables x, y, h, and k. We thus obtain the *second differential*[1] of the function,

$$\begin{aligned} d^2f = d(df) &= \frac{\partial}{\partial x}\left(\frac{\partial f}{\partial x} h + \frac{\partial f}{\partial y} k\right)h + \frac{\partial}{\partial y}\left(\frac{\partial f}{\partial x} h + \frac{\partial f}{\partial y} k\right)k \\ &= \frac{\partial^2 f}{\partial x^2} h^2 + 2\frac{\partial^2 f}{\partial x\,\partial y} hk + \frac{\partial^2 f}{\partial y^2} k^2 \\ &= \frac{\partial^2 f}{\partial x^2} dx^2 + 2\frac{\partial^2 f}{\partial x\,\partial y} dx\,dy + \frac{\partial^2 f}{\partial y^2} dy^2. \text{ [2]} \end{aligned}$$

Similarly, we may form the *higher differentials*

$$d^2f = d(d^2f) = \frac{\partial^3 f}{\partial x^3} dx^3 + 3\frac{\partial^3 f}{\partial x^2\,\partial y} dx^2\,dy + 3\frac{\partial^3 f}{\partial x\,\partial y^2} dx\,dy^2 + \frac{\partial^3 f}{\partial y^3} dy^3,$$

$$\begin{aligned} d^4f = \frac{\partial^4 f}{\partial x^4} dx^4 &+ 4\frac{\partial^4 f}{\partial x^3\,\partial y} dx^3\,dy + 6\frac{\partial^4 f}{\partial x^2\,\partial y^2} dx^2\,dy^2 \\ &+ 4\frac{\partial^4 f}{\partial x\,\partial y^3} dx\,dy^3 + \frac{\partial^4 f}{\partial y^4} dy^4, \end{aligned}$$

and, as is easily shown by induction, in general

$$d^nf = \frac{\partial^n f}{\partial x^n} dx^n + \binom{n}{1}\frac{\partial^n f}{\partial x^{n-1}\,\partial y} dx^{n-1}\,dy + \cdots$$

---

[1]We shall later see (p. 68) that the differentials of higher order introduced formally here correspond exactly to the terms of the same order in the expansion of the function.

[2]Traditionally, one writes the powers $(dx)^2$, $(dx)^3$, $(dy)^2$, $(dy)^3$ of differentials simply as dx^2, dx^3, dy^2, dy^3. This is, of course, somewhat misleading, since they might be confused with $d(x^2) = 2x\,dx$, $d(x^3) = 3x^2\,dx$, and so on.

$$\cdots + \binom{n}{k} \frac{\partial^n f}{\partial x^{n-k}\, \partial y^k}\, dx^{n-k}\, dy^k + \cdots + \frac{\partial^n f}{\partial y^n}\, dy^n.$$

The last formula can be expressed symbolically by the equation

$$d^n f = \left(\frac{\partial}{\partial x}\, dx + \frac{\partial}{\partial y}\, dy\right)^n f$$

where the expression on the right is first to be expanded formally by the binomial theorem, and then the terms

$$\frac{\partial^n f}{\partial x^n}\, dx^n,\ \frac{\partial^n f}{\partial x^{n-1}\, \partial y}\, dx^{n-1}\, dy, \ldots, \frac{\partial^n f}{\partial y^n}\, dy^n$$

are to be substituted for

$$\left(\frac{\partial}{\partial x}\, dx\right)^n f,\ \left(\frac{\partial}{\partial x}\, dx\right)^{n-1}\left(\frac{\partial}{\partial y}\, dy\right) f, \ldots, \left(\frac{\partial}{\partial y}\, dy\right)^n f.$$

For calculations with differentials the rule

$$d(fg) = f\, dg + g\, df$$

holds good; this follows immediately from the rule for the differentiation of a product.

In conclusion, we remark that the discussion in this section can immediately be extended to functions of more than two independent variables.

Exercises 1.5d

1. Find the total differentials for the following functions:

(a) $z = x^2y^2 + 3xy^3 - 2y^4$

(b) $z = \dfrac{xy}{x^2 + 2y^2}$

(c) $z = \log(x^4 - y^3)$

(d) $z = \dfrac{x}{y} + \dfrac{y}{x}$

(e) $z = \cos(x + \log y)$

(f) $z = \dfrac{x - y}{x + y}$

(g) $z = \arctan(x + y)$

(h) $z = x^y$

(i) $w = \cosh(x + y - z)$

(j) $w = x^2 - 2xz + y^3$.

2. Evaluate the total differential of $f(x) = x - y + (x^2 + y^2)^{1/3}$, for $x = 1$, $y = 2$, $dx = .1$, $dy = .3$.
3. Find $d^3f(x, y)$ for $f(x, y) = e^{x^2 + y^2}$.

e. *Application to the Calculus of Errors*

The differential $df = hf_x + kf_y$ is often used in practice as a convenient approximation to the increment of the function $f(x, y)$, $\Delta f = f(x + h, y + k) - f(x, y)$ as we pass from (x, y) to $(x + h, y + k)$. This use is exhibited particularly well in the so-called "calculus of errors" (cf. Volume I, p. 490). Suppose, for example, that we wish to find the possible error in the determination of the density of a solid body by the method of displacement. If m is the weight of the body in air and $\bar{m}$ its weight when submerged in water, then by Archimedes's principle, the loss of weight $(m - \bar{m})$ is the weight of the water displaced. If we are using the cgs (centimeter-gram-second) system of units, the weight of the water displaced is numerically equal to its volume and hence to the volume of the solid. The density s of the body is thus given in terms of the independent variables m and $\bar{m}$ by the formula $s = m/(m - \bar{m})$. The error in the measurement of the density s caused by an error dm in the measurement of m, and an error $d\bar{m}$ in the measurement of $\bar{m}$ is given approximately by the total differential

$$ds = \frac{\partial s}{\partial m}\, dm + \frac{\partial s}{\partial \bar{m}}\, d\bar{m}.$$

By the quotient rule, the partial derivatives are

$$\frac{\partial s}{\partial m} = -\frac{\bar{m}}{(m - \bar{m})^2} \quad \text{and} \quad \frac{\partial s}{\partial \bar{m}} = \frac{m}{(m - \bar{m})^2};$$

hence, the differential is

$$ds = \frac{-\bar{m}\, dm + m\, d\bar{m}}{(m - \bar{m})^2}.$$

Thus the error in s is greatest if dm and $d\bar{m}$ have opposite sign, say, if instead of m we measure too small an amount $m + dm$ and instead of $\bar{m}$ too large an amount $\bar{m} + d\bar{m}$. For example, if a piece of brass weighs about 100 gm in air, with a possible error 0.005 gm, and in water weighs about 88 gm, with a possible error of 0.008 gm, the density is given by our formula to within an error of about

$$\frac{88 \cdot 5 \cdot 10^{-3} + 100 \cdot 8 \cdot 10^{-3}}{12^2} \sim 9 \cdot 10^{-3},$$

or about 1 percent.

Exercises 1.5e

1. Find the approximate variation of the function $z = (x + y)/(x - y)$, as x varies from $x = 2$ to $x = 2.5$, and y, from $y = 4$ to $y = 4.5$.
2. Approximate the value of $\log [(1.02)^{1/4} + (0.96)^{1/6} - 1]$.
3. The base length x and height y of a right triangle are known to within errors of h, k, respectively. What is the possible error in the area?
4. If dz is the error of measurement in a quantity z, the *relative error* is defined as dz/z. Show that the relative error in a product $z = xy$ is the sum of the relative errors in the factors.
5. The acceleration g of gravity is to be determined by timing the fall in seconds of a body dropped from rest through a fixed distance x. If the measured time is t, we have $g = 2x/t^2$. If x is about 1 m and t about .45 sec show that the relative error of measurement in g is more sensitive to a relative error in t than a relative error in x.

1.6 Functions of Functions (Compound Functions) and the Introduction of New Independent Variables

a. Compound Functions. The Chain Rule

Frequently a function u of the independent variables x, y is given in the form

$$u = f(\xi, \eta, \ldots)$$

where the arguments $\xi, \eta, \ldots$ of f are themselves functions of x and y

$$\xi = \phi(x, y), \qquad \eta = \psi(x, y), \ldots .$$

We then say that

$$u = f(\xi, \eta, \ldots) = f(\phi(x, y), \psi(x, y), \ldots) = F(x, y) \tag{16}$$

is a *compound function* of x and y (compare Volume I, pp. 52 ff.).

For example, the function

$$u = F(x, y) = e^{xy} \sin (x + y) \tag{16a}$$

may be written as a compound function by means of the relations

$$u = f(\xi, \eta) = e^{\xi} \sin \eta, \tag{16b}$$

where $\xi = xy$ and $\eta = x + y$. Similarly, the function

$$u = F(x, y) = \log (x^4 + y^4) \cdot \text{arc sin} \sqrt{1 - x^2 - y^2} \tag{16c}$$

can be expressed in the form

$$u = f(\xi, \eta) = \eta \text{ arc sin } \xi, \tag{16d}$$

where $\xi = \sqrt{1 - x^2 - y^2}$ and $\eta = \log (x^4 + y^4)$.

In order to make the concept of compound function meaningful we assume that the functions $\xi = \phi(x, y)$, $\eta = \psi(x, y)$, . . . have the common domain R and map any points (x, y) of R into points $(\xi, \eta, \ldots)$ for which the function $u = f(\xi, \eta, \ldots)$ is defined, that is, into points of the domain S of f. The compound function

$$u = f(\phi(x, y), \psi(x, y), \ldots) = F(x, y)$$

is then defined in the region R.

A detailed examination of the regions R and S is often unnecessary, as in (16b), in which the argument point (x, y) can traverse the entire x, y-plane and the function $u = e^{\xi} \sin \eta$ is defined throughout the ξ, η-plane. On the other hand, (16d) shows the necessity for examining the domains R and S in the definition of compound functions. For the functions $\xi = \sqrt{1 - x^2 - y^2}$ and $\eta = \log (x^4 + y^4)$ are defined only in the region R consisting of the points $0 < x^2 + y^2 \leqq 1$, that is, the closed unit disk with center at the origin, the origin being deleted. Within this region we have $|\xi| < 1, \eta \leqq 0$. The corresponding points (ξ, η) all lie in the domain of the function η arc sin ξ, and thus the compound function $F(x, y)$ is defined in R.

A continuous function of continuous functions is itself continuous. More precisely, *if the function $u = f(\xi, \eta, \ldots)$ is continuous in the region S, and the functions $\xi = \phi(x, y)$, $\eta = \psi(x, y)$, . . . are continuous in the region R, then the compound function $u = F(x, y)$ is continuous in R.*

The proof follows immediately from the definition of continuity. Let (x_0, y_0) be a point of R, and let $\xi_0, \eta_0, \ldots$ be the corresponding values of $\xi, \eta, \ldots$. Now for any positive ε the absolute value of the difference

$$f(\xi, \eta, \ldots) - f(\xi_0, \eta_0, \ldots)$$

is less than ε, provided only that the inequality

$$\sqrt{(\xi - \xi_0)^2 + (\eta - \eta_0)^2 + \cdots} < \delta$$

is satisfied, where δ is a sufficiently small positive number. But by the continuity of $\phi(x, y)$, $\psi(x, y)$, . . . this inequality is satisfied if

$$\sqrt{(x - x_0)^2 + (y - y_0)^2} < \gamma,$$

where γ is a sufficiently small positive quantity. This establishes the continuity of the compound function.

Similarly, *a differentiable function of differentiable functions is itself differentiable.* This statement is formulated more precisely in the following theorem, which at the same time gives the rule for the differentiation of compound functions, the so-called *chain rule*:

If $\xi = \phi(x, y)$, $\eta = \psi(x, y)$, . . . *are differentiable functions of* x *and* y *in the region* R *and if* $f(\xi, \eta, \ldots)$ *is a differentiable function of* ξ, η, . . . *in the region* S, *then the compound function*

$$u = f(\phi(x, y), \psi(x, y), \ldots) = F(x, y) \tag{17}$$

is also a differentiable function of x *and* y*; its partial derivatives are given by the formulae*

$$\begin{aligned} F_x &= f_\xi\, \phi_x + f_\eta\, \psi_x + \cdots, \\ F_y &= f_\xi\, \phi_y + f_\eta\, \psi_y + \cdots, \end{aligned} \tag{18}$$

or, briefly, by

$$\begin{aligned} u_x &= u_\xi\, \xi_x + u_\eta\, \eta_x + \cdots, \\ u_y &= u_\xi\, \xi_y + u_\eta\, \eta_y + \cdots, \end{aligned} \tag{19}$$

Thus, in order to form the partial derivative with respect to x, we must first differentiate the compound function with respect to each of the variables $\xi, \eta, \ldots$, multiply each of these derivatives by the derivative of the corresponding variable with respect to x, and add all the products thus formed. This is the generalization of the chain rule for functions of one variable discussed in Volume I (p. 218).

Our statement can be written in a particularly simple and suggestive form if we use the notation of differentials, namely,

(20)
$$\begin{aligned} du &= u_\xi\, d\xi + u_\eta\, d\eta + \cdots \\ &= u_\xi\, (\xi_x\, dx + \xi_y\, dy) + u_\eta\, (\eta_x\, dx + \eta_y\, dy) + \cdots \\ &= (u_\xi\, \xi_x + u_\eta\, \eta_x + \cdots)\, dx + (u_\xi\, \xi_y + u_\eta\, \eta_y + \cdots)\, dy \\ &= u_x\, dx + u_y\, dy. \end{aligned}$$

This equation shows that we obtain the linear part of the increment of the compound function $u = f(\xi, \eta, \ldots) = F(x, y)$ by first writing this linear part as if $\xi, \eta, \ldots$ were the independent variables and then replacing $d\xi, d\eta, \ldots$ by the linear parts of the increments of the functions $\xi = \phi(x, y)$, $\eta = \psi(x, y), \ldots$. This fact exhibits the convenience and flexibility of the differential notation.

In order to prove our statement (18) we have merely to make use of the assumption that the functions concerned are differentiable. From this it follows that corresponding to the increments Δx and Δy of the independent variables x and y the quantities $\xi, \eta, \ldots$ change by the amounts

(20a)
$$\Delta\xi = \xi_x\, \Delta x + \xi_y\, \Delta y + \varepsilon_1 \sqrt{(\Delta x)^2 + (\Delta y)^2}$$

(20b)
$$\Delta\eta = \eta_x\, \Delta x + \eta_y\, \Delta y + \varepsilon_2 \sqrt{(\Delta x)^2 + (\Delta y)^2}, \ldots$$

where the numbers $\varepsilon_1, \varepsilon_2, \ldots$ tend to 0 for $\Delta x \to 0$ and $\Delta y \to 0$ or for $\sqrt{(\Delta x)^2 + (\Delta y)^2} \to 0$. The derivatives $\phi_x, \phi_y, \psi_x, \psi_y$ are taken for the arguments x, y. Moreover, if the quantities $\xi, \eta, \ldots$ undergo changes $\Delta\xi, \Delta\eta, \ldots$, the function $u = f(\xi, \eta, \ldots)$ changes by the amount

(21)
$$\Delta u = f_\xi \Delta\xi + f_\eta \Delta_\eta + \cdots + \delta\sqrt{(\Delta\xi)^2 + (\Delta\eta)^2 + \cdots}$$

where the quantity δ tends to 0 for $\Delta\xi \to 0$ and $\Delta\eta \to 0$, and f_ξ, f_η have the arguments ξ, η. Using here for $\Delta\xi, \Delta\eta, \ldots$ the amounts given by formulae (20a, b) corresponding to increments Δx and Δy in x and y, we find an equation of the form

(22)
$$\begin{aligned} \Delta u = (f_\xi\phi_x + f_\eta\psi_x + \cdots)\, \Delta x + (f_\xi\phi_y + f_\eta\psi_y + \cdots)\, \Delta y \\ + \varepsilon\sqrt{(\Delta x)^2 + (\Delta y)^2}. \end{aligned}$$

Here, for $\Delta x = \rho \cos \alpha$, $\Delta y = \rho \sin \alpha$, $\rho = \sqrt{(\Delta x)^2 + (\Delta y)^2}$, the quantity ε is given by

$$\varepsilon = \varepsilon_1 f_\xi + \varepsilon_2 f_\eta + \delta\sqrt{(\phi_x \cos\alpha + \phi_y \sin\alpha + \varepsilon_1)^2 + (\psi_x \cos\alpha + \psi_y \sin\alpha + \varepsilon_2)^2 + \cdots}$$

For $\rho \to 0$ the quantities Δx, Δy, ε_1, ε_2 tend to 0 and, hence, so do $\Delta\xi$, $\Delta\eta$, and δ. On the other hand, $f_\xi, f_\eta, \ldots, \phi_x, \phi_y, \psi_x, \psi_y, \ldots$ stay fixed. Consequently,

$$\lim_{\rho\to 0} \varepsilon = 0.$$

It follows from (22) that u considered as a function of the independent variables x, y is differentiable at the point (x, y) and that du is given by equation (20). From this expression for du we find that the partial derivatives u_x, u_y have the expressions (19) or (18).

Clearly this result is independent of the number of independent variables $x, y, \ldots$. It remains valid, for example, if quantities $\xi, \eta, \ldots$ depend on only one independent variable x, so that u is a compound function of the single variable x.

To calculate the higher partial derivatives, we need only differentiate the right-hand sides of our equations (19) with respect to x and y, treating $f_\xi, f_\eta, \ldots$ as compound functions. Confining ourselves for the sake of simplicity to the case of three functions ξ, η, and ζ, we obtain[1]

$$u_{xx} = f_{\xi\xi}\xi_x^2 + f_{\eta\eta}\eta_x^2 + f_{\zeta\zeta}\zeta_x^2 + 2f_{\xi\eta}\xi_x\eta_x + 2f_{\eta\zeta}\eta_x\zeta_x + 2f_{\xi\zeta}\xi_x\zeta_x + f_\xi\xi_{xx} + f_\eta\eta_{xx} + f_\zeta\zeta_{xx}, \tag{23a}$$

$$u_{xy} = f_{\xi\xi}\xi_x\xi_y + f_{\eta\eta}\eta_x\eta_y + f_{\zeta\zeta}\zeta_x\zeta_y + f_{\xi\eta}(\xi_x\eta_y + \xi_y\eta_x) + f_{\eta\zeta}(\eta_x\zeta_y + \eta_y\zeta_x) + f_{\xi\zeta}(\xi_x\zeta_y + \xi_y\zeta_x) + f_\xi\xi_{xy} + f_\eta\eta_{xy} + f_\zeta\zeta_{xy}, \tag{23b}$$

$$u_{yy} = f_{\xi\xi}\xi_y^2 + f_{\eta\eta}\eta_y^2 + f_{\zeta\zeta}\zeta_y^2 + 2f_{\xi\eta}\xi_y\eta_y + 2f_{\eta\zeta}\eta_y\zeta_y + 2f_{\xi\zeta}\xi_y\zeta_y + f_\xi\xi_{yy} + f_\eta\eta_{yy} + f_\zeta\zeta_{yy}. \tag{23c}$$

Exercises 1.6a

1. Find all partial derivatives of first and second order with respect to x and y for the following:

   (a) $z = u \log v$, where $u = x^2$, $v = \dfrac{1}{1+y}$

[1]It is assumed here that f is a function of ξ, η of class C^2 and that ξ, η, ζ are functions of x, y of class C^2. It follows that the compound function u of x and y again is of class C^2.

(b) $z = e^{uv}$, where $u = ax$, $v = \cos y$

(c) $z = u \arctan v$, where $u = \dfrac{xy}{x-y}$, $v = x^2y + y - x$

(d) $z = g(x^2 + y^2, e^{x-y})$

(e) $z = \tan (x \arctan y)$.

2. Calculate the partial derivatives of the first order for

(a) $w = \dfrac{1}{\sqrt{(x^2 + y^2 + 2xy \cos z)}}$

(b) $w = \arcsin \dfrac{x}{z + y^2}$

(c) $w = x^2 + y \log (1 + x^2 + y^2 + z^2)$

(d) $w = \arctan \sqrt{(x + yz)}$

3. Calculate the derivatives of

(a) $z = x^{(x^x)}$,

(b) $z = \left(\left(\dfrac{1}{x}\right)^{1/x}\right)^{1/x}$

4. Prove that if $f(x, y)$ satisfies Laplace's equation

$$\frac{\partial^2 f}{\partial x^2} + \frac{\partial^2 f}{\partial y^2} = 0,$$

so does $\phi(x, y) = f\left(\dfrac{x}{x^2 + y^2}, \dfrac{y}{x^2 + y^2}\right)$.

5. Prove that the functions

(a) $f(x, y) = \log \sqrt{x^2 + y^2}$,

(b) $g(x, y, z) = \dfrac{1}{\sqrt{x^2 + y^2 + z^2}}$,

(c) $h(x, y, z, w) = \dfrac{1}{x^2 + y^2 + z^2 + w^2}$,

satisfy the respective Laplace's equations,

(a) $f_{xx} + f_{yy} = 0$,

(b) $g_{xx} + g_{yy} + g_{zz} = 0$,

(c) $h_{xx} + h_{yy} + h_{zz} + h_{ww} = 0$.

Problems 1.6a

1. Prove that if $f(x, y)$ satisfies Laplace's equation

$$\frac{\partial^2 f}{\partial x^2} + \frac{\partial^2 f}{\partial y^2} = 0,$$

and if $u(x, y)$ and $v(x, y)$ satisfy the Cauchy-Riemann equations,

$$\frac{\partial u}{\partial x} = \frac{\partial v}{\partial y}, \quad \frac{\partial u}{\partial y} = -\frac{\partial v}{\partial x},$$

then the function $\phi(x, y) = f(u(x, y), v(x, y))$ is also a solution of Laplace's equation.

2. Prove if $z = f(x, y)$ is the equation of a cone, then

$$f_{xx}f_{yy} - f_{xy}^2 = 0$$

3. Let $f(x, y, z) = g(r)$, where $r = \sqrt{x^2 + y^2 + z^2}$.
   (a) Calculate $f_{xx} + f_{yy} + f_{zz}$.
   (b) Prove that if $f_{xx} + f_{yy} + f_{zz} = 0$, then $f(x, y, z) = \dfrac{a}{r} + b$, where a and b are constants.
4. Let $f(x_1, x_2, \ldots, x_n) = g(r)$, where

$$r = \sqrt{x_1^2 + x_2^2 + \cdots + x_n^2}$$

   (a) Calculate $f_{x_1x_1} + f_{x_2x_2} + \cdots + f_{x_nx_n}$(compare 1.4.a, Exercise 10).
   (b) Solve $f_{x_1x_1} + f_{x_2x_2} + \cdots + f_{x_nx_n} = 0$.

b. *Examples*[1]

1. Let us consider the function

$$u = \exp(x^2 \sin^2 y + 2xy \sin x \sin y + y^2).$$

We put

$$u = e^{\xi+\eta+\zeta}, \quad \xi = x^2 \sin^2 y, \quad \eta = 2xy \sin x \sin y, \quad \zeta = y^2$$

and obtain

$$\xi_x = 2x \sin^2 y, \qquad \eta_x = 2y \sin x \sin y + 2xy \cos x \sin y, \qquad \zeta_x = 0;$$
$$\xi_y = 2x^2 \sin y \cos y, \quad \eta_y = 2x \sin x \sin y + 2xy \sin x \cos y, \quad \zeta_y = 2y;$$
$$u_\xi = u_\eta = u_\zeta = e^{\xi+\eta+\zeta}.$$

Hence

$$u_x = 2 \exp(x^2 \sin^2 y + 2xy \sin x \sin y + y^2)(x \sin^2 y + y \sin x \sin y + xy \cos x \sin y)$$

and

$$u_y = 2 \exp(x^2 \sin^2 y + 2xy \sin x \sin y + y^2)(x^2 \sin y \cos y + x \sin x \sin y + xy \sin x \cos y + y).$$

[1]We note that the following differentiations can also be carried out directly, without using the chain rule for functions of several variables.

2. For the function

$$u = \sin(x^2 + y^2)$$

we put $\xi = x^2 + y^2$ and obtain

$$u_x = 2x\cos(x^2 + y^2), \qquad u_y = 2y\cos(x^2 + y^2)$$
$$u_{xx} = -4x^2\sin(x^2 + y^2) + 2\cos(x^2 + y^2),$$
$$u_{xy} = -4xy\sin(x^2 + y^2)$$
$$u_{yy} = -4y^2\sin(x^2 + y^2) + 2\cos(x^2 + y^2).$$

3. For the function

$$u = \arctan(x^2 + xy + y^2),$$

the substitution $\xi = x^2$, $\eta = xy$, $\zeta = y^2$ leads to

$$u_x = \frac{2x + y}{1 + (x^2 + xy + y^2)^2},$$
$$u_y = \frac{x + 2y}{1 + (x^2 + xy + y^2)^2}.$$

c. *Change of the Independent Variables*

The application of the chain rule (19) to a change of the independent variables is particularly important. For example, let $u = f(\xi, \eta)$ be a function of the two independent variables ξ, η, which we interpret as rectangular coordinates in the ξ,η-plane. We can introduce new rectangular coordinates x, y in that plane (see Volume I, p. 361) related to ξ, η by the formulae

(24a) $$\xi = \alpha_1 x + \beta_1 y, \qquad \eta = \alpha_2 x + \beta_2 y$$

or

(24b) $$x = \alpha_1 \xi + \alpha_2 \eta, \qquad y = \beta_1 \xi + \beta_2 \eta$$

Here,

$$\alpha_1 = \cos\gamma, \qquad \alpha_2 = -\sin\gamma, \qquad \beta_1 = \sin\gamma, \qquad \beta_2 = \cos\gamma,$$

where γ denotes the angle the positive ξ-axis forms with the positive

x-axis. The function $u = f(\xi, \eta)$ is then "transformed" into a new function

$$u = f(\xi, \eta) = f(\alpha_1 x + \beta_1 y, \alpha_2 x + \beta_2 y) = F(x,y),$$

which is formed from $f(\xi, \eta)$ by a process of compounding as described on p. 53. We say that the dependent variable u is "referred to the new independent variables x and y instead of ξ and η."

The rules of differentiation (19) on p. 55 at once yield

$$u_x = u_\xi \alpha_1 + u_\eta \alpha_2, \qquad u_y = u_\xi \beta_1 + u_\eta \beta_2, \tag{25}$$

where u_x, u_y denote the partial derivatives of the function $F(x, y)$, and u_ξ, u_η the partial derivatives of the function $f(\xi, \eta)$. Thus the partial derivatives of any function are transformed according to the same law (24b) as the independent variables when the coordinate axes are rotated. This is true for rotation of the axes in space as well.[1]

Another important change of the independent variables is that from rectangular coordinates (x, y) to *polar coordinates* (r, θ). The polar coordinates are connected with the rectangular coordinates by the equations

$$x = r\cos\theta, \qquad y = r\sin\theta \tag{26a}$$

$$r = \sqrt{x^2 + y^2}, \qquad \theta = \arccos\frac{x}{\sqrt{x^2+y^2}} = \arcsin\frac{y}{\sqrt{x^2+y^2}}. \tag{26b}$$

Referring a function $u = f(x, y)$ to polar coordinates, we have

$$u = f(x, y) = f(r\cos\theta, r\sin\theta) = F(r, \theta),$$

and u appears as a compound function of the independent variables r and θ. Hence, by the chain rule (19) we obtain

$$\begin{aligned} u_x &= u_r r_x + u_\theta \theta_x = u_r \frac{x}{r} - u_\theta \frac{y}{r^2} = u_r \cos\theta - u_\theta \frac{\sin\theta}{r}, \\ u_y &= u_r r_y + u_\theta \theta_y = u_r \frac{y}{r} + u_\theta \frac{x}{r^2} = u_r \sin\theta + u_\theta \frac{\cos\theta}{r}. \end{aligned} \tag{27}$$

These yield the useful equation

$$u_x^2 + u_y^2 = u_r^2 + \frac{1}{r^2} u_\theta^2, \tag{28}$$

[1] But, in general, not for other types of coordinate transformation.

By the rules (23a, b, c), the higher derivatives are given by

$$u_{xx} = u_{rr}\cos^2\theta + u_{\theta\theta}\frac{\sin^2\theta}{r^2} - 2u_{r\theta}\frac{\cos\theta\sin\theta}{r}$$
$$+ u_r\frac{\sin^2\theta}{r} + 2u_\theta\frac{\cos\theta\sin\theta}{r^2},$$

$$u_{xy} = u_{xy} = u_{rr}\cos\theta\sin\theta - u_{\theta\theta}\frac{\cos\theta\sin\theta}{r^2} + u_{r\theta}\frac{\cos^2\theta - \sin^2\theta}{r}$$
$$+ u_\theta\frac{\sin^2\theta - \cos^2\theta}{r^2} - u_r\frac{\sin\theta\cos\theta}{r},$$

$$u_{yy} = u_{rr}\sin^2\theta + u_{\theta\theta}\frac{\cos^2\theta}{r^2} + 2u_{r\theta}\frac{\cos\theta\sin\theta}{r}$$
$$+ u_r\frac{\cos^2\theta}{r} - 2u_\theta\frac{\cos\theta\sin\theta}{r^2}.$$

This leads to the expression in polar coordinates of the so-called Laplacian Δu, which appears in the important "Laplace," or "potential," equation $\Delta u = 0$ (see p. 33):

$$\Delta u = u_{xx} + u_{yy} = u_{rr} + u_{\theta\theta}\frac{1}{r^2} + u_r\frac{1}{r} \tag{29}$$
$$= \frac{1}{r^2}\left\{r\frac{\partial}{\partial r}\left(r\frac{\partial u}{\partial r}\right) + \frac{\partial^2 u}{\partial \theta^2}\right\}.$$

Conversely, we can apply the chain rule to express u_r and u_θ in terms of u_x and u_y. We find in this way

$$u_r = u_x x_r + u_y y_r = u_x\cos\theta + u_y\sin\theta, \tag{30a}$$

$$u_\theta = u_x x_\theta + u_y y_\theta = -u_x r\sin\theta + u_y r\cos\theta. \tag{30b}$$

We can also derive these equations by solving relations (27) for u_r and u_θ. Incidentally, equation (30a) has been encountered already as the expression for the derivative of u in the direction of the radius vector r on p. 45.

In general, whenever we are given relations defining a compound function,

$$u = f(\xi, \eta, \ldots),$$
$$\xi = \phi(x, y), \qquad \eta = \psi(x, y), \ldots$$

we may regard these as referring u to new independent variables x, y

instead of $\xi, \eta, \ldots$. Corresponding sets of values x, y and $\xi, \eta, \ldots$ of the independent variables assign the same value to u, whether it is regarded as a function $f(\xi, \eta \ldots)$ of $\xi, \eta, \ldots$ or as a function $F(x, y) = f(\phi(x, y), \psi(x, y), \ldots)$ of x, y.

In differentiations of a compound function $u = f(\xi, \eta, \ldots)$, we must distinguish clearly between the dependent variable u and the function $f(\xi, \eta, \ldots)$, which assigns values of u to values of the independent variables $\xi, \eta, \ldots$. The symbols of differentiation $u_\xi, u_\eta, \ldots$ have no meaning until the functional connection between u and the independent variables is specified. When dealing with compound functions $u = f(\xi, \eta, \ldots) = F(x, y)$, therefore, one really ought not to write u_ξ, u_η or u_x, u_y but instead $f_\xi(\xi, \eta)$, $f_\eta(\xi, \eta)$ or $F_x(x, y)$, $F_y(x, y)$, respectively. Yet, for the sake of brevity the simpler symbols u_ξ, u_η, u_x, u_y are often used when there is no risk of confusion. The chain rule is then written in the form

$$u_x = u_\xi \xi_x + u_\eta \eta_x, \qquad u_y = u_\xi \xi_y + u_\eta \eta_y, \tag{31}$$

which makes it unnecessary to give "names" f or F for the functional relation between u and ξ, η or x, y.

The following example illustrates the fact that the derivative of a quantity u with respect to a given variable depends on the nature of the functional connection between u and *all* of the independent variables; in particular, it depends on which of the independent variables are kept fixed during the differentiation. With the "identity transformation" $\xi = x$, $\eta = y$ the function $u = 2\xi + \eta$ becomes $u = 2x + y$, and we have $u_x = 2$, $u_y = 1$. If, however, we introduce the new independent variables $\xi = x$ (as before) and $\xi + \eta = v$, we find that $u = x + v$, so that $u_x = 1$, $u_v = 1$. Thus, differentiation with respect to the same independent variable x gives different results for different choices of the other variable.

Exercises 1.6c

1. Let $u = f(x, y)$, where $x = r\cos\theta$, $y = r\sin\theta$. Express $\sqrt{u_x^2 + u_y^2}$ in terms of u_r and u_θ.
2. Prove that the expression $f_{xx} + f_{yy}$ is unchanged by rotation of the coordinate system.
3. Show that the linear changes of variables $x = \alpha\xi + \beta\eta$, $y = \gamma\xi + \delta\eta$ transform the derivatives $f_{xx}(x, y)$, $f_{xy}(x, y)$, $f_{yy}(x, y)$ by the same rule as the coefficients a, b, c, respectively, of the polynominal

$$ax^2 + 2bxy + cy^2$$

4. Given $z = r^2 \cos\theta$, where r and θ are polar coordinates, find z_x and z_y at the point $\theta = \pi/4$, $r = 2$. Express z_r and z_θ in terms of z_x and z_y.
5. By the transformation $\xi = a + \alpha x + \beta y$, $\eta = b - \beta x + \alpha y$, in which a, b, α, β are constants and $\alpha^2 + \beta^2 = 1$, the function $u(x, y)$ is transformed into a function $U(\xi, \eta)$ of ξ and η. Prove that
$$U_{\xi\xi}U_{\eta\eta} - U_{\xi\eta}^{\;2} = u_{xx}u_{yy} - u_{xy}^2$$
6. Show how the expression $T_y - T_{xx}$ is transformed under the introduction of a variable $z = x/\sqrt{y}$ in place of y.
7. (a) Prove that the function
$$h(x, y) = f(x - y) + g(x + y)$$
for any twice continuously differentiable functions f, g, satisfies the condition $h_{xx} = h_{yy}$.
(b) Similarly, show that
$$H(x, y) = f(x - iy) + g(x + iy),$$
with $i^2 = -1$, satisfies the condition $H_{xx} = -H_{yy}$.

Problems 1.6c

1. Transform the Laplacian $u_{xx} + u_{yy} + u_{zz}$ into three-dimensional polar coordinates r, θ, ϕ defined by
$$\begin{aligned} x &= r\sin\theta\cos\phi \\ y &= r\sin\theta\sin\phi \\ z &= r\cos\theta. \end{aligned}$$
Compare with 1.6.a, Problem 3.
2. Find values a, b, c, d such that under the transformation $\xi = ax + by$, $\eta = cx + dy$, where $ad - bc \neq 0$, equation $Af_{xx} + 2Bf_{xy} + Cf_{yy} = 0$ becomes

(a) $f_{\xi\xi} + f_{\eta\eta} = 0$

(b) $f_{\xi\eta} = 0$ (A, B, C, constants)

Is this always possible?

1.7 The Mean Value Theorem and Taylor's Theorem for Functions of Several Variables

a. *Preliminary Remarks About Approximation by Polynomials*

We have already seen in Volume I (Chapter V, p. 451) how a function of a single variable can be approximated in the neighborhood of a given point with an accuracy higher than the nth order by means of a polynomial of degree n, the Taylor polynomial, provided that the function possesses derivatives up to the $(n + 1)$th order. Approximation by means of the linear part of the function, as given by the differential, is only the first step toward this closer approximation. In the case of functions of several variables, for example, of two independent variables, we may also seek an approximate representation in the neighborhood of a given point by means of a polynomial of degree n. In other words, we wish to approximate $f(x+h, y+k)$ by means of a "Taylor expansion" in terms of the increments h and k.

By a simple device this problem can be reduced to one for functions of only one variable. Instead of just considering $f(x+h, y+k)$, we introduce an additional variable t and regard the expression

$$F(t) = f(x + ht, y + kt) \tag{31}$$

as a function of t, keeping x, y, h, and k fixed for the moment. As t varies between 0 and 1, the point with coordinates $(x + ht, y + kt)$ traverses the line segment joining (x, y) and $(x + h, y + k)$. The Taylor expansion of $F(t)$ according to powers of t will yield for $t = 1$ an approximation to $f(x + h, y + k)$ of the desired kind.

We begin by calculating the derivatives of $F(t)$. If we assume that all the derivatives of the function $f(x, y)$ that we are about to write down are continuous in a region *entirely containing the line segment*, the chain rule (18) at once gives[1]

$$F'(t) = hf_x + kf_y, \tag{32a}$$

$$F''(t) = h^2 f_{xx} + 2hkf_{xy} + k^2 f_{yy}, \tag{32b}$$

. . . .

and, in general, we find by mathematical induction that the nth derivative is given by the expression

$$F^{(n)}(t) = h^n f_{x^n} + \binom{n}{1} h^{n-1} k f_{x^{n-1}y} + \binom{n}{2} h^{n-2} k^2 f_{x^{n-2}y^2} + \cdots + k^n f_{y^n}, \tag{32c}$$

[1]We have from the chain rule

$$F'(t) = \frac{d}{dt} f(x + ht, y + kt) = hf_\xi(\xi, \eta) + kf_\eta(\xi, \eta)$$

where $\xi = x + ht$, $\eta = y + kt$. We write here $f_x(x + ht, y + kt)$ for $f_\xi(x + ht, y + kt)$ since (again by the chain rule)

$$\frac{\partial}{\partial x} f(x + ht, y + kt) = f_\xi(x + ht, y + kt)$$

if x, y, h, k are considered independent variables.

which, as on p. 51, can be written symbolically in the form

$$F^{(n)}(t) = \left(h\frac{\partial}{\partial x} + k\frac{\partial}{\partial y}\right)^n f.$$

In this formula the symbolic power on the right is to be expanded by the binomial theorem and then the powers of $\partial/\partial x$, $\partial/\partial y$ multiplied by f are to be replaced by the corresponding nth derivatives $\partial^n f/\partial x^n$, $\partial^n f/\partial x^{n-1}\partial y, \ldots$. In all these derivatives the arguments $x + ht$ and $y + kt$ are to be written in place of x and y.

Exercises 1.7a

1. For $F(t) = f(x + ht,\ y + kt)$ find $F'(1)$ for:
   (a) $f(x, y) = \sin(x + y)$
   (b) $f(x, y) = \dfrac{y}{x}$
   (c) $f(x, y) = x^2 + 2xy^2 - y^4$
2. Find the slope of the curve $z(t) = F(t) = f(x + ht, y + kt)$ at $t = 1$, for $x = 0$, $y = 1$, $h = \frac{1}{2}$, $k = \frac{1}{4}$, and
   (a) $f(x, y) = x^2 + y^2$
   (b) $f(x, y) = \exp[x^2 + (y - 1)^2]$
   (c) $f(x, y) = \cos\pi(y - 1)\sin\pi x^2$

b. The Mean Value Theorem

Before taking up higher order approximations by polynomials, we derive a *mean value theorem* analogous to the one we already know for functions of one variable. This theorem relates the *difference* $f(x + h, y + k) - f(x, y)$ to the *partial derivatives* f_x and f_y. We expressly assume that these derivatives are continuous. On applying the ordinary mean value theorem to the function $F(t)$ we obtain

$$\frac{F(t) - F(0)}{t} = F'(\theta t),$$

where θ is a number between 0 and 1; using (31) and (32a) it follows that

$$\frac{f(x + ht, y + kt) - f(x, y)}{t} = hf_x(x + \theta ht, y + \theta kt) + kf_y(x + \theta ht, y + \theta kt).$$

Setting $t = 1$, we obtain the required *mean value theorem for functions of two variables* in the form

$$(33) \qquad \begin{aligned} &f(x + h, y + k) - f(x, y) \\ &\quad = hf_x(x + \theta h, y + \theta k) + kf_y(x + \theta h, y + \theta k) \\ &\quad = hf_x(\xi, \eta) + kf_y(\xi, \eta). \end{aligned}$$

Thus, *the difference between the values of the function at the points $(x + h, y + k)$ and (x, y) is equal to the differential at an intermediate point (ξ, η) on the line segment joining the two points.* It is worth noting that the *same* value of θ occurs in both f_x and f_y.

Just as for functions of a single variable (Volume I, p. 178), the mean value theorem can be used to obtain a modulus of continuity for a function $f(x, y)$ and, more precisely, to show that a function f as above is Lipschitz continuous. In order to apply the mean value theorem we must be able to join two points by a straight line segment along which f is defined. Assume then that the domain R of $f(x, y)$ is *convex*, that is, that the line segment joining any two points of R lies completely in R. Let f be continuously differentiable in R and let M be a bound for the absolute value of the derivatives of f:

$$|f_x(x, y)| < M, \qquad |f_y(x, y)| < M$$

for (x, y) in R. Then formula (33) can be applied and yields the inequality

$$(34) \qquad \begin{aligned} |f(x + h, y + k) - f(x, y)| &\leq |h|\,|f_x(\xi, \eta)| + |k|\,|f_y(\xi, \eta)| \\ &\leq |h| M + |k| M \leq 2M\sqrt{h^2 + k^2} \end{aligned}$$

Hence, the numerical value of the difference in the values of f at two points whose distance $\rho = \sqrt{h^2 + k^2}$ does not exceed a fixed multiple of the distance (namely, $2M\rho$). This is exactly what is meant by Lipschitz continuity of f. In particular we have

$$|f(x + h, y + k) - f(x, y)| < \varepsilon$$

for $\sqrt{h^2 + k^2} < \varepsilon/2M$. Thus f is uniformly continuous in R with the "modulus of continuity" $\delta = \varepsilon/2M$.

The following fact, the proof of which we leave to the reader, is a simple consequence of the mean value theorem. A function $f(x, y)$ whose partial derivatives f_x and f_y exist and have the value 0 at every point of a convex set is constant.

Exercises 1.7b

1. Interpret the mean value theorem geometrically.
2. Find a value θ for which

$$hf_x(x + \theta h, y + \theta k) + kf_y(x + \theta h, y + \theta k) = f(x + h, y + k) - f(x, y)$$

in each of the following cases:

(a) $f(x, y) = xy + y^2$, $x = y = 0$, $h = \frac{1}{2}$, $k = \frac{1}{4}$

(b) $f(x, y) = \sin \pi (x + y)$, $x = y = \frac{1}{4}$, $h = \frac{1}{8}$, $k = \frac{1}{4}$.

3. Show that there is a number θ, $0 < \theta < 1$ such that

$$\frac{2}{\pi} = \cos \frac{\pi\theta}{2} + \sin\left[\frac{\pi}{2}(1 - \theta)\right]$$

using the mean value theorem for the function

$$f(x, y) = \sin \pi x + \cos \pi y.$$

4. Derive the mean value theorem for a function $f(x, y, z)$ of three variables.
5. Find a number θ, $0 \leq \theta \leq 1$, for which

$$f\left(1, \frac{1}{2}, \frac{1}{3}\right) = f_x\left(\theta, \frac{\theta}{2}, \frac{\theta}{3}\right) + \frac{1}{2} f_y\left(\theta, \frac{\theta}{2}, \frac{\theta}{3}\right) + \frac{1}{3} f_z\left(\theta, \frac{\theta}{2}, \frac{\theta}{3}\right)$$

where

(a) $f(x, y, z) = xyz$

(b) $f(x, y, z) = x^2 + y^2 + 2xz$

Problems 1.7b

1. Let the domain of $f(x, y)$ be a polygonally connected region; that is, suppose that any two points P, Q of the domain can be connected within the domain by a sequence of segments $\overline{P_0P_1}, \overline{P_1P_2}, \ldots, \overline{P_{n-1}P_n}$, where $P_0 = P$ and $P_n = Q$. Prove that if the partial derivatives f_x and f_y have the value 0 at every point of the domain, then f is constant.

c. *Taylor's Theorem for Several Independent Variables*

If we apply Taylor's formula with Lagrange's form of the remainder (cf. Volume I, p. 452) to the function $F(t) = f(x + ht, y + kt)$, use the expressions (32a, b, c) for the derivatives of F, and put $t = 1$, we obtain *Taylor's theorem* for functions of two independent variables,

$$(35) \quad f(x + h, y + k) = f(x, y) + \{hf_x(x, y) + kf_y(x, y)\} + \frac{1}{2!}\{h^2 f_{xx}(x, y) + 2hkf_{xy}(x, y) + k^2 f_{yy}(x, y)\}$$

$$+\cdots+\frac{1}{n!}\left\{h^n f_{x^n}(x, y)+\binom{n}{1} h^{n-1} k f_{x^{n-1}y}(x, y)\right.$$

$$\left.+\cdots+k^n f_{y^n}(x, y)\right\}+R_n,$$

where R_n denotes the remainder term

$$R_n=\frac{1}{(n+1)!}\{h^{n+1} f_{x^{n+1}}(x+\theta h, y+\theta k)+\cdots$$
$$+k^{n+1} f_{y^{n+1}}(x+\theta h, y+\theta k)\}, \tag{36}$$

where $0<\theta<1$. The increment $f(x+h, y+k)-f(x, y)$ is thus written as a sum of homogeneous polynomials of degree 1, 2, . . . , $n+1$, which, apart from the factors

$$\frac{1}{1!}, \frac{1}{2!}, \ldots, \frac{1}{n!}, \frac{1}{(n+1)!},$$

are the first, second, . . ., nth differentials

$$df=hf_x+kf_y=\left(h \frac{\partial}{\partial x}+k \frac{\partial}{\partial y}\right) f$$

$$d^2 f=\left(h \frac{\partial}{\partial x}+k \frac{\partial}{\partial y}\right)^2 f=h^2 f_{xx}+2hk f_{xy}+k^2 f_{yy},$$

$$d^n f=\left(h \frac{\partial}{\partial x}+k \frac{\partial}{\partial y}\right)^n f=h^n f_{x^n}+\binom{n}{1} h^{n-1} k f_{x^{n-1}y}+\cdots+k^n f_{y^n}$$

of $f(x, y)$ at the point (x, y) and the $(n+1)$th differential $d^{n+1} f$ at an intermediate point on the line segment joining (x, y) and $(x+h, y+k)$. Hence, Taylor's theorem can be written more compactly as

$$f(x+h, y+k)=f(x, y)+df(x, y)+\frac{1}{2!} d^2 f(x, y)+\cdots$$
$$+\frac{1}{n!} d^n f(x, y)+R_n, \tag{37}$$

where

$$R_n=\frac{1}{(n+1)!} d^{n+1} f(x+\theta h, y+\theta k), \qquad 0<\theta<1. \tag{38}$$

In general the remainder R_n vanishes to a *higher* order than the term $d^n f$ just before it; that is, as $h \rightarrow 0$ and $k \rightarrow 0$, we have $R_n = o\{\sqrt{(h^2+k^2)^n}\}$.

From Taylor's theorem for functions of one variable the passage ($n \to \infty$) to *infinite Taylor series* led us to the expansions of many functions in power series. With functions of several variables such a process, even when possible, is in general too complicated. For us the importance of Taylor's theorem lies rather in the fact that the increment $f(x + h, y + k) - f(x, y)$ of a function is split up into increments df, d^2f, . . . of different orders.

Exercises 1.7c

1. Find the polynomial of second degree that best approximates $\sin x \sin y$ in the neighborhood of the origin.
2. For $f(x, y) = x^3 + 4y^2x$, approximate the value of $f(2.1, 2.9)$.
3. For $f(x, y) = x/y + y/x$, estimate the error in approximating the value of $f(.9, .9)$ by $f(1, 1)$.
4. Expand the function $f(x + h, y + k)$ in powers of h, k, for

   (a) $f(x, y) = x^3 - 2x^2y + y^2$

   (b) $f(x, y) = \cos(x + 2y)$ at $x = 0, y = \dfrac{\pi}{2}$

   (c) $f(x, y) = x^4y + 2y^2x - \sqrt{3x^2}$.
5. Expand $f(x, y, z) = xyz^2$ in powers of x, $y - 1$, $z + 1$.
6. Obtain the first few terms of the Taylor expansions of the following functions in a neighborhood of the origin (0, 0):

   (a) $z = \arc\tan \dfrac{y}{(x^2 + 1)}$

   (b) $z = \cosh x \sinh y$

   (c) $z = \cos x \cosh(x + y)$

   (d) $z = e^x \cos y$

   (e) $z = \dfrac{\sin x}{\cos y}$

   (f) $z = \log(1 - x) \log(1 - y)$

   (g) $z = e^{x^2 - y^2}$

   (h) $z = \cos(x + y)\, e^{-x^2}$

   (i) $z = \cos(x \cos y)$

   (j) $z = \sin(x^2 + y^2)$
7. Estimate the error in replacing $\cos x/\cos y$ by

$$1 - \frac{1}{2}(x^2 - y^2) \qquad \text{for} \qquad |x|, |y| < \frac{\pi}{6}.$$

Problems 1.7c

1. Find the Taylor series for the following functions and indicate their range of validity.

   (a) $\dfrac{1}{1 - x - y}$

(b) e^{x+y}.

2. Show that the law of cosines in spherical trigonometry,
$$\cos z = \cos x \cos y + \sin x \sin y \cos \theta,$$
reduces to the euclidean law of cosines,
$$z^2 = x^2 + y^2 - 2xy \cos \theta$$
in the neighborhood of the origin.

3. If $f(x, y)$ is a continuous function with continuous first and second derivatives, then
$$f_{xx}(0, 0) = \lim_{h \to +0} \frac{f(2h, e^{-1/2h}) - 2f(h, e^{-1/h}) + f(0, 0)}{h^2}$$

4. Prove that the function $f(x, y) = \exp(-y^2 + 2xy)$ can be expended in a series of the form
$$\sum_{n=0}^{\infty} \frac{H_n(x)}{n!} y^n,$$
that converges for all values of x and y and that the polynominals $H_n(x)$, the so-called *Hermite polynomials*, satisfy

(a) $H_n(x)$ is a polynomial of degree n.

(b) $H_n'(x) = 2nH_{n-1}(x)$

(c) $H_{n+1} - 2xH_n + 2nH_{n-1} = 0$

(d) $H_n'' - 2xH_n' + 2nH_n = 0$.

1.8 Integrals of a Function Depending on a Parameter

The concept of multiple integral of a function of several variables will be taken up in Chapters IV and V. For the moment we shall only study the *single* integrals arising in connection with such functions.

a. Examples and Definitions

If $f(x, y)$ is a continuous function of x and y in the rectangular region $\alpha \leqq x \leqq \beta$, $a \leqq y \leqq b$, we may think of the quantity x as fixed and integrate the function $f(x, y)$, considered as a function of y alone, over the interval $a \leqq y \leqq b$. We thus arrive at the expression
$$\int_a^b f(x, y)\, dy$$
which still depends on the choice of the quantity x. Thus, we are considering not just one integral but the family of integrals $\int_a^b f(x, y)\, dy$ obtained for different values of x. The quantity x, which is kept fixed

during the integration and to which we can assign any value in its interval, we call a *parameter*. Our ordinary *integral* therefore appears as a *function of the parameter x*.

Integrals that are functions of a parameter frequently occur in analysis and its applications. For example, as the substitution $xy = u$ readily shows, we have

$$\int_0^1 \frac{x\,dy}{\sqrt{1 - x^2y^2}} = \arc\sin x$$

for $-1 < x < 1$. Again, in integrating the general power function we may regard the exponent as a parameter and write accordingly

$$\int_0^1 y^x\,dy = \frac{1}{x+1},$$

where we assume that $x > -1$.

We can represent the region of definition of the function $f(x, y)$ geometrically and consider the parallel to the y-axis corresponding to the fixed value of the parameter x, as in Fig. 1.15. We obtain the function of y that is to be integrated by considering the values of the function $f(x, y)$ as a function of y along the line of intersection AB of the parallel with the rectangle. We may also speak of integrating the function $f(x, y)$ *along the segment AB*.

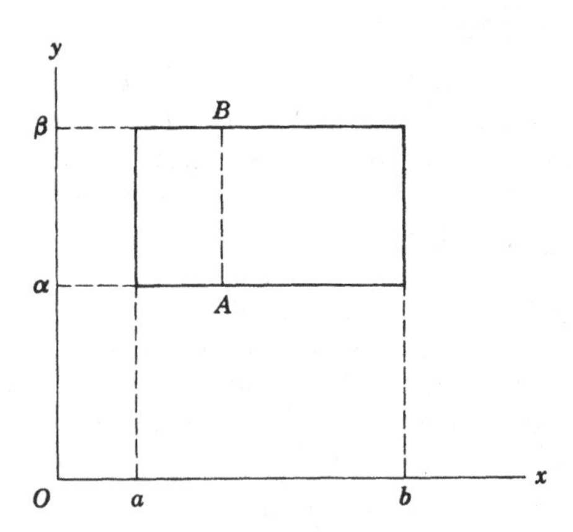

Figure 1.15

This geometrical point of view suggests a generalization. If the domain of definition R of the function $f(x, y)$ has the shape shown in

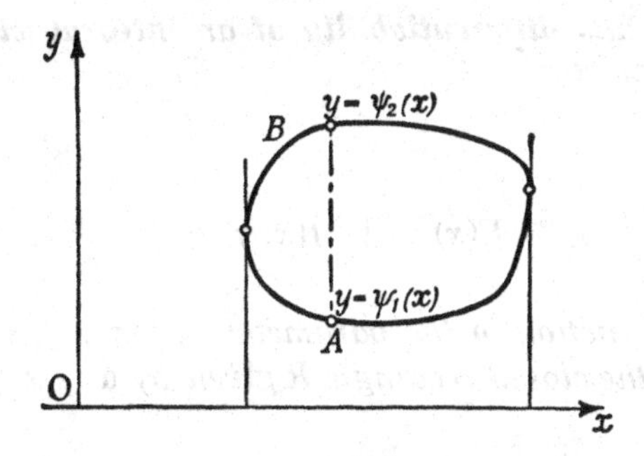

Figure 1.16

Fig. 1.16. such that any parallel to the y-axis cuts the boundary in at most two points, then for a fixed value of x we can again integrate the values of the function $f(x, y)$ along the line AB in which the parallel to the y-axis intersects the region R. The initial and final points of the interval of integration will themselves vary with x. We then have to consider an integral of the type

$$\int_{\psi_1(x)}^{\psi_2(x)} f(x, y)\, dy = F(x), \tag{39}$$

that is, an integral with the variable of integration y in which the parameter x is present both in the integrand and in the limits of integration. If we represent the function $f(x, y)$ by the surface $z = f(x, y)$ in x, y, z-space, then for a positive function f we can consider the cylinder with generators parallel to the z-axis having as its base the domain R of f in the x, y-plane and bounded on top by the surface $z = f(x, y)$. A fixed value of x corresponds to a plane parallel to the y, z-plane, which intersects the solid cylinder in a certain plane region. The area of that region is given by the integral in formula (39). For example, the integral

$$\int_{-\sqrt{1-x^2}}^{\sqrt{1-x^2}} \sqrt{1 - x^2 - y^2}\, dy$$

represents the area of the intersection of the hemisphere

$$0 < z < \sqrt{1 - x^2 - y^2}$$

with a plane $x =$ constant.

b. Continuity and differentiability of an integral with respect to the parameter

The integral

$$F(x) = \int_a^b f(x, y)\, dy$$

is a continuous function of the parameter x, for $\alpha \leq x \leq \beta$, if $f(x, y)$ is continuous in the closed rectangle R given by $\alpha \leqq x \leqq \beta$, $a \leqq y \leqq b$.

For

$$\left| F(x+h) - F(x) \right| = \left| \int_a^b (f(x+h, y) - f(x, y))\, dy \right|$$
$$\leqq \int_a^b \left| f(x+h, y) - f(x, y) \right| dy.$$

In virtue of the uniform continuity of $f(x, y)$, for sufficiently small values of h the integrand on the right, considered as a function of y, may be made uniformly as small as we please, and the statement follows immediately.

We next investigate the possibility of differentiating $F(x)$. We first consider the case in which the limits of integration are fixed and assume that the function $f(x, y)$ has a continuous partial derivative f_x in the closed rectangle R.[1] We shall prove that instead of first integrating with respect to y and then differentiating with respect to x we may reverse the order of these two processes:

THEOREM. *If in the closed rectangle $\alpha \leq x \leq \beta$, $a \leqq y \leq b$ the function $f(x, y)$ is continuous and has a continuous derivative with respect to x, we may differentiate the integral with respect to the parameter under the integral sign, that is,*

$$\frac{d}{dx} F(x) = \frac{d}{dx}\int_a^b f(x, y)\, dy = \int_a^b f_x(x, y)\, dy. \tag{40}$$

Moreover, $F'(x)$ is a continuous function of x.

Before proving this theorem, we remark that it yields a simple proof of the fact (already established on p. 37) that in the formation of the mixed derivative g_{xy} of a function $g(x, y)$ the order of differentiation can be changed, provided that g_y and g_{xy} are continuous and g_x exists. For if we put $f(x, y) = g_y(x, y)$, we have

[1]This means that f_x exists in the open rectangle and can be extended into the closed rectangle as a continuous function (see. p. 42).

$$g(x, y) = g(x, a) + \int_a^y f(x, \eta)\, d\eta.$$

Since $f(x, y)$ has a continuous derivative with respect to x in the rectangle $\alpha \leqq x \leqq \beta$, $a \leqq y \leqq b$, it follows that

$$g_x(x, y) = g_x(x, a) + \int_a^y f_x(x, \eta)\, d\eta,$$

and therefore by the fundamental theorem of calculus

$$g_{yx}(x, y) = f_x(x, y).$$

Since also $f_x(x, y) = g_{xy}(x, y)$ from the definition of f, we see that $g_{yx} = g_{xy}$.

PROOF. If both x and $x + h$ belong to the interval $\alpha \leqq x \leqq \beta$, we can write

$$\begin{aligned} F(x + h) - F(x) &= \int_a^b f(x + h, y)\, dy - \int_a^b f(x, y)\, dy \\ &= \int_a^b [f(x + h, y) - f(x, y)]\, dy. \end{aligned}$$

Since we have assumed that $f(x, y)$ is differentiable with respect to x, the mean value theorem of differential calculus in its usual form gives

$$f(x + h, y) - f(x, y) = hf_x(x + \theta h, y), \qquad 0 < \theta < 1.$$[1]

Moreover, since the derivative f_x is assumed to be continuous in the closed rectangle and therefore uniformly continuous, the absolute value of the difference

$$f_x(x + \theta h, y) - f_x(x, y)$$

is less than any positive quantity ε for all h with $|h| < \delta$ where $\delta = \delta(\varepsilon)$ is independent of x and y. Thus,

$$\left| \frac{F(x + h) - F(x)}{h} - \int_a^b f_x(x, y)\, dy \right|$$

[1]Here the quantity θ depends on y and may even vary discontinuously with y. This does not matter, for by the equation $f_x(x + \theta h, y) = h^{-1}[f(x + h, y) - f(x, y)]$ we see at once that $f_x(x + \theta h, y)$ is a continuous function of x and y and is therefore integrable.

$$= \left| \int_a^b f_x(x + \theta h, y)\, dy - \int_a^b f_x(x, y)\, dy \right|$$

$$\leqq \int_a^b \varepsilon\, dy = \varepsilon(b - a),$$

for $|h| < \delta(\varepsilon)$, provided $h \neq 0$. This means, however, that the relation

$$\lim_{h \to 0} \frac{F(x + h) - F(x)}{h} = \int_a^b f_x(x, y)\, dy = F'(x)$$

holds. This proves the existence of $F'(x)$ and formula (40). The continuity of F' follows from that of the integrand $f_x(x, y)$ (see p. 74).

In a similar way we can establish the continuity of the integral and the rule for differentiating the integral with respect to a parameter when the parameter occurs in the *limits of integration*.

For example, if we wish to differentiate

$$F(x) = \int_{\psi_1(x)}^{\psi_2(x)} f(x, y)\, dy,$$

we start with the expression

$$F(x) = \int_u^v f(x, y)\, dy = \phi(u, v, x),$$

where $u = \psi_1(x)$, $v = \psi_2(x)$. Here we assume that $\psi_1(x)$ and $\psi_2(x)$ have continuous first derivatives in an interval $\alpha \leqq x \leqq \beta$ and that

$$a < \psi_1(x) < \psi_2(x) < b$$

for $\alpha < x < \beta$. Let, moreover, $f(x, y)$ and $f_x(x, y)$ be continuous in the set

$$\alpha \leqq x \leqq \beta, \qquad a \leqq y \leqq b.$$

The function ϕ of the three independent variables u, v, x is defined then for

$$\alpha \leqq x \leqq \beta, \qquad a \leqq u \leqq b, \qquad a \leqq v \leqq b.$$

Moreover, it has continuous partial derivatives, since by formula (40)

$$\phi_x(u, v, x) = \frac{\partial}{\partial x} \int_u^v f(x, y)\, dy = \int_u^v f_x(x, y)\, dy$$

and by the fundamental theorem of calculus (Volume I, p. 185)

$$\phi_v(u, v, x) = \frac{\partial}{\partial v}\int_u^v f(x, y)\, dy = f(x, v)$$

$$\phi_u(u, v, x) = \frac{\partial}{\partial u}\int_u^v f(x, y)\, dy = -\frac{\partial}{\partial u}\int_v^u f(x, y)\, dy = -f(x, u).$$

We can apply the chain rule of differentiation (18) p. 55 to the compound function

$$F(x) = \phi\, [\psi_1(x), \psi_2(x), x]$$

and find

$$F'(x) = \phi_u \psi_1'(x) + \phi_v \psi_2'(x) + \phi_x.$$

This proves the existence of a continuous derivative of $F(x)$ for $\alpha < x < \beta$ and yields the formula

$$(41) \qquad \frac{d}{dx}\int_{\psi_1(x)}^{\psi_2(x)} f(x, y)\, dy$$

$$= \int_{\psi_1(x)}^{\psi_2(x)} f_x(x, y)\, dy - \psi_1'(x)\, f(x, \psi_1(x)) + \psi_2'(x)\, f(x, \psi_2(x)).$$

Taking, for example, for $F(x)$ the function

$$F(x) = \int_0^x \sin(xy)\, dy$$

we obtain

$$\frac{dF(x)}{dx} = \int_0^x y \cos(xy)\, dy + \sin(x^2).$$

For the example

$$F(x) = \int_0^1 \frac{x\, dy}{\sqrt{1 - x^2y^2}} = \text{arc sin } x,$$

for $-1 < x < +1$, we obtain the relation

$$F'(x) = \int_0^1 \frac{dy}{\sqrt{(1 - x^2y^2)^3}} = \frac{1}{\sqrt{1 - x^2}}$$

as the reader may verify directly.

Other examples are given by the sequence of integrals

$$(42) \qquad F_n(x) = \int_0^x \frac{(x-y)^n}{n!} f(y)\, dy, \qquad F_0(x) = \int_0^x f(y)\, dy,$$

where n is any positive integer and $f(y)$ is a continuous function of y alone, in the interval under consideration. Since the expression arising from differentiation with respect to the upper limit x vanishes, rule (41) yields the recursion formula

$$F_n'(x) = F_{n-1}(x)$$

for $n = 1, 2, 3, \ldots$. Since $F_0'(x) = f(x)$, this gives at once

$$(42a) \qquad F_n^{(n+1)}(x) = f(x).$$

Therefore $F_n(x)$ is that function whose $(n + 1)$th derivative is equal to $f(x)$ and which, together with its first n derivatives, vanishes for $x = 0$; it arises from $F_{n-1}(x)$ by integration from 0 to x. Hence, $F_n(x)$ is the function obtained from $f(x)$ by integrating $n + 1$ times between the limits 0 and x:

$$(42b) \qquad F_0(x) = \int_0^x f(y)\, dy, \qquad F_1(x) = \int_0^x F_0(y)\, dy,$$

$$F_2(x) = \int_0^x F_1(y)\, dy, \ldots, \qquad F_n(x) = \int_0^x F_{n-1}(y)\, dy.$$

This repeated integration can therefore be replaced by a single integration of the function $\frac{(x-y)^n}{n!} f(y)$ with respect to y.

The rules for differentiating an integral with respect to a parameter often remain valid even when differentiation under the integral sign yields a function that is not continuous everywhere. In such cases, instead of applying general criteria, it is more convenient to verify directly whether such a differentiation is permissible in each special case.

As an example, we consider the elliptic integral (cf. Volume I, p. 299).

$$F(k) = \int_{-1}^{+1} \frac{dx}{\sqrt{(1-x^2)(1-k^2x^2)}}; \qquad k^2 < 1.$$

The function

$$f(k, x) = \frac{1}{\sqrt{(1 - x^2)(1 - k^2x^2)}}$$

is discontinuous at $x = +1$ and at $x = -1$, but the integral (as an improper integral) has a meaning. Formal differentiation with respect to the parameter k gives

$$F'(k) = \int_{-1}^{+1} \frac{kx^2\,dx}{\sqrt{(1 - x^2)(1 - k^2x^2)^3}}$$

To investigate whether this equation is correct, we repeat the argument by which we obtained our differentiation formula. This gives

$$\frac{F(k + h) - F(k)}{h} = \int_{-1}^{+1} f_k(k + \theta h, x)\,dx$$
$$= \int_{-1}^{+1} \frac{(k + \theta h)x^2\,dx}{\sqrt{(1 - x^2)\,[1 - (k + \theta h)^2x^2]^3}}.$$

The difference between this expression and the integral obtained by formal differentiation is

$$\Delta = \int_{-1}^{+1} \frac{x^2}{\sqrt{1 - x^2}} \left(\frac{k + \theta h}{\sqrt{[1 - (k + \theta h)^2x^2]^3}} - \frac{k}{\sqrt{(1 - k^2x^2)^3}}\right) dx.$$

We must show that this integral tends to 0 with h. For this purpose we mark off about k an interval $k_0 \leqq k \leqq k_1$ not containing the values ± 1, and we choose h so small that $k + \theta h$ lies in this interval. The function

$$\frac{k}{\sqrt{(1 - k^2x^2)^3}}$$

is continuous in the closed region $-1 \leqq x \leqq 1$, $k_0 \leqq k \leqq k_1$, and is therefore uniformly continuous. The difference

$$\left|\frac{k + \theta h}{\sqrt{[1 - (k + \theta h)^2x^2]^3}} - \frac{k}{\sqrt{(1 - k^2x^2)^3}}\right|$$

consequently remains below a bound ε that is independent of x and k and which tends to 0 with h. Hence,

$$|\Delta| \leq \int_{-1}^{+1} \frac{x^2\,dx}{\sqrt{1 - x^2}}\,\varepsilon = M\varepsilon,$$

where M is a constant independent of ε. That is, the integral Δ tends to 0 as h does, which is what we wished to show.

Differentiation under the integral sign is therefore permissible in this case. Similar considerations apply in other cases.

Improper integrals with an infinite range of integration and depending on a parameter will be discussed on p. 462.

Exercises 1.8b

1. Let

$$F(k) = \int_a^b \alpha(x)\, \beta(x, k)\, dx,$$

where $\beta(x, k)$ and $\beta_k(x, k)$ are continuous for $a \leq x \leq b$, $k_0 < k < k_1$, and $\alpha(x)$ is continuous for $a < x < b$, and $\int_a^b |\alpha(x)|\, dx$ exists as an improper integral. Prove that

$$F'(k) = \int_a^b \alpha(x)\, \beta_k(x, k)\, dx \qquad \textit{for} \qquad k_0 < k < k_1.$$

2. Let

$$F(k) = \int_0^1 (x - 1)x^k \log^{-1} x\, dx \qquad \text{for} \qquad -1 < k.$$

Prove

(a) $\lim\limits_{k\to\infty} k\, F(k) = 1$

(b) $F(k) = \log \dfrac{2 + k}{1 + k}$.

c. *Interchange of Integrations. Smoothing of Functions*

The theorem on p. 74 about differentiation under the integral sign has the important consequence that we can interchange orders of integration.

Let $f(x, y)$ be continuous in the rectangle R given by

$$a \leqq x \leqq b, \qquad \alpha \leqq y \leqq \beta. \tag{42c}$$

Then the integrals

$$I = \int_a^b d\xi \int_\alpha^\beta f(\xi, \eta)\, d\eta \qquad \text{and} \qquad J = \int_\alpha^\beta d\eta \int_a^b f(\xi, \eta)\, d\xi \tag{42d}$$

have the same value. We call this value the *double integral of f over the rectangle* (42c).

As an example we consider the function $f(x, y) = y \sin(xy)$ in the rectangle $0 \leq x \leq 1,\ 0 \leq y \leq \frac{\pi}{2}$. Here

$$I = \int_0^1 d\xi \int_0^{\pi/2} \eta \sin(\xi\eta)\, d\eta = \int_0^1 \left(-\frac{\pi \cos(\pi\,\xi/2)}{2\xi} + \frac{\sin(\pi\,\xi/2)}{\xi^2}\right) d\xi$$
$$= \frac{\pi}{2} - 1$$

$$J = \int_0^{\pi/2} d\eta \int_0^1 \eta \sin(\xi\eta)\, d\xi = \int_0^{\pi/2} (1 - \cos\eta)\, d\eta = \frac{\pi}{2} - 1.$$

For the general proof of the identity $I = J$, we introduce the indefinite integrals

$$v(x, y) = \int_\alpha^y f(x, \eta)\, d\eta, \qquad u(x, y) = \int_a^x v(\xi, y)\, d\xi.$$

Applying formula (40) we find

$$u_y(x, y) = \int_a^x v_y(\xi, y)\, d\xi = \int_a^x f(\xi, y)\, d\xi$$

and thus

$$u(x, y) = u(x, \alpha) + \int_\alpha^y u_y(x, \eta)\, d\eta = \int_\alpha^y d\eta \int_a^x f(\xi, \eta)\, d\xi$$

For $x = b$, $y = \beta$ it follows that $I = J$.

We have associated here with a continuous function $f(x, y)$ in the rectangle R a function $u(x, y)$, which has continuous first derivatives

$$u_x(x, y) = \int_\alpha^y f(x, \eta)\, d\eta, \qquad u_y(x, y) = \int_a^x f(\xi, y)\, d\xi$$

and a continuous mixed second derivative

$$u_{xy}(x, y) = f(x, y).$$

We shall use the function for the purpose of "smoothing" f, that is, for constructing uniform approximations to f that have continuous partial derivatives.

For technical applications it often is essential to replace a continuous function f (itself perhaps only an approximation to an imperfectly known physical quantity) by a smooth function nearby. We know from the Weierstrass approximation theorem (Volume I, p. 569) that functions of one independent variable, continuous in an interval, can be approximated uniformly by polynomials, which even have derivatives of all orders. The analogous theorem holds for functions $f(x, y)$ continuous in a rectangle.

We can construct simpler approximations with a more moderate degree of smoothness by the process of "averaging" the function $f(x, y)$. It is convenient here to have extended the definition of f from its rectangular domain (42c) to the whole x, y-plane so that f is continuous everywhere.[1] For any $h > 0$ we form the *average* of f over the square of center (x, y) and sides of length $2h$ parallel to the axes:

$$F_h(x, y) = \frac{1}{4h^2}\int_{x-h}^{x+h} d\xi \int_{y-h}^{y+h} f(\xi, \eta)\, d\eta \tag{42e}$$

$$= \frac{u(x+h, y+h) - u(x+h, y-h) - u(x-h, y+h) + u(x-h, y-h)}{4h^2}$$

It is clear that $F_h(x, y)$ has continuous first derivatives and a continuous mixed second derivative.[2] In order to see that $F_h(x, y)$ approximates $f(x, y)$ for small h, we note that

$$F_h(x, y) - f(x, y) = \frac{1}{4h^2}\int_{x-h}^{x+h} d\xi \int_{y-h}^{y+h} [f(\xi, \eta) - f(x, y)]\, d\eta. \tag{42f}$$

Since f is uniformly continuous in some rectangle R' containing R in its interior, we know that f for given ε and sufficiently small h will vary by less than ε in every square of side $2h$ contained in R'. Then $|f(\xi, \eta) - f(x, y)| < \varepsilon$ in (42f), and $|F_h(x, y) - f(x, y)| < \varepsilon$. Hence

$$\lim_{h \to 0} F_h(x, y) \Rightarrow f(x, y) \text{ uniformly for } (x, y) \text{ in } R.$$

Thus we can find a smooth function $F_h(x, y)$ arbitrarily close to $f(x, y)$.

1.9 Differentials and Line Integrals

a. Linear Differential Forms

In Section 1.5d we defined the total differential du of a function $u = f(x, y, z)$ as the expression

[1]This can be achieved by continuing f as constant along rays perpendicular to one of the four sides of the rectangle and by continuing f into the remaining points of the plane as constant along rays from one of the four corners.

[2]In order to have $F_h(x, y)$ defined for all points of the rectangle R, we have to have f defined somewhat beyond R.

$$du = \frac{\partial f(x, y, z)}{\partial x} dx + \frac{\partial f(x, y, z)}{\partial y} dy + \frac{\partial f(x, y, z)}{\partial z} dz. \tag{43}$$

This definition for the differential of a function of several variables is suggested by the *chain rule of differentiation*. For if x, y, z are given functions of a variable t,

$$x = \varphi(t), \qquad y = \psi(t), \qquad z = \chi(t), \tag{44}$$

then the derivative of the compound function $u = f[\varphi(t), \psi(t), \chi(t)]$ according to the chain rule (19) is

$$\frac{du}{dt} = \frac{\partial f}{\partial x}\frac{dx}{dt} + \frac{\partial f}{\partial y}\frac{dy}{dt} + \frac{\partial f}{\partial z}\frac{dz}{dt}. \tag{45}$$

For functions u of a single variable t the differential has been defined as $du = \frac{du}{dt} dt$. Hence, here by (45)

$$du = \left(\frac{\partial f}{\partial x}\frac{dx}{dt} + \frac{\partial f}{\partial y}\frac{dy}{dt} + \frac{\partial f}{\partial z}\frac{dz}{dt}\right) dt$$

$$= \frac{\partial f}{\partial x}\frac{dx}{dt} dt + \frac{\partial f}{\partial y}\frac{dy}{dt} dt + \frac{\partial f}{\partial z}\frac{dz}{dt} dt,$$

which formally agrees with (43) if we remember that x, y, z (as functions of t) have the differentials

$$dx = \frac{dx}{dt} dt, \qquad dy = \frac{dy}{dt} dt, \qquad dz = \frac{dz}{dt} dt.$$

Thus the differential $du = df(x, y, z)$ as given by (43) furnishes immediately the differential $du = \frac{du}{dt} dt$ of u "along any curve" represented parametrically in the form (44).

The differential du as defined by (43) is a function of the six variables x, y, z, dx, dy, dz that is linear and homogeneous[1] in the variables dx, dy, dz, with coefficients that are functions of x, y, z. (There is, of course, no requirement that the differentials dx, dy, dz have to be "small" in any sense; such a restriction only arises if we want to use du as an approximation to the *increment*

[1]The most general linear function of three variables ξ, η, ζ is $A\xi + B\eta + C\zeta + D$ with coefficients A, B, C, D not depending on ξ, η, ζ; the linear function is called "homogeneous" or is said to be a "linear form" when $D = 0$ (see p. 13).

$$\Delta u = f(x + dx, y + dy, z + dz) - f(x, y, z)$$

as explained on p. 42).

The most general *linear differential form* in x,y,z-space is represented by the expression

$$L = A(x, y, z)\,dx + B(x, y, z)\,dy + C(x, y, z)\,dz. \tag{46}$$

It is a function L of the six variables x, y, z, dx, dy, dz that is a linear form in the "differential" variables dx, dy, dz, with coefficients depending on x, y, z. The total differentials du of functions are the special linear differential forms L that have coefficients of the form

$$A = \frac{\partial f(x, y, z)}{\partial x}, \qquad B = \frac{\partial f(x, y, z)}{\partial y}, \qquad C = \frac{\partial f(x, y, z)}{\partial z}, \tag{47}$$

for a suitable function $f = f(x, y, z)$. If a differential form L is the total differential of a function, we say it is an *exact* differential form or is *integrable*. Not every differential form is integrable; it is necessary that the coefficients A, B, C of L satisfy certain "integrability conditions":

If the coefficients A, B, C of the differential form L are of class C^1 (that is, have continuous first derivatives; see p. 42) and if L is exact, then the equations

$$\frac{\partial B}{\partial z} - \frac{\partial C}{\partial y} = 0, \qquad \frac{\partial C}{\partial x} - \frac{\partial A}{\partial z} = 0, \qquad \frac{\partial A}{\partial y} - \frac{\partial B}{\partial x} = 0 \tag{48}$$

hold.

Equations (48) simply are consequences of the rules for interchangeability of second derivatives. If A, B, C have continuous first derivatives and can be written in the form (47), then f has continuous second derivatives. Hence, by the theorem on p. 36, the order of differentiation does not matter. Thus, for example,

$$\frac{\partial A}{\partial y} = \frac{\partial}{\partial y}\frac{\partial f}{\partial x} = \frac{\partial}{\partial x}\frac{\partial f}{\partial y} = \frac{\partial B}{\partial x},$$

and similarly for the other identities in (48).

Hence, for example, the linear differential form

$$L = y\,dx + z\,dy + x\,dz$$

is not integrable, since here

$$\frac{\partial B}{\partial z} - \frac{\partial C}{\partial y} = \frac{\partial z}{\partial z} - \frac{\partial x}{\partial y} = 1 \neq 0.$$

On the other hand, the integrability conditions (48) are satisfied for the differential form

$$L = yz\,dx + zx\,dy + xy\,dz,$$

which, as a matter of fact, is the total differential du of the function $u = xyz$. To what extent the conditions (48) also are *sufficient* for expressing L as a total differential will be discussed in Section 1.10.

Similar conditions for integrability are obtained when the number of dimensions is other than three. For two independent variables x, y the general linear differential form is $L = A(x, y)\,dx + B(x, y)\,dy$. If L is the differential du of a function $u = f(x, y)$ the coefficients A, B satisfy the equation

$$\frac{\partial A}{\partial y} - \frac{\partial B}{\partial x} = 0.$$

In four dimensions, on the other hand, we obtain corresponding to equations (48) *six* integrability conditions by forming all possible mixed second derivatives of a function f of four variables.

The reason why it makes sense to consider a differential form L even when it is not an exact differential is that, along any curve C given parametrically in the form

$$x = \varphi(t), \qquad y = \psi(t), \qquad z = \chi(t),$$

L becomes the differential

$$L = \left(A\frac{dx}{dt} + B\frac{dy}{dt} + C\frac{dz}{dt}\right)dt$$

of a function of a single variable. This function is simply the one given by the indefinite integral

$$\int L = \int\left(A\frac{dx}{dt} + B\frac{dy}{dt} + C\frac{dz}{dt}\right)dt.$$

b. Line Integrals of Linear Differential Forms

For the purpose of discussing integration of linear differential forms over lines, it is important to have a clear picture of the concepts and properties of oriented arcs and closed curves. The reader is advised to reread Volume I, pp. 333–340, where all the relevant remarks are made for the case of *plane curves*. These apply equally well to curves in spaces of any number of dimensions.[1] Without restriction of generality we shall talk about integrals over curves in three-dimensional x, y, z-space.

A *simple arc* Γ is a set of points $P = (x, y, z)$ that can be represented parametrically in the form

$$(49) \qquad x = \varphi(t), \qquad y = \psi(t), \qquad z = \chi(t); \qquad a \leqq t \leqq b,$$

where φ, ψ, χ are continuous functions of t for $a \leqq t \leqq b$, and different t in that interval correspond to different points P. The parametric representation (49) constitutes a 1–1 continuous mapping of the interval on the t-axis onto the set Γ in space.[2] The same simple arc Γ has many different parametric representations. The most general one is obtained from the particular representation (49) by taking any continuous monotone function $\mu(\tau)$, mapping the interval $\alpha \leqq \tau \leqq \beta$ onto the interval $a \leqq t \leqq b$, and setting

$$(50) \qquad x = \phi[\mu(\tau)], \qquad y = \psi[\mu(\tau)], \qquad z = \chi[\mu(\tau)]; \quad \alpha \leqq \tau \leqq \beta.$$

There are two ways of ordering the points of Γ, which in any particular parametric representation (49) correspond to ordering according to either increasing or decreasing t. The choice of one of these two orderings converts Γ into an *oriented simple arc* Γ^*. We say that Γ^* is oriented *positively* with respect to the parameter t if the orientation of Γ^* corresponds to increasing t and *negatively* if it corresponds to decreasing t. The oriented simple arc with the opposite orientation is denoted by $-\Gamma^*$. The orientation is fixed completely if we know the order of any two points P_0, P_1 on Γ. If

[1]Specifically two-dimensional are only the notions of "positive and negative side" of a curve and of "clockwise and counterclockwise sense."

[2]The continuity of the mapping from t onto P is obvious from the assumed continuity of the functions φ, ψ, χ. It is important to realize that the inverse mapping $P \to t$ also is continuous. This means that given a sequence of points P_n on Γ converging to a point P the corresponding parameter values t_n converge to the parameter value for P. For the proof we observe that by the *compactness property of closed and bounded intervals* (Volume I, p. 95) a subsequence of the t_n converges to some value t with $a \leqq t \leqq b$. By the continuity of the original mapping, t is mapped on the limit P of the P_n. Because of the assumed 1–1 character of the mapping, t is determined uniquely by P. Hence, every convergent subsequence of the t_n has as limit the parameter value t corresponding to P. This proves, however, that the whole sequence of the t_n converges to t.

Γ^* is oriented positively with respect to the parameter t and if t_0 and t_1 are the parameter values for P_0, P_1, then $t_0 < t_1$ means that P_1 *follows* P_0 or P_0 *precedes* P_1 on Γ^* (Fig. 1.17).

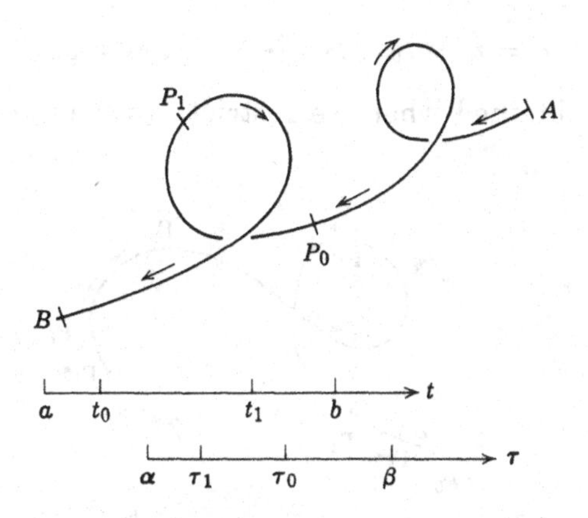

Figure 1.17 Simple arc in space oriented negatively with respect to parameter τ, positively with respect to parameter $t = \mu(\tau)$, where $\mu(\alpha) = b$, $\mu(\beta) = a$.

The end points of the oriented simple arc Γ^* correspond in the parametric representation (49) to the values $t = a, b$ in some order. We distinguish them respectively as "initial" and "final" point of Γ^*, the initial end point being the one that precedes the other one. If Γ^* has the initial point A and final point B we write

$$\Gamma^* = \widehat{AB}$$

The oppositely oriented arc is then

$$-\Gamma^* = \widehat{BA}$$

If Γ^* is oriented positively with respect to t, the initial point has parameter value a, and the final point, parameter value b.

An oriented simple arc $\Gamma^* = \widehat{AB}$ can be divided into oriented simple subarcs $\Gamma_1^*, \ldots, \Gamma_n^*$ by points $P_1, \ldots, P_{n-1}$ on Γ^* following each other according to the orientation. We put $P_0 = A$, $P_n = B$ and define for $i = 1, \ldots, n$ the arc Γ_i^* as the set of points on Γ^* consisting of P_{i-1}, P_i and all points preceding P_i and following P_{i-1}, ordered in the same way as on Γ^*. We write symbolically

$$\Gamma^* = \Gamma_1^* + \Gamma_2^* + \cdots + \Gamma_n^* \tag{51}$$

If Γ^* is oriented positively with respect to the parameter t in the representation (49) and if t_i is the parameter value corresponding to P_i, we have

$$a = t_0 < t_1 < t_2 < \cdots < t_n = b.$$

The arc Γ_i^* is obtained when we restrict t to the interval $t_{i-1} \leqq t \leqq t_i$ (Fig. 1.18).

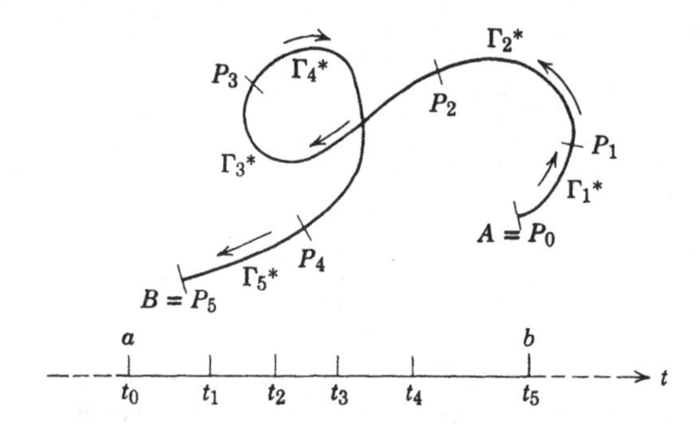

Figure 1.18 Oriented arc $\Gamma^* = AB$ represented as sum of arcs $\Gamma_{i+1}^* = P_i P_{i+1}$ such that $\Gamma^* = \Gamma_1^* + \Gamma_2^* + \Gamma_3^* + \Gamma_4^* + \Gamma_5^*$.

We are able now to define the integral $\int L$ of the linear differential form

$$L = A(x, y, z)\,dx + B(x, y, z)\,dy + C(x, y, z)\,dz \tag{52}$$

over a simple oriented arc Γ^*. We assume that the coefficients A, B, C of L are continuous in a neighborhood of Γ^*. We make the further assumption that the arc Γ^* not only is continuous but *sectionally smooth*, that is, that it can be represented parametrically by functions

$$x = \varphi(t), \qquad y = \psi(t), \qquad z = \chi(t); \qquad a \leqq t \leqq b, \tag{53}$$

which are sectionally smooth.[1]

[1]This means that φ, ψ, χ are continuous for $a \leqq t \leqq b$ and have continuous first derivatives in that interval except possibly for a finite number of jump-discontinuities of the derivatives. Notice that we require only the existence of *some* sectionally smooth parametric representation of Γ^*, while other representations need not be smooth.

Let $P_0, P_1, \ldots, P_n$ be any $n+1$ points of Γ^* following each other in the order determined by the orientation of Γ^*, where P_0 is the initial, and P_n the final, point of Γ^*.

We form the *Riemann sum*

$$F_n = \sum_{\nu=0}^{n-1}(A_\nu\,\Delta x_\nu + B_\nu\,\Delta y_\nu + C_\nu\,\Delta z_\nu). \tag{54}$$

Here A_ν, B_ν, C_ν are the values of A, B, C at some point Q_ν that precedes $P_{\nu+1}$ and follows P_ν on Γ^*, and Δx_ν, Δy_ν, Δz_ν stand for

$$x(P_{\nu+1}) - x(P_\nu), \qquad y(P_{\nu+1}) - y(P_\nu), \qquad z(P_{\nu+1}) - z(P_\nu).$$

We shall show that for $n \to \infty$ the sequence of F_n converges to a limit F, provided that the largest distance between successive points P_ν, $P_{\nu+1}$ tends to 0. The value of F does not depend on the particular choice of the points P_ν or of the intermediate points Q_ν. We call F the integral of the form L over the oriented arc Γ^*, and write

$$F = \int_{\Gamma^*} L = \int_{\Gamma^*} A\,dx + B\,dy + C\,dz \tag{55}$$

Since the definition of the integral does not refer to parametric representations, it is clear that the integral does not depend on the choice of parameters. The existence proof will imply that the integral is represented by the ordinary Riemann integral

$$\int_{\Gamma^*} L = \varepsilon\int_a^b \left(A\frac{dx}{dt} + B\frac{dy}{dt} + C\frac{dz}{dt}\right) dt \tag{56}$$

Here the integrand is the function of the single variable t obtained by substituting for the arguments x, y, z of A, B, C their expressions (53); moreover, $\varepsilon = +1$ when Γ^* is oriented positively with respect to t and $\varepsilon = -1$ when oriented negatively. Without distinguishing cases we can also write (56) as

$$\int_{\Gamma^*} L = \int_{t_i}^{t_f} \left(A\frac{dx}{dt} + B\frac{dy}{dt} + C\frac{dz}{dt}\right) dt, \tag{57}$$

where t_i is the parameter value for the initial point and t_f that of the final point of the oriented arc Γ^*; that is, $t_i = a$, $t_f = b$ when $\varepsilon = +1$, and $t_i = b$, $t_f = a$ when $\varepsilon = -1$.

To prove convergence of the Riemann sums F_n, we make use of the sectionally smooth parametric representation (53) of Γ^*. Let t_ν be the parameter value corresponding to the point P_ν. Since the correspondence between parameter values and points on the curve is continuous both ways for simple arcs (see footnote on p. 86), we see that as the largest distance between successive points tends to 0, the largest value of $|t_{\nu+1} - t_\nu|$ tends to 0 for $n \to \infty$. The functions $\varphi'(t)$, $\psi'(t)$, $\chi'(t)$ may have jump-discontinuities at a finite number of points. We can assume that all those points of discontinuity occur among our subdivision points $t_0, t_1, \ldots, t_n$, for since the A, B, C are bounded and the largest of the Δx_ν, Δy_ν, Δz_ν tend to 0 for $n \to \infty$, the effects of adding or subtracting contributions from a fixed finite number of subdivision points in the Riemann sum, F_n, disappear in the limit.

Since $\varphi(t)$, $\psi(t)$, $\chi(t)$ are now differentiable in the interior of each subinterval, we can apply the *mean value* theorem *of differential calculus* (see Volume I, p. 174) and find

$$\Delta x_\nu = \varphi(t_{\nu+1}) - \varphi(t_\nu) = \varphi'(\tau_\nu)(t_{\nu+1} - t_\nu)$$

$$\Delta y_\nu = \psi'(\tau_\nu')(t_{\nu+1} - t_\nu). \qquad \Delta z_\nu = \chi'(\tau_\nu'')(t_{\nu+1} - t_\nu),$$

with values τ_ν, τ_ν', τ_ν'' intermediate between t_ν and $t_{\nu+1}$. The point Q_ν on Γ^* also corresponds to a parameter value σ_ν intermediate between t_ν and $t_{\nu+1}$. Hence, the Riemann sum F_n in (54) takes the form

$$F_n = \sum_{\nu=0}^{n-1} [A(\sigma_\nu)\varphi'(\tau_\nu) + B(\sigma_\nu)\,\psi'(\tau_\nu') + C(\sigma_\nu)\,\chi'(\tau_\nu'')]\,[t_{\nu+1} - t_\nu].$$

Here the points $t_0, t_1, \ldots, t_n$ form a subdivision of the parameter interval $[a, b]$. If Γ^* is oriented positively with respect to t, the t_ν form an increasing sequence with $t_0 = a$, $t_n = b$, and $\Delta t_\nu = t_{\nu+1} - t_\nu > 0$. Otherwise, the t_ν are decreasing, $t_0 = b$, $t_n = a$, and $\Delta t_\nu < 0$. In our notation for the parameter interval, a always stands for the *smaller one* of the values a, b and thus may correspond to either the initial or the final point of the arc Γ^*.

If we now use the fundamental existence theorem for definite integrals as limits of Riemann sums (see Volume I, pp. 192 ff.), we find that $F = \lim_{n\to\infty} F_n$ exists and is given by formula (56).[1] The factor $\varepsilon = \pm 1$ arises from the assumption made in that theorem that the points of subdivision t_ν used in forming the Riemann sum constitute an *increasing* sequence. When the orientation of Γ^* corresponds to

[1]The intermediate values τ_ν, τ_ν', τ_ν'', σ_ν need not be the same for convergence (see the remarks on p. 195, Volume I).

decreasing t, we have to run through the values t_ν in opposite order, starting with t_n and ending with t_0, and change the sign of Δt_ν.

It is clear that the definition of line integral and the formula (56) can be extended to the case where Γ^* is an *oriented simple closed curve.*[1] In this case we form the Riemann sum by selecting n points $P_1, P_2, \ldots, P_n$ on Γ^* that follow each other in the order determined by the orientation, and we put $P_0 = P_n$ in the expression (54) for F_n.

Instances of integrals over curves in the x, y-plane have been encountered already in Volume I. Thus, the oriented area bounded by a closed oriented curve Γ^* had been represented in the form

$$A = \frac{1}{2}\int_a^b \left(x\frac{dy}{dt} - y\frac{dx}{dt}\right) dt$$

(see Volume I, p. 365); that is, as the line integral

$$A = \frac{1}{2}\int_{\Gamma^*} x\,dy - y\,dx$$

Another example is furnished by the work W done by a field of force with components ρ, σ in moving from a point P_0 to a point P_1 along a curve $\Gamma^* = \widehat{P_0P_1}$ referred to arc length s as parameter. Here (see Volume I, p. 420)

$$W = \int_{s_0}^{s_1} \left(\rho\frac{dx}{ds} + \sigma\frac{dy}{ds}\right) ds,$$

which can be written as

$$W = \int_{\Gamma^*} \rho\,dx + \sigma\,dy.$$

In the same way we can define the *work* done by forces in space with components ρ, σ, τ, in moving along an arc Γ^* in the direction given by its orientation as a line integral

$$W = \int_{\Gamma^*} \rho\,dx + \sigma\,dy + \tau\,dz.$$

---

[1]Such a curve has a continuous parametric representation (53), with different t corresponding to different points, except that $t = a$ and $t = b$ yield the same point. Moreover a cyclic order is specified on Γ^*, corresponding to either increasing or decreasing t (see Volume I, p. 339). We can always represent Γ^* as sum of oriented simple arcs Γ_i^* in the form (51), where for $i = 2, \ldots, n$ the final point of Γ_{i-1}^* is the initial point of Γ_i^* and where the final point of Γ_n^* is the initial point of Γ_1^*.

Exercises 1.9b

1. Find

$$\int z\,dx + x\,dy + y\,dz$$

(a) over the arc of the helix

$$x = \cos t, \qquad y = \sin t, \qquad z = t$$

joining the points (1, 0, 0) and (1, 0, 2 π);

(b) over the parabolic arc

$$x = x_0(1 - t^2), \qquad y = y_0(1 - t^2), \qquad z = t$$

joining the points (0, 0, 1) and (0, 0, −1) (for constant x_0, y_0).

c. *Dependence of Line Integrals on End Points*

We return to the general differential form L given by (52). Let Γ be a simple arc (not yet oriented) with a sectionally smooth parameter representation (53).

For any two points P_0, P_1 on Γ corresponding to the values t_0, t_1 of the parameter t, we can form the integral

$$I = \int_{t_0}^{t_1} \left(A\frac{dx}{dt} + B\frac{dy}{dt} + C\frac{dz}{dt}\right)dt.$$

By formula (57), I is equal to $\int L$ extended over the oriented subarc $\widehat{P_0P_1}$ of Γ that has P_0 as initial and P_1 as final point. It follows that I does not depend on the particular parameter representation. We write

$$I = \int_{P_0}^{P_1} L$$

The value of I is determined by the ordered pair of points P_0, P_1 and the simple arc of which they are end points.

For fixed P_0 we can define a function $f = f(P)$ *along the* arc Γ by the indefinite integral

$$f(P) = \int_{P_0}^{P} L = \int_{t_0}^{t} \left(A\frac{dx}{dt} + B\frac{dy}{dt} + C\frac{dz}{dt}\right) dt. \tag{58}$$

Taking f as a function of the independent variable t, we then have

$$\frac{df}{dt} = A\frac{dx}{dt} + B\frac{dy}{dt} + C\frac{dz}{dt}. \tag{59}$$

Writing this equation as

$$df = \frac{df}{dt}\,dt = A\,dx + B\,dy + C\,dz = L\,,$$

we thus express the linear differential form L (which need not be exact) as the differential of a function f; but we have to remember that this relation holds only along a special curve Γ on which f is defined.

For any points P and P' of Γ

$$\int_P^{P'} L = f(P') - f(P). \tag{60}$$

This follows immediately if we express the line integrals as integrals over the variable t and apply the fundamental connection between definite and indefinite integrals (see Volume I, p. 190). If Γ^*, the arc Γ with a certain orientation, has the initial point A and the final point B, we find, in particular, that

$$\int_{\Gamma^*} L = \int_A^B L = f(B) - f(A). \tag{61}$$

If $P_0, \ldots, P_n$ are points on Γ^* in the order determined by the orientation of Γ^*, with $P_0 = A$, $P_n = B$, we have

$$L = f(B) - f(A) = \sum_{\nu=0}^{n-1} [f(P_{\nu+1}) - f(P_\nu)]$$

$$= \sum_{\nu=0}^{n-1} \int_{P_\nu}^{P_{\nu+1}} L\,.$$

If we denote by $\Gamma_{\nu+1}^*$ the subarc with initial point P_ν and final point $P_{\nu+1}$, we have

$$\int_{P_\nu}^{P_{\nu+1}} L = \int_{\Gamma_{\nu+1}^*} L$$

Here the orientation of Γ_ν^* agrees with that of Γ so that

$$\Gamma^* = \Gamma_1^* + \Gamma_2^* + \cdots + \Gamma_n^*.$$

Therefore, *line integrals are additive*:

$$\int_{\Gamma_1^*+\cdots+\Gamma_n^*} L = \int_{\Gamma_1^*} L + \cdots + \int_{\Gamma_n^*} L \tag{62}$$

Similarly, if we interchange the end points of Γ^*,

$$\int_{-\Gamma^*} L = -\int_{\Gamma^*} L \tag{63}$$

These rules are of particular interest when applied to oriented closed curves represented as sums of oriented simple arcs. Consider a number of oriented simple closed curves $C_1^*, \ldots, C_n^*$(see Fig. 1.19),

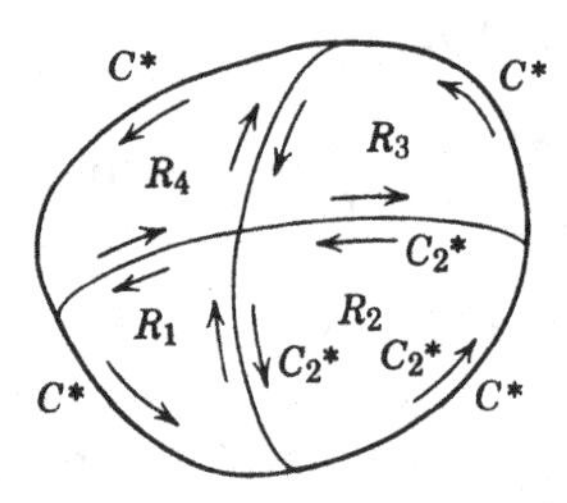

Figure 1.19 Additivity of line integrals over closed curves.

which may have portions in common. Assume that a simple arc Γ common to two of the curves, C_i^* and C_k^*, receives opposite orientations from C_i^* and C_k^* and that the portions of the curves not common to any two of them add up to an oriented closed curve C^*. Writing each line integral over a curve C_i^* as the sum of integrals over simple arcs and adding all these integrals, the contributions of the common arcs cancel out and we are left with the formula

$$\int_{C^*} L = \int_{C_1^*} L + \cdots + \int_{C_n^*} L \tag{64}$$

This situation arises, in particular, when the C_i^* are plane curves forming the boundaries of nonoverlapping two-dimensional regions R_i that together form a region R with boundary curve C^*, all C_i^* and C^* having the same orientation. More generally, the region R and its boundary C^* may lie on a surface, and R may be subdivided by arcs into subregions R_i with boundary curves C_i^* whose orientations fit together in the manner described.

A somewhat different application of the same principle occurs in the following theorem. Let two oriented closed curves C^* and C'^* (see Fig. 1.20) be subdivided by the points $A_1, \ldots, A_n$ and $A_1', \ldots, A_n'$, respectively, in the order of the sense of orientation, and let each pair of corresponding points A_i and A_i' be joined by a curved line. If

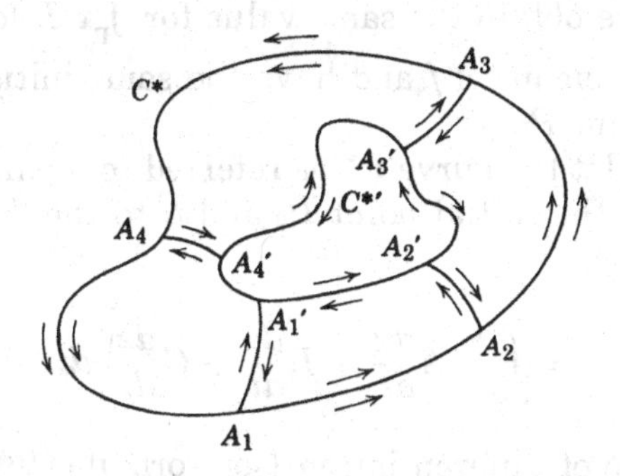

Figure 1.20

by C_i^* we denote the closed oriented curve $A_iA_{i+1}A_{i+1}'A_i'$ (identifying A_{n+1} with A_1 and A_{n+1}' with A_1'), then

$$\sum_{i=1}^{n} \int_{C_i^*} L = \int_{C'^*} L - \int_{C^*} L. \tag{65}$$

1.10 The Fundamental Theorem on Integrability of Linear Differential Forms

a. *Integration of Total Differentials*

A particularly important class of differential forms

$$L = A\,dx + B\,dy + C\,dz \tag{66}$$

are the total differentials of functions $u = f(x, y, z)$, with A, B, C of the form

$$A = \frac{\partial f}{\partial x}, \qquad B = \frac{\partial f}{\partial y}, \qquad C = \frac{\partial f}{\partial z}, \tag{67}$$

where f is a function with continuous first derivatives. While in general the value of $\int_{\Gamma^*} L$ depends not only on the end points but on the entire course of the curve, the following theorem is valid here:

The integral of a linear differential form L, which is the total differential of a function f, is equal to the difference of the values of f at the end points and does not depend on the course of Γ^* between those *points*. That is, we obtain the same value for $\int_{\Gamma^*} L$ for all curves Γ^* which lie in the domain of f and have the same initial point P_0 and the same final point P_1.

For the proof, let the curve Γ^* be referred to a parameter t where t_0 corresponds to the initial point P_0 and t_1 to the final point P_1. By (57), p. 89

$$\int_{\Gamma^*} L = \int_{t_0}^{t_1} \left(A\frac{dx}{dt} + B\frac{dy}{dt} + C\frac{dz}{dt}\right) dt.$$

By the chain rule of differentiation [see formula (18) p. 55] we then have

$$\int_{\Gamma^*} L = \int_{t_0}^{t_1} \frac{df}{dt}\, dt = f\Big|_{t_0}^{t_1} = f(P_1) - f(P_0), \tag{68}$$

where we write

$$f(P_i) = f(x(t_i), y(t_i), z(t_i))$$

for $i = 0,1$.

We observe that instead of requiring that the integral is independent of the path, we might just as well require that the integral over a simple closed curve Γ^* has the value 0, for if we divide the curve Γ^* by means of two points P_0 and P_1 into two oriented arcs Γ_1^* and Γ_2^*, we have

$$\Gamma^* = \Gamma_1^* + \Gamma_2^*,$$

where, say, Γ_1 has initial point P_0 and final point P_1, while Γ_2^* has initial point P_1 and final point P_0 (see p. 94). Then

$$\int_{\Gamma^*} L = \int_{\Gamma_1^*} L + \int_{\Gamma_2^*} L = \int_{\Gamma_1^*} L - \int_{-\Gamma_2^*} L$$

Here $-\Gamma_2^*$ has the same initial point P_0 and the same final point P_1 as Γ_1^*. The vanishing of $\int L$ over the closed curve Γ^* means exactly the same thing as the equality of L taken over the two simple arcs that have P_0 as initial point and P_1 as final point.

b. Necessary Conditions for Line Integrals to Depend Only on the End Points

Only under very special conditions is a line integral independent of the path or, what is equivalent, is the line integral round a closed path 0. For example, if a closed curve C^* in the x,y-plane forms the boundary of a region of positive area, then the line integral $\int(x\,dy - y\,dx)$ over C^* is not 0. We proved in the preceding section that for the independence of $\int L$ from the path joining the end points, it is sufficient that L is a total differential. The chief task of the theory of line integrals is to show that this condition is also necessary and then to express this necessary and sufficient condition in a form convenient for applications.

We shall investigate this question of independence for integrals over curves in three-space. But the results and proofs are exactly analogous in any number of dimensions. We make the assumption that $L = A\,dx + B\,dy + C\,dz$ is a linear differential form with coefficients A, B, C that are continuous functions of x, y, z in an open set R of space. The following theorem then holds:

The line integral $\int L$ taken over a simple oriented arc Γ^* in R is independent of the particular choice of Γ^* and determined solely by the initial and final point of Γ^* if and only if L is the total differential of a function $f(x, y, z)$ in R.

We have already proved on p. 95 that this condition is sufficient; that is, for an exact differential $L = A\,dx + B\,dy + C\,dz$ the integral $\int L$ is independent of the path. It is easy to see that the condition is necessary. Assume that $\int_{\Gamma^*} L$ depends only on the end points of Γ^*. We want to show that there exists a function $u(x, y, z)$ defined in R for which $du = L$. With no loss of generality we can assume that every two points of R can be connected by a simple polygonal arc that lies completely in R.[1] We pick a fixed point P_0 in R and define the function $u = u(x, y, z) = u(P)$ at any point P of R as $\int L$ extended over any simple arc with initial point P_0 and final point P. In order to compute the partial derivatives of u, we consider any point $(x, y, z) = P$ of R (Fig. 1.21). Since R is open, all points $(x + h, y, z) = P'$ will then also belong to R provided $|h|$ is sufficiently small. Let γ^* denote the oriented straight line segment joining P and P', while Γ^* shall denote a simple polygonal path joining P_0 to P. We can always modify Γ^* slightly to bring about that the last side of this polygonal arc, which has P as final point, is not parallel to the x-axis. Then Γ^* and γ^* have no point in common besides P (at least for $|h|$ sufficiently

[1]The open set R can always be decomposed into connected subsets that have this property (see Appendix 112). We then define u in each of these subsets by the construction indicated.

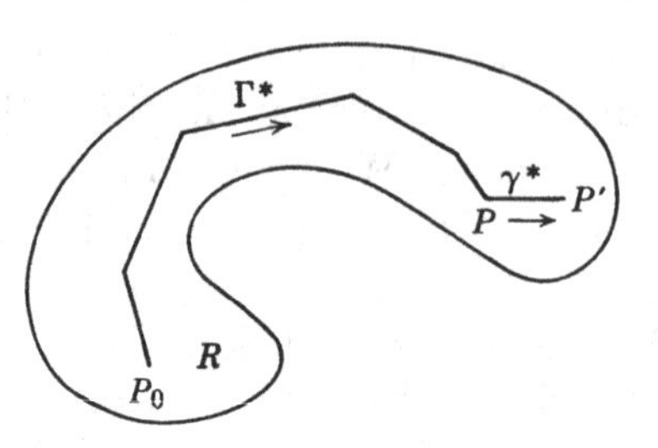

Figure 1.21

small), and $\Gamma^* + \gamma^*$ represents a simple arc with initial point P_0 and final point P'. It follows [see (62, p. 93)] that

$$u(x + h, y, z) - u(x, y, z) = u(P') - u(P) = \int_{\Gamma^* + \gamma^*} L - \int_{\Gamma^*} L = \int_{\gamma^*} L$$

$$= \int_x^{x+h} A(t, y, z)\, dt$$

Dividing by h and passing to the limit with $h \to 0$, we find that indeed

$$\frac{\partial u(x, y, z)}{\partial x} = A,$$

and similarly $\partial u/\partial y = B$ and $\partial u/\partial z = C$. This shows that $du = L$.

c. *Insufficiency of the Integrability Conditions*

The theorem on independence of line integrals we just proved is, however, of no great value unless we have some way of finding out whether a given differential L is a total differential or not. It is desirable to have some condition that involves only the coefficients A, B, C of $L = A\, dx + B\, dy + C\, dz$ and is easily verified. We have already recognized the integrability conditions

$$(69) \qquad \frac{\partial B}{\partial z} - \frac{\partial C}{\partial y} = 0, \qquad \frac{\partial C}{\partial x} - \frac{\partial A}{\partial z} = 0, \qquad \frac{\partial A}{\partial y} - \frac{\partial B}{\partial x} = 0$$

as necessary for the existence of a function $u = f(x, y, z)$ with the property that $L = du$. A form L satisfying (69) is called *closed*. Hence every exact form is closed. Since line integrals can be independent of the particular path joining any two points only when L is a total differential, we see that *conditions (69) are necessary, if L is to depend only on the end points of the path of integration.* Are these conditions also sufficient? They are sufficient if they permit us to construct a function $u = f(x, y, z)$ for which

$$A = \frac{\partial f}{\partial x}, \qquad B = \frac{\partial f}{\partial y}, \qquad C = \frac{\partial f}{\partial z}. \tag{70}$$

The surprising result is that the integrability conditions (69) suffice almost, but not quite, to ensure that L is the total differential of a function u and, hence, to ensure the independence of $\int L$ from the path. The identities (69) in themselves are not sufficient but become so if we add an assumption of quite a different character, one that concerns a *geometrical property* of the region in space in which L is considered.

A simple counterexample shows that conditions (69) alone are not sufficient to guarantee that $\int L$ taken over any closed curve is 0. We consider the differential

$$L = \frac{x\,dy - y\,dx}{x^2 + y^2} \tag{71}$$

corresponding to the choice of coefficients

$$A = \frac{-y}{x^2 + y^2}, \qquad B = \frac{x}{x^2 + y^2}, \qquad C = 0,$$

which are defined except for points on the line $x = y = 0$ (the z-axis). One verifies easily that the integrability conditions (69) are satisfied and thus that L is closed. When we integrate around the unit circle C^*: $x = \cos t$, $y = \sin t$, $z = 0$ in the x,y-plane, oriented positively with respect to t, we find

$$\int_{C^*} L = \int_0^{2\pi} \left(A\frac{dx}{dt} + B\frac{dy}{dt}\right) dt = \int_0^{2\pi} (\sin^2 t + \cos^2 t)\, dt$$
$$= 2\pi \neq 0.$$

As a matter of fact, it is easy to calculate $\int L$ around any closed curve C for the L given by (71). We introduce the polar angle θ of a point $P = (x, y, z)$ by

$$\cos\theta = \frac{x}{\sqrt{x^2 + y^2}}, \qquad \sin\theta = \frac{y}{\sqrt{x^2 + y^2}} \tag{72}$$

that is, the angle formed with the x, z-plane by the plane through P passing through the z-axis (see Fig. 1.22). Then

$$d\theta = d \arc\tan \frac{y}{x} = L, \tag{73}$$

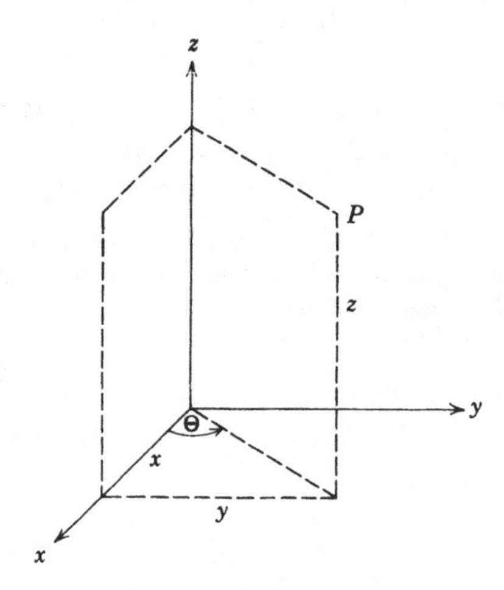

Figure 1.22

so that L is represented as total differential of the function $u = \theta$. The complications arise from the fact that formulae (72) define the values of θ only within whole multiples of 2π. Starting with some possible values θ_0 for θ at a point P_0, we can define θ in any point P by joining P to P_0 by a continuous curve and taking

$$\theta(P) = \theta_0 + \int_{P_0}^{P} d\theta = \theta_0 + \int L$$

(See Volume I, p. 434). But $\theta(P)$ defined in this way is multiple-valued depending on the choice of the curve: for a closed curve C^* the expression

$$\frac{1}{2\pi} \int_C d\theta$$

represents the number of times C *winds around* the z-axis in the clockwise sense (see Fig. 1.23). Hence, the value of

$$\int_{P_0}^{P} d\theta \tag{74}$$

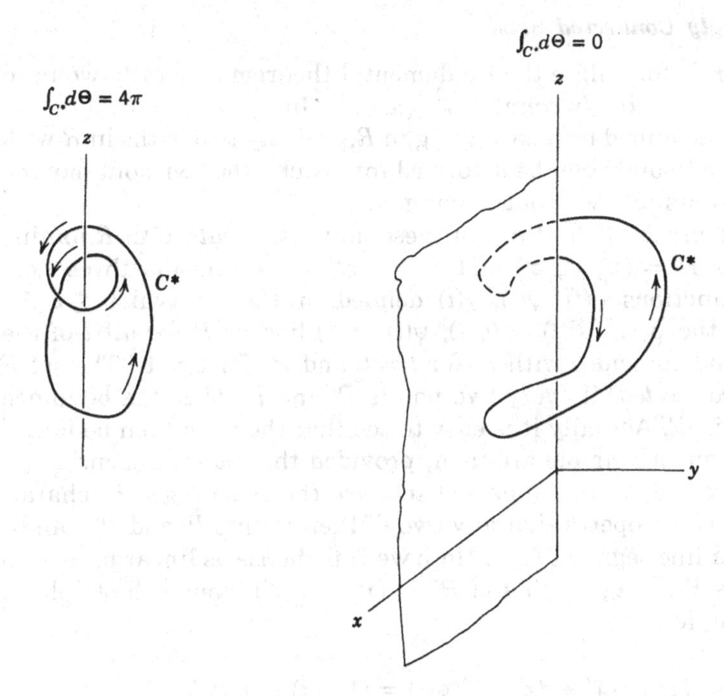

Figure 1.23

taken for two different paths with end points P_0, P is the same only if going along one path from P_0 to P and returning along the other path to P_0 we go zero-times around the z-axis. We can prevent any path from going around the z-axis by considering only points (x, y, z) with either $y \neq 0$ or with $y = 0$ and $x > 0$, erecting, in a manner of speaking, a wall along the half-plane

$$y = 0, \qquad x \leqq 0$$

which is not to be crossed. The points not excluded form a region R in which we can assign to θ a unique value with

$$-\pi < \theta < \pi$$

that constitutes a continuously differentiable function $\theta = \theta(x, y, z)$ with differential L. The integral (74) extended over any path in the region that joins P and P_0 has then a unique value $\theta(P) - \theta(P_0)$, which does not depend on the particular path. Similarly, the integral over a closed path in this region has the value 0.

d. Simply Connected Sets

In order to formulate the fundamental theorem generally we need the notion of a *simply connected*[1] *open set.* In such a set R, any two points can be joined by a path lying in R, and any two paths in R with the same end points can be deformed into each other without moving the end points and without leaving R.

We give precise definitions of these notions. A *path* C in R joining two points $P' = (x', y', z')$ and $P'' = (x'', y'', z'')$ means three continuous functions $\varphi(t)$, $\psi(t)$, $\chi(t)$ defined in the interval $0 \leqq t \leqq 1$ such that the point $P(t) = (\varphi(t), \psi(t), \chi(t))$ lies in R for all t of the interval and coincides with P' for $t = 0$ and P'' for $t = 1$.[2] The set R is called *connected*[3] if every two points P' and P'' of R can be joined by a path in R. Actually it is easy to see that they can then be joined also by a smooth simple arc in R, provided the set R is open.[4]

Trivial examples of connected sets are the *convex sets* R, characterized by the property that any two of their points P' and P'' can be joined by a line segment in R. Here we can choose as linear path with end points $P' = (x', y', z')$ and $P'' = (x'', y'', z'')$ simply the triple of linear functions

$$\varphi(t) = (1 - t)\,x' + tx'', \qquad \psi(t) = (1 - t)\,y' + ty'',$$
$$\chi(t) = (1 - t)\,z' + tz''$$

for $0 \leqq t \leqq 1$. Examples of such convex sets are solid spheres or cubes. Examples of connected, but not convex, sets are a solid torus, a spherical shell (i.e., the space between two concentric spheres), and the outside of a sphere or cylinder. Any set R whatsoever in space if it is not connected consists of connected subsets called the *components* of R. Disconnected are, for example, the set of points *not*

---

[1]More precisely "pathwise simply connected."

[2]Different t need not correspond to different $P(t)$. Notice that the description of a path does not only include the set of the points $P(t)$ in space (the "support" of the path) but also the choice of corresponding parameters t. Every simple arc in space determines many different paths corresponding to different parameter representations of the arc. We can always bring about by a linear substitution that the parameter values vary over the particular interval $0 \leq t \leq 1$.

[3]More precisely "pathwise connected."

[4]Taking a sufficiently fine subdivision of the parameter interval and joining corresponding points $P(t)$ by line segments, we first obtain a polygonal arc in R joining P' and P''. Omitting loops we get a simple polygonal arc. Replacing small portions near a corner by suitable parabolic arcs, we get a smooth simple arc in R joining P' and P''. See also p. 112.

belonging to a spherical shell or the set of points none of whose coordinates is an integer.

Let C_0 and C_1 be any two paths in R, given respectively by $(\varphi_0(t), \psi_0(t), \chi_0(t))$ and $(\varphi_1(t), \psi_1(t), \chi_1(t))$. Their end points P', P'', corresponding to $t = 0$ and $t = 1$, shall be the same. The connected set R is simply connected, if we can "deform C_0 into C_1" or "join C_0 and C_1" by means of a continuous family of paths C_λ with common end points P', P''. This shall mean that there exist continuous functions $(\varphi(t, \lambda), \psi(t, \lambda), \chi(t, \lambda)$ of the two variables t, λ for $0 \leqq t \leqq 1, 0 \leqq \lambda \leqq 1$, such that the point $P = (\varphi, \psi, \chi)$ always lies in R and such that P coincides with $(\varphi_0, \psi_0, \chi_0)$ for $\lambda = 0$, with $(\varphi_1, \psi_1, \chi_1)$ for $\lambda = 1$, with P' for $t = 0$ and with P'' for $t = 1$.[1] For each fixed λ the functions φ, ψ, χ determine a path C_λ in R that joins the points P' and P''. As λ varies from 0 to 1, the path C_λ changes continuously from C_0 to C_1, and in this sense represents a "continuous deformation" of C_0 into C_1 (see Fig. 1.24).

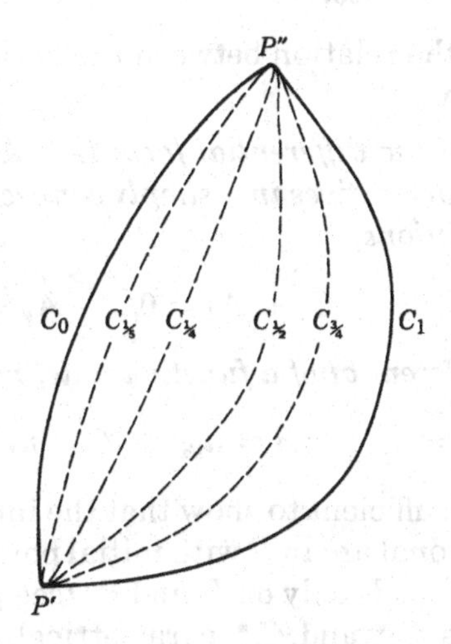

Figure 1.24

As is easily seen, *convex* sets R are simply connected. We only have to associate with the two curves C_0, C_1 having common end points P', P'' the curves C_λ given by

$$\begin{aligned} \varphi(t, \lambda) &= (1-\lambda)\,\varphi_0(t) + \lambda\varphi_1(t) \\ \psi(t, \lambda) &= (1-\lambda)\,\psi_0(t) + \lambda\psi_1(t) \\ \chi(t, \lambda) &= (1-\lambda)\,\chi_0(t) + \lambda\chi_1(t). \end{aligned}$$

[1]The paths C and C_1 are called *homotopic* relative to P', P''.

Here C_λ is obtained geometrically by joining points of C_0 and C_1 that belong to the same t by a line segment and taking the point that divides the segment in the ratio $\lambda/(1-\lambda)$. The points obtained in this way all lie in R because of the convexity of R. A different type of pathwise simply connected set is represented by a spherical shell. Not simply connected, on the other hand, is the set R obtained by removing the z-axis from x, y, z-space. Here the two paths (semicircles)

$$x = \cos \pi t, \qquad y = \sin \pi t, \qquad z = 0; \qquad 0 \leqq t \leqq 1$$

and

$$x = \cos \pi t, \qquad y = -\sin \pi t, \qquad z = 0; \qquad 0 \leqq t \leqq 1$$

have the same end points but cannot be deformed into each other without crossing the z-axis, which does not belong to R.[1]

e. *The Fundamental Theorem*

We can now state the relation between the notions of *closed* and of *exact* differential forms:

If the coefficients of the differential form $L = A\,dx + B\,dy + C\,dz$ *have continuous first derivatives in a simply connected set* R *and satisfy the integrability conditions*

$$B_z - C_y = 0, \qquad C_x - A_z = 0, \qquad A_y - B_x = 0, \tag{75a}$$

then L *is the total differential of a function* u *defined in* R*:*

$$A = u_x, \qquad B = u_y, \qquad C = u_z. \tag{75b}$$

For the proof, it is sufficient to show that the integral of L extended over any simple polygonal arc in R with initial point P' and final point P'' has a value that depends only on P' and P'' (see p. 97). We represent the two oriented arcs $C_0{}^*$ and $C_1{}^*$ parametrically by, respectively,

$$x = \phi_0(t), \qquad y = \psi_0(t), \qquad z = \chi_0(t), \qquad 0 \leqq t \leqq 1 \tag{76a}$$

and

$$x = \phi_1(t), \qquad y = \psi_1(t), \qquad z = \chi_1(t); \qquad 0 \leqq t \leqq 1 \tag{76b}$$

with $t = 0$ yielding P' and $t = 1$ yielding P''. Using the simple con-

[1]This follows from the fundamental theorem below and the fact that there exists a closed differential form, the one given by (71), whose integral over the whole circle does not vanish.

nectivity of R, we can "imbed" the paths (75a, b) into a continuous family[1]

(76c) $$x = \phi(t, \lambda), \qquad y = \psi(t, \lambda), \qquad z = \chi(t, \lambda)$$

reducing to (76a, b) for $\lambda = 0, 1$ and to P', P'' for $t = 0, 1$. We have by formula (56), p. 89.

(76d) $$\int_{C_1{}^*} L - \int_{C_0{}^*} L$$
$$= \int_0^1 [(Ax_t + By_t + Cz_t)|_{\lambda=1} - (Ax_t + By_t + Cz_t)|_{\lambda=0}]\, dt$$

where x, y, z are the functions of t, λ given by (76c). We assume, to begin with, that those functions have continuous first derivatives with respect to t, λ and a continuous mixed second derivative for $0 \leqq t \leqq 1$, $0 \leqq \lambda \leqq 1$. Then by (76d)

(76e) $$\int_{C_1{}^*} L - \int_{C_0{}^*} L = \int_0^1 dt \int_0^1 (Ax_t + By_t + Cz_t)_\lambda \, d\lambda$$

Now using the chain rule of differentiation and the integrability conditions (76a), we have the identity

$$\begin{aligned}(Ax_t + By_t + Cz_t)_\lambda &= Ax_{\lambda t} + By_{\lambda t} + Cz_{\lambda t} + A_x x_\lambda x_t + A_y y_\lambda x_t + A_z z_\lambda x_t \\ &\quad + B_x x_\lambda y_t + B_y y_\lambda y_t + B_z z_\lambda y_t + C_x x_\lambda z_t \\ &\quad + C_y y_\lambda z_t + C_z z_\lambda z_t \\ &= (Ax_\lambda + By_\lambda + Cz_\lambda)_t\end{aligned}$$

Interchanging orders of integration (see p. 80), we find that

$$\int_{C_1{}^*} L - \int_{C_0{}^*} L = \int_0^1 d\lambda \int_0^1 (Ax_\lambda + By_\lambda + Cz_\lambda)_t \, dt = 0,$$

since x_λ, t_λ, z_λ vanish for $t = 0, 1$, because the end points are independent of λ.

One sees the important part played in the proof by the assumption that R is simply connected. It enables us to convert the difference of the line integrals into a double integral over some intermediate region.

It is easy to remove the restrictions on the existence of derivatives of the functions ϕ, ψ, χ. Assume only that the arcs $C_0{}^*$ and $C_1{}^*$ are

[1]The paths of the family need not to be simple for $\lambda \neq 0,1$.

smooth, that is, that the functions $\phi(t, \lambda)$, $\psi(t, \lambda)$, $\chi(t, \lambda)$ have a continuous t-derivative when λ has one of the values 0 or 1 while being continuous for other values of λ. We can then (see p. 82) approximate these functions uniformly by functions $\bar{\phi}$, $\bar{\psi}$, $\bar{\chi}$, which have continuous first derivatives with respect to t and λ and a continuous mixed second derivative. In order that the smoother functions obtained represent a deformation of the paths C_0^* and C_1^* into each other, they have to agree with ϕ, ψ, χ for $\lambda = 0, 1$ and for $t = 0, 1$. This can always be brought about by a slight modification of $\bar{\phi}$, $\bar{\psi}$, $\bar{\chi}$, by adding suitable terms so that

$$\begin{aligned} x = {} & \bar{\phi}(t, \lambda) - (1 - \lambda)\,[\bar{\phi}(t, 0) - \phi_0(t)] - \lambda[\bar{\phi}(t, 1) - \phi_1(t)] \\ & - (1 - t)\,[\bar{\phi}(0, \lambda) - \phi_0(0)] - t[\bar{\phi}(1, \lambda) - \phi_0(1)] \\ & + (1 - t)(1 - \lambda)\,[\bar{\phi}(0, 0) - \phi_0(0)] + (1 - t)\lambda\,[\bar{\phi}(0, 1) - \phi_0(0)] \\ & + t(1 - \lambda)\,[\bar{\phi}(1, 0) - \phi_0(1)] + t\lambda\,[\bar{\phi}(1, 1) - \phi_0(1)] \end{aligned}$$

with analogous expressions for y and z. These functions have the correct values for $\lambda = 0, 1$, and for $t = 0, 1$, have continuous first derivatives and mixed second derivatives, and can be made to approximate the original ϕ, ψ, χ so closely that the corresponding points (x, y, z) still lie in the open set R.

Finally, the equality of the integrals of L can be extended to arcs C_0^* C_1^* that are only *sectionally* smooth, e.g. to polygonal arcs, by approximating these arcs by smooth ones with the same end points. The integrals over the approximating smooth arcs all have the same values, and the same follows then in the limit for the integrals over C_0^* and C_1^*.

Appendix

Geometrical intuition and physical reality always have provided powerful motivation and guiding ideas for constructive mathematical thought. Nevertheless, with the advance of analysis since the beginning of the nineteenth century, it has become a compelling necessity to cease invoking intuition as the prime justification of mathematical considerations. More and more, one has turned to rigorous proofs based on axiomatically hardened precision and clearly formulated concepts and procedures. In this development the notion of *set*, in particular of *point set*, has played a major role and by now has been absorbed into the fabric of analysis. Of some of these developments this appendix gives a simple introductory account.

A.1. The Principle of the Point of Accumulation in Several Dimensions and Its Applications

To establish the theory of functions of several variables on a firm basis, we can proceed in exactly the same way as in the case of functions of one variable. It is sufficient to discuss these matters in the case of two variables only, since the methods are essentially the same for functions of more than two independent variables.

a. The Principle of the Point of Accumulation

We base our discussion on Bolzano's and Weierstrass's principle of the point of accumulation. A pair of numbers (x, y) may be represented in the usual way by means of a point with the rectangular coordinates x and y in an x,y-plane. We now consider a bounded infinite set of such points $P(x, y)$, that is, a set containing an infinite number of distinct points, all of them lying in a bounded part of the plane, so that $|x| < C$ and $|y| < C$, where C is a constant. The principle of the point of accumulation states that *every bounded infinite set S of points has at least one point of accumulation.* That is, there exists a point Q with coordinates (ξ, η) such that an infinite number of points of S lie in every neighborhood of Q, say, in every region

$$(x - \xi)^2 + (y - \eta)^2 < \delta^2,$$

where δ is any positive number. It follows that, out of the infinite bounded set of points we can choose a sequence of distinct points $P_1, P_2, P_3, \ldots$ that converges to a limit Q. The sequence of the P_i can be constructed by induction, giving δ successively the values $1, \frac{1}{2}, \frac{1}{3}, \ldots$; we choose P_1 arbitrarily in S; if $P_1, \ldots, P_n$ have been defined, we take for P_{n+1} any one of the infinitely many points in the set S that have distance $< 1/(n + 1)$ from Q and are different from Q and from $P_1, \ldots, P_n$.

This principle of the point of accumulation for several dimensions can be proved analytically by the method used in the corresponding proof in Volume I (p. 95), merely by substituting rectangular regions for the intervals used there. An easier proof is obtained if we make use of the principle for one dimension. To do this we notice that by hypothesis every point $P(x, y)$ of the set S has an abscissa x for which the inequality $|x| < C$ holds. Either there is an $x = x_0$ that is the abscissa of an infinite number of points P (which therefore lie vertically above one another) or else each x belongs only to a finite number of points P. In the first case, we fix upon x_0 and consider the infinite number of values of y such that (x_0, y) belongs to our set. These values of y have a point of accumulation for one dimension. Hence, we can find a sequence of values of y, say $y_1, y_2, \ldots$, such that $y_n \to \eta_0$, from which it follows that the points (x_0, y_n) of the set tend to the limit point (x_0, η_0), which is thus a point of accumulation of the set. In the second case, there must be an infinite number of distinct values of x that are the abscissae of points of the set, and we can choose a sequence $x_1, x_2, \ldots$ of these abscissae tending to a limit ξ. For each x_n, let $P_n = (x_n, y_n)$ be a point of the set with abscissa x_n. The y_n form an infinite bounded set of numbers; hence, we can choose a subsequence $y_{n_1}, y_{n_2}, \ldots$ tending to a limit η. The corresponding subsequence of abscissae $x_{n_1}, x_{n_2}, \ldots$ still tends to the limit ξ; hence, the points $P_{n_1}, P_{n_2}, \ldots$ tend to the limit point (ξ, η). Thus, in either case, we can find a sequence of points of the set tending to a limit point, and the theorem is proved.

b. Cauchy's Convergence Test. Compactness

A consequence of the Bolzano-Weierstrass theorem is that *every bounded infinite sequence of points $P_1, P_2, \ldots$ has a convergent subsequence.* Indeed, if the sequence contains an infinite number of *distinct* elements, they form an infinite set of distinct points from which, according to the Weierstrass principle, we can choose a sequence converging to a point Q. If the sequence does not contain an infinite number of distinct elements, then at least one of its elements must be repeated infinitely often; there exists then a point Q that appears infinitely often in the sequence, and the subsequence formed by elements that equal Q converges to the point Q.

An important consequence is *Cauchy's convergence test:*

A sequence of points $P_1, P_2, \ldots$ in the plane (and similarly a sequence in n-dimensional euclidean space) converges to a limit if and only if for every $\varepsilon > 0$ there exists a number $N = N(\varepsilon)$ such that the distance between P_n and P_m is less than ε whenever both n and m are greater than N.

The proof proceeds exactly like the corresponding one for sequences of real numbers given in Volume I (p. 97). One sees immediately that a sequence satisfying the Cauchy condition is bounded; hence, by the preceding theorem, it contains a convergent subsequence with a limit Q, and it then follows immediately that the whole sequence converges to Q.

A set S of points in the plane was called *closed* if all boundary points of S belong to S. The limit Q of every convergent sequence of points of a closed set S is again a point of S (see p. 9). Since every bounded infinite sequence has been seen to contain a convergent subsequence of points, we find that *every infinite sequence formed from points of a bounded and closed set S of points in the plane contains a subsequence that converges to a point of S.* Generally we call a set S *compact*[1] if every sequence formed from elements of S contains a convergent subsequence with a limit in S. Hence, *a closed and bounded set of points in the plane (or in n-dimensional euclidean space) is compact.* The reader can easily verify the converse: Every compact set of points in the plane is closed and bounded. In the future we shall often refer to *closed and bounded* sets simply as *compact* sets.

c. The Heine-Borel Covering Theorem

A striking consequence of the Bolzano-Weierstrass principle is the *Heine-Borel theorem:*

Let there be given a compact (i.e., closed and bounded) set S and a system $\sum$ of infinitely many open sets that cover S in the sense that euery point of S belongs to at least one of the open sets in $\sum$. Then we can find a finite number of sets in $\sum$ that already cover S.

As an illustration consider the infinite set S of points on the x-axis consisting of the points $P_n = (1/n, 0)$ for $n = 1, 2, \ldots$ and of the origin $P_0 = (0, 0)$. This is a closed set. For $n = 1, 2, \ldots$, let S_n denote the open disk

$$\sqrt{(x-1/n)^2 + y^2} < \frac{1}{3n^2}$$

with center P_n and radius $1/3n^2$, and let S_0 denote the disk

$$\sqrt{x^2 + y^2} < \frac{1}{100}$$

Clearly the infinite system of all sets $S_0, S_1, S_2, \ldots$ covers S. In agreement with the Heine-Borel theorem we can pick a *finite* subsystem that covers S, for example the system consisting of $S_0, S_1, \ldots, S_{100}$. Here we immediately see the importance of the assumption that S be *closed.* The set T of points consisting of $P_1, P_2, \ldots$ alone, without P_0, is covered by the system consisting of $S_1, S_2, \ldots$, but no finite sub-

[1]Sometimes more precisely "sequentially compact."

system of these sets, each of which contains only a single point of T, can cover T.

To prove the Heine-Borel theorem, we use an indirect argument. Suppose that the theorem is false. The set S, being bounded, lies in a square Q. This square we subdivide into four equal squares. The part of S lying in at least one of these four squares or on its boundary cannot be covered by a finite number of the sets in $\sum$; for if each of the four parts of S could be covered in this way, S itself would be covered. This part of Q we call Q_1. We now subdivide Q_1 into four equal parts. By the same argument one of the four parts of Q_1 is a square Q_2 such that the points of S lying in Q_2 or on its boundary cannot be covered by a finite number of the open sets in $\sum$. Continuing in this way, we obtain an infinite sequence of squares Q_1, Q_2, Q_3, . . . each contained in the preceding one, their size shrinking to 0, and such that the points of S in the closure of any Q_n cannot be covered by a finite number of the sets in $\sum$. Clearly, for each n we can find a point P_n of S that lies in the interior or on the boundary of Q_n. Then P_1, P_2, . . . is a sequence of points of S. Since S is bounded, the sequence is bounded and must have a subsequence converging to some point A. Since S is closed, A is a point of S and hence contained in an open set Ω belonging to $\sum$. But then a whole neighborhood of A belongs to that open set Ω, say, the neighborhood consisting of the points having distance less than ε from A. We can choose an n so large that P_n has distance less than $\varepsilon/2$ from A and that the diagonal of Q_n has length less than $\varepsilon/2$. Then the whole square Q_n is contained in the ε-neighborhood of A and hence also in Ω. We see that the single set Ω of the system $\sum$ contains a whole square Q_n and its boundary, contrary to the assumption for the sequence Q_n. This completes the proof.

d. *An Application of the Heine-Borel Theorem to Closed Sets Contained in Open Sets*

Let R be an open set in the plane.[1] By definition every point P of R has a neighborhood that lies completely in R. For points P close to the boundary of R the neighborhood has to be very small. It is remarkable that for P confined to a closed subset S of R we can find a *uniform* size for the neighborhoods that are contained in R:

If a closed and bounded set S is contained in an open set R, there exists a positive ε such that the ε-neighborhood of every point P of S

[1]Everything said in this paragraph applies equally well to higher dimensions if we substitute the term "ball" for "disk."

is contained in R. In other words, the points not in R lie at least a distance ε away from all points of S.[1]

For the proof we make use of the assumption that R is open. For every point P in R there exists a disk with center P that is contained in R. The radius of this disk, call it r, depends on P; that is, $r = r(P)$. We take now for any P in S the open disk of radius $\frac{1}{2}r(P)$ and center P. By the Heine-Borel theorem a finite number of these disks can be found that cover the compact set S. Thus, we can find a finite number of points $P_1, \ldots, P_n$ in S such that every point P of S is contained in one of the disks of center P_k and radius $\frac{1}{2}r(P_k)$ for $k = 1, \ldots, n$. Let be the smallest of the positive numbers $\frac{1}{2}r(P_1), \ldots, \frac{1}{2}r(P_n)$. Then, f every P in S, the ε-neighborhood of P lies in R, for P lies in some d of center P_k and radius $\frac{1}{2}r(P_k)$. By construction the concentric D of radius $r(P_k)$ lies completely in R. Since $\overline{PP_k} < \frac{1}{2}r(P_k)$ and $\varepsilon \leq \frac{1}{2}r(P_k)$, the disk D contains the disk of radius ε about P. This shows that the disk of radius ε and center P lies in R.

As an example, we consider a curve S lying in the open set R. Such a curve is a set of points $P = (x, y)$ that can be represented in the form

$$x = \phi(t), \qquad y = \psi(t)$$

with the help of two continuous functions ϕ and ψ, where the parameter t varies over a closed interval $0 \leq t \leq 1$.[2] Such a curve S is a *closed* point set, for let $P_1, P_2, \ldots$ be a sequence of points on S converging to a point P. We consider the corresponding parameter values $t_1, t_2, \ldots$, which all lie in the closed interval $a \leq t \leq b$. Since a closed bounded interval is compact, a subsequence of the t_n converges to a value t in the interval. Since ϕ and ψ are continuous, the corresponding P_n converge to the point $Q = (x(t), y(t))$ on S. Thus, a subsequence of the sequence $P_1, P_2, \ldots$ converges to a point Q of S. Since the whole sequence converges to P, we have $P = Q$. and hence, P lies in S. Thus, S contains all limits of sequences of points of S and hence is closed.

If the curve lies in the open set R, we can find a positive number ε such that all disks of radius ε with centers on S lie in R. Since f and g are continuous, and hence uniformly continuous, we can find a positive number δ such that two points on S have distance less than ε if their parameter values t differ by less than δ. We can divide the

[1]It is essential that S is bounded. If, for example, R is the open half-plane $y > 0$ and S the closed set consisting of the points in the x,y-plane with $y \geq 1/x$, $x > 0$, the boundary of R comes arbitrarily close to points of S.

[2]The curve need not be *simple*; that is, different t may correspond to the same point P. The pair of functions defines a "path," and S is the *support* of that path.

parameter interval by points $t_1, \ldots, t_{n-1}$ such that

$$a = t_0 < t_1 < t_2 < \cdots < t_{n-1} < t_n = b$$

where the length of every subinterval is less than δ. Let $P_0, P_1, \ldots, P_n$ be the corresponding points on S. Then P_{i+1} always lies in the disk of radius ε about P_i. Also, the straight line segment joining P_i and P_{i+1} lies completely in the disk of radius ε and center P_i, and hence is contained in R. If we join successive points P_i by straight line segments, we obtain a *polygonal curve* that lies completely in R and has the same end points P_0, P_n as the continuous curve S. We can formulate this result as follows:

If two points of an open set R can be joined by a curve that lies in R, then they can also be joined by a polygonal curve in R.

A.2. Basic Properties of Continuous Functions

For functions f defined and continuous in a closed and bounded set S we can state the following two fundamental theorems:

The function f assumes a greatest value ("maximum") and a least value ("minimum") in S.

The function f is uniformly continuous in S.

The proofs of these theorems are like the corresponding proofs for functions of one variable (see Volume I, pp. 100–101) and need not be repeated.

The second theorem can also be obtained as an immediate consequence of the Heine-Borel theorem. Prescribe an $\varepsilon > 0$. If f is continuous at every point of S, there exists for every point P in S a δ-neighborhood of P of a certain radius $\delta = \delta(P)$ such that $|f(Q) - f(P)| < \varepsilon/2$ for any Q in S that lies in that neighborhood. Now for each P in S choose a neighborhood Ω_P of radius $\frac{1}{2}\delta(P)$. The Ω_P clearly cover S. We can select a finite number of them, say those with centers $P_1, \ldots, P_n$ that also cover S. Let Δ be the smallest of the numbers $\frac{1}{2}\delta(P_1), \ldots, \frac{1}{2}\delta(P_n)$. If then P and Q are any two points of S whose distance is less than Δ, the point P has distance less than $\frac{1}{2}\delta(P_k)$ from one of the points P_k with $k = 1, \ldots, n$. Since $\Delta \leq \frac{1}{2}\delta(P_k)$, we see that both P and Q lie in the $\delta(P_k)$-neighborhood of P_k. Hence,

$$|f(P) - f(P_k)| < \frac{1}{2}\varepsilon, \qquad |f(Q) - f(P)| < \frac{1}{2}\varepsilon,$$

and thus

$$|f(P) - f(Q)| < \varepsilon.$$

This establishes the uniform continuity of f since Δ is independent of the particular location of P and Q.

A.3. Basic Notions of the Theory of Point Sets

a. Sets and Subsets

In more complicated arguments involving sets of points (particularly in the theory of integration) it is convenient to use some standard notations for operations with sets. The sets of interest to us are always sets of numbers, of points, of functions, or of sets of these types. For example a "disk" in the plane is defined as a set of points (x, y) for which

$$(x - x_0)^2 + (y - y_0)^2 < r^2$$

for fixed x_0, y_0, r. An example of a set of sets (or *family* of sets) would be that consisting of all disks that contain the origin; that would be those disks for which $x_0^2 + y_0^2 < r^2$.

We shall refrain from trying to reduce the basic notion of *set* to still more fundamental ones or to analyze the logical difficulties involved in this notion. For us a set S is defined if for every object α exactly one of the two following statements is correct: (1) α belongs to S; (2) α *does not belong* to S. In case (1) one also says that α *is an element* of S or that α *is contained in* S; symbolically[1] one denotes this by

$$\alpha \in S,$$

and case (2) by

$$\alpha \notin S.$$

For example, if S is the disk given by the inequality $x^2 + y^2 < r^2$, then $\alpha \in S$ means that α is a point in the plane with coordinates x, y that has the property that $x^2 + y^2 < r^2$. Generally the elements of a set S can be characterized by some common *properties* (e.g., by the property of belonging to S). We write the set S of elements α that have the properties $A, B, \ldots$ symbolically as

$$S = \{\alpha\colon \alpha \text{ has the properties } A, B, \ldots\}.$$

[1]The symbol $\in$ must not be confused with the Greek letter ε.

For example, the disk S with center (x_0, y_0) and radius r can be described as

$$S = \{(x, y): x, y = \text{real numbers}; (x - x_0)^2 + (y - y_0)^2 < r^2\}.$$

The set described by

$$S = \{n : n = \text{integer}; \ 2 < n < 5\}$$

consists of the two elements $n = 3$ and $n = 4$.

For many purposes it is convenient to introduce the "empty" (or "null") set with the special symbol $\emptyset$. This set has no elements: $\alpha \notin \emptyset$ for all α. For example an open disk of radius 0 and center at the origin coincides with $\emptyset$:

$$\{(x, y) : x, y = \text{real numbers}; x^2 + y^2 < 0\} = \emptyset.$$

Two sets S and T are equal when they have the same elements, regardless of the different descriptions or properties used in their definition: $S = T$ means that $x \in S$ if and only if $x \in T$.

A set S is said to be a subset of a set T ("S is contained in T") if T contains all the elements that are contained in S, that is, if $\alpha \in S$ implies $\alpha \in T$. We write this symbolically:

$$S \subset T$$

or, more rarely,

$$T \supset S.$$

Thus, if S is the disk of radius 1 about the origin and T the disk of radius 4 about the point (1, 1), then $S \subset T$. Similarly, $\emptyset \subset S$ and $S \subset S$ for all sets S.

The symbols $\subset$ and $\supset$ are chosen, of course, for their similarity to the $<$ and $>$ signs of arithmetic (or more precisely to the $\leq$ and $\geq$ signs). They share with the latter symbols the basic properties:

$$S \subset T \text{ and } T \subset S \quad \text{implies} \quad S = T$$
$$S \subset T \text{ and } T \subset R \quad \text{implies} \quad S \subset R.[1]$$

[1]This is the common syllogism from logic: If all objects with the property A have the property B and all objects with the property B have the property C, then all objects with the property A have the property C.

A basic difference between the "contained in" signs for sets and the order signs for numbers is that for real numbers we always have either $x \leq y$ or $y \leq x$, whereas for sets neither of the propositions $S \subset T$ or $T \subset S$ has to hold. The symbol $\subset$ defines only a "partial" ordering between sets; of two sets neither may contain the other one.

b. Union and Intersection of Sets

During the last decades a great number of logical symbols have found wide acceptance in mathematics, so that it is now customary to express many mathematical theorems completely in symbolic notations without the use of ordinary words or sentence structure.[1] Use of proper symbolic notation has been essential for the development of mathematics from the very beginning; in fact, in rare instances, progress in some field may have slowed down for centuries just for lack of a suitable notation, as was perhaps the case with algebra in antiquity. On the other hand, too concentrated a notation may prove a great strain to the reader who tries to relate the information in the "dehydrated" form to his ordinary experience. Authors of books not primarily devoted to logic and foundations of mathematics compromise on the use of logical abbreviations in accordance with their tastes and the requirements of the special subjects under consideration.

There are two further set-theoretical symbols that we shall find almost indispensable later in this book, namely, the symbols for the operations of "union" and "intersection" of sets. Given two sets S and T we write $S \cup T$ for the "union" of the two sets, that is, for the set of elements that are "either" in S "or" in T:

$$S \cup T = \{a : a \in S \text{ or } a \in T\}.$$ [2]

Similarly, the "intersection" $S \cap T$ of S and T is defined as the set of elements that belong to both S and T:

$$S \cap T = \{a : a \in S \text{ and } a \in T\}.$$

---

[1]Examples of frequently used symbols follow:
$\{x_1, x_2, \ldots, x_n\}$: the set whose members are precisely $x_1, \ldots, x_n$
$S \times T$: the set of ordered pairs (a, b) with $a \in S$ and $b \in T$ ("Cartesian product" of the sets S, T)
$\rightarrow$: "implies"
$\exists\, x$: "there exists an x"
$\forall x$: "for all x."

[2]Here the word "or" like the Latin *vel* is not exclusive. $S \cup T$ consists of the elements that belong to *at least one* of the two sets S, T but may belong to both.

For example, if S and T are intervals on the real number axis and if

$$S = \{x : 3 < x < 5\},$$
$$T = \{x : 4 \leq x < 6\},$$

then

$$S \cup T = \{x : 3 < x < 6\}$$
$$S \cap T = \{x : 4 \leq x < 5\}$$

The operations $\cup$ and $\cap$ apply to any two sets S and T, provided we use the symbol for the empty set, writing

$$S \cap T = \emptyset$$

when S and T are *disjoint*, that is, have no common element. Notice that $S \cup \emptyset = S$, $S \cap \emptyset = \emptyset$ for any S.

The operation $\cup$ has many properties in common with addition. In particular, if S and T are "disjoint" sets—that is, sets without common elements—and have finitely many elements, then the number of elements in $S \cup T$ is just the sum of the numbers of elements in S and in T. There is, however, generally no unique inverse operation to union. Only if S and T are assumed to be disjoint and $S \subset R$, does the equation

$$S \cup T = R$$

have a unique solution T. For disjoint sets S, T the union is often denoted by $S + T$, and for $S \subset R$, the solution T of the equation $S + T = R$ by $R - S$ ("the complement of S relative to R"). We shall use the symbol $R - S$ more generally for any sets R, S to denote the set of elements of R that do not belong to S. Then $S + (R - S) = R \cup S$.

The union of n sets $S_1, \ldots, S_n$ is defined as the set of elements belonging to at least one of the sets $S_1, \ldots, S_n$ and is variously denoted by

$$\{a : a \in S_1 \text{ or } a \in S_2 \text{ or} \ldots \text{ or } a \in S_n\}$$
$$= S_1 \cup S_2 \cup \cdots \cup S_n$$
$$= \bigcup_{k=1}^{n} S_k$$

in analogy to the summation and product symbols. Similarly, the intersection of the sets $S_1, \ldots, S_n$, defined as the set of elements common to all of them, is

$$\{a : a \in S_1 \text{ and } a \in S_2 \text{ and } \ldots \text{ and } a \in S_n\}$$
$$= S_1 \cap S_2 \cap \cdots \cap S_n = \bigcap_{k=1}^{n} S_k.$$

We can with equal ease form unions and intersections of an infinite number of sets $S_1, S_2, \ldots, S_n, \ldots$, which we write respectively as

$$\bigcup_{k=1}^{\infty} S_k = \{a : a \in S_n \text{ for some } n\}$$
$$\bigcap_{k=1}^{\infty} S_k = \{a : a \in S_n \text{ for all } n\}.$$

For example, if S_n is the set of real numbers $x < n$

$$S_n = \{x : x \text{ real}, \ x < n\},$$

we have

$$\bigcup_{k=1}^{\infty} S_k = \{x : x \text{ real}\}$$
$$\bigcap_{k=1}^{\infty} S_k = \{x : x \text{ real}, \ x < 1\}.$$

In fact, union and intersection can be formed for arbitrary large families F of sets S even where the different sets S in F are not, or cannot be, distinguished by a subscript n with $n = 1, 2, 3, \ldots$. We write

$$\bigcup_{S \in F} S = \{a : a \in S \text{ for some } S \text{ with } S \in F\}$$
$$\bigcap_{S \in F} S = \{a : a \in S \text{ for all } S \text{ with } S \in F\}.$$

Thus the union of all disks in the x, y-plane containing the point $(1, 0)$ but not the point $(-1, 0)$ is the set of all (x, y) for which either $y \neq 0$ or $y = 0$ and $x > -1$. The intersection of the same family of disks contains the single point $(1, 0)$.

c. *Applications to Sets of Points in the Plane*

Some of our earlier results and definitions (see pp. 6–8) can be rewritten more compactly in the notation introduced in the last sections. Thus, given a set S of points in the plane, we obtain a decomposition of the whole plane π into three disjoint sets, namely, the set S^0

of interior points of S, the set ∂S of boundary points of S, and the set S_e of exterior points of S:

$$\pi = S^0 \cup \partial S \cup S_e$$

or more precisely,

$$\pi = S^0 + \partial S + S_e$$

Since the sets are disjoint:

$$S^0 \cap \partial S = \partial S \cap S_e = S_e \cap S^0 = \emptyset.$$

Here

$$S^0 \subset S \subset S^0 + \partial S.$$

The set $\bar{S}$ defined by

$$\bar{S} = S^0 + \partial S = S \cup \partial S \tag{1}$$

is the *closure* of S. We have $S^0 = S$ for open S and $\bar{S} = S$ for closed S.

The reader may verify as exercises the following propositions:

$\overline{\partial S} = \partial S$ ("The boundary of a set is always closed.")
$\bar{\bar{S}} = \bar{S}$ ("The closure of a set is always closed.")
$(S^0)^0 = S^0$, $(S_e)^0 = S_e$ ("The sets S^0 and S_e are open.")

2(a) $\quad S^0 \cup T^0 \subset (S \cup T)^0, \quad \overline{S \cup T} \subset \bar{S} \cup \bar{T}$

2(b) $\quad \partial(S \cup T) \subset \partial S \cup \partial T$

The union of open sets is open.
The union of a finite number of closed sets is closed.
The intersection of a finite number of open sets is open.
The intersection of closed sets is closed.

The last statements indicate a kind of symmetry ("duality") between the notions "open" and "closed," "union" and "intersection." This becomes more precise if we introduce the *complement* $C(S)$ of a set S, that is, the set of points in the plane π not belonging to S:[1]

$$C(S) = \{P : P \in \pi,\ P \notin S\} = \pi - S.$$

[1]For sets S of points on three-space Σ the complement of S is defined as $\Sigma - S$, the set of points of Σ not belonging to S.

We have

$$C(S^0) = \bar{S}_e, \qquad \partial C(S) = \partial S, \qquad C(S_e) = \bar{S}^0.$$

If S is open, $C(S)$ is closed, and vice versa. The complement of the intersection of several sets is the union of their complements.

In this notation the theorem of Heine-Borel takes a particularly simple form. "A family F of sets covers a set S" means simply that S is contained in the union of the sets of F. The theorem then simply states:

If F is a family of open sets in the plane and if S is a bounded and closed set such that

$$S \subset \bigcup_{T \in F} T,$$

then we can find a finite number of sets $T_1, T_2, \ldots, T_n \in F$ such that

$$S \subset \bigcup_{k=1}^{n} T_k.$$

A.4. Homogeneous Functions

The simplest homogeneous functions occurring in analysis and its applications are the *forms* or homogeneous polynomials in several variables (see p. 13). We say that a function of the form $ax + by$ is a homogeneous function of the first degree in x and y, that a function of the form $ax^2 + bxy + cy^2$ is a homogeneous function of the second degree, and in general that *a polynomial in x and y (or in a greater number of variables) is a homogeneous function of degree h if in each term the sum of the exponents of the independent variables is equal to h*, that is, if the terms (apart from constant coefficients) are of the form $x^h, x^{h-1}y, x^{h-2}y^2, \ldots, y^h$. These homogeneous polynomials have the property that the equation

$$f(tx, ty) = t^h f(x, y)$$

holds for every value of t. More generally, we say that *a function $f(x, y, \ldots)$ is homogeneous of degree h if it satisfies the equation*

$$f(tx, ty, \ldots) = t^h f(x, y, \ldots).$$

Examples of homogeneous functions that are not polynomials are

$$\tan\left(\frac{y}{x}\right) \qquad (h = 0),$$

$$x^2 \sin\frac{x}{y} + y\sqrt{x^2 + y^2}\log\frac{x + y}{x} \qquad (h = 2).$$

Another example is the cosine of the angle between two vectors with the respective components x, y, z and u, v, w:

$$\frac{xu + yv + zw}{\sqrt{x^2 + y^2 + z^2}\sqrt{u^2 + v^2 + w^2}} \qquad (h = 0).$$

The length of the vector with components x, y, z,

$$\sqrt{x^2 + y^2 + z^2}$$

is an example of a function that is *positively homogeneous* and of the first degree; that is, the equation defining homogeneous functions does not hold for this function unless t is positive or 0.

Homogeneous functions that are also differentiable satisfy Euler's partial differential equation

$$xf_x + yf_y + zf_z + \cdots = hf(x, y, z, \ldots).$$

To prove this we differentiate both sides of the equation $f(tx, ty, \ldots) = t^h f(x,y, \ldots)$ with respect to t; this is permissible, since the equation is an identity in t. Applying the chain rule to the function on the left, we obtain

$$xf_x(tx, ty, \ldots) + yf_y(tx, ty, \ldots) + \cdots = ht^{h-1}f(x, y, \ldots).$$

If we substitute $t = 1$ in this, the statement follows.

Conversely, it is easy to show that the homogeneity of the function $f(x, y, \ldots)$ is a consequence of Euler's relation, so that *Euler's relation is a necessary and sufficient condition for the homogeneity of the function.* The fact that a function is homogeneous of degree h can also be expressed by saying that the value of the function divided by x^h depends only on the ratios $y/x, z/x, \ldots$. It is therefore sufficient to show that it follows from the Euler relation that if new variables

$$\xi = x, \qquad \eta = \frac{y}{x}, \qquad \zeta = \frac{z}{x}, \ldots$$

are introduced, the function

$$\frac{1}{x^h} f(x, y, z, \ldots) = \frac{1}{\xi^h} f(\xi, \eta\xi, \zeta\xi, \ldots) = g(\xi, \eta, \zeta, \ldots)$$

no longer depends on the variable ξ (i.e., that the equation $g_\xi = 0$ is an identity). In order to prove this, we use the chain rule:

$$\begin{aligned} g_\xi &= (f_x + \eta f_y + \cdots) \frac{1}{\xi^h} - \frac{h}{\xi^{h+1}} f \\ &= (x f_x + y f_y + \cdots) \frac{1}{x^{h+1}} - \frac{h}{x^{h+1}} f. \end{aligned}$$

The expression on the right vanishes in virtue of Euler's relation, and our statement is proved.

This last statement can also be proved in a more elegant, but less direct, way. We wish to show that from Euler's relation it follows that the function

$$g(t) = t^h f(x, y, \ldots) - f(tx, ty, \ldots)$$

has the value 0 for all values of t. It is obvious that $g(1) = 0$. Again,

$$g'(t) = h t^{h-1} f(x, y, \ldots) - x f_x(tx, ty, \ldots) - y f_y(tx, ty, \ldots) - \ldots$$

On applying Euler's relation to the arguments $tx, ty, \ldots$ we find that

$$x f_x(tx, ty, \ldots) + y f_y(tx, ty, \ldots) + \cdots = \frac{h}{t} f(tx, ty, \ldots),$$

and thus $g(t)$ satisfies the differential equation

$$g'(t) = g(t) \frac{h}{t}.$$

If we write $g(t) = \gamma(t) t^h$, we obtain $g'(t) = \frac{h}{t} g(t) + t^h \gamma'(t)$, so that $\gamma(t)$ satisfies the differential equation

$$t^h \gamma'(t) = 0,$$

which has the unique solution $\gamma = \text{constant} = c$. Since for $t = 1$ it is obvious that $\gamma(t) = 0$, the constant c is 0, and so $g(t) = 0$ for all values of t, as was to be proved.

CHAPTER 2

Vectors, Matrices, Linear Transformations

Vectors in two dimensions have already been studied in Volume I, Chapter 4. Geometric concepts in higher dimensions make the use of vectors even more essential. Vectors serve to express many complicated equations concisely in a manner clearly exhibiting those features that do not depend on a particular choice of coordinate systems.

2.1 Operations with Vectors

a. Definition of Vectors

We introduce vectors in n-dimensional space as entities that can be added to each other and multiplied by scalars. Specifically, a vector $\mathbf{A}$ is a set of n real numbers[1] $a_1, \ldots, a_n$ in a definite order

$$\mathbf{A} = (a_1, \ldots, a_n)$$

(We always employ boldface type to denote vectors.) The numbers $a_1, \ldots, a_n$ are called the *components* of $\mathbf{A}$. Two vectors $\mathbf{A} = (a_1, \ldots, a_n)$ and $\mathbf{B} = (b_1, \ldots, b_n)$ are equal if and only if they have the same components.

The sum of any two vectors $\mathbf{A} = (a_1, \ldots, a_n)$ and $\mathbf{B} = (b_1, \ldots, b_n)$ is defined by

$$\text{(1a)} \qquad \mathbf{A} + \mathbf{B} = (a_1 + b_1,\ a_2 + b_2, \ldots, a_n + b_n);$$

[1]For our purposes it is sufficient to consider only *real* numbers as components, although vectors over other number fields also are used in other contexts.

we define the *product* of the vector $\mathbf{A} = (a_1, \ldots, a_n)$ by the scalar (i.e., real number) λ as

(1b) $$\lambda\mathbf{A} = (\lambda a_1, \lambda a_2, \ldots, \lambda a_n).$$[1]

More generally, we can form from any finite number of vectors $\mathbf{A} = (a_1, a_2, \ldots, a_n)$, $\mathbf{B} = (b_1, b_2, \ldots, b_n)$, $\ldots$, $\mathbf{D} = (d_1, d_2, \ldots, d_n)$ and an equal number of scalars $\lambda, \mu, \ldots, \gamma$ the *linear combination* $\lambda\mathbf{A} + \mu\mathbf{B} + \cdots + \gamma\mathbf{D} = (\lambda a_1 + \mu b_1 + \cdots + \gamma d_1, \ldots, \lambda a_n + \mu b_n + \cdots + \gamma d_n)$. In particular, any vector $\mathbf{A} = (a_1, \ldots, a_n)$ can be represented as a linear combination of the n "coordinate vectors"

(2a) $$\mathbf{E}_1 = (1, 0, 0, \ldots, 0), \qquad \mathbf{E}_2 = (0, 1, 0, \ldots, 0), \ldots,$$

$$\mathbf{E}_n = (0, 0, 0, \ldots, 1).$$

Obviously,

(2b) $$\mathbf{A} = a_1\mathbf{E}_1 + a_2\mathbf{E}_2 + \cdots + a_n\mathbf{E}_n.$$

We use the symbol $\mathbf{0}$ for the "zero vector," all of whose components vanish: $\mathbf{0} = (0, 0, \ldots, 0)$. We write $-\mathbf{A}$ for the vector $(-1)\mathbf{A} = (-a_1, -a_2, \ldots, -a_n)$.

It follows trivially from these definitions that sums of vectors and products with scalars obey all the usual algebraic laws, as far as they are meaningful.[2] Examples of objects conveniently represented by vectors are furnished by functions that are linear combinations of a finite number of suitably chosen functions. Thus, the general *polynomial* of degree $\leq n$ in the variable x

---

[1]Vectors differ from other objects that can be described by an ordered set of n real numbers (e.g., points in n-dimensional euclidean space or on a sphere in $n + 1$ dimensions) just by the fact that they permit the "linear operations" $\mathbf{A} + \mathbf{B}$ and $\lambda\mathbf{A}$. *Addition of points* defined similarly in terms of their coordinates would have no geometric meaning, at least no meaning independent of the special coordinate system used. Vectors will be represented later by pairs of points (see p. 109).

[2]These laws are the following:

(1) $\mathbf{A} + \mathbf{B} = \mathbf{B} + \mathbf{A}$, $\mathbf{A} + (\mathbf{B} + \mathbf{C}) = (\mathbf{A} + \mathbf{B}) + \mathbf{C}$

(2) $\lambda(\mathbf{A} + \mathbf{B}) = \lambda\mathbf{A} + \lambda\mathbf{B}$, $(\lambda + \mu)\mathbf{A} = \lambda\mathbf{A} + \mu\mathbf{A}$, $(\lambda\mu)\mathbf{A} = \lambda(\mu\mathbf{A})$

(3) There exists a unique element $\mathbf{O}$ such that $\mathbf{A} + \mathbf{O} = \mathbf{A}$ for all $\mathbf{A}$

(4) There exists a unique element $-\mathbf{A}$ for given $\mathbf{A}$ such that $\mathbf{A} + (-\mathbf{A}) = 0$

(5) $0\mathbf{A} = \mathbf{O}$, $1\mathbf{A} = \mathbf{A}$ for all $\mathbf{A}$.

Generally, sets of objects for which addition of the objects and multiplication by scalars are defined, and obey these laws, are called *vector spaces*.

$$P(x) = a_0 + a_1 x + a_2 x^2 + \cdots + a_n x^n,$$

can be represented by the single vector $\mathbf{A} = (a_0, a_1, \ldots, a_n)$ in $(n + 1)$-dimensional space. Addition of vectors and multiplication by scalars correspond then to the same operations carried out for the polynomials. Similarly, the general nth degree *trigonometric polynomial*

$$f(x) = \frac{1}{2} a_0 + \sum_{k=1}^{n} (a_k \cos kx + b_k \sin kx)$$

(see Volume I, p. 577) can be represented by the vector $(a_0, a_1, \ldots, a_n, b_1, b_2, \ldots, b_n)$ in $(2n + 1)$-dimensional space. The general *linear homogeneneous function* of three variables

$$u = a_1 x_1 + a_2 x_2 + a_3 x_3$$

is represented by the vector (a_1, a_2, a_3) in three-dimensional space, and the *general quadratic form* in three variables

$$u = a_1 x_1^2 + a_2 x_2^2 + a_3 x_3^2 + 2a_4 x_2 x_3 + 2a_5 x_3 x_1 + 2a_6 x_1 x_2$$

by the vector $(a_1, a_2, a_3, a_4, a_5, a_6)$ in six-dimensional space.

b. Geometric Representation of Vectors

Vectors in n-dimensional space, just as in the plane, can be visualized geometrically as certain mappings of space, the *translations* or *parallel displacements*. The vector $\mathbf{A} = (a_1, a_2, \ldots, a_n)$ may be depicted as the translation of n-dimensional euclidean space R^n that maps any point $P = (x_1, x_2, \ldots, x_n)$ into the point $P' = (x_1', x_2', \ldots, x_n')$ with coordinates

$$x_1' = x_1 + a_1,\ x_2' = x_2 + a_2,\ \ldots,\ x_n' = x_n + a_n.^1 \tag{3a}$$

The translation or the corresponding vector $\mathbf{A}$ is determined uniquely if for a single point $P = (x_1, x_2, \ldots, x_n)$ we give the image $P' = (x_1', x_2', \ldots, x_n')$; obviously by (3a)

$$\mathbf{A} = (x_1' - x_1, x_2' - x_2, \ldots, x_n' - x_n). \tag{3b}$$

[1]It is understood that both points P and P' lie in R^n and that their coordinates are taken with respect to the same coordinate system.

We shall denote this translation by $\mathbf{A} = \overrightarrow{PP'}$ and say that the vector **A** is *represented* by the ordered pair of points P and P' We call P the *initial point* and P' the *end point* or *final point* in this representation. In drawings the vector $\mathbf{A} = \overrightarrow{PP'}$ usually is indicated by an arrow extending from P to P'. The same vector **A** has many representations $\mathbf{A} = \overrightarrow{PP'}$ by a pair of points P and P'. The initial point P is completely arbitrary, since the mapping defined by **A** can act on any point and then determine an image P'.[1] The zero vector **0** corresponds to the "identity mapping" in which each point is mapped onto itself: $\mathbf{0} = \overrightarrow{PP}$.

As in the planar case (Volume I, p. 384) the sum of two vectors $\mathbf{A} = (a_1, \ldots, a_n)$, $\mathbf{B} = (b_1, \ldots, b_n)$ yields the *symbolic product* of the corresponding mappings. If **A** takes the point $P = (x_1, \ldots, x_n)$ into the point $P' = (x_1', \ldots, x_n')$ and **B** takes the point P' into $P'' = (x_1'', \ldots, x_n'')$, then $\mathbf{C} = \mathbf{A} + \mathbf{B}$ corresponds to the translation that takes P into P'', since

$$x_i'' = x_i' + b_i = (x_i + a_i) + b_i = x_i + (a_i + b_i)$$

for $i = 1, \ldots, n$. In vector notation we have

$$\mathbf{A} + \mathbf{B} = \overrightarrow{PP'} + \overrightarrow{P'P''} = \overrightarrow{PP''}. \tag{4}$$

If we represent **B** in the form $\overrightarrow{PP'''}$ giving it the same initial point P as **A**, we find that $\mathbf{A} + \mathbf{B} = \overrightarrow{PP''}$ is represented by the *diagonal of the parallelogram* with vertices P, P', P'', P''' (see Fig. 2.1).

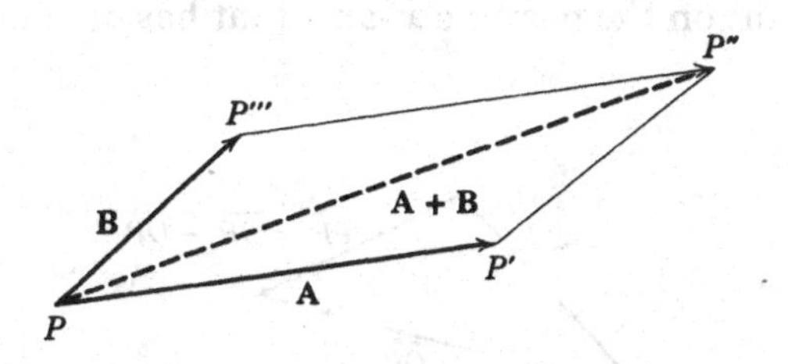

Figure 2.1 Addition of vectors.

[1]Occasionally the notation $P' - P$ is used for the vector $\overrightarrow{PP'}$, which, in accordance with formula (3b), suggests the notion of *vectors as differences of points*.

Interchanging initial and end point of the vector $\mathbf{A} = \overrightarrow{PP'} = (x_1' - x_1, x_2' - x_2, \ldots, x_n' - x_n)$ leads to the *opposite vector*

$$\overrightarrow{P'P} = (x_1 - x_1', x_2 - x_2', \ldots, x_n - x_n') = (-1)\,\mathbf{A} = -\mathbf{A}.$$

The mapping $P' \to P$ corresponding to $-\mathbf{A}$ is the inverse to the mapping $\mathbf{A}$; carrying out first $\mathbf{A}$ and then $-\mathbf{A}$ results in the identity mapping in accordance with the formula

$$(-\mathbf{A}) + \mathbf{A} = (-1 + 1)\,\mathbf{A} = 0\mathbf{A} = \mathbf{0}.$$

Corresponding to (4) we have the often used formula for the difference of two vectors $\mathbf{A} = \overrightarrow{PP'}$ and $\mathbf{B} = \overrightarrow{PP''}$ with common initial point:

$$\text{(4a)} \qquad \mathbf{B} - \mathbf{A} = \overrightarrow{PP''} - \overrightarrow{PP'} = \overrightarrow{PP''} + \overrightarrow{P'P} = \overrightarrow{P'P} + \overrightarrow{PP''} = \overrightarrow{P'P''}.$$

The difference of the vectors $\overrightarrow{PP''}$ and $\overrightarrow{PP'}$ is here represented by the third side of the triangle with vertices P, P', P''.

We can associate with every point $P = (x_1, \ldots, x_n)$ a unique vector that has the origin as initial point and P as end point; this is the vector

$$\overrightarrow{OP} = (x_1, \ldots, x_n),$$

the so-called *position vector* of P. The components of the position vector of P are just the coordinates of P. For example, the coordinate vector $\mathbf{E}_i = (0, \ldots, 0, 1, 0, \ldots, 0)$ in formula (2a) is the position vector of the point on the positive x_i-axis that has distance 1 from the

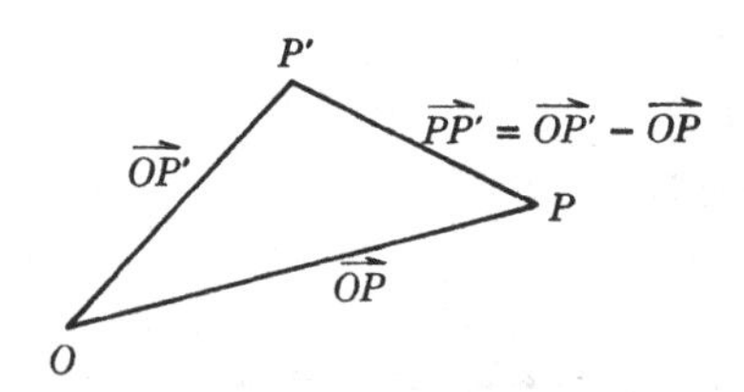

Figure 2.2 The vector $\overrightarrow{PP'}$ as difference of position vectors.

origin. Any vector $\mathbf{A} = \overrightarrow{PP'}$ can always be written as the difference of the position vectors of its end point and initial point:

$$\overrightarrow{PP'} = \overrightarrow{OP'} - \overrightarrow{OP} \tag{5}$$

(see Fig. 2.2).

c. *Length of Vectors, Angles Between Directions*

The distance between two points $P = (x_1, \ldots, x_n)$ and $P' = (x_1', \ldots, x_n')$ in n-dimensional euclidean space R^n is given by the formula[1]

$$r = \sqrt{(x_1' - x_1)^2 + (x_2' - x_2)^2 + \cdots + (x_n' - x_n)^2}. \tag{6}$$

Since only the differences of corresponding coordinates of P, P' enter into the expression for r, we see that the distance is the same for all pairs of points P, P' that represent the same vector $\mathbf{A} = \overrightarrow{PP'}$. We call r the *length of the vector* $\mathbf{A}$ and write $r = |\mathbf{A}|$. The vector $\mathbf{A} = (a_1, \ldots, a_n)$ has the length

$$|\mathbf{A}| = \sqrt{a_1^2 + a_2^2 + \cdots + a_n^2} \tag{6a}$$

The zero vector $\mathbf{0} = (0, 0, \ldots, 0)$ has length 0. The length of any other vector is a positive number.

In euclidean geometry, angles can be expressed in terms of lengths. This is achieved by the trigonometric formula ("law of cosines") that gives in a triangle with sides a, b, c the angle γ between the sides a and b:

$$\cos\gamma = \frac{a^2 + b^2 - c^2}{2ab}. \tag{6b}$$

We apply this formula to a triangle with vertices P, P', P''. (Fig. 2.3a). The sides a and b of the triangle are the lengths of the vectors $\mathbf{A} = \overrightarrow{PP'}$, $\mathbf{B} = \overrightarrow{PP''}$, while side c is the length of the vector

---

[1]In two or three dimensions the formula can be derived geometrically by applying the theorem of Pythagoras. In higher dimensions the expression for r can be considered as the *definition of distance* between two points in n-dimensional euclidean space, when referred to a Cartesian coordinate system.

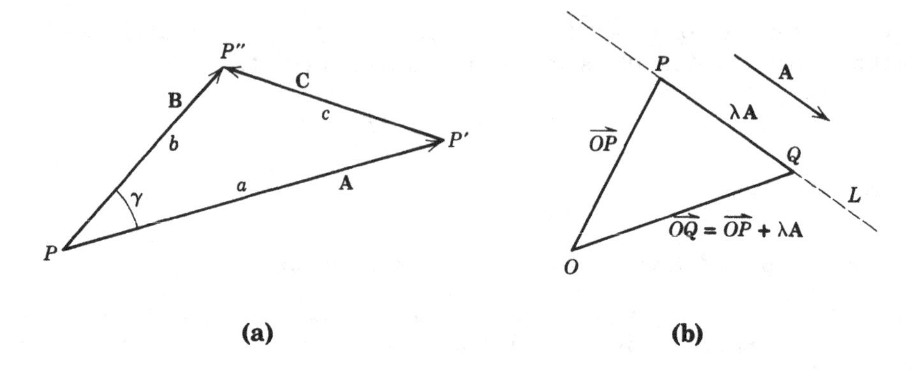

Figure 2.3 Vector representation of a line through a given point with a given direction.

$$\mathbf{C} = \overrightarrow{P'P''} = \overrightarrow{PP''} - \overrightarrow{PP'} = \mathbf{B} - \mathbf{A}.$$

For

$$\mathbf{A} = (a_1, \ldots, a_n), \qquad \mathbf{B} = (b_1, \ldots, b_n)$$

we have

$$\mathbf{C} = (c_1, \ldots, c_n) = (b_1 - a_1, \ldots, b_n - a_n).$$

By (6b)

$$\cos\gamma = \frac{|\mathbf{A}|^2 + |\mathbf{B}|^2 - |\mathbf{C}|^2}{2|\mathbf{A}|\ |\mathbf{B}|},$$

where

$$|\mathbf{A}|^2 = \sum_{i=1}^{n} a_i^2, \qquad |B|^2 = \sum_{i=1}^{n} b_i^2, \qquad |\mathbf{C}|^2 = \sum_{i=1}^{n} (b_i - a_i)^2.$$

Thus, for $\mathbf{A} \neq \mathbf{0}$, $\mathbf{B} \neq \mathbf{0}$,

$$\text{(7)} \qquad \cos\gamma = \frac{a_1b_1 + a_2b_2 + \cdots + a_nb_n}{\sqrt{a_1^2 + \cdots + a_n^2}\ \sqrt{b_1^2 + \cdots + b_n^2}}.$$

We see that the angle γ in the triangle $PP'P''$ depends only on the vectors $\mathbf{A} = \overrightarrow{PP'}$ and $\mathbf{B} = \overrightarrow{PP''}$. Accordingly, we call the quantity $\cos\gamma$

given by formula (7) *the cosine of the angle*[1] *between the vectors* $\mathbf{A} = (a_1, \ldots, a_n)$ and $\mathbf{B} = (b_1, \ldots, b_n)$.

Formula (7) for $\cos \gamma$ actually always defines *real* angles γ between any two nonzero vectors **A**, **B**, since it always yields a value with $|\cos \gamma| \leqq 1$. This is an immediate consequence of the *Cauchy-Schwarz inequality* (Volume I, p. 15)

$$(8) \qquad (a_1b_1 + a_2b_2 + \cdots + a_nb_n)^2 \leqq (a_1^2 + a_2^2 + \cdots + a_n^2)(b_1^2 + b_2^2 + \cdots + b_n^2).$$

In computing the angles between the vector **A** and any other vector **B** from (7), we need to know only the quantities

$$(9) \qquad \xi_i = \frac{a_i}{\sqrt{a_1^2 + \cdots + a_n^2}} \qquad (i = 1, \ldots, n)$$

which are called the *direction cosines* of **A**. All nonzero vectors with the same direction cosines form the same angles with other vectors and thus can be said to have the *same direction.* It follows from (7) that the direction cosines of A can be interpreted as cosines of certain angles:

$$(10) \qquad \xi_i = \cos \alpha_i,$$

where α_i is the angle between **A** and the ith "coordinate vector" $\mathbf{E}_i = (0, \ldots, 0, 1, 0, \ldots, 0)$. The n direction cosines of the vector **A** satisfy the identity[2]

$$(11) \qquad \cos^2 \alpha_1 + \cos^2 \alpha_2 + \cdots \cos^2 \alpha_n = 1.$$

The only vector without direction cosines (and thus without a direction) *is the zero vector.*

Two vectors **A** and **B** not equal to **0** have the same direction if and only if they have the same direction cosines, that is, if

---

[1]The angle γ itself is determined uniquely only if we confine γ to lie in the interval $0 \leqq \gamma \leqq \pi$. Replacing γ by $2n\pi \pm \gamma$ (where n is an integer), we obtain all other angles with the same value of $\cos \gamma$, and any of these will be considered as an angle between **A** and **B**.

[2]In two dimensions the relation $\cos^2 \alpha_1 + \cos^2 \alpha_2 = 1$ permits us to choose for α_2 the value $\pi/2 - \alpha_1$. In three or higher dimensions the relation (11) between the direction cosines does not correspond to any simple linear relation between the angles α_i themselves.

$$\frac{1}{|\mathbf{A}|}\mathbf{A} = \frac{1}{|\mathbf{B}|}\mathbf{B}.$$

Clearly, this is the case if and only if **A** and **B** satisfy a relation $\mathbf{A} = \lambda\mathbf{B}$, where λ is positive. Here $\lambda = |\mathbf{A}|/|\mathbf{B}|$ is the ratio of the lengths of the vectors. A vector of length 1 is called a *unit vector*. The vector

$$(\xi_1, \ldots, \xi_n) = \frac{1}{|\mathbf{A}|}\mathbf{A}$$

whose components are the direction cosines of **A** is the *unit vector in the direction of* **A**.

The vector $-\mathbf{A} = (-a_1, \ldots, -a_n)$ opposite to **A** has the direction cosines $-\xi_i$. We call its direction *opposite* to that of **A**. Two vectors **A** and **B** neither of which is the zero vector will be called *parallel* if they either have the same or the opposite directions. It is necessary for parallelism then that $\mathbf{A} = \lambda\mathbf{B}$ where λ is any number $\neq 0$. The components $a_1, \ldots, a_n$ of any vector $\mathbf{A} \neq \mathbf{0}$ parallel to a given direction are called *direction numbers* for that direction.

If we assign to a unit vector $(\xi_1, \ldots, \xi_n)$ the origin O as initial point, the end point $P = (\xi_1, \ldots, \xi_n)$ is a point on the "unit sphere" (i.e., the sphere of radius 1 and center at the origin O) $\xi_1^2 + \xi_2^2 + \cdots + \xi_n^2 = 1$. Since there exists exactly one unit vector in any given direction, we see that the different directions in n-dimensional space can be represented by the points of the unit sphere. The points on the sphere corresponding to opposite directions are diametrically opposite.

Intuitively a straight line can be thought of as a curve of "constant direction". This suggests that a *straight line* in n-dimensional space be defined as a locus of points with the property that all vectors $\neq \mathbf{0}$ with initial and end point on the line are parallel. This definition leads immediately to a *vector representation for lines*. For any distinct points P, Q on the line L the vector $\overrightarrow{PQ}$ is parallel to a fixed vector **A**, that is,

$$\overrightarrow{PQ} = \lambda\mathbf{A} \qquad (\lambda \neq 0).$$

If we keep P and **A** fixed and let Q run through all points of the line L we have for the position vector of Q the formula (see Fig. 2.3b)

$$\overrightarrow{OQ} = \overrightarrow{OP} + \overrightarrow{PQ} = \overrightarrow{OP} + \lambda\mathbf{A}. \tag{12}$$

Here the parameter λ varies over all real values; the value $\lambda = 0$ corresponds to the point $Q = P$. If Q has coordinates $x_1, \ldots, x_n$; P, the coordinates $y_1, \ldots, y_n$; and $\mathbf{A}$, the components $a_1, \ldots, a_n$, formula (12) corresponds to the *parametric representation* of the line

$$x_i = y_i + \lambda a_i \qquad (i = 1, \ldots, n)$$

where the parameter λ varies over all real λ. The point P divides the line L into two half-lines, or "rays," distinguished by the sign of λ. For $\lambda > 0$ the vector $\overrightarrow{PQ}$ has the same direction as $\mathbf{A}$ ("points" in the direction of $\mathbf{A}$); for $\lambda < 0$ the vector $\overrightarrow{PQ}$ points in the opposite direction.

d. Scalar Products of Vectors

The quantity appearing in the numerator of formula (7) for the angle γ between two vectors $\mathbf{A} = (a_1, \ldots, a_n)$ and $\mathbf{B} = (b_1, \ldots, b_n)$ is called the *scalar product* of $\mathbf{A}$ and $\mathbf{B}$ and denoted by $\mathbf{A} \cdot \mathbf{B}$:

$$\mathbf{A} \cdot \mathbf{B} = a_1 b_1 + a_2 b_2 + \cdots + a_n b_n. \tag{13}$$

Expressed in terms of geometric entities it can be written as

$$\mathbf{A} \cdot \mathbf{B} = |\mathbf{A}|\ |\mathbf{B}| \cos \gamma. \tag{14}$$

The scalar product of two vectors is the product of their lengths multiplied with the cosine of the angle between their directions. If $\mathbf{A} = \overrightarrow{PP'}$, $\mathbf{B} = \overrightarrow{PP''}$, we can interpret $p = |\mathbf{A}| \cos \gamma$ geometrically as the (signed) *projection* of the segment PP' onto the line PP'' (see Fig. 2.4). We call p the *component of the vector* $\mathbf{A}$ *in the direction of* $\mathbf{B}$. By formula (14) we have

$$\mathbf{A} \cdot \mathbf{B} = p|\mathbf{B}|. \tag{14a}$$

Thus the scalar product of the vectors $\mathbf{A}$, $\mathbf{B}$ is equal to the component of $\mathbf{A}$ in the direction of $\mathbf{B}$ multiplied by the length of $\mathbf{B}$.[1] If $\mathbf{B}$ is the coordinate vector $\mathbf{E}_i = (0, \ldots, 1, \ldots 0)$ in the direction of the positive x_i-axis, the component of $\mathbf{A}$ in the direction of $\mathbf{B}$ is simply a_i, the ith component of the vector $\mathbf{A}$. One easily verifies from the

[1]It is, of course, also equal to the component of $\mathbf{B}$ in the direction of $\mathbf{A}$ multiplied by the length of $\mathbf{A}$.

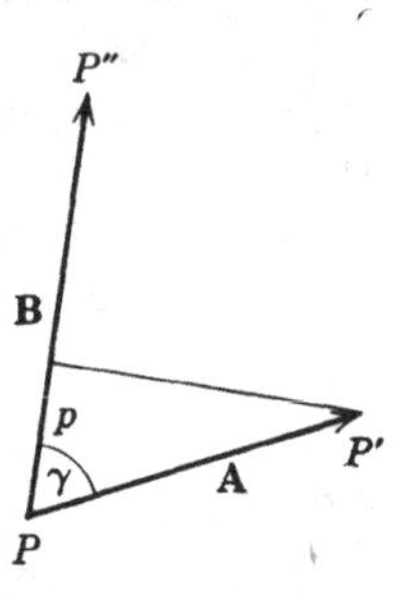

Figure 2.4 Scalar product of the vectors $\mathbf{A} = \overrightarrow{PP'}$ and $\mathbf{B} = \overrightarrow{PP''}$.

definition (13) that the scalar product satisfies the usual algebraic laws

$$\mathbf{A} \cdot \mathbf{B} = \mathbf{B} \cdot \mathbf{A} \qquad \text{(commutative law)} \tag{15a}$$

$$\lambda(\mathbf{A} \cdot \mathbf{B}) = (\lambda\mathbf{A}) \cdot \mathbf{B} = \mathbf{A} \cdot (\lambda\mathbf{B}) \qquad \text{(associative law)[1]} \tag{15b}$$

$$\mathbf{A} \cdot (\mathbf{B} + \mathbf{C}) = \mathbf{A} \cdot \mathbf{B} + \mathbf{A} \cdot \mathbf{C}, \qquad (\mathbf{A} + \mathbf{B}) \cdot \mathbf{C} = \mathbf{A} \cdot \mathbf{C} + \mathbf{B} \cdot \mathbf{C} \qquad \text{(distributive laws).} \tag{15c}$$

The fundamental importance of the scalar product stems from the fact that, expressed in terms of the components of the vectors **A** and **B**, it has the simple *algebraic* expression (13), while at the same time it has a purely geometric interpretation represented by formula (14), which makes no mention of the components of the vectors in any specific coordinate system. Scalar products are not only useful in describing angles but form the basis for deriving analytic expressions for areas and volumes as well.

We conclude from the Cauchy-Schwarz inequality (8) that the scalar product satisfies the inequality

$$|\mathbf{A} \cdot \mathbf{B}| \leqq |\mathbf{A}|\ |\mathbf{B}|, \tag{16}$$

which just expresses that $|\cos \gamma| \leqq 1$. We shall see (p. 191) that the

[1]Since the scalar product of two vectors is not a vector but a scalar, there is no associative law involving *scalar* products of three vectors.

equality in (16) holds only if the vectors **A** and **B** are parallel or if at least one of them is the zero vector.

We notice that by (6a), (13) for $\mathbf{B} = \mathbf{A}$

$$\mathbf{A} \cdot \mathbf{A} = |\mathbf{A}|^2, \tag{17a}$$

That is, *the scalar product of a vector with itself is the square of its length.* This also follows from (14), since the vector **A** forms the angle $\gamma = 0$ with itself. The important relation

$$\mathbf{A} \cdot \mathbf{B} = 0 \tag{17b}$$

for nonzero vectors **A**, **B** corresponds to $\cos\gamma = 0$ or $\gamma = \pi/2$. It characterizes the vectors **A**, **B** as "perpendicular" or "orthogonal" or "normal" to each other. On the other hand, $\mathbf{A} \cdot \mathbf{B} > 0$ means $\cos\gamma > 0$; that is, we can assign to γ a value with $0 \leqq \gamma < \pi/2$; the directions of the vectors form an *acute* angle. Similarly, $\mathbf{A} \cdot \mathbf{B} < 0$ means that the vectors form an angle with $\pi/2 < \gamma \leqq \pi$, an *obtuse* angle, with each other.

For example, the two coordinate vectors (see p. 123)

$$\mathbf{E}_1 = (1, 0, 0, \ldots, 0) \quad \text{and} \quad \mathbf{E}_2 = (0, 1, 0, \ldots, 0)$$

are orthogonal to each other, since

$\mathbf{E}_1 \cdot \mathbf{E}_2 = 1\cdot 0 + 0\cdot 1 + 0\cdot 0 + \cdots + 0\cdot 0 = 0$. More generally, *any two distinct coordinate vectors* $\mathbf{E}_i$ and $\mathbf{E}_k$ *are orthogonal*:

$$\mathbf{E}_i \cdot \mathbf{E}_k = 0 \qquad (i \neq k). \tag{17c}$$

For $k = i$, we have, of course,

$$\mathbf{E}_i \cdot \mathbf{E}_i = |\mathbf{E}_i|^2 = 1; \tag{17d}$$

the coordinate vectors have length 1.

e. *Equation of Hyperplanes in Vector Form*

The locus of the points $P = (x_1, \ldots, x_n)$ in n-dimensional space R^n satisfying a linear equation of the form

$$a_1x_1 + a_2x_2 + \cdots + a_nx_n = c \tag{18}$$

(where $a_1, a_2, \ldots, a_n$ do not all vanish) is called a *hyperplane*. The prefix "hyper-" is needed because n-dimensional space contains "planes," or "linear manifolds," of various dimensions; the hyperplanes can be identified with the $(n-1)$-dimensional euclidean spaces contained in the n-dimensional space R^n. They are the ordinary two-dimensional planes in three-dimensional space, the straight lines in the plane, the points on a line.

Introducing the vector $\mathbf{A} = (a_1, a_2, \ldots, a_n)$ and the position vector $\mathbf{X} = (x_1, \ldots, x_n) = \overrightarrow{OP}$ of the point P, we can write equation (18) in vector notation as

$$\mathbf{A} \cdot \mathbf{X} = c \qquad (\mathbf{A} \neq 0). \tag{18a}$$

Let $\mathbf{Y} = (y_1, \ldots, y_n) = \overrightarrow{OQ}$ be the position vector of *a* particular point Q of the hyperplane, so that $\mathbf{A} \cdot \mathbf{Y} = c$. Subtracting this equation from (18a), we find that the points P of the hyperplane satisfy

$$0 = \mathbf{A} \cdot \mathbf{X} - \mathbf{A} \cdot \mathbf{Y} = \mathbf{A} \cdot (\mathbf{X} - \mathbf{Y}) = \mathbf{A} \cdot \overrightarrow{PQ}. \tag{19}$$

Hence the vector $\mathbf{A}$ is perpendicular to the line joining any two points of the hyperplane. The hyperplane consists of those points obtained by proceeding from any one of its points Q in all directions perpendicular to $\mathbf{A}$. We call the direction of $\mathbf{A}$ "normal" to the hyperplane (see Fig. 2.5).

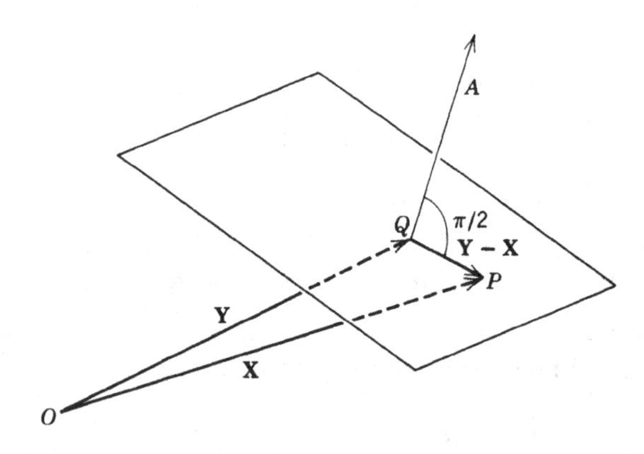

Figure 2.5 Law of formation of third-order determinant.

The hyperplane with equation (18a) divides space into the two open half-spaces given by $\mathbf{A} \cdot \mathbf{X} < c$ and $\mathbf{A} \cdot \mathbf{X} > c$. The vector **A** *points into* the half-space $\mathbf{A} \cdot \mathbf{X} > c$. By this we mean that a ray from a point Q of the hyperplane in the direction of **A** consists of points whose position vectors **X** satisfy $\mathbf{A} \cdot \mathbf{X} > c$. Indeed the position vectors **X** of points P of such a ray are given by

$$\mathbf{X} = \overrightarrow{OP} = \overrightarrow{OQ} + \lambda \mathbf{A} = \mathbf{Y} + \lambda \mathbf{A}$$

[see (12)], where **Y** is the position vector of Q and λ is a positive number. Then obviously

$$\mathbf{A} \cdot \mathbf{X} = \mathbf{A} \cdot \mathbf{Y} + \mathbf{A} \cdot \lambda \mathbf{A} = c + \lambda |\mathbf{A}|^2 > c.$$

More generally, any vector **B** forming an acute angle with **A** points into the half-space $\mathbf{A} \cdot \mathbf{X} > c$, since $\mathbf{A} \cdot \mathbf{B} > 0$ implies that

$$\mathbf{A} \cdot \mathbf{X} = \mathbf{A} \cdot (\mathbf{Y} + \lambda \mathbf{B}) = \mathbf{A} \cdot \mathbf{Y} + \lambda \mathbf{A} \cdot \mathbf{B} > c.$$

If the constant c is positive, the half-space $\mathbf{A} \cdot \mathbf{X} < c$ will be the one containing the origin, since $\mathbf{A} \cdot \mathbf{O} = 0 < c$. Then **A** has the normal direction "away from the origin".

The linear equation (18a) describing a given hyperplane is not unique. For we can multiply the equation with an arbitrary constant factor $\lambda \neq 0$, which amounts to replacing the vector **A** by the parallel vector $\lambda \mathbf{A}$ and the constant c by λc. If $c \neq 0$—that is, if the hyperplane does not pass through the origin—we can choose

$$\lambda = \frac{\operatorname{sgn} c}{|\mathbf{A}|} \cdot$$

Multiplying (18a) by λ, we obtain the *normal form* of the *equation of the hyperplane*

$$\mathbf{B} \cdot \mathbf{X} = p \tag{20}$$

Here p is a positive constant, and **B** is the unit normal vector pointing away from the origin. The constant p in equation (20) is simply the *distance of the hyperplane from the origin 0*, that is, the shortest distance of any point of the hyperplane from 0. For let P be any point of the hyperplane and let **X** be the position vector of P. Then the distance of P from the origin 0 is given by

$$|\overrightarrow{OP}| = |\mathbf{X}| = |\mathbf{X}|\ |\mathbf{B}|.$$

It follows from (16), (20) that

$$|\overrightarrow{OP}| \geqq \mathbf{B} \cdot \mathbf{X} = p.$$

Equality holds for the special point P of the hyperplane with position vector

$$\overrightarrow{OP} = \mathbf{X} = p\mathbf{B}.$$

The line joining this point to the origin has the direction of the normal to the hyperplane. More generally we can find the distance d of any point Q in space with position vector $\mathbf{Y}$ from the hyperplane. As the reader may verify by himself,

$$d = |\mathbf{B} \cdot \mathbf{Y} - p|. \tag{20a}$$

f. Linear Dependence of Vectors and Systems of Linear Equations

Many problems in mathematical analysis can be reduced to the study of linear relations between a number of vectors in n-dimensional space. A vector $\mathbf{Y}$ is called *dependent*[1] *on the vectors* $\mathbf{A}_1, \mathbf{A}_2, \ldots, \mathbf{A}_m$ if $\mathbf{Y}$ can be represented as a "linear combination" of $\mathbf{A}_1, \ldots, \mathbf{A}_m$, that is, if there exist scalars $x_1, \ldots, x_m$ such that

$$\mathbf{Y} = x_1\mathbf{A}_1 + x_2\mathbf{A}_2 + \cdots + x_m\mathbf{A}_m. \tag{21}$$

Here m is any natural number. The zero vector is always dependent, since it can be represented in the form (21) choosing for all the scalars x_i the value 0. Dependence of $\mathbf{Y}$ on a single vector $\mathbf{A}_1 \neq 0$ means that either $\mathbf{Y} = 0$ or that $\mathbf{Y}$ is parallel to $\mathbf{A}_1$. Choosing for $\mathbf{A}_1, \ldots, \mathbf{A}_m$ the n coordinate vectors

$$\mathbf{E}_1 = (1, 0, \ldots, 0), \qquad \mathbf{E}_2 = (0, 1, \ldots, 0), \ldots, \tag{22}$$

$$\mathbf{E}_n = (0, 0, \ldots, 1)$$

we see that the relation (21) holds for any vector $\mathbf{Y} = (y_1, \ldots, y_n)$ if we choose $x_1 = y_1, x_2 = y_2, \ldots, x_n = y_n$:

$$Y = y_1\mathbf{E}_1 + y_2\mathbf{E}_2 + \cdots + y_n\mathbf{E}_n. \tag{23}$$

[1]What we call here "dependent" is often called "linearly dependent" in the literature. Since we do not consider any other kind of dependence between vectors, we drop the word "linear."

Thus, every vector in space is dependent on the coordinate vectors. On the other hand, none of the n coordinate vectors $\mathbf{E}_i$ is dependent on any of the others, as is easily seen. More generally, *a vector* $\mathbf{Y} \neq \mathbf{0}$ *cannot be dependent on vectors* $\mathbf{A}_1, \mathbf{A}_2, \ldots, \mathbf{A}_m$ *if* $\mathbf{Y}$ *is orthogonal to each of the vectors* $\mathbf{A}_1, \ldots, \mathbf{A}_m$. For multiplying relation (21) scalarly by itself yields that

$$\begin{aligned}|\mathbf{Y}|^2 = \mathbf{Y}\cdot\mathbf{Y} &= \mathbf{Y}\cdot(x_1\mathbf{A}_1 + x_2\mathbf{A}_2 + \cdots + x_m\mathbf{A}_m)\\ &= x_1\mathbf{Y}\cdot\mathbf{A}_1 + x_2\mathbf{Y}\cdot\mathbf{A}_2 + \cdots + x_m\mathbf{Y}\cdot\mathbf{A}_m = 0,\end{aligned}$$

and hence that $\mathbf{Y} = \mathbf{0}$.

We call the vectors $\mathbf{A}_1, \ldots, \mathbf{A}_m$ *dependent* if there exist scalars $x_1, x_2, \ldots, x_m$ *that do not all vanish,* such that

$$x_1\mathbf{A}_1 + x_2\mathbf{A}_2 + \cdots + x_m\mathbf{A}_m = 0. \tag{24}$$

If $\mathbf{A}_1, \ldots, \mathbf{A}_m$ are not dependent — that is, if (24) holds only for $x_1 = x_2 = \cdots = x_m = 0$ — we call $\mathbf{A}_1, \ldots, \mathbf{A}_m$ *independent.* For example, the coordinate vectors $\mathbf{E}_1, \ldots, \mathbf{E}_n$ are independent, since

$$\mathbf{0} = x_1\mathbf{E}_1 + x_2\mathbf{E}_2 + \cdots + x_n\mathbf{E}_n = (x_1, x_2, \ldots, x_n)$$

obviously implies that $x_1 = x_2 = \cdots = x_n = 0$.

The two notions of "dependence of a vector on a set of vectors" and "dependence of a set of vectors" are closely related. *A number of vectors are dependent if and only if we can find one of them that is dependent on the others.* For, obviously, relation (21) expressing that $\mathbf{Y}$ is dependent on $\mathbf{A}_1, \ldots, \mathbf{A}_m$ can be written in the form

$$x_1\mathbf{A}_1 + \cdots + x_m\mathbf{A}_m + (-1)\mathbf{Y} = 0,$$

which shows that the $m + 1$ vectors $\mathbf{A}_1, \mathbf{A}_2, \ldots, \mathbf{A}_m, \mathbf{Y}$ are dependent. Conversely, if $\mathbf{A}_1, \ldots, \mathbf{A}_m$ are dependent, we have a relation of the form (24) where not all coefficients x_i vanish. If, say, x_k does not vanish, we can solve equation (24) for $\mathbf{A}_k$, expressing $\mathbf{A}_k$ as a linear combination of the other vectors.

Dependence of the vector $\mathbf{Y}$ on the vectors $\mathbf{A}_1, \ldots, \mathbf{A}_m$ means that a certain system of linear equations has solutions $x_1, \ldots, x_m$. For let $\mathbf{Y} = (y_1, \ldots, y_n)$, and let the vector $\mathbf{A}_k$ be given by

$$\mathbf{A}_k = (a_{1k}, a_{2k}, \ldots, a_{nk}).$$

Then the vector equation (21), written out by components, is equivalent to the system of n linear equations

$$
\begin{aligned}
&a_{11}x_1 + a_{12}x_2 + \cdots + a_{1m}x_m = y_1 \\
&a_{21}x_1 + a_{22}x_2 + \cdots + a_{2m}x_m = y_2 \\
&\cdots\cdots\cdots\cdots\cdots\cdots \\
&a_{n1}x_1 + a_{n2}x_2 + \cdots + a_{nm}x_m = y_n
\end{aligned}
\tag{25}
$$

for the unknown quantities $x_1, \ldots, x_m$. Obviously, $\mathbf{Y}$ is dependent on $\mathbf{A}_1, \ldots, \mathbf{A}_m$ if and only if the system (25) posesses at least one solution $x_1, \ldots, x_m$. Similarly, the vectors $\mathbf{A}_1, \ldots, \mathbf{A}_m$ are dependent if and only if the "homogeneous" system of equations

$$
\begin{aligned}
&a_{11}x_1 + a_{12}x_2 + \cdots + a_{1m}x_m = 0 \\
&a_{21}x_1 + a_{22}x_2 + \cdots + a_{2m}x_m = 0 \\
&\cdots\cdots\cdots\cdots\cdots\cdots \\
&a_{n1}x_1 + a_{n2}x_2 + \cdots + a_{nm}x_m = 0.
\end{aligned}
\tag{25a}
$$

has a "nontrivial" solution $x_1, \ldots, x_m$, that is, has a solution different from the trivial solution[1]

$$x_1 = x_2 = \cdots = x_m = 0.$$

We found one set of n vectors in n-dimensional space that are independent, namely, the coordinate vectors $\mathbf{E}_1, \ldots, \mathbf{E}_n$. Basic for the theory of vectors is the fact that n is the maximum number of independent vectors:

FUNDAMENTAL THEOREM OF LINEAR DEPENDENCE. *Every $n + 1$ vectors in n-dimensional space are dependent.*

Before proving this theorem we consider some of its far-reaching implications. We can conclude immediately that any set of more than n vectors in n-dimensional space is dependent. For any dependence (24) between the first $n + 1$ of m vectors can be considered a dependence of all m vectors, if to the remaining vectors we assign the coefficient 0. The fundamental theorem then implies: *The system of homogeneous linear equations* (25a) *always has a nontrivial solution if $m > n$, that is, if the number of unknowns exceeds the number of equations.*

We can formulate the last statement geometrically in a different way, if we interprete each of the equations (25a) as stating that a

[1]Equations of the type $P(x_1, x_2, \ldots, x_m) = 0$ where P is a homogeneous polynomial (see p. 13) are called *homogeneous*. They always have the trivial solution $x_1 = x_2 = \cdots = x_m = 0$. Moreover any solution $x_1, \ldots, x_m$ stays a solution if we multiply all of the x_i by the same factor λ.

certain scalar product of two vectors in m-dimensional space vanishes. A nontrivial solution $x_1, \ldots, x_m$ then corresponds to a vector $\mathbf{X} = (x_1, \ldots, x_m) \neq \mathbf{0}$. The vanishing of the scalar product of two nonvanishing vectors means that the vectors are perpendicular to each other. Equations (25a) state that $\mathbf{X}$ is perpendicular to the n vectors $(a_{11}, a_{12}, \ldots, a_{1m}), (a_{21}, a_{22}, \ldots, a_{2m}), \ldots, (a_{n1}, a_{n2}, \ldots, a_{nm})$. We have then: *Given a set of nonvanishing vectors whose number is less than the dimension of the space, we can find a vector that is perpendicular to all of them* (and hence, by p. 137, is independent of them).

Returning to vectors in n-dimensional space, we observe a further consequence of the fundamental theorem: *Every vector* $\mathbf{Y}$ *in n-dimensional space is dependent on n given vectors* $\mathbf{A}_1, \ldots, \mathbf{A}_n$, *provided* $\mathbf{A}_1, \ldots, \mathbf{A}_n$ are *independent*. For since the $n + 1$ vectors $\mathbf{A}_1, \ldots, \mathbf{A}_n, \mathbf{Y}$ must be dependent, we have a relation of the form

$$z_1\mathbf{A}_1 + z_2\mathbf{A}_2 + \cdots + z_n\mathbf{A}_n + z_{n+1}\mathbf{Y} = 0,$$

where not all of the quantities $z_1, \ldots, z_{n+1}$ vanish. Then $z_{n+1} \neq 0$, since otherwise $\mathbf{A}_1, \ldots, \mathbf{A}_n$ would be dependent, contrary to assumption. It follows that

$$\mathbf{Y} = x_1\mathbf{A}_1 + x_2\mathbf{A}_2 + \cdots + x_n\mathbf{A}_n \tag{26}$$

where

$$x_i = -\frac{z_i}{z_{n+1}} \qquad (i = 1, \ldots, n).$$

Incidentally, the coefficients x_k in the representation (26) of $\mathbf{Y}$ as a linear combination of the independent vectors $\mathbf{A}_1, \ldots, \mathbf{A}_n$ are uniquely determined, for if there were a second representation

$$\mathbf{Y} = y_1\mathbf{A}_1 + y_2\mathbf{A}_2 + \cdots + y_n\mathbf{A}_n$$

it would follow by subtracting that

$$(x_1 - y_1)\mathbf{A}_1 + (x_2 - y_2)\mathbf{A}_2 + \cdots + (x_n - y_n)\mathbf{A}_n = 0.$$

Here for independent vectors $\mathbf{A}_1, \ldots, \mathbf{A}_n$ we conclude that all coefficients vanish and hence that $x_1 = y_1, \ldots, x_n = y_n$.

On the other hand, if $\mathbf{A}_1, \ldots, \mathbf{A}_n$ are dependent, we certainly can find a vector $\mathbf{Y}$ that does not depend on $\mathbf{A}_1, \ldots, \mathbf{A}_n$, for in that case, one of the vectors $\mathbf{A}_1, \ldots, \mathbf{A}_n$ is dependent on the others, say $\mathbf{A}_n$ on $\mathbf{A}_1, \ldots, \mathbf{A}_{n-1}$; a vector $\mathbf{Y}$ dependent on $\mathbf{A}_1, \ldots, \mathbf{A}_n$ is then also dependent on $\mathbf{A}_1, \ldots, \mathbf{A}_{n-1}$. There are, however, vectors $\mathbf{Y}$ in n-dimensional space that do not depend on $n-1$ given vectors (see p. 139).

Since independence of $\mathbf{A}_1, \ldots, \mathbf{A}_n$ is equivalent to the fact that the corresponding system of homogeneous linear equations (25a) has only the trivial solution, we have deduced the following basic theorem on solvability of systems of linear equations from the fundamental theorem:

The system of n linear equations

$$\begin{aligned} a_{11}x_1 + a_{12}x_2 + \cdots + a_{1n}x_n &= y_1 \\ a_{21}x_1 + a_{22}x_2 + \cdots + a_{2n}x_n &= y_2 \\ \cdots\cdots\cdots\cdots\cdots\cdots & \\ a_{n1}x_1 + a_{n2}x_2 + \cdots + a_{nn}x_n &= y_n \end{aligned} \tag{27}$$

has a unique solution $x_1, \ldots, x_n$ for any given numbers $y_1, \ldots, y_n$ provided the homogeneous equations

$$\begin{aligned} a_{11}x_1 + a_{12}x_2 + \cdots + a_{1n}x_n &= 0 \\ a_{21}x_1 + a_{22}x_2 + \cdots + a_{2n}x_n &= 0 \\ \cdots\cdots\cdots\cdots\cdots\cdots & \\ a_{n1}x_1 + a_{n2}x_2 + \cdots + a_{nn}x_n &= 0 \end{aligned} \tag{27a}$$

have only the trivial solution $x_1 = x_2 = \cdots = x_n = 0$. If the system (27a) has a nontrivial solution we can find values $y_1, \ldots, y_n$ for which the system (27) has no solution.

We have here a pure existence theorem, that gives no indication, how the solution $x_1, x_2 \ldots, x_n$, if it exists, can actually be obtained. This can be achieved by means of determinants, as discussed in Section 2.3 below.

We proceed to the proof of the fundamental theorem, using induction over the dimension n. The theorem states that any $n+1$ vectors $\mathbf{A}_1, \ldots, \mathbf{A}_n, \mathbf{Y}$ in n-dimensional space are dependent. For $n = 1$, vectors become scalars, and the statement to be proved is the following: For any two *numbers* Y and $\mathbf{A}$ we can find numbers x_0, x_1, which do not both vanish, such that

$$x_0 Y + x_1 \mathbf{A} = 0.$$

This is trivial. If $Y = \mathbf{A} = 0$, we take $x_0 = x_1 = 1$; in all other cases, we take $x_0 = \mathbf{A}$, $x_1 = -Y$.

Assume that we have proved that any n vectors in $(n-1)$-dimensional space are dependent. Let $\mathbf{A}_1, \ldots, \mathbf{A}_n, \mathbf{Y}$ be vectors in n-dimensional space. We want to prove that $\mathbf{A}_1, \ldots, \mathbf{A}_n, \mathbf{Y}$ are dependent. This is certainly the case, if $\mathbf{A}_1, \ldots, \mathbf{A}_n$ alone are already dependent. Thus we restrict ourselves to the case that $\mathbf{A}_1, \ldots, \mathbf{A}_n$ are independent; we shall prove that then $\mathbf{Y}$ is dependent on $\mathbf{A}_1, \ldots, \mathbf{A}_n$. It is sufficient to prove that each of the coordinate vectors $\mathbf{E}_1, \ldots \mathbf{E}_n$ in (22) is dependent on $\mathbf{A}_1, \ldots, \mathbf{A}_n$, for any vector $\mathbf{Y}$ is, by (23), a linear combination of the $\mathbf{E}_i$ and hence also of the $\mathbf{A}_k$ if the $\mathbf{E}_i$ can be expressed in terms of the $\mathbf{A}_k$. We shall prove only that $\mathbf{E}_n$ is dependent on $\mathbf{A}_1, \ldots, \mathbf{A}_n$, since the proof for the other $\mathbf{E}_i$ is similar. We only have to show that the system of equations

$$\begin{aligned} a_{i1}x_1 + a_{i2}x_2 + \cdots + a_{in}x_n &= 0 \qquad (i = 1, \ldots, n-1) \\ a_{n1}x_1 + a_{n2}x_2 + \cdots + a_{nn}x_n &= 1 \end{aligned} \tag{28}$$

has a solution $x_1, \ldots, x_n$. Now the first $n-1$ equations, which are homogeneous, have a nontrivial solution $x_1, \ldots, x_n$ as a consequence of the induction assumption that n vectors in $(n-1)$-dimensional space are dependent. For that solution, let

$$a_{n1}x_1 + a_{n2}x_2 + \cdots + a_{nn}x_n = c.$$

Here $c \neq 0$, since otherwise the vectors $\mathbf{A}_1, \ldots, \mathbf{A}_n$ would be dependent. Dividing $x_1, x_2, \ldots, x_n$ by c, we obtain then the desired solution of the system (28). This completes the proof of the fundamental theorem.

Exercises 2.1

1. Give the coordinate representation of the line passing through the point $P = (-2, 0, 4)$ and in the direction of the vector $\mathbf{A} = (2, 1, 3)$.
2. (a) What is the equation of the line passing through the points $P = (3, -2, 2)$ and $Q = (6, -5, 4)$?
   (b) Give the equation of the line passing through any two distinct points P and Q.
3. If $\mathbf{A}$ and $\mathbf{B}$ are two vectors with initial point O and final points P and Q, then the vector with O as initial point and the point dividing PQ in the ratio $\lambda : (1-\lambda)$ as final point is given by
$$(1-\lambda)\,\mathbf{A} + \lambda\mathbf{B}.$$
4. In Exercise 3, for what values of λ does the position vector correspond to a point on the ray in the direction of Q from P?
5. The center of mass of the vertices of a tetrahedron $PQRS$ may be

defined as the point dividing MS in the ratio 1:3, where M is the center of mass of the vertices PQR. Show that this definition is independent of the order in which the vertices are taken and that it agrees with the general definition of the center of mass (Volume I, p. 373).

6. Two edges of a tetrahedron are called opposite if they have no vertex in common. For example, the edges PQ and RS of the tetrahedron of Exercise 5 are opposite. Show that the segment joining the midpoints of opposite edges of a tetrahedron passes through the center of mass of the vertices.

7. Let $A_1, \ldots, A_n$ be n arbitrary particles in space, with masses, $m_1, m_2, \ldots, m_n$, respectively. Let G be their center of mass and let $\mathbf{A}_1 \ldots, \mathbf{A}_n$ denote the vectors with initial point G and final points $A_1, \ldots, A_n$. Prove that

$$m_1\mathbf{A}_1 + m_2\mathbf{A}_2 + \cdots + m_n\mathbf{A}_n = 0.$$

8. The real numbers form a one-dimensional vector space where addition of "vectors" is ordinary addition and multiplication by scalars is ordinary multiplication. Show that the positive real numbers also form a vector space where addition of vectors is ordinary multiplication and scalar multiplication is appropriately defined.

9. Verify that the complex numbers form a two-dimensional vector space where addition is ordinary addition and the scalars are real numbers.

10. Let P and Q be diametrically opposite points and R any other point on a sphere. Show that PR meets QR at right angles.

11. (a) Obtain the normal form of the plane through the point $P = (-3, 2, 1)$ and perpendicular to the vector $\mathbf{A} = (1, 2, -2)$.
    (b) What is the distance of the point $Q = (1, -1, -1)$ from the plane?
    (c) Do O and Q lie on the same or opposite sides of the plane?

12. (a) Let the equation of a hyperplane be given in the form (18). Determine the coordinates of the foot of the perpendicular from a point P to the hyperplane.
    (b) In Exercise 11, give the feet of the perpendiculars from O and Q on the plane.

13. Let $\mathbf{A}$ and $\mathbf{B}$ be nonparallel vectors. Show that

$$\mathbf{C} = \mathbf{A} - \frac{\mathbf{A} \cdot \mathbf{B}}{|\mathbf{B}|^2}\mathbf{B}$$

is perpendicular to $\mathbf{B}$. The vector $\mathbf{C}$ is called the component of $\mathbf{A}$ perpendicular to $\mathbf{B}$.

14. Find the angle ϕ between the plane

$$Ax + By + Cz + D = 0.$$

and the line

$$x = x_0 + \alpha t, \; y = y_0 + \beta t, \qquad z = z_0 + \gamma t.$$

2.2 Matrices and Linear Transformations

a. Change of Base. Linear Spaces

Every vector $\mathbf{Y}$ in n-dimensional space R^n can be written as a linear combination of the coordinate vectors $\mathbf{E}_1, \ldots, \mathbf{E}_n$ defined by (22); namely,

$$\mathbf{Y} = y_1\mathbf{E}_1 + \cdots + y_n\mathbf{E}_n, \tag{29}$$

where the y_i are the components of $\mathbf{Y}$. We can generalize the notion of coordinate vector and of components by considering any m independent vectors $\mathbf{A}_1, \ldots, \mathbf{A}_m$ in S_n. If $\mathbf{Y}$ is a vector dependent on the $\mathbf{A}_i$, we have

$$\mathbf{Y} = x_1\mathbf{A}_1 + \cdots + x_m\mathbf{A}_m \tag{30}$$

where the coefficients x_i are determined uniquely by $\mathbf{Y}$. We call $x_1, \ldots, x_m$ the *components of* $\mathbf{Y}$ *with respect to the base* $\mathbf{A}_1, \ldots, \mathbf{A}_m$. With respect to this base, the base vector $\mathbf{A}_1$ has the components $1, 0, \ldots, 0$; the base vector $\mathbf{A}_2$, the components $0, 1, \ldots, 0$; and so on. For any scalar λ the vector

$$\lambda\mathbf{Y} = \lambda x_1\mathbf{A}_1 + \cdots + \lambda x_m\mathbf{A}_m$$

also is dependent on the $\mathbf{A}_i$ and has components $\lambda x_1, \ldots, \lambda x_m$. Similarly, if

$$\mathbf{Y}' = x_1'\mathbf{A}_1 + \cdots + x_m'\mathbf{A}_m$$

is a second vector depending on the $\mathbf{A}_i$, the sum

$$\mathbf{Y} + \mathbf{Y}' = (x_1 + x'_1)\mathbf{A}_1 + \cdots + (x_m + x_m')\mathbf{A}_m$$

has the components $x_1 + x_1', \ldots, x_m + x_m'$ with respect to our base.

For $m < n$ not all vectors $\mathbf{Y}$ in n-dimensional space are dependent on $\mathbf{A}_1, \ldots, \mathbf{A}_m$. The vectors dependent on m independent vectors are said to form an *m-dimensional vector space.* We can visualize such a space by choosing an arbitrary point P_0 with position vector $\mathbf{B} = \overrightarrow{OP_0}$ as initial point for all the vectors $\mathbf{A}_1, \ldots, \mathbf{A}_m$. Let

$$\mathbf{A}_i = \overrightarrow{P_0P_i} \qquad (i = 1, \ldots, m) \tag{31a}$$

and let $\mathbf{Y} = \overrightarrow{P_0P}$ be the vector given by (30). Then the point P has the position vector

(31b) $$\overrightarrow{OP} = \overrightarrow{OP_0} + \overrightarrow{P_0P} = \mathbf{B} + x_1\mathbf{A}_1 + \cdots + x_m\mathbf{A}_m.$$

The points P in relation (31b) are said to form the *m-dimensional linear manifold* S_m *through* P_0 *spanned by the vectors* $\mathbf{A}_1, \ldots, \mathbf{A}_m$. Every point P in S_m uniquely determines values $x_1, \ldots, x_m$, which we call *affine coordinates* for P. In this affine coordinate system for S_m the "origin" — that is, the point with $x_1 = x_2 = \cdots = x_m = 0$ — is the point P_0; the point with affine coordinates $x_1 = 1, x_2 = \cdots = x_m = 0$ is P_1, the end point of the vector $\mathbf{A}_1 = \overrightarrow{P_0P_1}$, and so on. For two points P and P' of S_m with position vectors

$$\overrightarrow{OP} = \mathbf{B} + x_1\mathbf{A}_1 + \cdots + x_m\mathbf{A}_m, \qquad \overrightarrow{OP'} = \mathbf{B} + x_1'\mathbf{A}_1 + \cdots + x_m'\mathbf{A}_m,$$

the vector

$$\overrightarrow{PP'} = \overrightarrow{OP'} - \overrightarrow{OP} = (x_1' - x_1)\mathbf{A}_1 + \cdots + (x_m' - x_m)\mathbf{A}_m$$

has as components with respect to the base $\mathbf{A}_1, \ldots, \mathbf{A}_m$ the differences of the affine coordinates of the points P and P'.

According to our definition a one-dimensional linear manifold S_1 through the point P_0 is the locus of points P with position vectors of the form

$$\overrightarrow{OP} = \mathbf{B} + x_1\mathbf{A}_1$$

where $\mathbf{B}$ and $\mathbf{A}_1$ are fixed vectors, ($\mathbf{A}_1 \neq 0$) and x_1 ranges over all real numbers. Of course, S_1 is merely the straight line through P_0 parallel to the direction of the vector $\mathbf{A}_1$ (see p. 130). A two-dimensional linear manifold or two-dimensional *plane* S_2 consists of the points P with position vectors

$$\overrightarrow{OP} = \mathbf{B} + x_1\mathbf{A}_1 + x_2\mathbf{A}_2$$

where $\mathbf{B}$, $\mathbf{A}_1$, $\mathbf{A}_2$ are fixed vectors ($\mathbf{A}_1$ and $\mathbf{A}_2$ independent) and x_1 and x_2 range over all real numbers. The n-dimensional linear spaces S_n are identical with the whole space R^n; for any vector $\mathbf{Y}$ is dependent on n linearly independent vectors $\mathbf{A}_1, \ldots, \mathbf{A}_n$ (see p. 133), and hence the position vector of any point P is representable in the form

$$\overrightarrow{OP} = \mathbf{B} + x_1\mathbf{A}_1 + \cdots + x_n\mathbf{A}_n.$$

The $(n-1)$-dimensional linear manifolds can be seen to be identical with the hyperplanes defined on p. 133. For given any $n-1$ vectors $\mathbf{A}_1, \ldots, \mathbf{A}_{n-1}$ in n-dimensional space, we can find a vector $\mathbf{A}$ perpendicular to all of them (see page 139.) Then for

$$\overrightarrow{OP} = \mathbf{B} + x_1\mathbf{A}_1 + \cdots + x_{n-1}\mathbf{A}_{n-1}$$

we have the relation

$$\mathbf{A} \cdot \overrightarrow{OP} = \mathbf{B} \cdot \mathbf{A} + x_1 \cdot \mathbf{A}_1 \cdot \mathbf{A} + \cdots + x_{n-1}\mathbf{A}_{n-1} \cdot \mathbf{A} = \mathbf{B} \cdot \mathbf{A}$$
$$= \text{constant},$$

which is just a linear equation for the coordinates of P.

In general, the determination of the components x_i of a vector $\mathbf{Y}$ with respect to a base $\mathbf{A}_1, \ldots, \mathbf{A}_m$ requires the solution of a system of linear equations of the type (25). In one important special case, the x_i can be found directly, namely, when the base vectors form an *orthonormal* system. We call the vectors $\mathbf{A}_1, \ldots, \mathbf{A}_m$ orthonormal if each of them has length 1 and any two are orthogonal to each other, that is, if

$$\mathbf{A}_i \cdot \mathbf{A}_k = \begin{cases} 1 \text{ for } i = k \\ 0 \text{ for } i \neq k. \end{cases} \tag{32}$$

If a vector $\mathbf{Y}$ is of the form

$$\mathbf{Y} = x_1\mathbf{A}_1 + x_2\mathbf{A}_2 + \cdots + x_m\mathbf{A}_m,$$

we find, using the *orthogonality relations* (32), that

$$\mathbf{Y} \cdot \mathbf{A}_i = x_1\mathbf{A}_1 \cdot \mathbf{A}_i + x_2\mathbf{A}_2 \cdot \mathbf{A}_i + \cdots + x_m\mathbf{A}_m \cdot \mathbf{A}_i = x_i \qquad (i = 1, \ldots, m). \tag{33}$$

In particular, $\mathbf{Y} = \mathbf{0}$ implies $x_i = 0$ for $i = 1, \ldots, m$; thus *orthonormal vectors always are independent.* Formula (33) shows that the component x_i of the vector $\mathbf{Y}$ with respect to an orthonormal base $\mathbf{A}_1, \ldots, \mathbf{A}_m$ is equal to the *component* $\mathbf{Y} \cdot \mathbf{A}_i$ of the vector $\mathbf{Y}$ *in the direction of* $\mathbf{A}_i$. The coordinate vectors $\mathbf{E}_1, \ldots, \mathbf{E}_n$ defined by equations (22) form just such an orthonormal base, and the components of the vector $\mathbf{Y} = (y_1, \ldots, y_n)$ with respect to this base are the quantities $\mathbf{Y} \cdot \mathbf{E}_i = y_i$.

An orthonormal base is also distinguished by the fact that the length of a vector and the scalar product of two vectors is given by the same formulae as in the original base $\mathbf{E}_1, \ldots, \mathbf{E}_n$. Given any two vectors $\mathbf{Y}$ and $\mathbf{Y}'$ of the form

$$\text{(34a)} \quad \mathbf{Y} = x_1\mathbf{A}_1 + \cdots + x_m\mathbf{A}_m, \qquad \mathbf{Y}' = x_1'\mathbf{A}_1 + \cdots + x_m'\mathbf{A}_m$$

we have

$$\begin{aligned} \text{(34b)} \quad \mathbf{Y} \cdot \mathbf{Y}' &= (x_1\mathbf{A}_1 + \cdots + x_m\mathbf{A}_m) \cdot (x_1'\mathbf{A}_1 + \cdots + x_m'\mathbf{A}_m) \\ &= x_1\mathbf{A}_1 \cdot (x_1'\mathbf{A}_1 + \cdots + x_m'\mathbf{A}_m) + \cdots \\ &\quad + x_m\mathbf{A}_m \cdot (x_1'\mathbf{A}_1 + \cdots + x_m'\mathbf{A}_m) \\ &= x_1x_1' + x_2x_2' + \cdots + x_mx_m'.\text{[1]} \end{aligned}$$

In the particular case $\mathbf{Y}' = \mathbf{Y}$ we find for the length of the vector $\mathbf{Y}$ the formula

$$\text{(34c)} \qquad |\mathbf{Y}| = \sqrt{\mathbf{Y} \cdot \mathbf{Y}} = \sqrt{x_1^2 + \cdots + x_m^2}.$$

If the m-dimensional linear manifold S_m through the point P_0 is spanned by m orthonormal vectors $\mathbf{A}_1, \ldots, \mathbf{A}_m$, the corresponding affine coordinate system is called a *Cartesian coordinate system* for the space S_m. The coordinate vectors $\mathbf{A}_1, \ldots, \mathbf{A}_m$ are mutually perpendicular and of length 1. The distance d between any two points with Cartesian coordinates $(x_1, \ldots, x_m)$ and $(x_1', \ldots, x_m')$ is given by the formula

$$d = \sqrt{(x_1' - x_1)^2 + \cdots + (x_m' - x_m)^2}$$

More generally any geometric relation based on the notion of distance (such as angle, area, volume) has the same analytic expression in any Cartesian coordinate system.

b. Matrices

The relation

$$\text{(35a)} \qquad \mathbf{Y} = x_1\mathbf{A}_1 + \cdots + x_m\mathbf{A}_m$$

between vectors $\mathbf{A}_1, \ldots, \mathbf{A}_m, \mathbf{Y}$ in n-dimensional space can be written as a system of linear equations [see (25), p. 138]

[1]Without the orthogonality relations we could only conclude that $\mathbf{Y} \cdot \mathbf{Y}'$ is given by the more complicated expression

$$\mathbf{Y} \cdot \mathbf{Y}' = \sum_{i,k} c_{ik}x_ix_k \qquad \text{where} \qquad c_{ik} = \mathbf{A}_i \cdot \mathbf{A}_k.$$

$$
\begin{aligned}
a_{11}x_1 + a_{12}x_2 + \cdots + a_{1m}x_m &= y_1 \\
a_{21}x_1 + a_{22}x_2 + \cdots + a_{2m}x_m &= y_2 \\
\cdots\cdots\cdots\cdots\cdots\cdots\cdots\cdots \\
a_{n1}x_1 + a_{n2}x_2 + \cdots + a_{nm}x_m &= y_n
\end{aligned} \tag{35b}
$$

connecting the components $y_1, \ldots, y_n$ of the vector $\mathbf{Y}$ in the original coordinate system with the components $x_1, \ldots, x_m$ of $\mathbf{Y}$ with respect to the base vectors $\mathbf{A}_i = (a_{1i}, a_{2i}, \ldots, a_{ni})$ for $i = 1, \ldots, m$. The linear relations (35b) between the quantities x_i and y_j are completely described by the system of $n \times m$ coefficients a_{ji}. The system of coefficients arranged in a rectangular array

$$
\mathbf{a} = \begin{pmatrix}
a_{11} & a_{12} & \cdots & a_{1m} \\
a_{21} & a_{22} & \cdots & a_{2m} \\
\vdots & \vdots & & \vdots \\
a_{n1} & a_{n2} & \cdots & a_{nm}
\end{pmatrix}, \tag{36}
$$

as they appear in (35b) is called a *matrix*.
(We shall usually denote matrices by boldface lower-case letters).

The matrix $\mathbf{a}$ in (36) has mn "elements"

$$a_{ji}; \qquad j = 1, \ldots, n; \qquad i = 1, \ldots, m.$$

These elements are arranged in m "columns"

$$
\begin{pmatrix} a_{11} \\ a_{21} \\ \vdots \\ a_{n1} \end{pmatrix}, \begin{pmatrix} a_{12} \\ a_{22} \\ \vdots \\ a_{n2} \end{pmatrix}, \ldots, \begin{pmatrix} a_{1m} \\ a_{2m} \\ \vdots \\ a_{nm} \end{pmatrix}
$$

or in n "rows"

$$
\begin{aligned}
&(a_{11} \quad a_{12} \quad \cdots \quad a_{1m}), \\
&(a_{21} \quad a_{22} \quad \cdots \quad a_{2m}), \\
&\cdots\cdots\cdots\cdots \\
&(a_{n1} \quad a_{n2} \quad \cdots \quad a_{nm}).
\end{aligned}
$$

Two matrices are considered equal only if they agree in the number of rows and columns and if corresponding elements are the same.

The columns of the matrix **a** can be identified respectively with the set of components of the vectors $\mathbf{A}_1, \mathbf{A}_2, \ldots, \mathbf{A}_m$. We shall often write the matrix **a** whose columns are formed from the components of the vectors $\mathbf{A}_1, \mathbf{A}_2, \ldots, \mathbf{A}_m$ as

$$\mathbf{a} = (\mathbf{A}_1, \mathbf{A}_2, \ldots, \mathbf{A}_m). \tag{37}$$

The system of equations (35b) expressing the n quantities $y_1, \ldots, y_n$ as linear functions of the m quantities $x_1, \ldots, x_m$ can be compressed into the single symbolic equation

$$\mathbf{aX} = \mathbf{Y}, \tag{38}$$

where $\mathbf{X}$ stands for the vector $(x_1, \ldots, x_m)$ and $\mathbf{Y}$ for the vector $(y_1, \ldots, y_n)$. If the column vectors $\mathbf{A}_1, \ldots, \mathbf{A}_m$ of the matrix **a** are independent, we can interpret (38) as describing a *change of base* or of coordinate system for vectors.

The equation connects the components $x_1, \ldots, x_m$ of the vector with respect to the base $\mathbf{A}_1, \ldots, \mathbf{A}_m$ in the subspace S_m with the components $y_1, \ldots, y_n$ of the same vector with respect to the base $\mathbf{E}_1, \ldots, \mathbf{E}_n$ for the whole space S_n. This might be called the "passive" interpretation of (38), in which the geometrical objects—the vectors—stay fixed and only the reference system is switched.

There is another, "active" interpretation, in which the vectors change rather than the coordinate system. Equations (36) then describe a *mapping* of vectors $(x_1, \ldots, x_m)$ in an m-dimensional space onto vectors $(y_1, \ldots, y_n)$ in an n-dimensional space. A mapping given by equation (38), or in more detail by the equivalent system of equations (35b), is called *linear*, or *affine*.[1]

[1]In an affine mapping of vectors the components y_j of the image vector $\mathbf{Y}$ are *homogeneous* linear functions of components x_i of the original vector $\mathbf{X}$, as in formulae (35b). If we identify $\mathbf{X}$ and $\mathbf{Y}$ with *position vectors* of points, formulae (35b) define a mapping of points $(x_1, \ldots, x_m)$ in the space R^m onto points $(y_1, \ldots, y_n)$ in the space R^n. The point mappings obtained in this way are the special affine mappings that take the origin of R^m into the origin of R^n. The most general affine mapping of points is given by *inhomogeneous* linear equations

$$y_j = \sum_{i=1}^{m} a_{ji}x_i + b_j \qquad (j = 1, \ldots, n) \tag{*}$$

(It can be obtained from a special mapping taking the origin into the origin by a translation with components b_j). Applying the mapping (*) to two points $P' = (x_1', \ldots, x_m')$, $P'' = (x_1'', \ldots, x_m'')$ with images $Q' = (y_1', \ldots, y_n')$, $Q'' = (y_1'', \ldots, y_n'')$, we see that the corresponding mapping of the vectors $\overrightarrow{P'P''} = (x_1'' - x_1', \ldots, x_m'' - x_m') = (x_1, \ldots, x_m)$ onto the vectors $\overrightarrow{Q'Q''} = (y_1'' - y_1', \ldots, y_n'' - y_n') = (y_1, \ldots, y_n)$ is given by the *homogeneous* equations (35b).

For example the system of equations

(38a) $$y_1 = \frac{2}{3}x_1 - \frac{1}{3}x_2, \qquad y_2 = -\frac{1}{3}x_1 + \frac{2}{3}x_2,$$

$$y_3 = -\frac{1}{3}x_1 - \frac{1}{3}x_2$$

corresponding to the matrix

$$\mathbf{a} = \begin{pmatrix} \frac{2}{3} & -\frac{1}{3} \\ -\frac{1}{3} & \frac{2}{3} \\ -\frac{1}{3} & -\frac{1}{3} \end{pmatrix}$$

can be interpreted as a mapping of vectors $\mathbf{X} = (x_1, x_2)$ in the plane onto vectors $\mathbf{Y} = (y_1, y_2, y_3)$ in three-dimensional space. Here the image vectors all satisfy the relation

(38b) $$y_1 + y_2 + y_3 = 0$$

and hence are orthogonal to the vector $\mathbf{N} = (1, 1, 1)$. Identifying the vectors $\mathbf{X}, \mathbf{Y}$ with position vectors of points, we have in (38a) a mapping of the $x_1\,x_2$-plane onto the plane π in $y_1\,y_2\,y_3$-space with equation (38b). Geometrically the point (y_1, y_2, y_3) is obtained by projecting the point $(x_1, x_2, 0)$ perpendicularly onto the plane π.[1] Alternately, equations (38a) can be interpreted passively as a parametric representation for the plane π, with x_1 and x_2 playing the role of parameters.

Different matrices give rise to different linear mappings, for by (35b) the coordinate vectors

$$\mathbf{E}_1 = (1, 0, \ldots, 0), \qquad \mathbf{E}_2 = (0, 1, \ldots, 0), \ldots$$

are mapped onto the vectors

$$\mathbf{A}_1 = (a_{11}, a_{21}, \ldots, a_{n1}), \qquad \mathbf{A}_2 = (a_{12}, a_{22}, \ldots, a_{n2}), \ldots$$

Thus, the column vectors $\mathbf{A}_1, \mathbf{A}_2, \ldots, \mathbf{A}_n$ of the matrix $\mathbf{a}$ are just the images of the coordinate vectors $\mathbf{E}_1, \mathbf{E}_2, \ldots, \mathbf{E}_n$. Hence, the matrix $\mathbf{a}$ is determined uniquely by the mapping.

[1]The line joining $(x_1, x_2, 0)$ and (y_1, y_2, y_3) is parallel to the normal N of π.

Of particular importance are the linear mappings $\mathbf{Y} = \mathbf{aX}$ of the n-dimensional vector space *into itself*; they map a vector $\mathbf{X} = (x_1, \ldots, x_n)$ onto a vector $\mathbf{Y} = (y_1, \ldots, y_n)$ with the same number of components. Such mappings correspond to matrices $\mathbf{a}$ with as many rows as columns, so-called *square matrices*.[1] Written out by components, the mapping $\mathbf{Y} = \mathbf{aX}$ corresponding to a square matrix $\mathbf{a}$ with n rows and columns takes the form (27). p.140. The basic theorem of solvability of systems of n linear equations for n unknown quantities (p. 140) can now be stated alternatively as follows:

For a square matrix $\mathbf{a}$ *there are two mutually exclusive possibilities:*

(1) $\mathbf{aX} \neq \mathbf{0}$ *for every vector* $\mathbf{X} \neq \mathbf{0}$

(2) $\mathbf{aX} = \mathbf{0}$ *for some vector* $\mathbf{X} \neq \mathbf{0}$.

In case (1) there exists for every vector $\mathbf{Y}$ *a unique vector* $\mathbf{X}$ *such that* $\mathbf{Y} = \mathbf{aX}$. *In case (2) there exist vectors* $\mathbf{Y}$ *for which the equation* $\mathbf{Y} = \mathbf{aX}$ *holds for no vector* $\mathbf{X}$.[2]

We call the matrix $\mathbf{a}$ *singular* in case (2) and *nonsingular* in case (1). Since existence of a nontrivial solution $\mathbf{X}$ of the equation $\mathbf{aX} = \mathbf{0}$ is equivalent to dependence of the column vectors of the matrix $\mathbf{a}$, we see that a *square matrix* $\mathbf{a}$ *is singular if and only if its column vectors are dependent.*

c. Operations with Matrices

It is customary to denote the elements of a matrix $\mathbf{a}$ as in (36) by letters bearing two subscripts, such as a_{ji}. The subscripts indicate the *location* or *address* of the element in the matrix, the first subscript giving the row number, the second the column number. For a matrix with n rows and m columns having elements a_{ji} the subscript j ranges over $1, 2, \ldots, n$ and the subscript i over $1, 2, \ldots, m$. Equation (36) is often abbreviated into the formula

$$\mathbf{a} = (a_{ji}),$$

which only exhibits the elements of the matrix $\mathbf{a}$ but does not show the numbers of rows and columns, which have to be deduced from the context.[3] In the example

[1]The more general matrices with arbitrary numbers of rows and columns are referred to as *rectangular* matrices.

[2]In case (1) the equation $\mathbf{Y} = \mathbf{aX}$ represents a 1–1 mapping of the n-dimensional vector space onto itself. In case (2) the mapping is neither 1–1 nor onto.

[3]The letter a in a_{ji} is the name of a real-valued function of the independent variables j and i. The domain of this function consists of the points in the j, i-plane whose

$$\mathbf{a} = (a_{ji}) = \begin{pmatrix} 1! & 2! & 3! & \cdots & m! \\ 2! & 3! & 4! & \cdots & (m+1)! \\ 3! & 4! & 5! & \cdots & (m+2)! \\ \vdots & \vdots & \vdots & & \vdots \\ n! & (n+1)! & (n+2)! & \cdots & (m+n-1)! \end{pmatrix}$$

we have $a_{ji} = (i + j - 1)!$

Addition of matrices and multiplication of matrices by scalars are defined in the same way as for vectors. If $\mathbf{a} = (a_{ji})$ and $\mathbf{b} = (b_{ji})$ are matrices of the same "size"—that is, with the same numbers of rows and columns—we define $\mathbf{a} + \mathbf{b}$ as the matrix obtained by adding corresponding elements:

$$\mathbf{a} + \mathbf{b} = (a_{ji} + b_{ji}).$$

Similarly, for a scalar λ we define $\lambda\mathbf{a}$ as the matrix obtained by multiplying each element of $\mathbf{a}$ by the factor λ:

$$\lambda\mathbf{a} = (\lambda a_{ji}).$$

One verifies immediately the rules

$$(\mathbf{a} + \mathbf{b})\,\mathbf{X} = \mathbf{a}\mathbf{X} + \mathbf{b}\mathbf{X}, \qquad (\lambda\mathbf{a})\,\mathbf{X} = \lambda(\mathbf{a}\mathbf{X}) \tag{39}$$

for the mappings of vectors $\mathbf{X}$ determined by the matrices.

More significant is the fact that matrices of suitable sizes can be *multiplied* with each other. A natural definition of the product of two matrices $\mathbf{a}, \mathbf{b}$ is obtained by considering the *symbolic product,* or *composition,* of the corresponding mappings (see Volume I, p. 52). If $\mathbf{a} = (a_{ji})$ is a matrix with m columns and n rows, and if $\mathbf{X} = (x_1, \ldots, x_m)$ is a vector with m components, then $\mathbf{a}$ determines the mappings $\mathbf{Y} = \mathbf{a}\mathbf{X}$ of the vector $\mathbf{X}$ onto the vector $\mathbf{Y} = (y_1, \ldots, y_n)$ with the n components

$$y_j = \sum_{i=1}^{m} a_{ji} x_i \qquad (j = 1, \ldots, n).$$

If now $\mathbf{b} = (b_{kj})$ is a matrix with n columns and p rows, then the

---

coordinates are integers with $1 \leqq j \leqq n$, and $1 \leqq i \leqq m$. Ordinarily we write a function f of two independent variables x, y as $f(x, y)$, and a more consistent notation here would be $a(j, i)$ instead of the customary a_{ji}.

mapping $\mathbf{Z} = \mathbf{bY}$ will map $\mathbf{Y}$ onto the vector $\mathbf{Z} = (z_1, \ldots, z_p)$ with the p components

$$z_k = \sum_{j=1}^{n} b_{kj}\, y_j = \sum_{j=1}^{n} \sum_{i=1}^{m} b_{kj}\, a_{ji}\, x_i = \sum_{i=1}^{m} c_{ki}\, x_i,$$

where

$$c_{ki} = \sum_{j=1}^{n} b_{kj}\, a_{ji} \quad (k = 1, \ldots, p;\, i = 1, \ldots, m). \tag{40}$$

Thus $\mathbf{Z} = \mathbf{cX}$, where $\mathbf{c} = \mathbf{ba} = (c_{ki})$ is the matrix with p rows and m columns and with elements given by formula (40). Accordingly, we define the product $\mathbf{c} = \mathbf{ba}$ of the matrices $\mathbf{b}$ and $\mathbf{a}$ as the matrix with elements c_{ki} given by (40).

We observe that the product $\mathbf{ba}$ is defined only if the number of columns of $\mathbf{b}$ is the same as the number of rows of $\mathbf{a}$. This corresponds to the obvious fact that the symbolic product of two mappings can only be formed, if the domain of the first factor contains the range of the second one. Thus it could happen very well that the product $\mathbf{ba}$ is defined but not the product $\mathbf{ab}$ with the factors in the reverse order. But even where both $\mathbf{ba}$ and $\mathbf{ab}$ are defined the *commutative law of multiplication* $\mathbf{ab} = \mathbf{ba}$ *in general does not hold for matrices.* For example, for

$$\mathbf{a} = \begin{pmatrix} 0 & 1 \\ -1 & 0 \end{pmatrix}, \qquad \mathbf{b} = \begin{pmatrix} 1 & 0 \\ 0 & -1 \end{pmatrix}$$

we have

$$\mathbf{ab} = \begin{pmatrix} 0 & -1 \\ -1 & 0 \end{pmatrix}, \qquad \mathbf{ba} = \begin{pmatrix} 0 & 1 \\ 1 & 0 \end{pmatrix}.$$

However, one easily verifies from formula (40) that matrix multiplication obeys the associative and distributive laws

$$\mathbf{a}(\mathbf{bc}) = (\mathbf{ab})\mathbf{c}, \tag{41a}$$

$$\mathbf{a}(\mathbf{b} + \mathbf{c}) = \mathbf{ab} + \mathbf{ac}, \qquad (\mathbf{a} + \mathbf{b})\mathbf{c} = \mathbf{ac} + \mathbf{bc}, \tag{41b}$$

(for matrices of appropriate sizes). We might say that all algebraic manipulations for matrices are permitted as long as the products involved are defined and we do not interchange factors.

The mapping of vectors determined by the matrix **a**, which we had written as $\mathbf{Y} = \mathbf{aX}$, can be considered a special example of matrix multiplication *provided* we write **X** and **Y** as "column vectors," that is, as matrices with a single column and with m and n rows, respectively:

$$\mathbf{X} = \begin{pmatrix} x_1 \\ x_2 \\ \cdot \\ \cdot \\ \cdot \\ x_m \end{pmatrix}, \qquad \mathbf{Y} = \begin{pmatrix} y_1 \\ y_2 \\ \cdot \\ \cdot \\ \cdot \\ y_n \end{pmatrix}$$

d. Square Matrices. The Reciprocal of a Matrix. Orthogonal Matrices

Of particular importance in applications are the matrices with the same number of rows and columns, the so-called *square matrices* (the more general matrices with arbitrary numbers of rows and columns are referred to as *rectangular* matrices). The *order* of a square matrix is the number of its rows or columns. Any two square matrices of the same order n can be added or multiplied. In particular, we can form *powers* of such a matrix:

$$\mathbf{a}^2 = \mathbf{aa}, \qquad \mathbf{a}^3 = \mathbf{aaa}, \cdots.$$

The zero matrix **0** of order n is the matrix all of whose elements are 0, or all of whose columns are zero vectors:

(42a) $$\mathbf{0} = (0, 0, \ldots, 0).$$

It has the obvious properties

(42b) $$\mathbf{a} + \mathbf{0} = \mathbf{0} + \mathbf{a} = \mathbf{a}, \qquad \mathbf{a0} = \mathbf{0a} = \mathbf{0}$$

(for all n-th order matrices **a**),

(42c) $\mathbf{0X} = \mathbf{0}$ for all vectors **X** with n components.

The *unit matrix,* of order n, denoted by **e** is the matrix corresponding to the identity mapping of vectors **X**:

(43a) $$\mathbf{eX} = \mathbf{X}$$

for all vectors **X**. Since then in particular $\mathbf{eE}_k = \mathbf{E}_k$ for all coordinate vectors $\mathbf{E}_k$, we find that the unit matrix has the coordinate vectors as columns:

$$\mathbf{e} = (\mathbf{E}_1, \mathbf{E}_2, \ldots, \mathbf{E}_n) = \begin{pmatrix} 1 & 0 & 0 & \cdots & 0 \\ 0 & 1 & 0 & \cdots & 0 \\ \cdot & \cdot & \cdot & & \cdot \\ \cdot & \cdot & \cdot & & \cdot \\ \cdot & \cdot & \cdot & & \cdot \\ 0 & 0 & 0 & \cdots & 1 \end{pmatrix} \tag{43b}$$

One verifies immediately that $\mathbf{e}$ plays the role of a "unit" in matrix multiplication:

$$\mathbf{ae} = \mathbf{ea} = \mathbf{a} \tag{43c}$$

for all n-th order $\mathbf{a}$.

We call an nth order matrix $\mathbf{b}$ *reciprocal* to the nth order matrix $\mathbf{a}$ if

$$\mathbf{ab} = \mathbf{e}. \tag{44}$$

If $\mathbf{b}$ is reciprocal to $\mathbf{a}$, then $\mathbf{a}$ corresponds to the inverse of the mapping of vectors furnished by $\mathbf{b}$, for if $\mathbf{b}$ maps a vector $\mathbf{Y}$ onto $\mathbf{X}$ (i.e., if $\mathbf{X} = \mathbf{bY}$), then $\mathbf{a}$ maps $\mathbf{X}$ back onto $\mathbf{Y}$, since $\mathbf{aX} = \mathbf{abY} = \mathbf{eY} = \mathbf{Y}$. More concretely, if we know a reciprocal $\mathbf{b}$ of the matrix $\mathbf{a} = (a_{ji})$, we can write down a solution $\mathbf{X} = (x_1, x_2, \ldots, x_n)$ of the system of linear equations

$$\begin{aligned} a_{11}x_1 + a_{12}x_2 + \cdots + a_{1n}x_n &= y_1 \\ a_{21}x_1 + a_{22}x_2 + \cdots + a_{2n}x_n &= y_2 \\ \cdots\cdots\cdots\cdots\cdots\cdots \\ a_{n1}x_1 + a_{n2}x_2 + \cdots + a_{nn}x_n &= y_n \end{aligned}$$

for any given $(y_1, \ldots, y_n) = \mathbf{Y}$. Since $\mathbf{abY} = \mathbf{eY} = \mathbf{Y}$, we have indeed a solution given by $\mathbf{X} = \mathbf{bY}$, that is, by

$$\begin{aligned} x_1 &= b_{11}y_1 + \cdots + b_{1n}y_n \\ \cdots&\cdots\cdots\cdots\cdots \\ x_n &= b_{n1}y_1 + \cdots + b_{nn}y_n. \end{aligned}$$

Every real number a except zero has a reciprocal b for which $ab = 1$. However, there are matrices different from the zero matrix that have no reciprocal. If $\mathbf{a}$ has a reciprocal, the equation $\mathbf{aX} = \mathbf{Y}$ has for every vector $\mathbf{Y}$ the solution $\mathbf{X} = \mathbf{bY}$, since

$$\mathbf{abY} = \mathbf{eY} = \mathbf{Y}.$$

Hence (see p. 150) the matrix $\mathbf{a}$ must be nonsingular; that is, the columns of $\mathbf{a}$ are independent vectors. *Singular matrices have no reciprocal.* The condition $\mathbf{ab} = \mathbf{e}$ for the reciprocal matrix $\mathbf{b}$ of $\mathbf{a}$ can be written out in the form

$$\sum_{r=1}^{n} a_{jr} b_{rk} = e_{jk}, \tag{45}$$

where a_{jr}, b_{rk}, e_{jk} denote respectively the general elements of the matrices $\mathbf{a}$, $\mathbf{b}$, $\mathbf{e}$. For fixed k we have in (45) a system of n linear equations for the vector $\mathbf{B}_k = (b_{1k}, b_{2k}, \ldots, b_{nk})$, which represents the kth column of the matrix $\mathbf{b}$. If the matrix $\mathbf{a}$ is nonsingular, there exists a unique solution $\mathbf{B}_k$ of (45) for every k. *Hence, a nonsingular matrix* $\mathbf{a}$ *has one and only one reciprocal* $\mathbf{b}$.

Let $\mathbf{a}$ be any nonsingular matrix and $\mathbf{b}$ its reciprocal; that is, $\mathbf{ab} = \mathbf{e}$. Take an arbitrary vector $\mathbf{X}$ and put $\mathbf{Y} = \mathbf{aX}$. Since both $\mathbf{Z} = \mathbf{X}$ and $\mathbf{Z} = \mathbf{bY}$ are solutions of the equations $\mathbf{Y} = \mathbf{aZ}$ and since the solution is unique, we must have

$$\mathbf{bY} = \mathbf{X}$$

for every vector $\mathbf{X}$. Hence (see p.149) $\mathbf{a}$ is the reciprocal of $\mathbf{b}$:

$$\mathbf{ba} = \mathbf{e}.$$

The reciprocal of a nonsingular matrix $\mathbf{a}$ is usually denoted by $\mathbf{a}^{-1}$. We have

$$\mathbf{aa}^{-1} = \mathbf{a}^{-1}\mathbf{a} = \mathbf{e}, \tag{46}$$

where $\mathbf{e}$ is the unit matrix. The reciprocal can be calculated by solving the system of linear equations (45) for the b_{rk}. Since the elements e_{jk} of the unit matrix have the value 0 for $j \neq k$ and 1 for $j = k$, equations (45) state that the scalar product of the jth row of the matrix $\mathbf{a}$ with the kth column of the matrix $\mathbf{a}^{-1}$ has the value 0 for $j \neq k$ and 1 for $j = k$. Furthermore, since $\mathbf{a}^{-1}\,\mathbf{a} = \mathbf{e}$ we see that the scalar product of the jth row of $\mathbf{a}^{-1}$ with the kth column of $\mathbf{a}$ also has the value 0 for $j \neq k$ and 1 for $j = k$.

Multiplying by reciprocals enables us to "divide" an equation between matrices by a nonsingular matrix. For example, the matrix equation

$$\mathbf{ab} = \mathbf{c},$$

where $\mathbf{a}$ is a nonsingular matrix, can be solved for $\mathbf{b}$ by multiplying the equation *from the left* by $\mathbf{a}^{-1}$:

$$\mathbf{a}^{-1}\mathbf{c} = \mathbf{a}^{-1}(\mathbf{ab}) = (\mathbf{a}^{-1}\mathbf{a})\mathbf{b} = \mathbf{eb} = \mathbf{b}.$$

Similarly, the equation

$$\mathbf{ba} = \mathbf{c}$$

leads to

$$\mathbf{ca}^{-1} = \mathbf{b}.$$

From the point of view of euclidean geometry the most important square matrices are the so-called *orthogonal* matrices, which correspond to transitions from one Cartesian coordinate system to another such system or to linear transformations that preserve length. A square matrix $\mathbf{a}$ is called orthogonal if its column vectors $\mathbf{A}_1, \ldots, \mathbf{A}_n$ form an orthonormal system:

$$\mathbf{A}_i \cdot \mathbf{A}_k = \begin{cases} 0 & \text{for} \quad i \neq k \\ 1 & \text{for} \quad i = k \end{cases} \tag{47}$$

(see p. 145). Since vectors forming an orthonormal system are independent, it follows that *orthogonal matrices are always nonsingular.* The vector relation $\mathbf{aX} = \mathbf{Y}$ corresponding to the matrix $\mathbf{a}$, interpreted passively, describes how the components $y_1, \ldots, y_n$ of a vector with respect to the coordinate vectors $\mathbf{E}_1, \ldots, \mathbf{E}_n$ are connected with the components of the same vector with respect to the base $\mathbf{A}_1, \ldots, \mathbf{A}_n$. For an orthogonal matrix $\mathbf{a}$ the base $\mathbf{A}_1, \ldots, \mathbf{A}_n$ consists of n mutually orthogonal vectors of length 1, forming a "Cartesian" coordinate system, in which distance is given by the usual expression (see p. 146). Interpreted actively, $\mathbf{Y} = \mathbf{aX}$ represents a linear mapping in which the coordinate vectors $\mathbf{E}_i$ are mapped onto the vectors $\mathbf{A}_i$. This mapping takes a vector

$$\mathbf{X} = (x_1, \ldots, x_n) = x_1\mathbf{E}_1 + \cdots + x_n\mathbf{E}_n$$

into the vector

$$\mathbf{Y} = \mathbf{aX} = \mathbf{a}(x_1\mathbf{E}_1 + \cdots + x_n\mathbf{E}_n) = x_1\mathbf{aE}_1 + \cdots + x_n\mathbf{aE}_n$$
$$= x_1\mathbf{A}_1 + \cdots + x_n\mathbf{A}_n.$$

The mapping preserves the length of any vector, since by (47)

$$|\mathbf{Y}|^2 = \mathbf{Y}\cdot\mathbf{Y} = (x_1\mathbf{A}_1 + \cdots + x_n\mathbf{A}_n)\cdot(x_1\mathbf{A}_1 + \cdots + x_n\mathbf{A}_n)$$
$$= x_1^2 + \cdots + x_n^2 = |\mathbf{X}|^2.$$

More generally the mapping preserves the scalar product of any two vectors and hence also angles between directions, as is easily verified. Such length preserving mappings are known as *orthogonal transformations*, or *rigid motions*. In two dimensions they are easily identified with the changes of coordinate axes discussed in Volume I (p. 361). A vector $\mathbf{A}_1$ of length 1 in two dimensions is of the form $\mathbf{A}_1 = (\cos\gamma, \sin\gamma)$ with some suitable angle γ. The only vectors $\mathbf{A}_2$ of length 1 that are perpendicular to $\mathbf{A}_1$ are

$$\mathbf{A}_2 = \left(\cos\left(\gamma + \frac{\pi}{2}\right), \sin\left(\gamma + \frac{\pi}{2}\right)\right) = \left(-\sin\gamma, \cos\gamma\right)$$

and

$$\mathbf{A}_2 = \left(\cos\left(\gamma - \frac{\pi}{2}\right), \sin\left(\gamma - \frac{\pi}{2}\right)\right) = \left(\sin\gamma, -\cos\gamma\right).$$

Thus the general second-order orthogonal matrix is either of the form

$$\text{(48)}\qquad \mathbf{a} = \begin{pmatrix} \cos\gamma & -\sin\gamma \\ \sin\gamma & \cos\gamma \end{pmatrix} \quad \text{or} \quad \mathbf{a} = \begin{pmatrix} \cos\gamma & \sin\gamma \\ \sin\gamma & -\cos\gamma \end{pmatrix}.$$

The orthorgonality relations (47) permit one immediately to write down the inverse $\mathbf{a}^{-1}$ of an orthogonal matrix $\mathbf{a}$. We just take for $\mathbf{a}^{-1}$ the matrix that has the $\mathbf{A}_k$ as *row* vectors; the scalar product of the jth row of $\mathbf{a}^{-1}$ with the kth column of $\mathbf{a}$ is then 0 for $j \neq k$ and 1 for $j = k$, as required by the relation $\mathbf{a}^{-1}\,\mathbf{a} = \mathbf{e}$. Generally, for any matrix $\mathbf{a} = (a_{jk})$, one defines the *transpose* $\mathbf{a}^T = (b_{jk})$ as the matrix obtained from $\mathbf{a}$ by interchanging rows and columns. More precisely $b_{jk} = a_{kj}$.[1] For an orthogonal matrix we simply have

[1]Thinking of **a** as written out as a rectangular array, one defines the "main diagonal" of **a** as the line running from the upper left-hand corner downward at slope -1. It is the line containing the elements $a_{11}, a_{22}, a_{33}, \ldots$. The transpose of **a** is obtained by "reflecting" **a** in the main diagonal.

$$\mathbf{a}^{-1} = \mathbf{a}^T. \tag{49}$$

For example,

$$\begin{pmatrix} \cos\gamma & -\sin\gamma \\ \sin\gamma & \cos\gamma \end{pmatrix}^{-1} = \begin{pmatrix} \cos\gamma & \sin\gamma \\ -\sin\gamma & \cos\gamma \end{pmatrix}.$$

Following (46) we can write relation (49) as

$$\mathbf{a}^T\mathbf{a} = \mathbf{e}, \qquad \mathbf{a}\mathbf{a}^T = \mathbf{e}. \tag{49a}$$

The second relation shows that in an orthogonal matrix the scalar product of the jth row with the kth row is 0 for $j \neq k$ and 1 for $j = k$. Thus *in an orthogonal matrix the row vectors also form an orthonormal system.*

Exercises 2.2

1. In each case describe the space through P spanned by the vectors $\mathbf{A}_k$.

   (a) $P = (-1, 2, 1)$; $\mathbf{A}_1 = (4, 0, 3)$

   (b) $P = (2, 1, -4)$ $\mathbf{A}_1 = (3, -2, 1)$, $\mathbf{A}_2 = (1, 0, -1)$

   (c) $P = (2, 1, -4, 2)$, $\mathbf{A}_1 = (3, -2, 1, 2)$, $\mathbf{A}_2 = (1, 0, -1, 2)$.

2. Verify that $\mathbf{E}_1 = (2/3, 2/3, -1/3)$, $\mathbf{E}_2 = (1/\sqrt{2}, -1/\sqrt{2}, 0)$, $\mathbf{E}_3 = (\sqrt{2}/6, \sqrt{2}/6, 2\sqrt{2}/3)$ form an orthonormal base and obtain the representations of the given vectors in terms of this base:

   (a) $\mathbf{A}_1 = (\sqrt{2}, \sqrt{2}, \sqrt{2})$

   (b) $\mathbf{A}_2 = (3, -3, 3)$

   (c) $\mathbf{A}_3 = (1, 0, 0)$

3. Given linearly independent vectors $\mathbf{A}_1, \mathbf{A}_2, \ldots, \mathbf{A}_m$, construct mutually perpendicular unit vectors $\mathbf{E}_1, \mathbf{E}_2, \ldots, \mathbf{E}_m$ with the property that $\mathbf{E}_k$ is a linear combination of $\mathbf{A}_1, \mathbf{A}_2, \ldots, \mathbf{A}_k$, for $k = 1, 2, \ldots, m$.
4. From the result of Exercise 3, prove the fundamental theorem of linear dependence.
5. What is the distance of the point $P = (x_0, y_0, z_0)$ from the straight line given by

   $$x = at + b, \quad y = ct + d, \quad z = et + f?$$

   (*Hint*: Find the foot of the perpendicular from P to the line.)
6. Does the following system of equations have a nontrivial solution?

   $$x + 2y + 3z = 0$$

   $$2x + 3y + z = 0$$

$3x + y + 2z = 0$

7. Find the representation of the vector (a_1, a_2, a_3) with respect to the base $\mathbf{A}_1 = (1, 2, 3)$, $\mathbf{A}_2 = (2, 3, 1)$, $\mathbf{A}_3 = (3, 1, 2)$.
8. Determine the matrix for changing from Cartesian coordinates for the base $\mathbf{E}_1, \mathbf{E}_2, \mathbf{E}_3$ to affine coordinates for the base $\mathbf{A}_1, \mathbf{A}_2, \mathbf{A}_3$ given in Exercise 7.
9. Prove that if the matrix $\mathbf{a}$ is singular, there exist vectors $\mathbf{Y}$ for which $\mathbf{Y} = \mathbf{aX}$ has no solution.
10. Obtain the products $\mathbf{ab}$ and $\mathbf{ba}$ for the matrices

$$\mathbf{a} = \begin{pmatrix} 1 & 2 & 0 \\ 0 & 0 & 1 \\ 2 & 1 & 0 \end{pmatrix}, \quad \mathbf{b} = \begin{pmatrix} -2 & 1 & 0 \\ 0 & 1 & -2 \\ 1 & 0 & 1 \end{pmatrix}$$

11. Find conditions that the 2×2 matrix

$$\begin{pmatrix} a & b \\ c & d \end{pmatrix}$$

has a reciprocal and give that reciprocal if it exists.

12. Show that there is only one unit matrix.
13. Find the reciprocal of $\mathbf{ab}$, if neither $\mathbf{a}$ nor $\mathbf{b}$ is singular.
14. Sometimes a singular $n \times n$ matrix is defined as a matrix that maps n-dimensional space onto a space of lower dimension. Show that this definition is equivalent to the one given here.
15. Interpret the matrices in (48) geometrically.
16. Prove that $\mathbf{a}$ is orthogonal if and only if $\mathbf{a}^T = \mathbf{a}^{-1}$.
17. Show that the transpose of a product $\mathbf{ab}$ is the product $\mathbf{b}^T\mathbf{a}^T$ of the transposed matrices in reverse order.
18. Show that the product of orthogonal matrices is orthogonal.
19. Verify that mapping by an orthogonal matrix preserves scalar products; that is, if $\mathbf{a}$ is orthogonal, then $(\mathbf{aX}) \cdot (\mathbf{aY}) = \mathbf{X} \cdot \mathbf{Y}$
20. Show that any length-preserving matrix is orthogonal.
21. Prove that an affine transformation transforms the center of mass of a system of particles into the center of mass of the image particles.

2.3 Determinants

a. Determinants of Second and Third Order

Mathematical analysis includes the study of nonlinear mappings in spaces of several dimensions. Such a study, however, has to be preceded by one of the linear mappings $\mathbf{Y} = \mathbf{aX}$ where $\mathbf{X}$ and $\mathbf{Y}$ are vectors and $\mathbf{a}$ a matrix. In particular, it is of basic importance to analyze the structure of the inverse of such a mapping or—what amounts to the same thing—analyze the structure of the solutions of a system of n linear equations

$$(50) \qquad \begin{cases} a_{11}x_1 + a_{12}x_2 + \cdots + a_{1n}x_n = y_1 \\ a_{21}x_1 + a_{22}x_2 + \cdots + a_{2n}x_n = y_2 \\ \cdots\cdots\cdots\cdots\cdots\cdots \\ a_{n1}x_1 + a_{n2}x_2 + \cdots + a_{nn}x_n = y_n \end{cases}$$

for n unknown quantities $x_1, \ldots, x_n$.

The process of solving n linear equations in n variables leads to certain algebraic expressions called *determinants*, which have a great number of terms. In the beginning, the explicit definition and the properties of determinants appear somewhat mystifying. The mystery will disappear when we base the definition of determinant on one single property, that of being a multilinear alternating form of n vectors in n-dimensional space. From this conceptual approach all the important properties of determinants can easily be derived. We shall see in later chapters of this book that determinants are of the utmost importance in extending differential and integral calculus to higher dimensions.

It is instructive to write out the explicit solution of equations (50) for the first few values of n. For $n = 1$ we have the single equation

$$a_{11}x_1 = y_1$$

with the solution

$$(50a) \qquad x_1 = \frac{y_1}{a_{11}}.$$

For $n = 2$ we have the system

$$a_{11}x_1 + a_{12}x_2 = y_1$$

$$a_{21}x_1 + a_{22}x_2 = y_2.$$

Multiplying the first equation by a_{22}, the second by a_{12} and subtracting, we eliminate x_2 and find a single equation for x_1; similarly, multiplying the first equation by a_{21} and the second by a_{11} and subtracting eliminates x_1. In this way we find for x_1, x_2 the expressions

$$(50b) \qquad x_1 = \frac{a_{22}y_1 - a_{12}y_2}{a_{11}a_{22} - a_{12}a_{21}}, \qquad x_2 = \frac{a_{11}y_2 - a_{21}y_1}{a_{11}a_{22} - a_{12}a_{21}}.$$

For $n = 3$ we have the system

$$\left\{\begin{aligned} a_{11}x_1 + a_{12}x_2 + a_{13}x_3 &= y_1 \\ a_{21}x_1 + a_{22}x_2 + a_{23}x_3 &= y_2 \\ a_{31}x_1 + a_{32}x_2 + a_{33}x_3 &= y_3. \end{aligned}\right. \tag{50c}$$

We can reduce this system to two equations for x_1, x_2, thus eliminating x_3, by multiplying the second equation by a_{13}/a_{23} and subtracting it from the first and by multiplying the third equation by a_{13}/a_{33} and subtracting it from the from the first. The two resulting equations for x_1, x_2 alone can then be solved as before. After some algebraic manipulation we find that

(50d)

$$x_1 = \frac{a_{22}a_{33}y_1 + a_{12}a_{23}y_2 + a_{13}a_{32}y_2 - a_{13}a_{22}y_3 - a_{23}a_{32}y_1 - a_{12}a_{33}y_2}{a_{11}a_{22}a_{33} + a_{12}a_{23}a_{31} + a_{13}a_{21}a_{32} - a_{13}a_{22}a_{31} - a_{11}a_{23}a_{32} - a_{12}a_{21}a_{33}},$$

with similar formulae for x_2 and x_3. For $n = 4$, the computations become completely unwieldy and it is clear that only a systematic approach can bring order into the results.

We notice that in each case the solution x_i takes the form of a quotient, where the denominator is a function of the coefficients a_{ji} alone, that is, a function of the matrix $\mathbf{a} = (a_{ji})$. For $n = 1$ this function is simply the coefficient a_{11} itself. For $n = 2$, the denominator

$$a_{11}a_{22} - a_{12}a_{21},$$

formed from the elements of the matrix

$$\mathbf{a} = \begin{pmatrix} a_{11} & a_{12} \\ a_{21} & a_{22} \end{pmatrix},$$

is called the *determinant of the matrix* $\mathbf{a}$ and written

$$a_{11}a_{22} - a_{12}a_{21} = \det(\mathbf{a}) = \begin{vmatrix} a_{11} & a_{12} \\ a_{21} & a_{22} \end{vmatrix}. \tag{51a}$$

It is clear that the numerators in (50b) also can be written as determinants, giving rise to the expressions

$$x_1 = \frac{\begin{vmatrix} y_1 & a_{12} \\ y_2 & a_{22} \end{vmatrix}}{\begin{vmatrix} a_{11} & a_{12} \\ a_{21} & a_{22} \end{vmatrix}}; \qquad x_2 = \frac{\begin{vmatrix} a_{11} & y_1 \\ a_{12} & y_2 \end{vmatrix}}{\begin{vmatrix} a_{11} & a_{12} \\ a_{21} & a_{22} \end{vmatrix}} \tag{51b}$$

Of course, these formulae make sense only if the determinant in the denominator does not have the value 0.

Formula (50d) suggests introducing as determinant of the third-order matrix

$$\mathbf{a} = \begin{pmatrix} a_{11} & a_{12} & a_{13} \\ a_{21} & a_{22} & a_{23} \\ a_{31} & a_{32} & a_{33} \end{pmatrix}$$

the expression

$$\text{(52a)} \qquad \begin{aligned} & a_{11}a_{22}a_{33} + a_{12}a_{23}a_{31} + a_{13}a_{21}a_{32} - a_{13}a_{22}a_{31} \\ & \qquad - a_{11}a_{23}a_{32} - a_{12}a_{21}a_{33} \\ & = \det(\mathbf{a}) = \begin{vmatrix} a_{11} & a_{12} & a_{13} \\ a_{21} & a_{22} & a_{23} \\ a_{31} & a_{32} & a_{33} \end{vmatrix} \end{aligned}$$

The law of formation of such a third-order determinant can be expressed by the easily remembered "diagonal rule" (Fig. 2.5a). We repeat the first two columns after the third; form the product of each triad of numbers in the diagonal lines, multiplying the products associated with lines slanting downward to the right by $+1$ and to the left by -1; and add. (This rule holds only for third-order determinants!).

With the help of third-order determinants we can write the solution of the system (50c) in the more concise form

$$x_1 = \frac{\begin{vmatrix} y_1 & a_{12} & a_{13} \\ y_2 & a_{22} & a_{23} \\ y_3 & a_{32} & a_{33} \end{vmatrix}}{\begin{vmatrix} a_{11} & a_{12} & a_{13} \\ a_{21} & a_{22} & a_{23} \\ a_{31} & a_{32} & a_{33} \end{vmatrix}}, \quad x_2 = \frac{\begin{vmatrix} a_{11} & y_1 & a_{13} \\ a_{21} & y_2 & a_{23} \\ a_{31} & y_3 & a_{33} \end{vmatrix}}{\begin{vmatrix} a_{11} & a_{12} & a_{13} \\ a_{21} & a_{22} & a_{23} \\ a_{31} & a_{32} & a_{33} \end{vmatrix}}, \quad x_3 = \frac{\begin{vmatrix} a_{11} & a_{12} & y_1 \\ a_{21} & a_{22} & y_2 \\ a_{31} & a_{32} & y_{33} \end{vmatrix}}{\begin{vmatrix} a_{11} & a_{12} & a_{13} \\ a_{21} & a_{22} & a_{23} \\ a_{31} & a_{32} & a_{33} \end{vmatrix}}$$

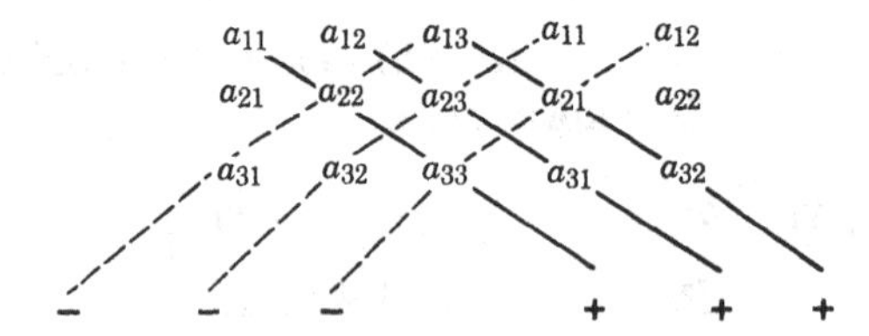

Figure 2.5a

By analogy we define the determinant of the first order matrix

$$\mathbf{a} = (a_{11})$$

on the basis of (50a) as

$$a_{11} = \det(\mathbf{a}).$$

We see then that in each of the cases $n = 1,2,3$ the solution $(x_1, \ldots, x_n)$ of the system (50) can be described as follows ("Cramer's rule"): *Each unknown x_i is the quotient of two determinants. In the denominator we have the determinant of the matrix $\mathbf{a} = (a_{jk})$; in the numerator we have the determinant of the matrix obtained by replacing the ith column of the matrix* $\mathbf{a}$ *by the quantities $y_1, y_2, \ldots, y_n$ appearing on the right-hand side of the equations.*

b. Linear and Multilinear Forms of Vectors

In order to define determinants of higher order and to formulate their principal properties, it is necessary to make use of some general algebraic notions.

A function $f(a_1, \ldots, a_n)$ of the n independent variables $a_1, \ldots, a_n$ can be considered as a *function of the vector* $\mathbf{A} = (a_1, \ldots, a_n)$ *and written* in the form $f(\mathbf{A})$. We call f a *linear form* in $\mathbf{A}$, if

$$f(\mathbf{A} + \mathbf{B}) = f(\mathbf{A}) + f(\mathbf{B}) \tag{53a}$$

for any two vectors $\mathbf{A}$, $\mathbf{B}$ and

$$f(\lambda\mathbf{A}) = \lambda f(\mathbf{A}) \tag{53b}$$

for any vector $\mathbf{A}$ and any scalar λ.

The two rules (53a, b) can be compressed into the single requirement that

$$f(\lambda\mathbf{A} + \mu\mathbf{B}) = \lambda f(\mathbf{A}) + \mu f(\mathbf{B}) \tag{54a}$$

for any vectors $\mathbf{A}$, $\mathbf{B}$ and scalars λ, μ. Written out in detail, the rule (54a) becomes

$$\begin{aligned} f(\lambda a_1 + \mu b_1, \ldots, \lambda a_n + \mu b_n) \\ = \lambda f(a_1, \ldots, a_n) + \mu f(b_1, \ldots, b_n). \end{aligned} \tag{54b}$$

For example, the function

$$f(\mathbf{A}) = 3a_2 - 27a_3$$

is a linear form, while

$$f(\mathbf{A}) = |\mathbf{A}| = \sqrt{a_1^2 + \cdots + a_n^2}$$

is not.

Relation (54a) immediately implies the more general rule for linear forms

$$f(\lambda_1\mathbf{A}_1 + \cdots + \lambda_m\mathbf{A}_m) = \lambda_1 f(\mathbf{A}_1) + \cdots + \lambda_m f(\mathbf{A}_m) \tag{54c}$$

valid for any m vectors $\mathbf{A}_1, \ldots, \mathbf{A}_m$ and scalars $\lambda_1, \ldots, \lambda_m$. This rule yields an explicit expression for the most general linear form in the vector $\mathbf{A}$. Using the coordinate vectors $\mathbf{E}_1, \ldots, \mathbf{E}_n$, we have by (2b) the representation

$$\mathbf{A} = (a_1, \ldots, a_n) = a_1\mathbf{E}_1 + a_2\mathbf{E}_2 + \cdots + a_n\mathbf{E}_n$$

for the vector $\mathbf{A}$. Hence, by (54c), f is of the form

$$\begin{aligned} f(\mathbf{A}) &= a_1 f(\mathbf{E}_1) + a_2 f(\mathbf{E}_2) + \cdots + a_n f(\mathbf{E}_n) \\ &= c_1a_1 + c_2a_2 + \cdots + c_na_n \end{aligned} \tag{55a}$$

where the c_i have the constant values

$$c_i = f(\mathbf{E}_i). \tag{55b}$$

Combining the coefficients c_i into the vector $\mathbf{C} = (c_1, \ldots, c_n)$, we have

$$f(\mathbf{A}) = \mathbf{C} \cdot \mathbf{A}. \tag{55c}$$

The most general linear form in a vector **A** *is the scalar product of* **A** *with a suitable constant vector* **C**.

A function $f(\mathbf{A}, \mathbf{B})$ of two vectors $\mathbf{A} = (a_1, \ldots, a_n)$, $\mathbf{B} = (b_1, \ldots, b_n)$ is called a *bilinear form* in $\mathbf{A}, \mathbf{B}$ if f is a linear form in $\mathbf{A}$ for fixed $\mathbf{B}$ and a linear form in $\mathbf{B}$ for fixed $\mathbf{A}$; this means that we require that

$$f(\lambda\mathbf{A} + \mu\mathbf{B}, \mathbf{C}) = \lambda f(\mathbf{A}, \mathbf{C}) + \mu f(\mathbf{B}, \mathbf{C}) \tag{56a}$$

$$f(\mathbf{A}, \lambda\mathbf{B} + \mu\mathbf{C}) = \lambda f(\mathbf{A}, \mathbf{B}) + \mu f(\mathbf{A}, \mathbf{C}) \tag{56b}$$

for any vectors $\mathbf{A}, \mathbf{B}, \mathbf{C}$ and scalars λ, μ. The simplest example of a bilinear form is the scalar product

$$f(\mathbf{A}, \mathbf{B}) = \mathbf{A} \cdot \mathbf{B}.$$

In this example, the rules (56a, b) just reduce to the associative and distributive laws (15b, c), p. 132 for scalar products.

We find more generally from (56a, b) that

$$\text{(56c)} \qquad \begin{aligned} f(\alpha\mathbf{A} + \beta\mathbf{B}, \gamma\mathbf{C} + \delta\mathbf{D}) &= \alpha f(\mathbf{A}, \gamma\mathbf{C} + \delta\mathbf{D}) + \beta f(\mathbf{B}, \gamma\mathbf{C} + \delta\mathbf{D}) \\ &= \alpha\gamma f(\mathbf{A}, \mathbf{C}) + \alpha\delta f(\mathbf{A}, \mathbf{D}) + \beta\gamma f(\mathbf{B}, \mathbf{C}) + \beta\delta f(\mathbf{B}, \mathbf{D}). \end{aligned}$$

Thus, we can operate with bilinear forms as with ordinary products in "multiplying out" expressions. Using again the decomposition

$$\mathbf{A} = (a_1, \ldots, a_n) = a_1\mathbf{E}_1 + \cdots + a_n\mathbf{E}_n$$
$$\mathbf{B} = (b_1, \ldots, b_n) = b_1\mathbf{E}_1 + \cdots + b_n\mathbf{E}_n$$

for the vectors **A**, **B**, we arrive at the formula

$$\begin{aligned} f(\mathbf{A}, \mathbf{B}) &= f(a_1\mathbf{E}_1 + a_2\mathbf{E}_2 + \cdots + a_n\mathbf{E}_n, \\ &\qquad\qquad b_1\mathbf{E}_1 + b_2\mathbf{E}_2 + \cdots + b_n\mathbf{E}_n) \\ &= \sum_{j,k=1}^{n} a_j b_k f(\mathbf{E}_j, \mathbf{E}_k) \end{aligned}$$

Hence, the most general bilinear form in **A**, **B** is given by

$$\text{(57a)} \qquad f(\mathbf{A}, \mathbf{B}) = \sum_{j,k=1}^{n} c_{jk} a_j b_k$$

with constant coefficients

$$\text{(57b)} \qquad c_{jk} = f(\mathbf{E}_j, \mathbf{E}_k).$$

For **B** = **A** the bilinear form f goes over into the *quadratic form*

$$\text{(57c)} \qquad f(\mathbf{A}, \mathbf{A}) = \sum_{j,k=1}^{n} c_{jk} a_j a_k.$$

In a similar way one defines *trilinear* forms $f(\mathbf{A}, \mathbf{B}, \mathbf{C})$ in three vectors **A**, **B**, **C** as functions that are linear forms in each vector separately. One finds, exactly as before, that the most general trilinear form is given by an expression

$$\text{(58a)} \qquad f(\mathbf{A}, \mathbf{B}, \mathbf{C}) = \sum_{j,k,r=1}^{n} c_{jkr} a_j b_k c_r,$$

where

(58b) $$c_{jkr} = f(\mathbf{E}_j, \mathbf{E}_k\, \mathbf{E}_r).$$

More general *multilinear* forms f in any number m of vectors can be defined in an obvious manner. It is only the matter of notation that injects a new element, since we can no longer associate different letters with different vectors. We denote the vectors by $\mathbf{A}_1, \mathbf{A}_2, \ldots,$ $\mathbf{A}_m$ and introduce their components a_{jk} by

$$\mathbf{A}_1 = (a_{11}, a_{21}, \ldots, a_{n1}), \qquad \mathbf{A}_2 = (a_{12}, a_{22}, \ldots, a_{n2}), \ldots,$$

$$\mathbf{A}_m = (a_{1m}, a_{2m}, \ldots, a_{nm}).$$

The function f is a multilinear form $f(\mathbf{A}_1, \ldots, \mathbf{A}_m)$ in $\mathbf{A}_1, \mathbf{A}_2, \ldots,$ $\mathbf{A}_m$ if it is a linear form in each vector when the others are held fixed. We can also consider f as function of the matrix

$$\mathbf{a} = (\mathbf{A}_1, \mathbf{A}_2, \ldots, \mathbf{A}_m) = (a_{jk})$$

that has $\mathbf{A}_1, \mathbf{A}_2, \ldots, \mathbf{A}_m$ as column vectors. In analogy to (58a) the most general multilinear form in $\mathbf{A}_1, \mathbf{A}_2, \ldots, \mathbf{A}_m$ is given by

(59a) $$f(\mathbf{A}_1, \mathbf{A}_2, \ldots, \mathbf{A}_m) = \sum_{\substack{j_1, j_2, \ldots, j_m \\ = 1, \ldots, n}} c_{j_1 j_2 \cdots j_m} a_{j_1 1} a_{j_2 2} \cdots a_{j_m m}$$

where[1]

(59b) $$c_{j_1 j_2 \cdots j_m} = f(\mathbf{E}_{j_1}, \mathbf{E}_{j_2}, \ldots, \mathbf{E}_{j_m}).$$

c. *Alternating Multilinear Forms. Definition of Determinants*

The determinants of second and third order defined in formulae (51a) and (52a) are special multilinear forms. The determinant of second order in (51a) p.161 is a bilinear form of the two 2-dimensional vectors

(60a) $$\mathbf{A}_1 = (a_{11}, a_{21}), \qquad \mathbf{A}_2 = (a_{12}, a_{22});$$

[1]The use of subscripts of subscripts in these formulae is somewhat cumbersome. Here $j_1, j_2, \ldots, j_m$ stands for any combination of m numbers selected from the set of numbers $1, 2, \ldots, n$. Such a combination could also be considered as a function $j(k)$ whose domain is the set of numbers $k = 1, 2, \ldots, m$ and whose range is in the set of numbers $j = 1, 2, \ldots, n$. Any one of these combinations or functions gives rise to a term in the sum in formula (59a).

the determinant of third order in (52a) is a trilinear function of the three 3-dimensional vectors

(60b) $$\mathbf{A}_1 = (a_{11}, a_{21}, a_{31}), \qquad \mathbf{A}_2 = (a_{12}, a_{22}, a_{32}),$$
$$\mathbf{A}_3 = (a_{13}, a_{23}, a_{33}).$$

(The linearity of determinants in each vector separately follows by inspection from the fact that each product in the explicit expansion contains exactly one factor with a given second subscript). The extra feature that sets the determinants apart from other multilinear forms, is their *alternating* character.

A function of several arguments (which could be vectors or scalars) is called *alternating* if it just changes in sign, when we interchange any two of the arguments. Examples of alternating functions of scalar arguments are

(61a) $$\phi(x, y) = y - x$$

(61b) $$\phi(x, y, z) = (z - y)(z - x)(y - x).$$

A function f of two n-dimensional vectors $\mathbf{A}_1$, $\mathbf{A}_2$ is alternating if

$$f(\mathbf{A}_1, \mathbf{A}_2) = -f(\mathbf{A}_2, \mathbf{A}_1)$$

for all $\mathbf{A}_1$, $\mathbf{A}_2$. This implies in particular for $\mathbf{A}_1 = \mathbf{A}_2 = \mathbf{A}$ that

$$f(\mathbf{A}, \mathbf{A}) = 0.$$

Let $n = 2$ and f be an alternating function of the vectors $\mathbf{A}_1$, $\mathbf{A}_2$ given by (60a), which is also a bilinear form. Then

$$f(\mathbf{E}_1, \mathbf{E}_1) = f(\mathbf{E}_2, \mathbf{E}_2) = 0, \qquad f(\mathbf{E}_2, \mathbf{E}_1) = -f(\mathbf{E}_1, \mathbf{E}_2).$$

It follows from (57a, b) that

(62a) $$f(\mathbf{A}_1, \mathbf{A}_2) = f(a_{11}\mathbf{E}_1 + a_{21}\mathbf{E}_2,\ a_{12}\mathbf{E}_1 + a_{22}\mathbf{E}_2)$$
$$= c(a_{11}a_{22} - a_{12}a_{21}) = c \begin{vmatrix} a_{11} & a_{12} \\ a_{21} & a_{22} \end{vmatrix} = c \det(\mathbf{A}_1, \mathbf{A}_2),$$

where the constant c has the value

(62b) $$c = f(\mathbf{E}_1, \mathbf{E}_2).$$

Thus, *every bilinear alternating form of two vectors* $\mathbf{A}_1$, $\mathbf{A}_2$ *in two-dimensional space differs from the determinant of the matrix with columns* $\mathbf{A}_1$, $\mathbf{A}_2$ *only by a constant factor c.*

More generally, an alternating bilinear form of two vectors in n dimensions can be written

$$f(\mathbf{A}_1, \mathbf{A}_2) = \sum_{j,k=1}^{n} c_{jk} a_{j1} a_{k2},$$

where

$$c_{jk} = -c_{kj}, \qquad c_{jj} = 0.$$

Combining the terms with subscripts differing only by a permutation, we can express f as a linear combination of second-order determinants:

$$\text{(62c)} \qquad f(\mathbf{A}_1, \mathbf{A}_2) = \sum_{\substack{j,k=1 \\ j<k}}^{n} c_{jk}(a_{j1}a_{k2} - a_{k1}a_{j2}) = \sum_{\substack{j,k=1 \\ j<k}}^{n} c_{jk} \begin{vmatrix} a_{j1} & a_{k1} \\ a_{j2} & a_{k2} \end{vmatrix}.$$

For an alternating function f of three vectors, we have the relations

$$\text{(63a)} \qquad f(\mathbf{A}, \mathbf{B}, \mathbf{C}) = -f(\mathbf{B}, \mathbf{A}, \mathbf{C}) = -f(\mathbf{A}, \mathbf{C}, \mathbf{B}) = -f(\mathbf{C}, \mathbf{B}, \mathbf{A}),$$

from which it follows that also

$$\text{(63b)} \qquad f(\mathbf{A}, \mathbf{B}, \mathbf{C}) = f(\mathbf{B}, \mathbf{C}, \mathbf{A}) = f(\mathbf{C}, \mathbf{A}, \mathbf{B}).$$

In particular, f vanishes whenever two of its arguments are equal. Let $\mathbf{A}_1$, $\mathbf{A}_2$, $\mathbf{A}_3$ be the three-dimensional vectors given by (60b). By (58a, b) the general alternating trilinear form f in $\mathbf{A}_1$, $\mathbf{A}_2$, $\mathbf{A}_3$ is

$$f(\mathbf{A}_1, \mathbf{A}_2, \mathbf{A}_3) = \sum_{j,k,r=1}^{3} c_{jkr} a_{j1} a_{k2} a_{r3}$$

Here, using (63a, b),

$$c_{jkr} = f(\mathbf{E}_j, \mathbf{E}_k, \mathbf{E}_r) = \varepsilon_{jkr} f(\mathbf{E}_1, \mathbf{E}_2, \mathbf{E}_3),$$

with $\varepsilon_{jkr} = 0$, if two of the numbers j, k, r are equal and

(64a) $\quad \varepsilon_{123} = \varepsilon_{231} = \varepsilon_{312} = 1, \qquad \varepsilon_{213} = \varepsilon_{132} = \varepsilon_{321} = -1.$

Using the fact that the function $\phi(x, y, z)$ in formula (61b) changes sign whenever two of its arguments are interchanged, we find for ε_{jkr} the concise expression

(64b)
$$\begin{aligned}\varepsilon_{jkr} &= \operatorname{sgn} \phi(j, k, r)\\ &= \operatorname{sgn} (r-k)(r-j)(k-j).\end{aligned}$$

Comparison with the expression (52a), p. 162 for a third-order determinant shows that

(64c)
$$f(\mathbf{A}_1, \mathbf{A}_2, \mathbf{A}_3) = c \begin{vmatrix} a_{11} & a_{12} & a_{13} \\ a_{21} & a_{22} & a_{23} \\ a_{31} & a_{32} & a_{33} \end{vmatrix},$$

where $c = f(\mathbf{E}_1, \mathbf{E}_2, \mathbf{E}_3)$ is a constant. We have the same result as in two dimensions: *The most general trilinear alternating form in three 3-dimensional vectors* $\mathbf{A}_1, \mathbf{A}_2, \mathbf{A}_3$ *differs from the determinant of the matrix with columns* $\mathbf{A}_1, \mathbf{A}_2, \mathbf{A}_3$, *only by a constant factor c.* Obviously, then, the third-order determinant of the matrix with columns $\mathbf{A}_1, \mathbf{A}_2, \mathbf{A}_3$ is that uniquely determined trilinear alternating form in the vectors $\mathbf{A}_1, \mathbf{A}_2, \mathbf{A}_3$ that has the value 1 when $\mathbf{A}_1, \mathbf{A}_2, \mathbf{A}_3$ are respectively equal to the coordinate vectors $\mathbf{E}_1, \mathbf{E}_2, \mathbf{E}_3$.[1]

It is clear now how we can define determinants of higher order. Let $\mathbf{a}$ be the matrix

(65a)
$$\mathbf{a} = \begin{pmatrix} a_{11} & a_{12} & \cdots & a_{1n} \\ a_{21} & a_{22} & \cdots & a_{2n} \\ \vdots & \vdots & & \vdots \\ a_{n1} & a_{n2} & \cdots & a_{nn} \end{pmatrix},$$

with column vectors $\mathbf{A}_1, \mathbf{A}_2, \ldots, \mathbf{A}_n$. Let f be a multilinear alternating form in $\mathbf{A}_1, \ldots, \mathbf{A}_n$. Then f is given by (59a). Here the coefficients $c_{j_1 j_2 \cdots j_n}$ have the form

(65b)
$$c_{j_1 j_2 \cdots j_n} = f(\mathbf{E}_{j_1}, \mathbf{E}_{j_2}, \ldots, \mathbf{E}_{j_n}).$$

They change sign, whenever we interchange any two of the numbers $j_1, j_2, \ldots, j_n$. Denote by $\phi(x_1, \ldots, x_n)$ the product

[1]The last condition expresses that the unit matrix $\mathbf{e}$ has the determinant 1.

(65c)
$$\begin{aligned}\phi(x_1, x_2, \ldots, x_n) &= (x_n - x_{n-1})\ (x_n - x_{n-2}) \cdots (x_n - x_2)\ (x_n - x_1)\\ &\qquad (x_{n-1} - x_{n-2}) \cdots (x_{n-1} - x_2)\,(x_{n-1} - x_1)\\ &\qquad \cdots\cdots\cdots\cdots\\ &\qquad (x_3 - x_2)\ (x_3 - x_1)\\ &\qquad (x_2 - x_1)\\ &= \prod_{\substack{j,k=1,\ldots,n\\ j<k}} (x_k - x_j).\end{aligned}$$

It is easily seen that ϕ is an alternating function of the scalars $x_1, \ldots, x_n$ that vanishes only when two of those scalars are equal. Then,

(65d)
$$\varepsilon_{j_1 j_2 \ldots j_n} = \operatorname{sgn} \phi(j_1, j_2, \ldots, j_n)$$

is an alternating function of $j_1, \ldots, j_n$, which only assumes the values $+1, 0, -1$. For $j_1, \ldots, j_n$ restricted to the values $1, 2, \ldots, n$, we have $\varepsilon_{j_1 j_2 \ldots j_n} = 0$, unless the numbers $j_1, \ldots, j_n$ are distinct, that is, unless they form a *permutation* of the numbers $1, 2, \ldots, n$. One calls $j_1, \ldots, j_n$ an *even permutation* of $1, 2, \ldots, n$ if $\varepsilon_{j_1 j_2 \ldots j_n} = +1$ and an *odd permutation* if $\varepsilon_{j_1 j_2 \ldots j_n} = -1$. An even permutation can be rearranged in the order $1, 2, \ldots, n$ by an even number of interchanges of two elements, an odd permutation by an odd number of such interchanges.

Obviously, by (65b),

(65e)
$$c_{j_1 j_2 \ldots j_n} = \varepsilon_{j_1 j_2 \ldots j_n} f(\mathbf{E}_1, \ldots, \mathbf{E}_n).$$

We define the determinant of the matrix $\mathbf{a}$ in (65a) as

(66a)
$$\det(\mathbf{a}) = \begin{vmatrix} a_{11} & a_{12} & \cdots & a_{1n}\\ a_{21} & a_{22} & \cdots & a_{2n}\\ \cdots & \cdots & \cdots & \cdots\\ a_{n1} & a_{n2} & & a_{nn}\end{vmatrix} = \sum_{j_1,\ldots,j_n=1}^{n} \varepsilon_{j_1 j_2 \ldots j_n} a_{j_1 1} a_{j_2 2} \ldots a_{j_n n}.$$

We have then the result: *The most general multilinear alternating form f in n n-dimensional vectors $\mathbf{A}_1, \ldots, \mathbf{A}_n$ differs from the determinant of the matrix with columns $\mathbf{A}_1, \ldots, \mathbf{A}_n$ only by the constant factor $c = f(\mathbf{E}_1, \ldots, \mathbf{E}_n)$.*

d. Principal Properties of Determinants

Formula (66a) gives the explicit expansion of an nth-order determinant in terms of its n^2 elements a_{jk}. Counting only the terms with nonvanishing coefficients $\varepsilon_{j_1 j_2 \ldots j_n}$, the determinant is an nth-degree form in the a_{jk} consisting of $n!$ terms. Each term (aside from the coefficient $\varepsilon_{j_1 j_2 \ldots j_n} = \pm 1$) is a product of n of the elements, one from each column and from each row. In principle, the expansion formula makes it possible to compute a determinant for any given values of the elements. In practice, the formula has too many terms to keep track of (120 in the case of fifth-order determinants; 3,628,800 in the case of tenth-order determinants) to be useful for numerical computations, and more efficient ways of evaluating determinants have been devised.

The basic properties of determinants already are incorporated in our definition as alternating multilinear forms of n vectors $\mathbf{A}_1, \mathbf{A}_2, \ldots, \mathbf{A}_n$ in n-dimensional space. If $\mathbf{a}$ is the matrix with these vectors as column vectors, we write

$$\det(\mathbf{a}) = \det(\mathbf{A}_1, \ldots, \mathbf{A}_n).$$

It follows immediately that *the determinant of the square matrix* $\mathbf{a}$ *changes sign if we interchange any two columns of* $\mathbf{a}$; *in particular, the determinant of a matrix* $\mathbf{a}$ *with two identical columns vanishes.* Using the linearity of the determinant in each of its column vectors separately, we find that *multiplying one column of the matrix* $\mathbf{a}$ *by a factor* λ *has the effect of multiplying the determinant of* $\mathbf{a}$ *by* λ.[1] For example,

$$\det(\lambda\mathbf{A}_1, \mathbf{A}_2, \ldots, \mathbf{A}_n) = \lambda\det(\mathbf{A}_1, \mathbf{A}_2, \ldots, \mathbf{A}_n). \tag{67a}$$

In particular, we find for $\lambda = 0$ and $\mathbf{A}_1$ arbitrary that

$$\det(\mathbf{0}, \mathbf{A}_2, \ldots, \mathbf{A}_n) = 0. \tag{67b}$$

The same considerations apply, of course, to any other column, and we find that *the determinant of a matrix* $\mathbf{a}$ *vanishes if any column of* $\mathbf{a}$ *is the zero vector.* From the multilinearity of determinants, we conclude more generally that

[1]Multiplying all elements of the nth order matrix $\mathbf{a}$ by the factor λ is equivalent to multiplying each of its n columns by λ and, hence, results in multiplying the determinant of $\mathbf{a}$ by λ^n. Thus, $\det(\lambda\mathbf{a}) = \lambda^n \det(\mathbf{a})$.

(67c)
$$\begin{aligned}&\det(\mathbf{A}_1 + \lambda\mathbf{A}_2, \mathbf{A}_2, \ldots, \mathbf{A}_n)\\ &\quad = \det(\mathbf{A}_1, \mathbf{A}_2, \ldots, \mathbf{A}_n) + \lambda \det(\mathbf{A}_2, \mathbf{A}_2, \ldots, \mathbf{A}_n)\\ &\quad = \det(\mathbf{A}_1, \mathbf{A}_2, \ldots, \mathbf{A}_n),\end{aligned}$$

since the matrix $(\mathbf{A}_2, \mathbf{A}_2, \ldots, \mathbf{A}_n)$ has two identical columns. Generally, *the value of the determinant of the matrix* **a** *does not change if we add a multiple of one column of* **a** *to a different column.*[1]

Of fundamental importance is the multiplication law for determinants:

The determinant of the product of two nth-order matrices **a** *and* **b** *is the product of their determinants:*

(68a)
$$\det(\mathbf{ab}) = \det(\mathbf{a}) \cdot \det(\mathbf{b}).$$

Written out by elements, the rule takes the form

(68b)
$$\begin{vmatrix} a_{11} & a_{12} & \cdots & a_{1n}\\ a_{21} & a_{22} & \cdots & a_{2n}\\ \vdots & \vdots & & \vdots\\ a_{n1} & a_{n2} & \cdots & a_{nn}\end{vmatrix} \times \begin{vmatrix} b_{11} & b_{12} & \cdots & b_{1n}\\ b_{21} & b_{22} & \cdots & b_{2n}\\ \vdots & \vdots & & \vdots\\ b_{n1} & b_{n2} & \cdots & b_{nn}\end{vmatrix} = \begin{vmatrix} c_{11} & c_{12} & \cdots & c_{1n}\\ c_{21} & c_{22} & \cdots & c_{2n}\\ \vdots & \vdots & & \vdots\\ c_{n1} & c_{n2} & \cdots & c_{nn}\end{vmatrix}$$

where

(68c)
$$c_{jk} = a_{j1}b_{1k} + a_{j2}b_{2k} + \cdots + a_{jn}b_{nk} = \sum_{r=1}^{n} a_{jr}b_{rk}.$$

This law is a simple consequence of our definition of determinants. Let $\mathbf{c} = \mathbf{ab}$ be the product matrix. We hold the matrix **a** fixed and consider the determinant of **c** in its dependence on **b**. By (68c) the kth-column vector of the matrix c

$$\mathbf{C}_k = (c_{1k}, c_{2k}, \ldots, c_{nk})$$

has elements c_{jk} which are linear forms in the kth-column vector $\mathbf{B}_k$

[1]Obviously multiplying a column by the factor λ and adding it to the *same* column changes the value of the determinant by the factor $1 + \lambda$.

of the matrix **b**. It follows that det (**c**) is a linear form in the vector $\mathbf{B}_k$ when the other columns of **b** are held fixed. It is also clear that interchanging two columns of **b** corresponds exactly to interchanging the corresponding columns of **c**. Hence, det(**c**) is an alternating multilinear form in the column vectors of the matrix **b**. Consequently (see p. 170),

$$\det(\mathbf{c}) = \gamma \det(\mathbf{b}),$$

where γ is the value of det (**c**) for the case where

$$\mathbf{B}_1 = \mathbf{E}_1, \mathbf{B}_2 = \mathbf{E}_2, \ldots, \mathbf{B}_n = \mathbf{E}_n$$

or where **b** is the unit matrix **e**. Now, if $\mathbf{b} = \mathbf{e}$, then obviously $\mathbf{c} = \mathbf{ab} = \mathbf{ae} = \mathbf{a}$, and consequently $\gamma = \det(\mathbf{a})$. This proves (68a).

On p. 157 we defined the transpose $\mathbf{a}^T$ of the matrix **a** as the matrix obtained from **a** by interchanging rows and columns. We have the surprising fact that a square matrix and its transpose have the same determinant:

$$\det(\mathbf{a}^T) = \det(\mathbf{a}) \tag{68d}$$

or

$$\begin{vmatrix} a_{11} & a_{21} & \cdots & a_{n1} \\ a_{12} & a_{22} & \cdots & a_{n2} \\ \cdot & \cdot & & \cdot \\ \cdot & \cdot & & \cdot \\ \cdot & \cdot & & \cdot \\ a_{1n} & a_{2n} & \cdots & a_{nn} \end{vmatrix} = \begin{vmatrix} a_{11} & a_{12} & \cdots & a_{1n} \\ a_{21} & a_{22} & \cdots & a_{2n} \\ \cdot & \cdot & & \cdot \\ \cdot & \cdot & & \cdot \\ \cdot & \cdot & & \cdot \\ a_{n1} & a_{n2} & \cdots & a_{nn} \end{vmatrix}. \tag{68e}$$

For $n = 2,3$ one easily verifies this identity from the explicit expressions (51a), (52a), pp. 161–2. We only indicate the proof for general n, which can be based on the expansion formula (66a) for det (**a**). In each term of the sum with nonvanishing coefficient, we can rearrange the factors according to the first subscripts, so that

$$a_{j_1 1} a_{j_2 2} \ldots a_{j_n n} = a_{1k_1} a_{2k_2} \ldots a_{nk_n},$$

where $k_1, k_2, \ldots, k_n$ form again a permutation of the numbers 1, 2, . . ., n.[1] One easily shows that

[1]Looking at $j_1, j_2, \ldots, j_n$ as a function mapping the set $1, 2, \ldots, n$ onto itself, we have in $k_1, k_2, \ldots, k_n$ just the inverse function; that is, the equation $j_r = s$ is equivalent to $k_s = r$.

$$\varepsilon_{j_1 j_2 \ldots j_n} = \varepsilon_{k_1 k_2 \ldots k_n}$$

(this is left as an exercise for the reader). Hence,

$$\det(\mathbf{a}) = \sum_{k_1, \ldots, k_n = 1}^{n} \varepsilon_{k_1 k_2 \ldots k_n} a_{1k_1} a_{2k_2} \ldots a_{nk_n} = \det(\mathbf{a}^T).$$

An immediate consequence of formula (68d) is that a determinant can be considered as an alternating multilinear function of its row vectors. In particular *a determinant changes sign if we interchange any two rows.*

The multiplication rule (68a) states that *the product of the determinants of two square matrices* **a**, **b** *is equal to the determinant of the matrix* **ab** *whose elements are the scalar products of the row vectors of* **a** *with the column vectors of* **b**. We use now that the determinant of a matrix **a** is equal to the determinant of its transpose $\mathbf{a}^T$, which is obtained by interchanging rows and columns of **a**. It follows then that

$$\det(\mathbf{a}) \cdot \det(\mathbf{b}) = \det(\mathbf{a}^T) \cdot \det(\mathbf{b}) = \det(\mathbf{a}^T\mathbf{b}).$$

Hence, *the product of the determinants of the matrices* **a** *and* **b** *is also equal to the determinant of the matrix* $\mathbf{a}^T\mathbf{b}$, *obtained by forming the scalar products of the columns of* **a** *with the columns of* **b**. If

$$\mathbf{a} = (\mathbf{A}_1, \ldots, \mathbf{A}_n) \qquad \text{and} \qquad \mathbf{b} = (\mathbf{B}_1, \ldots, \mathbf{B}_n),$$

we obtain the identity

$$\text{(68f)} \quad \det(\mathbf{A}_1, \ldots, \mathbf{A}_n) \cdot \det(\mathbf{B}_1, \ldots, \mathbf{B}_n)$$

$$= \begin{vmatrix} \mathbf{A}_1 \cdot \mathbf{B}_1 & \mathbf{A}_1 \cdot \mathbf{B}_2 & \ldots & \mathbf{A}_1 \cdot \mathbf{B}_n \\ \mathbf{A}_2 \cdot \mathbf{B}_1 & \mathbf{A}_2 \cdot \mathbf{B}_2 & \ldots & \mathbf{A}_2 \cdot \mathbf{B}_n \\ \vdots & \vdots & & \vdots \\ \mathbf{A}_n \cdot \mathbf{B}_1 & \mathbf{A}_n \cdot \mathbf{B}_2 & \ldots & \mathbf{A}_n \cdot \mathbf{B}_n \end{vmatrix}$$

A simple application of these rules to *orthogonal matrices* **a**, for which [see formula (49), p. 158] $\mathbf{a}^{-1} = \mathbf{a}^T$ or $\mathbf{a}^T\mathbf{a} = \mathbf{e}$, yields

$$\det(\mathbf{a}^T\mathbf{a}) = \det(\mathbf{a}^T) \cdot \det(\mathbf{a}) = [\det(\mathbf{a})]^2 = \det(\mathbf{e}) = 1.$$

Consequently, *the determinant of an orthogonal matrix can only have the values* $+1$ or -1. The geometric interpretation of this result will be given on p. 202.

e. Application of Determinants to Systems of Linear Equations

Determinants provide a convenient tool for deciding when n vectors $\mathbf{A}_1, \mathbf{A}_2, \ldots, \mathbf{A}_n$ in n-dimensional space are dependent or, equivalently, when the square matrix $\mathbf{a}$ with columns $\mathbf{A}_1, \ldots, \mathbf{A}_n$ is singular.

The necessary and sufficient condition for a square matrix to be singular is that its determinant vanishes.

Let indeed $\mathbf{a}$ be singular. Then the column vectors $\mathbf{A}_1, \mathbf{A}_2, \ldots, \mathbf{A}_n$ are dependent. Thus, one of the column vectors, say $\mathbf{A}_1$, is dependent on the others:

$$\mathbf{A}_1 = \lambda_2\mathbf{A}_2 + \lambda_3\mathbf{A}_3 + \cdots + \lambda_n\mathbf{A}_n.$$

It follows from the multilinearity of determinants that

$$\begin{aligned}\det(\mathbf{a}) &= \det(\lambda_2\mathbf{A}_2 + \lambda_3\mathbf{A}_3 \cdots + \lambda_n\mathbf{A}_n, \mathbf{A}_2, \mathbf{A}_3, \ldots, \mathbf{A}_n)\\ &= \lambda_2\det(\mathbf{A}_2, \mathbf{A}_2, \mathbf{A}_3, \ldots, \mathbf{A}_n) + \lambda_3 \det(\mathbf{A}_3, \mathbf{A}_2, \mathbf{A}_3, \mathbf{A}_n),\\ &\quad + \cdots + \lambda_n \det(\mathbf{A}_n, \mathbf{A}_2, \mathbf{A}_3, \ldots, \mathbf{A}_n)\\ &= 0,\end{aligned}$$

since each of the matrices has a repeated column.[1]

Conversely, if $\mathbf{a}$ is nonsingular, there exists (see p. 155) a reciprocal $\mathbf{b} = \mathbf{a}^{-1}$ of $\mathbf{a}$:

$$\mathbf{ab} = \mathbf{e},$$

where $\mathbf{e}$ is the unit matrix. By the multiplication rule for determinants, it follows that

$$\det(\mathbf{a}) \cdot \det(\mathbf{b}) = \det(\mathbf{e}) = 1$$

and, hence, that $\det(\mathbf{a}) \neq 0$. This proves that $\mathbf{a}$ is singular if and only if $\det(\mathbf{a}) = 0$.

We consider now the system of linear equations

[1]More generally, this argument shows that an alternating multilinear form in m vectors in n-dimensional space vanishes identically for $m > n$, since then the vectors are necessarily dependent.

$$
\text{(69a)} \qquad \begin{cases} a_{11}x_1 + a_{12}x_2 + \cdots + a_{1n}x_n = y_1 \\ a_{21}x_1 + a_{22}x_2 + \cdots + a_{2n}x_n = y_2 \\ \cdots\cdots\cdots\cdots\cdots\cdots\cdots\cdots \\ a_{n1}x_1 + a_{n2}x_2 + \cdots + a_{nn}x_n = y_n \end{cases}
$$

corresponding to the matrix **a**. Following the discussion on p. 150 we have to distinguish the two cases (1) det (**a**) $\neq 0$ and (2) det (**a**) $= 0$. In case (1) equations (69a) have a unique solution for every $y_1, \ldots, y_n$. In case (2) there does not always exist a solution, and it is never unique. We now have not only an explicit test to distinguish between the two cases with the help of determinants but also shall find the means to calculate the solution in case (1). Introducing the vector

$$
\mathbf{Y} = (y_1, y_2, \cdots, y_n),
$$

we can write the system (69a) in the form

$$
\text{(69b)} \qquad x_1\mathbf{A}_1 + x_2\mathbf{A}_2 + \cdots + x_n\mathbf{A}_n = \mathbf{Y},
$$

where the $\mathbf{A}_k$ are the column vectors of the matrix **a**. Then,

$$
\begin{aligned}
&\det(\mathbf{Y}, \mathbf{A}_2, \mathbf{A}_3, \ldots, \mathbf{A}_n) \\
&= \det(x_1\mathbf{A}_1 + x_2\mathbf{A}_2 + \cdots + x_n\mathbf{A}_n, \mathbf{A}_2, \mathbf{A}_3, \ldots, \mathbf{A}_n) \\
&= x_1 \det(\mathbf{A}_1, \mathbf{A}_2, \mathbf{A}_3, \ldots, \mathbf{A}_n) + x_2 \det(\mathbf{A}_2, \mathbf{A}_2, \mathbf{A}_3, \ldots, \mathbf{A}_n) \\
&\qquad\qquad + x_3 \det(\mathbf{A}_3, \mathbf{A}_2, \mathbf{A}_3, \ldots, \mathbf{A}_n) + \cdots \\
&\qquad\qquad + x_n \det(\mathbf{A}_n, \mathbf{A}_2\ \mathbf{A}_2, \ldots, \mathbf{A}_n) \\
&= x_1 \det(\mathbf{A}_1, \mathbf{A}_2, \ldots, \mathbf{A}_n)
\end{aligned}
$$

and similarly,

$$
\det(\mathbf{A}_1, \mathbf{Y}, \mathbf{A}_3, \ldots, \mathbf{A}_n) = x_2 \det(\mathbf{A}_1, \mathbf{A}_2, \ldots, \mathbf{A}_n)
$$

and so on. If the matrix **a** is nonsingular, we can divide by its determinant and obtain the solution $x_1, x_2, \ldots, x_n$ expressed by determinants:

$$
x_1 = \frac{\det(\mathbf{Y}, \mathbf{A}_2, \ldots, \mathbf{A}_n)}{\det(\mathbf{A}_1, \mathbf{A}_2, \ldots, \mathbf{A}_n)}, \qquad x_2 = \frac{\det(\mathbf{A}_1, \mathbf{Y}, \ldots, \mathbf{A}_n)}{\det(\mathbf{A}_1, \mathbf{A}_2, \ldots, \mathbf{A}_n)},
$$

$$
\ldots, x_n = \frac{\det(\mathbf{A}_1, \mathbf{A}_2, \ldots, \mathbf{Y})}{\det(\mathbf{A}_1, \mathbf{A}_2, \ldots, \mathbf{A}_n)}.
$$

This is *Cramer's rule* for the solution of n linear equations in n unknown quantities.

Exercises 2.3

1. Evaluate the following determinants:

(a) $\begin{vmatrix} 3 & 4 & 5 \\ 4 & 5 & 6 \\ 5 & 6 & 7 \end{vmatrix}$ (c) $\begin{vmatrix} 1 & 1 & 1 \\ 2 & 3 & 4 \\ 3 & -1 & 7 \end{vmatrix}$

(b) $\begin{vmatrix} 1 & 1 & 1 \\ 1 & 2 & 4 \\ 1 & 3 & 9 \end{vmatrix}$ (d) $\begin{vmatrix} 1 & x & x^3 \\ 1 & y & y^3 \\ 1 & z & z^3 \end{vmatrix}$

2. Find the relation that must exist between a, b, c in order that the system of equations

$$3x + 4y + 5z = a$$
$$4x + 5y + 6z = b$$
$$5x + 6y + 7z = c$$

may have a solution.

3. (a) Verify that the determinant of the unit matrix is 1.
(b) Show that if $\mathbf{a}$ is nonsingular, then $\det(\mathbf{a}^{-1}) = 1/\det(\mathbf{a})$.

4. Obtain the values of

(a) ε_{321}, (b) ε_{2143}, (c) ε_{4231}, (d) ε_{54321}

5. Show that the determinant

$$\begin{vmatrix} a & b & c \\ d & e & f \\ g & h & k \end{vmatrix}$$

can always be reduced to the form

$$\begin{vmatrix} \alpha & 0 & 0 \\ 0 & \beta & 0 \\ 0 & 0 & \gamma \end{vmatrix}$$

merely by repeated application of the following processes: (1) interchanging two rows or two columns, and (2) adding a multiple of one row (or column) to another row (or column).

6. A matrix is diagonal if $a_{ij} = 0$ whenever $i \neq j$. Show that the determinant of the $n \times n$ diagonal matrix (a_{ij}) is the product $a_{11} a_{22} \ldots a_{nn}$.

7. The matrix (a_{ij}) is upper-triangular if $a_{ij} = 0$ whenever $j < i$. Show that

$$\det(a_{ij}) = a_{11}a_{22} \cdots a_{nn}.$$

8. Evaluate

(a)
$$\begin{vmatrix} 1 & x & x^2 \\ 1 & y & y^2 \\ 1 & z & z^2 \end{vmatrix}$$

(b)
$$\begin{vmatrix} 1! & 2! & 3! \\ 2! & 3! & 4! \\ 3! & 4! & 5! \end{vmatrix}$$

(c)
$$\begin{vmatrix} 1! & 2! & 3! & 4! \\ 2! & 3! & 4! & 5! \\ 3! & 4! & 5! & 6! \\ 4! & 5! & 6! & 7! \end{vmatrix}$$

9. Solve the equations

$$\begin{aligned} 2x - 3y + 4z &= 4 \\ 4x - 9y + 16z &= 10 \\ 8x - 27y + 64z &= 34. \end{aligned}$$

10. Prove the identity

$$(a^2 + b^2)(c^2 + d^2) = (ac + bd)^2 + (bc - ad)^2$$

by forming the product of the determinants

$$\begin{vmatrix} a & b \\ -b & a \end{vmatrix} \text{ and } \begin{vmatrix} c & d \\ -d & c \end{vmatrix}$$

11. If $A = x^2 + y^2 + z^2$, $B = xy + yz + zx$, show that

$$D = \begin{vmatrix} B & A & B \\ B & B & A \\ A & B & B \end{vmatrix} = (x^3 + y^3 + z^3 - 3xyz)^2.$$

12. Show that

$$\Delta = \begin{vmatrix} t_1 + x & a + x & a + x & a + x \\ b + x & t_2 + x & a + x & a + x \\ b + x & b + x & t_3 + x & a + x \\ b + x & b + x & b + x & t_4 + x \end{vmatrix}$$

is of the form $A + Bx$, where A and B are independent of x. By giving particular values to x, prove that

$$A = \frac{af(b) - bf(a)}{a - b}, \qquad B = \frac{f(b) - f(a)}{b - a},$$

where

$$f(t) = (t_1 - t)(t_2 - t)(t_3 - t)(t_4 - t).$$

13. Prove that any bilinear form f in A and $\mathbf{B}$ may be written

$$\mathbf{A} \cdot (\mathbf{cB}) = (\mathbf{c}^T\mathbf{A}) \cdot \mathbf{B}$$

14. Prove that in a nonsingular affine transformation the image of a quadric

$$ax^2 + by^2 + cz^2 + dxy + exz + fyz + gx + hy + iz + j = 0$$

is another quadric.

15. If the three determinants

$$\begin{vmatrix} a_1 & a_2 \\ b_1 & b_2 \end{vmatrix}, \quad \begin{vmatrix} a_1 & a_2 \\ c_1 & c_2 \end{vmatrix}, \quad \begin{vmatrix} b_1 & b_2 \\ c_1 & c_2 \end{vmatrix}$$

do not all vanish, then the necessary and sufficient condition for the existence of a solution of the three equations

$$a_1x + a_2y = d$$

$$b_1x + b_2y = e$$

$$c_1x + c_2y = f$$

is

$$D = \begin{vmatrix} a_1 & a_2 & d \\ b_1 & b_2 & e \\ c_1 & c_2 & f \end{vmatrix} = 0.$$

16. State the condition that the two straight lines $x = a_1t + b_1$, $y = a_2t + b_2$, $z = a_3t + b_3$ and $x = c_1t + d_1$, $y = c_2t + d_2$, $z = c_3t + d_3$ either intersect or are parallel.

17. Prove (68d) by verifying that it does not matter whether the factors in each term of the expansion (66a) are ordered by their first or second subscripts, namely, with

$$a_{j_1 1}\, a_{j_2 2} \cdots a_{j_n n} = a_{1k_1}\, a_{2k_2} \cdots a_{nk_n},$$

that

$$\varepsilon_{j_1 j_2 \cdots j_n} = \varepsilon_{k_1 k_2 \cdots k_n}.$$

18. Prove that the affine transformation

$$x' = ax + by + cz$$

$$y' = dx + ey + fz$$

$$z' = gx + hy + kz$$

leaves at least one direction unaltered.

2.4 Geometrical Interpretation of Determinants

a. Vector Products and Volumes of Parallelepipeds in Three-Dimensional Space

In Volume I (p. 388) we defined the "cross product" of two vectors $\mathbf{A} = (a_1, a_2)$ and $\mathbf{B} = (b_1, b_2)$ in the plane as the scalar

$$\mathbf{A} \times \mathbf{B} = a_1 b_2 - a_2 b_1. \tag{70a}$$

Here $|\mathbf{A} \times \mathbf{B}|$ represents twice the area of the triangle with vertices P_0, P_1, P_2, where $\mathbf{A} = \overrightarrow{P_0P_1}$, $\mathbf{B} = \overrightarrow{P_0P_2}$. We call $|\mathbf{A} \times \mathbf{B}|$ the area of the parallelogram *spanned* by the vectors $\mathbf{A}$, $\mathbf{B}$, that is, of the parallelogram with successive vertices P_0, P_1, Q, P_2. The sign of $\mathbf{A} \times \mathbf{B}$ determines the orientation of the parallelogram.[1] In determinant notation the cross product takes the form

$$\mathbf{A} \times \mathbf{B} = \begin{vmatrix} a_1 & b_1 \\ a_2 & b_2 \end{vmatrix} = \det(\mathbf{A}, \mathbf{B}). \tag{70b}$$

Thus, $|\det(\mathbf{A}, \mathbf{B})|$ can be interpreted geometrically as the area of the parallelogram spanned by the vectors $\mathbf{A}, \mathbf{B}$. Analogous interpretations will be found for higher-order determinants.

For three vectors $\mathbf{A} = (a_1, a_2, a_3)$, $\mathbf{B} = (b_1, b_2, b_3)$, $\mathbf{C} = (c_1, c_2, c_3)$ in three-dimensional space, it is natural to form the determinant

$$\det(\mathbf{A}, \mathbf{B}, \mathbf{C}) = \begin{vmatrix} a_1 & b_1 & c_1 \\ a_2 & b_2 & c_2 \\ a_3 & b_3 & c_3 \end{vmatrix}$$

Written out as a linear form in the vector $\mathbf{C}$ we have, by (52a),

$$\begin{aligned} \det(\mathbf{A}, \mathbf{B}, \mathbf{C}) &= (a_2 b_3 - a_3 b_2) c_1 + (a_3 b_1 - a_1 b_3) c_2 + (a_1 b_2 - a_2 b_1) c_3 \\ &= \mathbf{Z} \cdot \mathbf{C}, \end{aligned} \tag{71a}$$

where $\mathbf{Z} = (z_1, z_2, z_3)$ is the vector with components

$$z_1 = a_2 b_3 - a_3 b_2 = \begin{vmatrix} a_2 & b_2 \\ a_3 & b_3 \end{vmatrix}, \tag{71b}$$

[1] We have $\mathbf{A} \times \mathbf{B} > 0$ if the sense (counterclockwise or clockwise) in which the vertices follow each other is the same as that for the "coordinate square" with successive vertices (0, 0), (1, 0), (1, 1,), (0, 1).

$$z_2 = a_3b_1 - a_1b_3 = \begin{vmatrix} a_3 & b_3 \\ a_1 & b_1 \end{vmatrix},$$

$$z_3 = a_1b_2 - a_2b_1 = \begin{vmatrix} a_1 & b_1 \\ a_2 & b_2 \end{vmatrix}.$$

We call the vector Z the "vector product," or "cross product," of the vectors **A**, **B** and write $Z = \mathbf{A} \times \mathbf{B}$.[1] Then, by definition,

(71c) $$\det(\mathbf{A},\mathbf{B},\mathbf{C}) = (\mathbf{A} \times \mathbf{B}) \cdot \mathbf{C}.$$

Because of this formula the scalar det (**A**, **B**, **C**) is sometimes referred to as the *triple vector product* of **A**, **B**, **C**.

The components z_i of the vector $\mathbf{Z} = \mathbf{A} \times \mathbf{B}$ are themselves second-order determinants and, hence, are bilinear alternating forms of the vectors **A**, **B**. This leads immediately to the laws for vector multiplication:

(72a) $$(\lambda\mathbf{A}) \times \mathbf{B} = \mathbf{A} \times (\lambda\mathbf{B}) = \lambda(\mathbf{A} \times \mathbf{B});$$

(72b) $$(\mathbf{A}' + \mathbf{A}'') \times \mathbf{B} = \mathbf{A}' \times \mathbf{B} + \mathbf{A}'' \times \mathbf{B};$$

$$\mathbf{A} \times (\mathbf{B}' + \mathbf{B}'') = \mathbf{A} \times \mathbf{B}' + \mathbf{A} \times \mathbf{B}''$$

(72c) $$\mathbf{A} \times \mathbf{B} = -\mathbf{B} \times \mathbf{A}$$

Relation (72c) could be called the "anticommutative" law of multiplication. It has the important consequence that

(72d) $$\mathbf{A} \times \mathbf{A} = 0 \quad \text{for all vectors } \mathbf{A}.$$

More generally, *the vector product of two vectors* **A**, **B** *vanishes if and only if* **A** *and* **B** *are dependent.* For by (71c) the relation $\mathbf{A} \times \mathbf{B} = 0$ is equivalent to

$$\det(\mathbf{A}, \mathbf{B}, \mathbf{C}) = 0 \quad \text{for all vectors } \mathbf{C},$$

or to the fact (see p. 175) that **A**, **B**, **C** are dependent for all **C**. Now we can always find a vector **C** that is independent of **A** and **B** (see p. 139) Then the dependence of **A**, **B**, **C** implies that **A** and **B** are dependent.

[1]The vector product of two vectors in three-dimensions is again a *vector*, in contrast to cross products of vectors in two dimensions and scalar products in any number of dimensions, which are *scalars*.

The vector product $\mathbf{A} \times \mathbf{B}$ is perpendicular to both of the vectors $\mathbf{A}$ and $\mathbf{B}$, since by (71c),

$$\text{(72e)} \quad (\mathbf{A} \times \mathbf{B}) \cdot \mathbf{A} = \det(\mathbf{A}, \mathbf{B}, \mathbf{A}) = 0, \; (\mathbf{A} \times \mathbf{B}) \cdot \mathbf{B} = \det(\mathbf{A}, \mathbf{B}, \mathbf{B}) = 0.$$

Hence, for $\mathbf{A} = \overrightarrow{P_0P_1}$ and $\mathbf{B} = \overrightarrow{P_0P_2}$ independent, the direction of $\mathbf{A} \times \mathbf{B}$ is one of the two directions perpendicular to any plane $P_0P_1P_2$ spanned by $\mathbf{A}$ and $\mathbf{B}$. The length of the vector $\mathbf{A} \times \mathbf{B}$ also has a simple geometric interpretation. We have, by (71b),

$$\begin{aligned} \text{(72f)} \qquad |\mathbf{A} \times \mathbf{B}|^2 &= (a_2b_3 - {}_3b_2)^2 + (a_3b_1 - a_1b_3)^2 + (a_1b_2 - a_2b_1)^2 \\ &= (a_1^2 + a_2^2 + a_3^2)(b_1^2 + b_2^2 + b_3^2) \\ &\quad - (a_1b_1 + a_2b_2 + a_3b_3)^2 \\ &= |\mathbf{A}|^2|\mathbf{B}|^2 - (\mathbf{A} \cdot \mathbf{B})^2. \end{aligned}$$ [1]

Using the fact [formula (14), p. 131] that

$$\mathbf{A} \cdot \mathbf{B} = |\mathbf{A}|\,|\mathbf{B}| \cos \gamma,$$

where γ is the angle between the directions of $\mathbf{A}$ and $\mathbf{B}$, we find from (72f) that

$$|\mathbf{A} \times \mathbf{B}| = \sqrt{|\mathbf{A}|^2|\mathbf{B}|^2 - |\mathbf{A}|^2|\mathbf{B}|^2 \cos^2 \gamma} = |\mathbf{A}|\mathbf{B}| \sin \gamma$$

For $\mathbf{A} = \overrightarrow{P_0P_1}$, $\mathbf{B} = \overrightarrow{P_0P_2}$ we have in $|\mathbf{B}| \sin \gamma$ (where γ is assigned a value between 0 and π) the distance of the point P_2 from the line P_0P_1 (Fig. 2.6). Hence (exactly as in two dimensions), the quantity $|\mathbf{A} \times \mathbf{B}|$ gives the area of the parallelogram with vertices P_0, P_1, Q, P_2 "spanned" by the vectors $\mathbf{A}$, $\mathbf{B}$ or twice the area of the triangle with vertices P_0, P_1, P_2.

The individual components of the product $\mathbf{A} \times \mathbf{B} = (z_1, z_2, z_3)$ also can be interpreted geometrically. For example, the expression

$$z_3 = a_1b_2 - a_2b_1$$

is just the cross product of the two-dimensional vectors (a_1, a_2) and

[1]This identity incidentally yields an immediate proof of the Cauchy-Schwarz inequality

$$|\mathbf{A} \cdot \mathbf{B}| \leqq |\mathbf{A}|\;|\mathbf{B}|$$

(see p. 132). It also supplies the additional piece of information that the equality sign holds if and only if the vectors $\mathbf{A}$ and $\mathbf{B}$ are dependent.

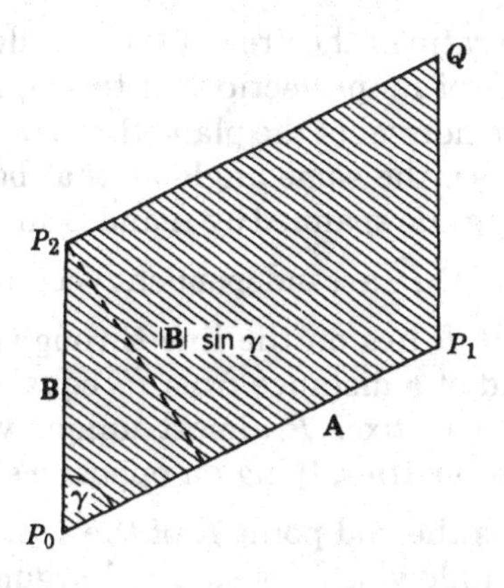

Figure 2.6 Area $|\mathbf{A} \times \mathbf{B}|$ of parallelogram spanned by two vectors **A**, **B**.

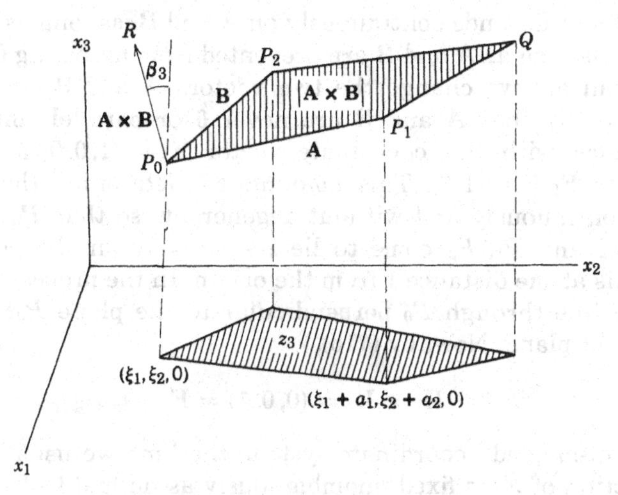

Figure 2.7 Components of vector product $\mathbf{A} \times \mathbf{B} = (z_1, z_2, z_3)$ interpreted as projected areas.

(b_1, b_2) [see (70a)]. If P_0 has the coordinates ξ_1, ξ_2, ξ_3, we have in $|z_3|$ the area of the parallelogram in the x_1, x_2-plane with vertices (ξ_1, ξ_2), $(\xi_1 + a_1, \xi_2 + a_2)$, $(\xi_1 + a_1 + b_1, \xi_2 + a_2 + b_2)$, $(\xi_1 + b_1, \xi_2 + b_2)$. This parallelogram is just the projection onto the x_1, x_2-plane of the parallelogram with vertices P_0, P_1, Q, P_2, spanned in space by the vectors **A**, **B** (see Fig. 2.7). If $\mathbf{A} \times \mathbf{B}$ has the direction cosines $\cos \beta_1$, $\cos \beta_2$, $\cos \beta_3$, we have [see (9), p. 129]

$$|z_3| = |\mathbf{A} \times \mathbf{B}|\,|\cos \beta_3|$$

Thus $|\cos \beta_3|$ gives the ratio of the area of the parallelogram spanned by **A** and **B** to the area of its projection on the x_1, x_2-plane. Here β_3 is the angle between the normal of the plane through P_0, P_1, P_2 and the x_3-axis. This is, of course, the same angle as that between the plane containing the parallelogram spanned by **A** and **B** and the x_1, x_2-plane.[1]

If $\mathbf{A} = \overrightarrow{P_0P_1}$ and $\mathbf{B} = \overrightarrow{P_0P_2}$ are independent vectors, we have $\mathbf{A} \times \mathbf{B} = \overrightarrow{P_0R}$, where the point R lies on the line through P_0 perpendicular to the plane $P_0P_1P_2$ and at a distance from P_0 equal to twice the area of the triangle $P_0P_1P_2$. This fixes R almost uniquely. There are only two points with these properties, lying on opposite sides of the plane. Which of these points is the end point R of the vector $\mathbf{A} \times \mathbf{B} = \overrightarrow{P_0R}$ can be decided by the following "continuity" argument. The vector product $\mathbf{A} \times \mathbf{B}$ depends continuously on the vectors A, B since its components are bilinear functions of those of **A**, **B**. Then the direction of $\mathbf{A} \times \mathbf{B}$ also depends continuously on **A** and **B**, as long as $\mathbf{A} \times \mathbf{B} \neq \mathbf{0}$, that is, as long as **A** and **B** are prevented from becoming **0** or parallel. We can always change the two vectors **A** and **B** continuously in such a way that **A** and **B** are never **0** or parallel until finally **A** coincides with the coordinate vector $\mathbf{E}_1 = (1, 0, 0)$ and **B** with the vector $\mathbf{E}_2 = (0, 1, 0)$. This amounts to deforming the triangle $P_0P_1P_2$ continuously and without degeneracy, so that P_0 goes into the origin and P_1, P_2 come to lie respectively on the positive x_1- and x_2-axis at the distance 1 from the origin. In the process, the point R on the line through P_0 perpendicular to the plane $P_0P_1P_2$ never crosses that plane. Now, by (71b),

$$\mathbf{E}_1 \times \mathbf{E}_2 = (0, 0, 1) = \mathbf{E}_3$$

In a "right-handed" coordinate system, the kind we usually employ, the direction of E_3 is fixed unambiguously as normal to $\mathbf{E}_1$ and $\mathbf{E}_2$ in such a way that the 90° rotation about the x_3-axis that takes $\mathbf{E}_1$ into $\mathbf{E}_2$ appears *counterclockwise* from the point $(0, 0, 1)$. Then, generally, if our coordinate system is right-handed, the direction of $\mathbf{A} \times \mathbf{B} = \overrightarrow{P_0R}$ is such that the rotation about the line $\overrightarrow{P_0R}$ of the vector $\mathbf{A} = \overrightarrow{P_0P_1}$ into the vector $\mathbf{B} = \overrightarrow{P_0P_2}$ by an angle γ between 0 and π appears counterclockwise when viewed from R (see Fig. 2.8). Similarly, in a left-handed coordinate system the 90° rotation from $\mathbf{E}_1$ into $\mathbf{E}_2$ appears

[1]In general, the area of the projection of a plane figure onto a second plane equals the product of the area of the original figure with the cosine of the angle between the two planes, as will become clear when we discuss transformations of integrals.

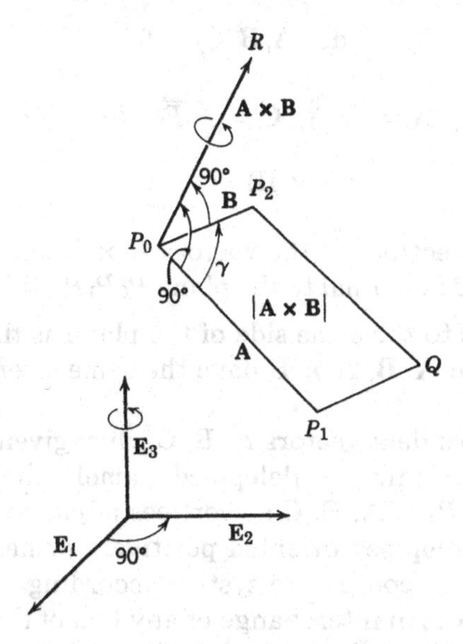

Figure 2.8 Vector product $\mathbf{A} \times \mathbf{B}$ in right-handed coordinate system.

clockwise from (0, 0, 1), and so also does then the rotation from **A** into **B** appear from the end point R of $\mathbf{A} \times \mathbf{B} = \overrightarrow{P_0R}$.

Generally, an ordered triple of three independent vectors **A**, **B**, **C** defines a certain *sense* or *orientation*. If $\mathbf{A} = \overrightarrow{P_0P_1}$, $\mathbf{B} = \overrightarrow{P_0P_2}$, and $\mathbf{C} = \overrightarrow{P_0P_3}$, we can rotate the direction of **A** into that of **B** by an angle between 0 and π in the plane $P_0P_1P_2$. The sense of the triple **A**, **B**, **C** by definition is the sense (counterclockwise or clockwise) that rotation appears to have, when viewed from that side of the plane to which **C** points.[1] The triple **B**, **A**, **C** has the *opposite* orientation. *The orientation of the triple* **A**, **B**, $\mathbf{A} \times \mathbf{B}$ *is always the same as that of the coordinate vectors* $\mathbf{E}_1$, $\mathbf{E}_2$, $\mathbf{E}_3$.

We call the triple **A**, **B**, **C** oriented positively with respect to the x_1, x_2, x_3-coordinate system if it has the same orientation as the triple of vectors $\mathbf{E}_1$, $\mathbf{E}_2$, $\mathbf{E}_3$, and oriented negatively if it has the opposite orientation. *For the triple* **A**, **B**, **C** *to be oriented positively with respect to the* x_1, x_2x_3,*-coordinates it is necessary and sufficient that*

[1]The same type of orientation determines the difference between left-handed and right-handed screws. The motion of a screw consists of a combination of translatory motion along an axis and rotation about that axis. The distinction between the two types of screws is defined by the sense of the rotation, clockwise or counterclockwise, when viewed from that direction of the axis in which the translation proceeds.

$$\det(\mathbf{A}, \mathbf{B}, \mathbf{C}) > 0 \tag{73}$$

For let $\mathbf{A} = \overrightarrow{P_0P_1}$, $\mathbf{B} = \overrightarrow{P_0P_1}$, $\mathbf{C} = \overrightarrow{P_0P_3}$. Relation (73) means that

$$(\mathbf{A} \times \mathbf{B}) \cdot \mathbf{C} > 0,$$

that is, that the directions of the vectors $\mathbf{A} \times \mathbf{B}$ and $\mathbf{C}$ form an acute angle. Since $\mathbf{A} \times \mathbf{B}$ is normal to the plane $P_0P_1P_2$, this implies that the vector $\overrightarrow{P_0P_3}$ points to the same side of the plane as the vector $\mathbf{A} \times \mathbf{B}$. Hence, $\mathbf{A}$, $\mathbf{B}$, $\mathbf{C}$ and $\mathbf{A}$, $\mathbf{B}$, $\mathbf{A} \times \mathbf{B}$ have the same orientation, which is that of $\mathbf{E}_1$, $\mathbf{E}_2$, $\mathbf{E}_3$.

The three independent vectors $\mathbf{A}$, $\mathbf{B}$, $\mathbf{C}$ when given the same initial point $\mathbf{P}_0$ "span" a certain parallelepiped, namely, the one that has the end points P_1, P_2, P_3 of $\mathbf{A}$, $\mathbf{B}$, $\mathbf{C}$ as vertices adjacent to the vertex P_0. We call the parallelepiped oriented positively or negatively with respect to the x_1, x_2, x_3-coordinate system according to the orientation of the triple $\mathbf{A}$, $\mathbf{B}$, $\mathbf{C}$. An interchange of any two of the vectors $\mathbf{A}$, $\mathbf{B}$, $\mathbf{C}$ reverses the orientation for the parallelepiped spanned by the vectors.[1]

Let θ be the angle formed by the direction of the vectors $\mathbf{C}$ and $\mathbf{A} \times \mathbf{B}$. By (71c),

$$\det(\mathbf{A}, \mathbf{B}, \mathbf{C}) = |\mathbf{A} \times \mathbf{B}|\,|\mathbf{C}| \quad \cos\theta \tag{74a}$$

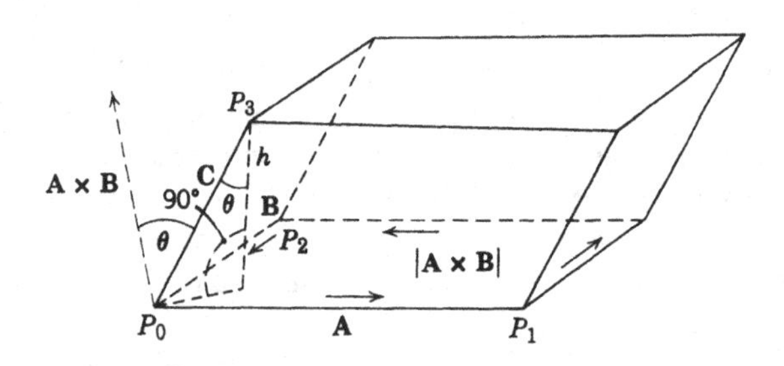

Figure 2.9 Volume $V = |\mathbf{A} \times \mathbf{B}|h$ of parallelepiped.

[1]The orientation of the parallelepiped can be visualized as an orientation ascribed to each face of the parallelepiped (i.e., as a sense assigned to the boundary polygon of the face) such that a common edge of two neighboring faces is assigned opposite senses in the orientation of the two faces. The orientation of all faces is determined uniquely if for a single face the sense of one edge is prescribed. For the orientation of the parallelepiped spanned by $\mathbf{A}$, $\mathbf{B}$, $\mathbf{C}$, the sense of the edge P_0P_1 in the face spanned by the vectors $\overrightarrow{P_0P_2}$ and $\overrightarrow{P_0P_1}$ is that of proceding from P_0 to P_1 (see Fig. 2.9).

Since $\mathbf{A} \times \mathbf{B}$ is perpendicular to the plane $P_0P_1P_2$, the angle between the line P_0P_3 and the plane $P_0P_1P_2$ is $\frac{1}{2}\pi - \theta$. Thus,

$$h = |\mathbf{C}|\,|\cos\theta| = |\mathbf{C}|\left|\sin\left(\frac{\pi}{2} - \theta\right)\right| \tag{74b}$$

is the distance of the point P_3 from the plane $P_0P_1P_2$, that is the *altitude* of the parallelepiped from P_3. Since the volume V of the parallelepiped is equal to the area $|\mathbf{A} \times \mathbf{B}|$ of one face multiplied with the corresponding altitude h, it follows from (74a, b) that

$$V = |\mathbf{A} \times \mathbf{B}|h = |\det(\mathbf{A}, \mathbf{B}, \mathbf{C})|. \tag{74c}$$

In words, *the volume of a parallelepiped spanned by three vectors* **A**, **B**, **C** *is the absolute value of the determinant of the matrix with columns* **A**, **B**, **C**. Thus, the value of det(**A**, **B**, **C**) determines both the volume and the orientation of the parallelepiped spanned by **A**, **B**, **C**. We express this fact by the formula

$$\det(\mathbf{A}, \mathbf{B}, \mathbf{C}) = \varepsilon V, \tag{74}$$

where V is the volume of the parallelepiped spanned by the vectors **A**, **B**, **C** and $\varepsilon = +1$ if the parallelepiped is oriented positively with respect to x_1, x_2x_3,-coordinates and $\varepsilon = -1$ if oriented negatively.

b. Expansion of a Determinant with Respect to a Column. Vector Products in Higher Dimensions

Only in three dimensions can we define a product $\mathbf{A} \times \mathbf{B}$ of two vectors **A**, **B** that again is a vector.[1] The closest analogue in n-dimensions would be a "vector product" of $n - 1$ vectors. Taking n vectors,

$$\mathbf{A}_1 = (a_{11}, \ldots, a_{n1}), \ldots, \mathbf{A}_n = (a_{1n}, \ldots, a_{nn})$$

in n-dimensional space, we can form the determinant of the matrix $(\mathbf{A}_1, \ldots, \mathbf{A}_n)$ with those vectors as columns. The determinant of this matrix is a linear form in the last vector $\mathbf{A}_n$ and can be written as a scalar product

$$\det(\mathbf{A}_1, \ldots, \mathbf{A}_n) = z_1a_1 + z_2a_2 + \cdots + z_na_n = \mathbf{Z} \cdot \mathbf{A}_n, \tag{75}$$

[1] In higher dimensions we cannot associate with two vectors **A**, **B** a third vector **C** outside the plane spanned by **A**, **B** in a *geometric* fashion, that is, by a construction that determines **C** uniquely and does not change under rigid motions.

where the vector $\mathbf{Z} = (z_1, \ldots, z_n)$ depends only on the $n - 1$ vectors $\mathbf{A}_1, \mathbf{A}_2, \ldots, \mathbf{A}_{n-1}$. Obviously, $\mathbf{Z}$ is linear in each of the vectors $\mathbf{A}_1, \ldots, \mathbf{A}_{n-1}$ separately and is alternating. We can call $\mathbf{Z}$ the *vector product* of $\mathbf{A}_1, \ldots, \mathbf{A}_{n-1}$ and denote it by

$$\mathbf{Z} = \mathbf{A}_1 \times \mathbf{A}_2 \times \cdots \times \mathbf{A}_{n-1}. \tag{76}$$

It is clear from (75) that

$$\mathbf{Z} \cdot \mathbf{A}_1 = \mathbf{Z} \cdot \mathbf{A}_2 = \ldots = \mathbf{Z} \cdot \mathbf{A}_{n-1} = 0;$$

we see that *the vector product of $n - 1$ vectors is orthogonal to each of the vectors,* as in three dimensions. The length of the vector product $\mathbf{Z}$ also can be interpreted geometrically as volume of the oriented $(n - 1)$-dimensional parallelepiped spanned by the vectors $\mathbf{A}_1, \ldots, \mathbf{A}_{n-1}$, as we shall see later.

Just as in three dimensions, the components of $\mathbf{Z}$ can be written as determinants in analogy to formulae (71b). We first derive such a determinant expression for the component z_n of $\mathbf{Z}$. By (75),

$$z_n = \mathbf{Z} \cdot \mathbf{E}_n = \det(\mathbf{A}_1, \ldots, \mathbf{A}_{n-1}, \mathbf{E}_n),$$

where

$$\mathbf{E}_n = (0,0, \ldots, 0,1)$$

is the n-th coordinate vector. Taking $\mathbf{A}_n = \mathbf{E}_n$ in the general expansion formula (66a) p.170 for determinants amounts to replacing the last factor $a_{j_n n}$ in each term by 1 for $j_n = n$ and by 0 for $j_n \neq n$. For $j_n = n$ the coefficient $\varepsilon_{j_1 \ldots j_{n-1} j_n}$ vanishes, unless $j_1, \ldots, j_{n-1}$ constitute a permutation of the numbers $1, 2, \ldots, n - 1$. In that case, the coefficient (65c, d) reduces to

$$\begin{aligned} \varepsilon_{j_1 \ldots j_{n-1} j_n} = \varepsilon_{j_1 \ldots j_{n-1} n} &= \operatorname{sgn} \phi(j_1, \ldots, j_{-1}, n) \\ &= \operatorname{sgn} (n - j_{n-1}) \cdots (n - j_1) \phi(j_1, \ldots, j_{n-1}) \\ &= \operatorname{sgn} \phi(j_1, \ldots, j_{n-1}) = \varepsilon_{j_1 \ldots j_{n-1}} \end{aligned}$$

It follows from (66a) that

$$z_n = \sum_{j_1, \ldots, j_{n-1} = 1}^{n-1} \varepsilon_{j_1 \cdots j_{n-1}} a_{j_1 1} a_{j_2 2} \cdots a_{j_{n-1} n-1} \tag{77a}$$

$$= \begin{vmatrix} a_{11} & a_{12} & \ldots & a_{1\,n-1} \\ a_{21} & a_{22} & \ldots & a_{2\,n-1} \\ \vdots & \vdots & & \vdots \\ a_{n-1\,1} & a_{n-1\,2} & \ldots & a_{n-1\,n-1} \end{vmatrix}.$$

We see that z_n is equal to the determinant of the matrix obtained from the matrix $(\mathbf{A}_1, \ldots, \mathbf{A}_n)$ by omitting the last row and column. Generally, one defines a *minor* of a matrix $\mathbf{a}$ as the determinant of a square matrix obtained from $\mathbf{a}$ by omitting some of the rows and columns, while preserving the relative positions of the remaining elements. The minor *complementary to an element* a_{jk} of a square matrix $\mathbf{a}$ is the one obtained by omitting from $\mathbf{a}$ the row and column containing the element a_{jk}. *Thus* z_n *is equal to the minor complementary to* a_{nn}.

The other components of the vector $\mathbf{Z}$ have similar representations. We have, for example, by (75),

$$z_{n-1} = \det(\mathbf{A}_1, \ldots, \mathbf{A}_{n-1}, \mathbf{E}_{n-1}).$$

To evaluate this determinant, we interchange the last two rows (see p. 174) which changes the sign of the determinant. The last column $\mathbf{E}_{n-1}$ then goes over into $\mathbf{E}_n$, and we find from our previous result that $-z_{n-1}$ is equal to the determinant obtained by omitting the last row and column of the new matrix or, equivalently, is equal to the minor complementary to the element $a_{n-1\,n}$ in the original matrix. Similarly, one finds that $\pm z_i$ for each $i = 1, \ldots, n$ is equal to the minor complementary to the element a_{in}, where the positive sign applies for $n - i$ even, the negative one for $n - i$ odd.

Formula (75) thus constitutes an expansion of an nth-order determinant in terms of $(n - 1)$-order determinants, the minors complementary to the elements in the last column. For example, for $n = 4$ we have the formula

(77b)
$$\begin{vmatrix} a_{11} & a_{12} & a_{13} & a_{14} \\ a_{21} & a_{22} & a_{23} & a_{24} \\ a_{31} & a_{32} & a_{33} & a_{34} \\ a_{41} & a_{42} & a_{43} & a_{44} \end{vmatrix}$$

$$= -a_{14} \begin{vmatrix} a_{21} & a_{22} & a_{23} \\ a_{31} & a_{32} & a_{33} \\ a_{41} & a_{42} & a_{43} \end{vmatrix} + a_{24} \begin{vmatrix} a_{11} & a_{12} & a_{13} \\ a_{31} & a_{32} & a_{33} \\ a_{41} & a_{42} & a_{43} \end{vmatrix}$$

$$-a_{34}\begin{vmatrix} a_{11} & a_{12} & a_{13} \\ a_{21} & a_{22} & a_{23} \\ a_{41} & a_{42} & a_{43} \end{vmatrix} + a_{44}\begin{vmatrix} a_{11} & a_{12} & a_{13} \\ a_{21} & a_{22} & a_{23} \\ a_{31} & a_{32} & a_{33} \end{vmatrix}$$

Interchanging columns, we can derive similar formulae for expanding a determinant in terms of the minors complementary to the elements of any given column. Expansions of this type play a role in many proofs that involve induction over the dimension of the space, as we shall see in the next sections.

c. *Areas of Parallelograms and Volumes of Parallelepipeds in Higher Dimensions*

Surfaces in space can be built up from infinitesimal parallelograms. Thus, formulae for areas of curved surfaces and for integrals over surfaces require knowledge of an expression for the area of a parallelogram in space. Similarly, formulae for volumes or volume integrals over curved manifolds have to be based on expressions for volumes of parallelepipeds in higher dimensions. Such expressions are easily derived in greatest generality with the help of determinants.

The basic quantity associated with vectors is the scalar product of two vectors

$$\mathbf{A} = (a_1, \ldots, a_n) \qquad \text{and} \qquad \mathbf{B} = (b_1, \ldots, b_n),$$

which in any Cartesian coordinate system is given by

$$\mathbf{A} \cdot \mathbf{B} = a_1 b_1 + \cdots a_n b_n.$$

While the individual components a_j and b_k of $\mathbf{A}$ and $\mathbf{B}$ depend on the special Cartesian coordinate system used, the scalar product has an independent *geometric* meaning:

$$\mathbf{A} \cdot \mathbf{B} = |\mathbf{A}|\,|\mathbf{B}| \cos \gamma,$$

where $|\mathbf{A}|$, $|\mathbf{B}|$ are the lengths of the vectors $\mathbf{A}$ and $\mathbf{B}$, and γ the angle between them. If follows that any quantity that can be expressed in terms of scalar products has an *invariant* geometric meaning and does not depend on the special Cartesian coordinate system used.

The simplest quantity expressible in terms of scalar products is the distance of two points P_0, P_1 which is the length of the vector $\mathbf{A} = \overrightarrow{P_0P_1}$. The square of that distance is given by

(78a) $$|\mathbf{A}|^2 = \mathbf{A} \cdot \mathbf{A}.$$

With two vectors **A**, **B** in n-dimensional space, we can associate *the area of a parallelogram spanned by the two vectors* if we give them a common initial point P_0. Let $\mathbf{A} = \overrightarrow{P_0P_1}$ and $\mathbf{B} = \overrightarrow{P_0P_2}$. The vectors then span a parallelogram P_0, P_1, Q, P_2 that has P_1 and P_2 as vertices adjacent to the vertex P_0. By elementary geometry the area α of the parallelogram is equal to the product of adjacent sides multiplied by the sine of the included angle γ:

$$\begin{aligned}\alpha &= |A|\,|B| \sin\gamma \\ &= \sqrt{|A|^2|B|^2 - |A|^2|B|^2\cos^2\gamma} \\ &= \sqrt{|A|^2|B|^2-(A\cdot B)^2}\end{aligned}$$

as we found already on p. 182 for the special case $n = 3$. We can write this formula for the area α more elegantly in the form of a determinant for the square of α:

(78b) $$\alpha^2 = (\mathbf{A}\cdot\mathbf{A})(\mathbf{B}\cdot\mathbf{B}) - (\mathbf{A}\cdot\mathbf{B})(\mathbf{B}\cdot\mathbf{A}) = \begin{vmatrix} \mathbf{A}\cdot\mathbf{A} & \mathbf{A}\cdot\mathbf{B} \\ \mathbf{B}\cdot\mathbf{A} & \mathbf{B}\cdot\mathbf{B}\end{vmatrix}$$

The determinant that appears here on the right-hand side is called the *Gram determinant* of the vectors **A**, **B** and denoted by $\Gamma(\mathbf{A}, \mathbf{B})$. It is clear from the derivation that

$$\Gamma(\mathbf{A}, \mathbf{B}) \geqq 0$$

for all vectors **A**, **B** and that equality holds only if **A** and **B** are dependent.[1]

We can derive a similar expression for the square of *the volume V of a parallelepiped spanned by three vectors* **A**, **B**, **C** in n-dimensional space. We represent the vectors in the form

$$\mathbf{A} = \overrightarrow{P_0P_1}, \qquad \mathbf{B} = \overrightarrow{P_0P_2}, \qquad \mathbf{C} = \overrightarrow{P_0P_3}$$

and consider the parallelepiped that has P_1, P_2, P_3 as vertices adjacent to the vertex P_0. Its volume V can be defined as the product of the area α of one of its faces multiplied by the corresponding altitude h. Choosing for α the area of the parallelogram spanned

[1] That is, if either one of the vectors vanishes ($|\mathbf{A}|$ or $|\mathbf{B}| = 0$) or if they are parallel ($\sin\gamma = 0$).

by the vectors **A** and **B**, we have to take for h the distance of the point P_3 from the plane through P_0, P_1, P_2. Thus,

$$V^2 = h^2\alpha^2 = h^2\Gamma(A,B) = h^2\begin{vmatrix} A \cdot A & A \cdot B \\ B \cdot A & B \cdot B \end{vmatrix}.$$

We interpret h to stand for the "perpendicular" distance of P_3 from the plane $P_0\ P_1\ P_2$, that is, the length of that vector $\mathbf{D} = \overrightarrow{PP_3}$ which is perpendicular to the plane and has its initial point P in the plane. For a point P in the plane $P_0P_1P_2$ the vector $\overrightarrow{P_0P}$ must be dependent on $\mathbf{A} = \overrightarrow{P_0P_1}$ and $\mathbf{B} = \overrightarrow{P_0P_2}$ (see p. 144):

$$\overrightarrow{P_0P} = \lambda\mathbf{A} + \mu\mathbf{B}.$$

Hence, the vector **D** has the form

$$\mathbf{D} = \overrightarrow{PP_3} = \overrightarrow{P_0P_3} - \overrightarrow{P_0P} = \mathbf{C} - \lambda\mathbf{A} - \mu\mathbf{B}$$

with suitable constants λ, μ. If **D** is to be perpendicular to the plane spanned by **A** and **B**, we must have

$$\text{(79a)} \qquad \mathbf{A} \cdot \mathbf{D} = 0, \qquad \mathbf{B} \cdot \mathbf{D} = 0.$$

This leads to a system of linear equations for determining λ and μ:

$$\text{(79b)} \quad \mathbf{A} \cdot \mathbf{C} = \lambda\mathbf{A} \cdot \mathbf{A} + \mu\mathbf{A} \cdot \mathbf{B}, \qquad \mathbf{B} \cdot \mathbf{C} = \lambda\mathbf{B} \cdot \mathbf{A} + \mu\mathbf{B} \cdot \mathbf{B}.$$

The determinant of these equations is just the Gram determinant $\Gamma(\mathbf{A}, \mathbf{B})$. Assuming **A** and **B** to be independent vectors, we have $\Gamma(\mathbf{A}, \mathbf{B}) \neq 0$. There exists, then, a uniquely determined solution λ, μ of equations (79) and, hence, a unique vector $\mathbf{D} = \overrightarrow{PP_3}$ perpendicular to the plane $P_0P_1P_2$ and with initial point in that plane. The length of that vector is equal to the distance h, so that by (79a)

$$\begin{aligned} h^2 &= |\mathbf{D}|^2 = \mathbf{D} \cdot \mathbf{D} = (\mathbf{C} - \lambda\mathbf{A} - \mu\mathbf{B}) \cdot \mathbf{D} \\ &= \mathbf{C} \cdot \mathbf{D} - \lambda\mathbf{A} \cdot \mathbf{D} - \mu\mathbf{B} \cdot \mathbf{D} \\ &= \mathbf{C} \cdot \mathbf{D} = \mathbf{C} \cdot \mathbf{C} - \lambda\mathbf{C} \cdot \mathbf{A} - \mu\mathbf{C} \cdot \mathbf{B}. \end{aligned}$$

This results in the expression

$$\text{(79c)} \qquad V^2 = (\mathbf{C} \cdot \mathbf{C} - \lambda\mathbf{A} \cdot \mathbf{C} - \mu\mathbf{B} \cdot \mathbf{C})\ \Gamma(\mathbf{A},\mathbf{B}).$$

This expression for the square of the volume of the parallelepiped spanned by **A**, **B**, **C** can be written more elegantly as the Gram determinant formed from the vectors **A**, **B**, **C**:

$$V^2 = \begin{vmatrix} \mathbf{A}\cdot\mathbf{A} & \mathbf{A}\cdot\mathbf{B} & \mathbf{A}\cdot\mathbf{C} \\ \mathbf{B}\cdot\mathbf{A} & \mathbf{B}\cdot\mathbf{B} & \mathbf{B}\cdot\mathbf{C} \\ \mathbf{C}\cdot\mathbf{A} & \mathbf{C}\cdot\mathbf{B} & \mathbf{C}\cdot\mathbf{C} \end{vmatrix} = \Gamma(\mathbf{A},\mathbf{B},\mathbf{C}). \tag{79d}$$

To show the identity of the expressions (79c) and (79d) for V^2, we make use of the fact that the value of the determinant $\Gamma(\mathbf{A}, \mathbf{B}, \mathbf{C})$ does not change if we subtract from the last column λ-times the first column and μ-times the second column:

$$\Gamma(\mathbf{A},\mathbf{B},\mathbf{C}) = \begin{vmatrix} \mathbf{A}\cdot\mathbf{A} & \mathbf{A}\cdot\mathbf{B} & \mathbf{A}\cdot\mathbf{C} - \lambda\mathbf{A}\cdot\mathbf{A} - \mu\mathbf{A}\cdot\mathbf{B} \\ \mathbf{B}\cdot\mathbf{A} & \mathbf{B}\cdot\mathbf{B} & \mathbf{B}\cdot\mathbf{C} - \lambda\mathbf{B}\cdot\mathbf{A} - \mu\mathbf{B}\cdot\mathbf{B} \\ \mathbf{C}\cdot\mathbf{A} & \mathbf{C}\cdot\mathbf{B} & \mathbf{C}\cdot\mathbf{C} - \lambda\mathbf{C}\cdot\mathbf{A} - \mu\mathbf{C}\cdot\mathbf{B} \end{vmatrix}.$$

It follows from (79b) that

$$\Gamma(\mathbf{A},\mathbf{B},\mathbf{C}) = \begin{vmatrix} \mathbf{A}\cdot\mathbf{A} & \mathbf{A}\cdot\mathbf{B} & 0 \\ \mathbf{B}\cdot\mathbf{A} & \mathbf{B}\cdot\mathbf{B} & 0 \\ \mathbf{C}\cdot\mathbf{A} & \mathbf{C}\cdot\mathbf{B} & \mathbf{C}\cdot\mathbf{C} - \lambda\mathbf{C}\cdot\mathbf{A} - \mu\mathbf{C}\cdot\mathbf{B} \end{vmatrix}.$$

Expanding this determinant in terms of the last column leads back immediately to the expression (79c).

Formula (79d) shows that the volume V of the parallelepiped spanned by the vectors **A**, **B**, **C** does not depend on the choice of the face and of the corresponding altitude used in the computation, for the value of $\Gamma(\mathbf{A}, \mathbf{B}, \mathbf{C})$ does not change when we permute **A**, **B**, **C**. For example, $\Gamma(\mathbf{B}, \mathbf{A}, \mathbf{C})$ can be obtained by interchanging in the determinant for $\Gamma(\mathbf{A}, \mathbf{B}, \mathbf{C})$ the first two rows and then the first two columns.

Formula (79c) can be written as

$$\Gamma(\mathbf{A},\mathbf{B},\mathbf{C}) = |\mathbf{D}|^2\Gamma(\mathbf{A},\mathbf{B}).$$

It follows that

$$\Gamma(\mathbf{A},\mathbf{B},\mathbf{C}) \geqq 0$$

for any vectors **A**, **B**, **C**. Here the equal sign can only hold if either $\Gamma(\mathbf{A}, \mathbf{B}) = 0$ or $\mathbf{D} = 0$. The relation $\Gamma(\mathbf{A}, \mathbf{B}) = 0$ would imply that **A** and **B** are dependent. If $\mathbf{D} = 0$, we would have $\mathbf{C} = \lambda\mathbf{A} + \mu\mathbf{B}$, so that **C** would depend on **A** and **B**. *Hence the Gram determinant* $\Gamma(\mathbf{A}, \mathbf{B}, \mathbf{C})$ *vanishes if and only if the vectors* **A**, **B**, **C** *are dependent.*

For $n = 3$ formula (79d) follows immediately from the formula (74c) for the volume V of an oriented parallelepiped spanned by three vectors **A**, **B**, **C** in three-dimensional space. This is a consequence of identity (68f) p. 174 according to which

$$\det(\mathbf{A}, \mathbf{B}, \mathbf{C}) \det(\mathbf{A}, \mathbf{B}, \mathbf{C}) = \Gamma(\mathbf{A}, \mathbf{B}, \mathbf{C}).$$

The expression for V^2 as a Gram determinant has the advantage of showing that V is independent of the special cartesian coordinate system used, and hence that V has a geometrical meaning.

We can proceed to "volumes" V of four-dimensional parallelepipeds spanned by four vectors $\mathbf{A} = \overrightarrow{P_0P_1}$, $\mathbf{B} = \overrightarrow{P_0P_2}$, $\mathbf{C} = \overrightarrow{P_0P_3}$, $\mathbf{D} = \overrightarrow{P_0P_4}$ in n-dimensional space ($n \geqq 4$). Defining V as the product of the volume of the three-dimensional parallelepiped spanned by the three vectors **A**, **B**, **C** with the distance of the point P_4 from the three-dimensional "plane" through the points P_0, P_1, P_2, P_3, we arrive by the exactly same steps as before at an expression for V^2 as a Gram determinant:

$$\text{(80a)} \qquad V^2 = \begin{vmatrix} \mathbf{A}\cdot\mathbf{A} & \mathbf{A}\cdot\mathbf{B} & \mathbf{A}\cdot\mathbf{C} & \mathbf{A}\cdot\mathbf{D} \\ \mathbf{B}\cdot\mathbf{A} & \mathbf{B}\cdot\mathbf{B} & \mathbf{B}\cdot\mathbf{C} & \mathbf{B}\cdot\mathbf{D} \\ \mathbf{C}\cdot\mathbf{A} & \mathbf{C}\cdot\mathbf{B} & \mathbf{C}\cdot\mathbf{C} & \mathbf{C}\cdot\mathbf{D} \\ \mathbf{D}\cdot\mathbf{A} & \mathbf{D}\cdot\mathbf{B} & \mathbf{D}\cdot\mathbf{C} & \mathbf{D}\cdot\mathbf{D} \end{vmatrix} = \Gamma(\mathbf{A}, \mathbf{B}, \mathbf{C}, \mathbf{D})$$

If here $n = 4$, the Gram determinant becomes the square of the determinant of the matrix with columns **A**, **B**, **C**, **D**, and we find that

$$\text{(80b)} \qquad V = |\det(\mathbf{A}, \mathbf{B}, \mathbf{C}, \mathbf{D})|.$$

More generally, m vectors $\mathbf{A}_1, \ldots, \mathbf{A}_m$ in n-dimensional space, to which we assign a common initial point P_0, span an m-dimensional parallelepiped. The square of the volume V of that parallelepiped is given by the Gram determinant

$$\text{(81a)} \quad V^2 = \begin{vmatrix} \mathbf{A}_1\cdot\mathbf{A}_1 & \mathbf{A}_1\cdot\mathbf{A}_2 & \cdots & \mathbf{A}_1\cdot\mathbf{A}_m \\ \mathbf{A}_2\cdot\mathbf{A}_1 & \mathbf{A}_2\cdot\mathbf{A}_2 & \cdots & \mathbf{A}_2\cdot\mathbf{A}_m \\ \vdots & \vdots & & \vdots \\ \mathbf{A}_m\cdot\mathbf{A}_1 & \mathbf{A}_m\cdot\mathbf{A}_2 & \cdots & \mathbf{A}_m\cdot\mathbf{A}_m \end{vmatrix} = \Gamma(\mathbf{A}_1, \ldots, \mathbf{A}_m)$$

For $m = n$ we obtain for the volume of the parallelepiped spanned by n vectors in n-space the formula

$$V = |\det(\mathbf{A}_1, \ldots, \mathbf{A}_n)|. \tag{81b}$$

One proves by induction over m that

$$\Gamma(\mathbf{A}_1, \ldots, \mathbf{A}_m) \geqq 0,$$

where equality holds if and only if the vectors $\mathbf{A}_1, \ldots, \mathbf{A}_m$ are dependent.[1]

d. Orientation of Parallelepipeds in n-Dimensional Space

Later on, in Chapter 5, when we need a consistent method to fix the sign of multiple integrals, we have to make use of signed volumes and orientations of parallelepipeds in n-dimensional space.

For the volume spanned by n vectors $\mathbf{A}_1, \ldots, \mathbf{A}_n$ in n-dimensional space we have by (81b) the expression

$$V = |\det \mathbf{A}_1, \ldots, \mathbf{A}_n)|.$$

We call $\det(\mathbf{A}_1, \ldots, \mathbf{A}_n)$ the volume in $(x_1 \cdots x_n)$-coordinates of the *oriented* parallelepiped spanned by $\mathbf{A}_1, \ldots, \mathbf{A}_n$. The parallelepiped or the set of vectors $\mathbf{A}_1, \ldots, \mathbf{A}_n$ is called positively oriented with respect to the coordinate system if $\det(\mathbf{A}_1, \ldots, \mathbf{A}_n)$ is positive, negatively if the determinant is negative. Thus,

$$\det(\mathbf{A}_1, \ldots, \mathbf{A}_n) = \varepsilon V, \tag{81c}$$

where V is the volume of the parallelepiped spanned by the vectors $\mathbf{A}_1, \ldots, \mathbf{A}_n$ and $\varepsilon = +1$ or -1 according to whether the parallelepiped is oriented positively or negatively with respect to the coordinate system.

While the square of $\det(\mathbf{A}_1, \ldots, \mathbf{A}_n)$ has a geometrical meaning independent of the Cartesian coordinate system, this is not the case for the sign of the determinant. Interchanging, for example, the x_1- and x_2-axes results in the interchange of the first two rows of the determinant and, hence, in a change of sign in $\det(\mathbf{A}_1, \ldots, \mathbf{A}_n)$. What has an independent geometric meaning, however, is the state-

[1] In the case of dependent vectors $\mathbf{A}_1, \ldots, \mathbf{A}_m$ with common initial point P_0 the parallelepiped spanned by these vectors "collapses" into a linear manifold of $m-1$ dimensions or less and has m-dimensional volume equal to 0.

ment that *two* n-dimensional parallelepipeds in n-dimensional space have the *same* or have the *opposite* orientation.

Consider two ordered sets of vectors $\mathbf{A}_1, \ldots, \mathbf{A}_n$ and $\mathbf{B}_1, \ldots, \mathbf{B}_n$ in n-dimensional space, where we assume that each set consists of independent vectors. Obviously, the two sets have the same orientation—that is, are both oriented positively or both negatively with respect to the $x_1 \cdots x_n$-system—if and only if the condition

$$\det(\mathbf{A}_1, \ldots, \mathbf{A}_n) \cdot \det(\mathbf{B}_1, \ldots, \mathbf{B}_n) > 0 \tag{82a}$$

is satisfied. Using the identity (68f), we can write this condition in the form

$$[\mathbf{A}_1, \ldots, \mathbf{A}_n;\ \mathbf{B}_1, \ldots, \mathbf{B}_n] > 0, \tag{82b}$$

where the symbol on the left denotes the function of $2n$ vectors defined by

$$[\mathbf{A}_1, \ldots, \mathbf{A}_n;\ \mathbf{B}_1, \ldots, \mathbf{B}_n] = \begin{vmatrix} \mathbf{A}_1 \cdot \mathbf{B}_1 & \mathbf{A}_1 \cdot \mathbf{B}_2 & \cdots & \mathbf{A}_1 \cdot \mathbf{B}_n \\ \mathbf{A}_2 \cdot \mathbf{B}_1 & \mathbf{A}_2 \cdot \mathbf{B}_2 & \cdots & \mathbf{A}_2 \cdot \mathbf{B}_n \\ \vdots & \vdots & & \vdots \\ \mathbf{A}_n \cdot \mathbf{B}_1 & \mathbf{A}_n \cdot \mathbf{B}_2 & \cdots & \mathbf{A}_n \cdot \mathbf{B}_n \end{vmatrix} \tag{82c}$$

Notice that for $\mathbf{B}_1 = \mathbf{A}_1, \ldots, \mathbf{B}_n = \mathbf{A}_n$ the symbol $[\mathbf{A}_1, \ldots, \mathbf{A}_n; \mathbf{B}_1, \ldots, \mathbf{B}_n]$ reduces to the Gram determinant $\Gamma(\mathbf{A}_1, \ldots, \mathbf{A}_n)$. Formulae (82b, c) make it evident that having the same orientation is a geometric property that does not depend on the specific Cartesian coordinate system used. We denote this property symbolically by

$$\Omega(\mathbf{A}_1, \ldots, \mathbf{A}_n) = \Omega(\mathbf{B}_1, \ldots, \mathbf{B}_n) \tag{82d}$$

and the property of having the opposite orientation[1] by

[1]The individual orientation Ω of an n-tuple of vectors does not stand for a "number." Formula (82f) only associates a value ± 1 with the ratio of two *orientations*, while formulae (82d, e) express equality or inequality of orientations. It is, of course, possible to describe the two different possible orientations of n-tuples completely by numerical values, say, giving the value $\Omega = +1$ to one orientation, the value $\Omega = -1$ to the other. This involves, however, the arbitrary selection of a "standard orientation" we call $+1$—for example, that given by the coordinate vectors—whereas the relations (82d, e, f) are meaningful independent of any numerical value assigned to Ω. Analogous situations are common throughout mathematics. For

(82e) $$\Omega(\mathbf{A}_1, \ldots, \mathbf{A}_n) = -\Omega(\mathbf{B}_1, \ldots, \mathbf{B}_n).$$

Then, generally, for two sets of n independent vectors in n-dimensional space,

(82f) $$\Omega(\mathbf{B}_1, \ldots, \mathbf{B}_n) = \operatorname{sgn}[\mathbf{A}_1, \ldots, \mathbf{A}_n; \mathbf{B}_1, \ldots, \mathbf{B}_n]\Omega(\mathbf{A}_1, \ldots, \mathbf{A}_n).$$

The set $\mathbf{A}_1, \ldots, \mathbf{A}_n$ is oriented positively or negatively with respect to $x_1 \cdots x_n$-coordinates according to whether

(83a) $$\Omega(\mathbf{A}_1, \ldots, \mathbf{A}_n) = \Omega(\mathbf{E}_1, \ldots, \mathbf{E}_n)$$

or

(83b) $$\Omega(\mathbf{A}_1, \ldots, \mathbf{A}_n) = -\Omega(\mathbf{E}_1, \ldots, \mathbf{E}_n),$$

where $\mathbf{E}_1, \ldots, \mathbf{E}_n$ are the coordinate vectors. On occasion, we shall denote the orientation $\Omega(\mathbf{E}_1, \ldots, \mathbf{E}_n)$ of the coordinate system by

$$\Omega(x_1, x_2, \ldots, x_n).$$

For two sets of n vectors in n-dimensional space $\mathbf{A}_1, \ldots, \mathbf{A}_n$ and $\mathbf{A}_1', \ldots, \mathbf{A}_n'$ we have by (82c), (81b)

(84a) $$[\mathbf{A}_1, \ldots, \mathbf{A}_n; \mathbf{A}_1', \ldots, \mathbf{A}_n'] = \varepsilon\varepsilon' VV'$$

Here V and V' are, respectively, the volumes of the parallelepipeds spanned by the two sets of vectors; the factors ε, ε' depend on their orientations and those of the coordinate vectors:

(84b) $$\varepsilon = \operatorname{sgn}[\mathbf{A}_1, \ldots, \mathbf{A}_n; \mathbf{E}_1, \ldots, \mathbf{E}_n]$$

(84c) $$\varepsilon' = \operatorname{sgn}[\mathbf{A}_1', \ldots, \mathbf{A}_n'; \mathbf{E}_1, \ldots, \mathbf{E}_n].$$

The product

(84d) $$\varepsilon\varepsilon' = \operatorname{sgn}[\mathbf{A}_1, \ldots, \mathbf{A}_n; \mathbf{A}_1', \ldots, \mathbf{A}_n']$$

---

example, in euclidean geometry, equality of distances and even the ratio of distances have a meaning even when no numerical values are assigned to the distances (as in Euclid's *Elements*). It is true that we can describe distances by real numbers, such that the ratio of distances is just that of the corresponding real numbers. This requires the arbitrary selection of a "standard distance" (e.g., a meter), to which all other distances are referred, and thus introduces in some sense a "nongeometrical" element.

is independent of the choice of the coordinate system and has the value $+1$ if the parallelepipeds have the same orientation but -1 if the opposite orientation.

Using the definition in terms of scalar products, we can form the expression

$$[\mathbf{A}_1, \ldots, \mathbf{A}_m; \mathbf{A}_1', \ldots, \mathbf{A}_m'] \tag{85a}$$

$$= \begin{vmatrix} \mathbf{A}_1 \cdot \mathbf{A}_1' & \mathbf{A}_1 \cdot \mathbf{A}_2' & \cdots \mathbf{A}_1 \cdot \mathbf{A}_m' \\ \mathbf{A}_2 \cdot \mathbf{A}_1' & \mathbf{A}_2 \cdot \mathbf{A}_2' & \cdots \mathbf{A}_2 \cdot \mathbf{A}_m' \\ \vdots & \vdots & \vdots \\ \mathbf{A}_m \cdot \mathbf{A}_1' & \mathbf{A}_m \cdot \mathbf{A}_2' & \cdots \mathbf{A}_m \cdot \mathbf{A}_m' \end{vmatrix}$$

for any $2m$ vectors $\mathbf{A}_1, \ldots, \mathbf{A}_m'$ in n-dimensional space. It is clear from the definition that this expression is a multilinear form in the $2m$ vectors. For example, the vector $\mathbf{A}_1'$ occurs only in the first column and the elements of that column are linear forms in $\mathbf{A}_1'$. Since the whole determinant is a linear form in the elements of the first column, it follows that it is a linear form in $\mathbf{A}_1'$. It also is evident from (85a) that the expression is an alternating function of the vectors $\mathbf{A}_1', \ldots, \mathbf{A}_m'$ for fixed $\mathbf{A}_1, \ldots, \mathbf{A}_m$ and an alternating function of $\mathbf{A}_1. \ldots, \mathbf{A}_m$ for fixed $\mathbf{A}_1', \ldots, \mathbf{A}_m'$. It follows (see the footnote on p. 000) that

$$[\mathbf{A}_1, \ldots, \mathbf{A}_m; \mathbf{A}_1', \ldots, \mathbf{A}_m'] = 0 \tag{85b}$$

whenever the m vectors $\mathbf{A}_1, \ldots, \mathbf{A}_m$ or the m vectors $\mathbf{A}_1', \ldots, \mathbf{A}_m'$ are dependent. In particular (85b) always holds when $m > n$.

Assume then that $m \leqq n$ and that the vectors $\mathbf{A}_1, \ldots, \mathbf{A}_m$ and the vectors $\mathbf{A}_1', \ldots, \mathbf{A}_m'$ are independent. We can assume that all these vectors are given the same initial point, say the origin O of n-dimensional space. Then $\mathbf{A}_1, \ldots, \mathbf{A}_m$ span an m-dimensional linear manifold π through O and $\mathbf{A}_1', \ldots, \mathbf{A}_m'$ another such plane π'. Introduce an orthonormal system of vectors $\mathbf{E}_1, \ldots, \mathbf{E}_m$ as coordinate vectors in π and another orthonormal system of vectors $\mathbf{E}_1', \ldots, \mathbf{E}_m'$ in π'.[1] For fixed $\mathbf{A}_1, \ldots, \mathbf{A}_m$ the function (85b) is an alternating multilinear form in the vectors $\mathbf{A}_1', \ldots, \mathbf{A}_m'$ and, hence (see p. 149), is given by

[1]These two systems of coordinate vectors in π and π' do not have to be related to each other in any way nor to the coordinate system to which the whole n-dimensional space containing π and π' is referred.

$$[\mathbf{A}_1, \ldots, \mathbf{A}_m; \mathbf{A}_1', \ldots, \mathbf{A}_m']$$
$$= [\mathbf{A}_1, \ldots, \mathbf{A}_m; \mathbf{E}_1', \ldots, \mathbf{E}_m'] \det(\mathbf{A}_1', \ldots, \mathbf{A}_m'),$$

where $\det(\mathbf{A}_1', \ldots, \mathbf{A}_m')$ is the determinant of the matrix formed by the components of the vectors $\mathbf{A}_1', \ldots, \mathbf{A}_m'$ referred to $\mathbf{E}_1', \ldots, \mathbf{E}_m'$ as coordinate vectors. Obviously the coefficient $[\mathbf{A}_1, \ldots, \mathbf{A}_m; \mathbf{E}_1', \ldots, \mathbf{E}_m']$ itself is an alternating multilinear form in $\mathbf{A}_1, \ldots, \mathbf{A}_m$ and, hence, given by

$$[\mathbf{E}_1, \ldots, \mathbf{E}_m; \mathbf{E}_1', \ldots, \mathbf{E}_m'] \det(\mathbf{A}_1, \ldots, \mathbf{A}_m),$$

where the last determinant is formed from the matrix of components of $\mathbf{A}_1, \ldots, \mathbf{A}_m$ referred to the coordinate vectors $\mathbf{E}_1, \ldots, \mathbf{E}_m$.

Using formula (81c), we obtain the identity

$$[\mathbf{A}_1, \ldots, \mathbf{A}_m; \mathbf{A}_1', \ldots, \mathbf{A}_m'] = \mu\varepsilon\varepsilon' VV'. \tag{85c}$$

Here V and V' are respectively the volumes of the parallelepipeds spanned by the vectors $\mathbf{A}_1, \ldots, \mathbf{A}_m$ and $A_1', \ldots, A_m'$. The factors ε, ε' relate the orientations of the parallelepipeds to those of the coordinate systems in π and π':

$$\varepsilon = \operatorname{sgn} [\mathbf{A}_1, \ldots, \mathbf{A}_m; \mathbf{E}_1, \ldots, \mathbf{E}_m],$$

$$\varepsilon' = \operatorname{sgn} \mathbf{A}_1', \ldots, \mathbf{A}_m'; \mathbf{E}_1', \ldots, \mathbf{E}_m'].$$

Finally, the coefficient

$$\mu = [\mathbf{E}_1, \ldots, \mathbf{E}_m; \mathbf{E}_1', \ldots, \mathbf{E}_m']$$

depends only on the spaces π and π' and the coordinate systems chosen in those spaces. If $\pi = \pi'$ we can choose

$$\mathbf{E}' = \mathbf{E}_1, \ldots, \mathbf{E}_m' = \mathbf{E}_m;$$

in that case $\mu = 1$, as in formula (84a).

For $\mu \neq 0$, we can use formula (85c) to relate orientations in two distinct m-dimensional linear manifolds π and π', both lying in the same n-dimensional space.[1] Replacing, if necessary, one of the coordinate

[1]One verifies easily that $\mu = 0$ only when π and π' are *perpendicular to each other*, that is, when π' contains a vector orthogonal to all vectors in π. More generally, the coefficient μ can be interpreted as cosine of the angle between the two manifolds (see problem 13, p. 203).

vectors by its opposite, we can always contrive that $\mu > 0$. Then, by (85c),

$$\text{(85d)} \qquad \operatorname{sgn}[\mathbf{A}_1, \ldots, \mathbf{A}_m; \mathbf{A}_1', \ldots, \mathbf{A}_m'] = \varepsilon\varepsilon'$$

Thus, the condition

$$[\mathbf{A}_1, \ldots, \mathbf{A}_m; \mathbf{A}_1', \ldots, \mathbf{A}_m'] > 0$$

for any $\mathbf{A}_1, \ldots, \mathbf{A}_m$ in π and $\mathbf{A}_1', \ldots, \mathbf{A}_m'$ in π' signifies that both sets of vectors are oriented positively or both oriented negatively with respect to the coordinate systems in those spaces.

e. Orientation of Planes and Hyperplanes

The choice of a particular Cartesian coordinate system in an m-dimensional linear manifold π determines a certain orientation

$$\Omega(\mathbf{E}_1, \ldots, \mathbf{E}_m),$$

where $\mathbf{E}_1, \ldots, \mathbf{E}_m$ are the coordinate vectors. This choice fixes which sets of m vectors $\mathbf{A}_1, \ldots, \mathbf{A}_m$ in π are called positively oriented, namely, those with the same orientation as $\mathbf{E}_1, \ldots, \mathbf{E}_m$. We denote by π^* the combination of the linear space π with the selection of a particular orientation in π and call π^* an *oriented linear* manifold. We write $\Omega(\pi^*)$ for the selected orientation and call m independent vectors $\mathbf{A}_1, \ldots, \mathbf{A}_m$ in π oriented positively if

$$\Omega(\mathbf{A}_1, \ldots, \mathbf{A}_m) = \Omega(\pi^*).$$

We call π^* *oriented positively with respect* to a particular Cartesian coordinate system if the orientation of the coordinate vectors is the same as that of π^*.

An oriented two-dimensional plane π^* can be visualized as a plane with a distinguished *positive sense of rotation.* If a pair of vectors $\mathbf{A}, \mathbf{B}$ is oriented "positively" with respect to π^*, the positive sense of rotation of η^* is the sense of the rotation by an angle less than 180° that takes the direction of $\mathbf{A}$ into that of $\mathbf{B}$.[1]

If the oriented two-dimensional plane π^* lies in an oriented three-dimensional plane σ^*, we can distinguish a *positive* and *negative* side

[1]Notice that the orientation of π^* can only be described by pointing out a specific positively oriented pair of vectors $\mathbf{B}, \mathbf{C}$ in π or a specific rotating object in π (e.g., a clock) that has the distinguished sense of rotation. There is no abstract way of deciding whether a given rotation is *clockwise* or *counterclockwise*, anymore than there is an abstract way of saying which is the *right* and which the *left* side. These questions can only be decided by reference to some *standard objects.*

of π^*. Let P_0 be any point of π^*. We take two independent vectors $\mathbf{B} = \overrightarrow{P_0P_1}$, $\mathbf{C} = \overrightarrow{P_0P_2}$ in π^* for which

$$\Omega(\mathbf{B}, \mathbf{C}) = \Omega(\pi^*). \tag{86a}$$

A third vector $\mathbf{A} = \overrightarrow{P_0P_3}$, independent of $\mathbf{B}$, $\mathbf{C}$ is said to point to the *positive side of* π^* if

$$\Omega(\mathbf{A}, \mathbf{B}, \mathbf{C}) = \Omega(\sigma^*). \tag{86b}$$

If σ^* is oriented positively with respect to a Cartesian coordinate system, we can replace condition (86b) by

$$\det(\mathbf{A}, \mathbf{B}, \mathbf{C}) > 0 \tag{86c}$$

in that system. If σ^* is oriented positively with respect to the usual right-handed coordinate system, then the positive side of an oriented plane π^* is the one from which the positive sense of rotation in π^* appears counterclockwise.

The same terminology applies to oriented hyperplanes π^* in n-dimensional oriented space σ^*. Given $n - 1$ vectors $\mathbf{A}_2, \ldots, \mathbf{A}_n$ in π^* with

$$\Omega(\mathbf{A}_2, \ldots, \mathbf{A}_n) = \Omega(\pi^*), \tag{87a}$$

a vector $\mathbf{A}_1$ is said to point to the positive side of π^*, if

$$\Omega(\mathbf{A}_1, \ldots, \mathbf{A}_{n-1}, \mathbf{A}_n) = \Omega(\sigma^*), \tag{87b}$$

f. Change of Volume of Parallelepipeds in Linear Transformations

A square matrix $\mathbf{a} = (a_{jk})$ with n rows and columns determines a linear transformation or mapping $\mathbf{Y} = \mathbf{aX}$ of vectors $\mathbf{X}$ in n-dimensional space into vectors $\mathbf{Y}$ of the same space. Here we assume that $\mathbf{X}$ and $\mathbf{Y}$ are referred to the same coordinate vectors $\mathbf{E}_1, \ldots, \mathbf{E}_n$. For $\mathbf{X} = (x_1, \ldots, x_n)$, $\mathbf{Y} = (y_1, \ldots, y_n)$ the transformation, written out by components, has the form

$$y_j = \sum_{r=1}^{n} a_{jr} x_r \qquad (j = 1, \ldots, n).$$

A set of n vectors $\mathbf{B}_1 = (b_{11}, \ldots, b_{n1}), \ldots, \mathbf{B}_n = (b_{1n}, \ldots, b_{nn})$ is transformed into the set of n vectors $\mathbf{C}_1 = (c_{11}, \ldots, c_{n1}), \ldots, \mathbf{C}_n = (c_{1n}, \ldots, c_{nn})$, where

$$c_{jk} = \sum_{r=1}^{n} a_{jr} b_{rk}$$

By the rule for the determinant of a product of matrices (p. 172), we have

$$\det(\mathbf{C}_1, \ldots, \mathbf{C}_n) = \det(\mathbf{a}) \cdot \det(\mathbf{B}_1, \ldots, \mathbf{B}_n) \tag{88a}$$

This formula contains the two formulae

$$|\det(\mathbf{C}_1, \ldots, \mathbf{C}_n)| = |\det(\mathbf{a})|\,|\det(\mathbf{B}_1, \ldots, \mathbf{B}_n)| \tag{88b}$$

$$\operatorname{sgn}\det(\mathbf{C}_1, \ldots, \mathbf{C}_n) = [\operatorname{sgn}\det(\mathbf{a})][\operatorname{sgn}\det(\mathbf{B}_1, \ldots, \mathbf{B}_n). \tag{88c}$$

These two rules can be formulated immediately in geometrical language:

The linear transformation of n-dimensional space into itself corresponding to a square matrix **a** *multiplies the volume of every parallelepiped spanned by n vectors by the same constant factor* $|\det(\mathbf{a})|$. *It preserves the orientation of all n-dimensional parallelepipeds, if det* $(\mathbf{a}) > 0$, *and changes the orientation of all of them if det* $(\mathbf{a}) < 0$.[1]

For a rigid motion, the matrix **a** is orthogonal and, hence (see p. 175), has determinant $+1$ or -1. Thus, *rigid motions preserve the volume of parallelepipeds*. Those for which $\det(\mathbf{a}) = +1$ preserve sense; the others invert it.

Exercises 2.4

1. Treat number 5 of Exercises 2.2 in terms of vector products.
2. In a uniform rotation let (α, β, γ) be the direction cosines of the axis of rotation, which passes through the origin, and ω the angular velocity. Find the velocity of the point (x, y, z).
3. Show that the plane through the three points (x_1, y_1, z_1), (x_2, y_2, z_2), (x_3, y_3, z_3) is given by

$$\begin{vmatrix} x_1 - x & y_1 - y & z_1 - z \\ x_2 - x & y_2 - y & z_2 - z \\ x_3 - x & y_3 - y & z_3 - z \end{vmatrix} = 0.$$

[1]It is important to emphasize the assumptions in this theorem. Only volumes of n-dimensional parallelepipeds are multiplied by the same factor; lower-dimensional ones are multiplied by factors that vary with their location. Also, we have to assume that image and original refer to the same coordinate system if the statement about orientations is to hold.

4. Find the shortest distance between two straight lines l and l' in space, given by the equations $x = at + b$, $y = ct + d$, $z = et + f$ and $x = a't + b'$, $y = c't + d'$, $z = e't + f'$.

5. Show that the area of a convex polygon with the successive vertices $P_1(x_1, y_1)$, $P_2(x_2, y_2)$, . . ., $P_n(x_n, y_n)$ is given by half the absolute value of

$$\begin{vmatrix} x_1 & x_2 \\ y_1 & y_2 \end{vmatrix} + \begin{vmatrix} x_2 & x_3 \\ y_2 & y_3 \end{vmatrix} + \cdots + \begin{vmatrix} x_{n-1} & x_n \\ y_{n-1} & y_n \end{vmatrix} + \begin{vmatrix} x_n & x_1 \\ y_n & y_1 \end{vmatrix}.$$

6. Prove that the area of the triangle with vertices (x_1, y_1), (x_2, y_2), and (x_3, y_3) is

$$\frac{1}{2}\begin{vmatrix} x_1 & y_1 & 1 \\ x_2 & y_2 & 1 \\ x_3 & y_3 & 1 \end{vmatrix}.$$

7. If the vertices of the triangle of the preceding exercise have rational coordinates, prove the triangle cannot be equilateral.

8. (a) Prove the inequality

$$D = \begin{vmatrix} a & b & c \\ a' & b' & c' \\ a'' & b'' & c'' \end{vmatrix} \leqq \sqrt{(a^2 + b^2 + c^2)(a'^2 + b'^2 + c'^2)(a''^2 + b''^2 + c''^2)}.$$

(b) When does the equality sign hold?

9. Prove the vector identities

(a) $\mathbf{A} \times (\mathbf{B} \times \mathbf{C}) = (\mathbf{A} \cdot \mathbf{C})\,\mathbf{B} - (\mathbf{A} \cdot \mathbf{B})\,\mathbf{C}$

(b) $(\mathbf{X} \times \mathbf{Y}) \cdot (\mathbf{X}' \times \mathbf{Y}') = (\mathbf{X} \cdot \mathbf{X}')(\mathbf{Y} \cdot \mathbf{Y}') - (\mathbf{X} \cdot \mathbf{Y}')(\mathbf{Y} \cdot \mathbf{X}')$

(c) $[\mathbf{X} \times (\mathbf{Y} \times \mathbf{Z})] \cdot \{[\mathbf{Y} \times (\mathbf{Z} \times \mathbf{X})] \times [\mathbf{Z} \times (\mathbf{X} \times \mathbf{Y})]\} = 0.$

10. Give the formula for a rotation through the angle ϕ about the axis $x:y:z = 1:0:-1$ such that the rotation of the plane $x = z$ is positive when looked at from the point $(-1, 0, 1)$.

11. If $\mathbf{A}$, $\mathbf{B}$, and $\mathbf{C}$ are independent, use the two representations of $\mathbf{X} = (\mathbf{A} \times \mathbf{B}) \times (\mathbf{C} \times \mathbf{D})$ obtained from Exercise 9a to express $\mathbf{D}$ as a linear combination of $\mathbf{A}$, $\mathbf{B}$, and $\mathbf{C}$.

12. Let Ox, Oy, Oz and Ox', Oy', Oz' be two right-handed coordinate systems. Assume that Oz and Oz' do not coincide; let the angle zOz' be θ $(0 < \theta < \pi)$. Draw the half-line Ox_1 at right angles to both Oz and Oz' and such that the system Ox_1, Oz, Oz' has the same orientation as Ox, Oy, Oz. The Ox_1 is the line of intersection of the planes Oxy and $Ox'y'$. Let the angle $x0x_1$ be ϕ and the angle x_1Ox' be ψ and let them be measured in the usual positive sense in their respective planes, Oxy and $Ox'y'$. Find the matrix for the change of coordinates.

13. Let π and π' be two m-dimensional linear subspaces of the same n-dimensional space with respective orthonormal bases $\mathbf{E}_1, \mathbf{E}_2, \ldots, \mathbf{E}_m$ and $\mathbf{E}_1', \mathbf{E}_2', \ldots, \mathbf{E}_m'$. Show that $\mu = [\mathbf{E}_1, \mathbf{E}_2, \ldots, \mathbf{E}_m; \mathbf{E}_1', \mathbf{E}_2', \ldots, \mathbf{E}_m'] = 0$ if and only if π and π' are orthogonal, that is, one space contains a vector perpendicular to all the vectors of the other.

2.5 Vector Notions in Analysis

a. Vector Fields

Mathematical analysis comes into play when we are concerned with a *vector manifold* depending on one or more continuously varying parameters.

If, for example, we consider a material occupying a portion of space and in a state of motion, then at a given instant each particle of the material will have a definite velocity represented by a vector $\mathbf{U} = (u_1, u_2, u_3)$. We say that these vectors form a *vector field* in the region in question. The three components of the field vector then appear as three functions

$$u_1(x_1, x_2, x_3), \quad u_2(x_1, x_2, x_3), \quad u_3(x_1, x_2, x_3)$$

of the three coordinates x_1, x_2, x_3 of the position of the particle at the instant in question. We would usually represent $\mathbf{U}$ as a vector with initial point (x_1, x_2, x_3).

The forces acting at different points of space likewise form a vector field. As an example of a *force field* we consider the gravitational force per unit mass exerted by a heavy particle, according to Newton's law of attraction. According to that law the field vector $\mathbf{F} = (f_1, f_2, f_3)$ at each point (x_1, x_2, x_3) is directed toward the attracting particle, and its magnitude is inversely proportional to the square of the distance from the particle.

Field vectors, like $\mathbf{U}$ or $\mathbf{F}$, have a physical meaning independent of coordinates. In a given Cartesian x_1, x_2, x_3-coordinate system the vector $\mathbf{U}$ has components u_1, u_2, u_3 that depend on the coordinate system. In a different Cartesian coordinate system the point that originally had coordinates x_1, x_2, x_3 receives the coordinates y_1, y_1, y_3 where the y_i and x_k are connected by equations of the form

(89a)
$$\begin{cases} y_1 = a_{11}x_1 + a_{12}x_2 + a_{13}x_3 + b_1 \\ y_2 = a_{21}x_1 + a_{22}x_2 + a_{23}x_3 + b_2 \\ y_3 = a_{31}x_1 + a_{32}x_2 + a_{33}x_3 + b_3 \end{cases}$$

or

(89b)
$$y_j = \sum_{k=1}^{3} a_{jk}x_k + b_j \qquad (i = 1, 2, 3).$$

The components v_1, v_2, v_3 of the vector $\mathbf{U}$ in the new coordinate system are then given by the corresponding *homogenenous* relations

$$v_j = \sum_{k=1}^{3} a_{jk} u_k \qquad (j = 1, 2, 3). \tag{89c}$$

The matrix $\mathbf{a} = (a_{jk})$ is orthogonal, so that (see p. 158) its reciprocal is equal to its transpose. Consequently, the solutions of equations (89b), (89c) for x_k and u_k take the form

$$x_k = \sum_{j=1}^{3} a_{jk}(y_j - b_j) \qquad (k = 1, 2, 3), \tag{89d}$$

$$u_k = \sum_{k=1}^{3} a_{jk} v_j \qquad (k = 1, 2, 3). \tag{89e}$$

Any three functions u_1, u_2, u_3 of the variables x_1, x_2, x_3 determine a field of vectors $\mathbf{U}$ with components u_1, u_2, u_3 in x_1, x_2, x_3-coordinates. If the field is to have a meaning independent of the choice of coordinate systems, the components v_i of $\mathbf{U}$ in a Cartesian y_1, y_2, y_3-coordinate system have to be given by formula (89c) whenever the y_i and x_i are connected by formulae (89a).

b. *Gradient of a Scalar*

A scalar is a function $s = s(P)$ of the points P in space. In any Cartesian coordinate system in which the point P is described by its coordinates x_1, x_2, x_3 the scalar s becomes a function $s = f(x_1, x_2, x_3)$. We may regard the three partial derivatives

$$u_1 = \frac{\partial s}{\partial x_1} = f_{x_1}(x_1, x_2, x_3),$$

$$u_2 = \frac{\partial s}{\partial x_2} = f_{x_2}(x_1, x_2, x_3),$$

$$u_3 = \frac{\partial s}{\partial x_3} = f_{x_3}(x_1, x_2, x_3).$$

as components in x_1, x_2, x_3-coordinates of a vector $\mathbf{U} = (u_1, u_2, u_3)$.

In any new Cartesian y_1, y_2, y_3-coordinate system connected with the original one by relations (89a) or (89d), the scalar s is represented by the function

$$\begin{aligned} s &= g(y_1, y_2, y_3) \\ &= f\left(\sum_{k=1}^{3} a_{k1}(y_k - b_k), \sum_{k=1}^{3} a_{k2}(y_k - b_k), \sum_{k=1}^{3} a_{k3}(y_k - b_k)\right) \end{aligned}$$

By the *chain rule of differentiation* (p. 55) we have

$$\begin{aligned} v_j &= \frac{\partial s}{\partial y_j} = g_{y_j}(y_1, y_2, y_3) \\ &= \sum_{k=1}^{3} \frac{\partial s}{\partial x_k}\frac{\partial x_k}{\partial y_j} \\ &= \sum_{k=1}^{3} u_k a_{jk}. \end{aligned}$$

Using the relations (89c), we see that the vector $\mathbf{U}$ has the components $v_j = \partial s/\partial y_j$ in the y_1, y_2, y_3-system. Thus the partial derivatives of the scalar s formed in any cartesian coordinate system form the components of a vector $\mathbf{U}$ that does not depend on the system. We call $\mathbf{U}$ the *gradient of the scalar* s and write

$$\mathbf{U} = \operatorname{grad} s.$$

By formula (14b), p. 45 the derivative of s in the direction with direction cosines $\cos\alpha_1$, $\cos\alpha_2$, $\cos\alpha_3$ is given in x_1, x_2, x_3-coordinates by

$$D_{(\alpha)}s = \frac{\partial s}{\partial x_1}\cos\alpha_1 + \frac{\partial s}{\partial x_2}\cos\alpha_2 + \frac{\partial s}{\partial x_3}\cos\alpha_3. \tag{90}$$

Introducing the unit vector $\mathbf{R} = (\cos\alpha_1, \cos\alpha_2, \cos\alpha_3)$ in the direction with direction angles $\alpha_1, \alpha_2, \alpha_3$, we can write the derivative of s in that direction in vector notation as

$$D_{(\alpha)}s = \mathbf{R}\cdot\operatorname{grad} s. \tag{90b}$$

We find from the Cauchy-Schwarz inequality (see p. 132) for $|\mathbf{R}| = 1$.

$$|D_{(\alpha)}s| \leqq |\mathbf{R}|\,|\operatorname{grad} s| = |\operatorname{grad} s|$$

Thus, *the derivative of s in any direction never exceeds the length of the gradient of s*. Taking for $\mathbf{R}$ the unit vector in the direction of grad s, we find for the directional derivative the value

$$D_{(\alpha)}s = \frac{1}{|\operatorname{grad} s|}(\operatorname{grad} s)\cdot(\operatorname{grad} s) = |\operatorname{grad} s|$$

Thus, *the length of the gradient vector of s is equal to the maximum rate of change of s in any direction. The direction of the gradient is the one in which the scalar s increases most rapidly, while in the opposite direction s decreases most rapidly.*

We shall return to the geometrical interpretation of the gradient in Chapter 3. We can, however, immediately give an intuitive idea of the *direction* of the gradient. Confining ourselves first to vectors in two dimensions, we have to consider the gradient of a scalar $s = f(x_1, x_2)$. We shall suppose that s is represented by its level lines (or contour lines)

$$s = f(x_1, x_2) = \text{constant} = c$$

in the x_1, x_2-plane. Then the derivative of s at a point P in the direction of the level line through P is obviously 0, for if Q is another point on the same level line, the equation $s(Q) - s(P) = 0$ holds; dividing by the distance ρ of Q and P and letting ρ tend to 0 we find in the limit (see p. 45) that the derivative of s in the direction tangential to the level line at P is 0. Thus, by (90b), $R \cdot \text{grad } s = 0$ if $\mathbf{R}$ is a unit vector in the direction of the tangent to the level line, and therefore, *at every point the gradient vector of s is perpendicular to the level line through that point.* An exactly analogous statement holds for the gradient in three dimensions. If we represent the scalar s by its *level surfaces*

$$s = f(x_1, x_2, x_3) = \text{constant} = c,$$

the gradient has component zero in every direction tangential to the level surface and is therefore perpendicular to the level surface.

In applications, we frequently meet with vector fields that represent the gradient of a scalar function. The gravitational field of force due to particle of mass M concentrated in a point $Q = (\xi_1, \xi_2, \xi_3)$ may be taken as an example. Let $\mathbf{F} = (f_1, f_2, f_3)$ denote the force exerted by the attractive mass M on a particle of mass m located at the point $P = (x_1, x_2, x_3)$. Denote by $\mathbf{R}$ the vector

$$\mathbf{R} = \overrightarrow{QP} = (x_1 - \xi_1, x_2 - \xi_2, x_3 - \xi_3).$$

By Newton's law of gravitation, $\mathbf{F}$ has the direction of $-\mathbf{R}$ and the magnitude $C/|\mathbf{R}|^2$, where $C = \gamma m M$ (here γ denotes the universal gravitational constant). Hence,

$$\mathbf{F} = -\frac{C}{|\mathbf{R}|^3}\mathbf{R}$$

or

$$f_j = C\frac{\xi_j - x_j}{\sqrt{(\xi_1 - x_1)^2 + (\xi_2 - x_2)^2 + (\xi_3 - x_3)^2}^3} \qquad (j = 1, 2, 3).$$

By differentiation, one verifies immediately that

$$f_j = \frac{\partial}{\partial x_j}\frac{C}{\sqrt{(\xi_1 - x_1)^2 + (\xi_2 - x_2)^2 + (\xi_3 - x_3)^2}} \qquad (j = 1, 2, 3).$$

Hence,

$$\mathbf{F} = \operatorname{grad}\frac{C}{r}, \tag{91}$$

where

$$r = \sqrt{(\xi_1 - x_1)^2 + (\xi_2 - x_2)^2 + (\xi_3 - x_3)^2} = |\mathbf{R}|$$

is the distance of the two particles at P and Q.

If a field of force is the gradient of a scalar function, this scalar function is often called the *potential function* of the field. We shall consider this concept from a more general point of view in the study of work and energy (pp. 657 and 714).

c. *Divergence and Curl of a Vector Field*

By differentiation we have assigned to every scalar a vector field, the gradient. Similarly, we can assign by differentiation to every vector field $\mathbf{U}$ a certain scalar, known as the *divergence* of the vector field $\mathbf{U}$. For a specific Cartesian x_1, x_2, x_3-coordinate system in which $\mathbf{U} = (u_1, u_2, u_3)$, we define the divergence of the vector $\mathbf{U}$ as the function

$$\operatorname{div}\mathbf{U} = \frac{\partial u_1}{\partial x_1} + \frac{\partial u_2}{\partial x_2} + \frac{\partial u_3}{\partial x_3}, \tag{92}$$

that is, as the sum of the partial derivatives of the three components with respect to the corresponding coordinates. We can show that the scalar div $\mathbf{U}$ defined in this way does not depend on the particular choice of Cartesian coordinate system.[1] Let the coordinates

[1]This would not be the case for other expressions formed from the first derivatives of the components of the vector $\mathbf{U}$, for example,

$$\frac{\partial u_1}{\partial x_1} + \frac{\partial u_2}{\partial x_2} - \frac{\partial u_3}{\partial x_3}$$

or

$$\frac{\partial u_1}{\partial x_2} \cdot \frac{\partial u_2}{\partial x_3} \cdot \frac{\partial u_3}{\partial x_1}.$$

y_1, y_2, y_3 of a point in a different Cartesian system be connected with x_1, x_2, x_3 by equations (89b); the components v_1, v_2, v_3 of $\mathbf{U}$ in the new system are then given by relations (89c). We have from the chain rule of differentiation

$$\begin{aligned}\operatorname{div} \mathbf{U} &= \sum_{k=1}^{3} \frac{\partial u_k}{\partial x_k} = \sum_{k,j=1}^{3} \frac{\partial u_k}{\partial y_j} \frac{\partial y_j}{\partial x_k} \\ &= \sum_{j,k=1}^{3} a_{jk} \frac{\partial u_k}{\partial y_j} = \sum_{j=1}^{3} \frac{\partial}{\partial y_j} \sum_{k=1}^{3} a_{jk} u_k \\ &= \sum_{j=1}^{3} \frac{\partial v_j}{\partial x_j},\end{aligned}$$

which shows that we are led to the same scalar div $\mathbf{U}$ in any other coordinate system.

Here we content ourselves with the formal definition of the divergence; its physical interpretation will be discussed later (Chapter V, Section 9).

We shall adopt the same procedure for the so-called *curl* of a vector field $\mathbf{U}$. The curl is itself a vector

$$\mathbf{B} = \operatorname{curl} \mathbf{U}.$$

If in a x_1, x_2, x_3-coordinate system the vector $\mathbf{U}$ has the components u_1, u_2, u_3, we define the components b_1, b_2, b_3 of curl $\mathbf{U}$ by

$$(93) \qquad b_1 = \frac{\partial u_3}{\partial x_2} - \frac{\partial u_2}{\partial x_3}, \quad b_2 = \frac{\partial u_1}{\partial x_3} - \frac{\partial u_3}{\partial x_1}, \quad b_3 = \frac{\partial u_2}{\partial x_1} - \frac{\partial u_1}{\partial x_2}.$$

We could verify as in the other cases that our definition of the curl of a vector $\mathbf{U}$ actually yields a vector independent of the particular coordinate system, provided the Cartesian coordinate systems considered all have the same orientation. However, we omit these computations here, since in Chapter V, p. 616 we shall give a physical interpretation of the curl that clearly brings out its vectorial character.

The three concepts of gradient, divergence, and curl can all be related to one another if we use a symbolic vector with the components

$$\frac{\partial}{\partial x_1}, \frac{\partial}{\partial x_2}, \frac{\partial}{\partial x_3}.$$

This *vector differential operator* is usually denoted by the symbol ∇, pronounced "del." The gradient of a scalar s is the product of the symbolic vector ∇ with the scalar quantity s; that is, it is the vector

$$\text{(94)} \qquad \operatorname{grad} s = \nabla s = \left(\frac{\partial}{\partial x_1} s,\ \frac{\partial}{\partial x_2} s,\ \frac{\partial}{\partial x_3} s\right).^{1}$$

The divergence of a vector $\mathbf{U} = (u_1, u_2, u_3)$ is the scalar product

$$\text{(94b)} \qquad \operatorname{div} \mathbf{U} = \nabla \cdot \mathbf{U} = \frac{\partial}{\partial x_1} u_1 + \frac{\partial}{\partial x_2} u_2 + \frac{\partial}{\partial x_3} u_3.$$

Finally the curl of the vector $\mathbf{U}$ is the vector product

$$\text{(94c)} \qquad \operatorname{curl} \mathbf{U} = \nabla \times \mathbf{U}$$

$$= \left(\frac{\partial}{\partial x_2} u_3 - \frac{\partial}{\partial x_3} u_2,\ \frac{\partial}{\partial x_3} u_1 - \frac{\partial}{\partial x_1} u_3,\ \frac{\partial}{\partial x_1} u_2 - \frac{\partial}{\partial x_2} u_1\right)$$

[see (71b), p. 180. The fact that the vector ∇ is independent of the Cartesian coordinate system used to define its components follows from the chain rule of differentiation; under the coordinate transformation (89d), we have by the chain rule

$$\frac{\partial}{\partial y_j} = \sum_{k=1}^{3} \frac{\partial x_k}{\partial y_j} \frac{\partial}{\partial x_k} = \sum_{k=1}^{3} a_{jk} \frac{\partial}{\partial x_k},$$

which shows that the components of ∇ transform according to the rule (89c) for vectors. This makes it obvious that also ∇s, $\nabla \cdot \mathbf{U}$ and $\nabla \times \mathbf{U}$ do not depend on coordinates.[2]

In conclusion, we mention a few relations that constantly recur. *The curl of a gradient is zero;* in symbols,

$$\text{(95a)} \qquad \operatorname{curl} \operatorname{grad} s = \nabla \times (\nabla s) = 0.$$

[1]We are forced here to write the vector in front of the scalar in the product ∇s, contrary to our usual habit, since the components of the symbolic vector ∇ do not commute with ordinary scalars.

[2]This statement has to be qualified in the case of the curl. Generally, magnitude and direction of the vector product of two vectors has a geometrical meaning, as explained on p. 185, except that the product changes into the opposite when we change the orientation of the Cartesian coordinate system used. This implies for a vector $\mathbf{U}$ that $\operatorname{curl} \mathbf{U} = \nabla \times \mathbf{U}$ behaves like a vector, as long as we do not change the orientation of the coordinate system (i.e., as long as only orthogonal transformations with determinant $+1$ are used). Changing the orientation of the coordinate system results in changing $\operatorname{curl} \mathbf{U}$ into its opposite.

The divergence of a curl is zero; in symbols,

(95b) $$\text{div curl } \mathbf{U} = = \nabla \cdot (\nabla \times \mathbf{U}) = 0.$$

As we easily see, these relations follow from the definitions of divergence, curl, and gradient, using the interchangeability of differentiations. Relations (95a, b) also follow formally if we apply the ordinary rules for vectors to the symbolic vector ∇, since then

$$\nabla \times (\nabla s) = (\nabla \times \nabla)s = \mathbf{0}, \quad \nabla \cdot (\nabla \times \mathbf{U}) = \det(\nabla, \nabla, \mathbf{U}) = 0.$$

Another extremely important combination of our vector differential operators is the *divergence of a gradient:*

(95c) $$\text{div grad } s = \nabla \cdot (\nabla s) = \frac{\partial^2 s}{\partial x_1^2} + \frac{\partial^2 s}{\partial x_2^2} + \frac{\partial^2 s}{\partial x_3^2} = \Delta s.$$

Here

(95d) $$\Delta = \nabla \cdot \nabla = \frac{\partial^2}{\partial x_1^2} + \frac{\partial^2}{\partial x_2^2} + \frac{\partial^2}{\partial x_3^2}$$

is known as the "Laplace operator" or the "Laplacian." The partial differential equation

(95e) $$\Delta s = \frac{\partial^2 s}{\partial x_1^2} + \frac{\partial^2 s}{\partial x_2^2} + \frac{\partial^2 s}{\partial x_3^2} = 0$$

satisfied by many important scalars s in mathematical physics is called the "Laplace equation" or "potential equation."

The terminology of "vector analysis" is often used also when the number of independent variables is other than three. A system of n functions $u_1, \ldots, u_n$ of n indenpendent variables $x_1, \ldots, x_n$ determines a *vector field* in n-dimensional space. The concepts of gradient of a scalar and of the Laplace operator then retain their meaning. Notions analogous to the curl of a vector become more complicated. The most satisfactory approach to analogues of relations (95a,b) in n dimensions is through the calculus of *exterior differential forms,* which will be described in the next chapter.

d. Families of Vectors. Application to the Theory of Curves in Space and to Motion of Particles

In addition to vector fields we also consider one-parametric manifolds of vectors, called *families of vectors*, where the vectors $\mathbf{U} = (u_1, u_2, u_3)$ do not correspond to each point of a region in space but to each value of a single parameter t. We write $\mathbf{U} = \mathbf{U}(t)$. The derivative of the vector $\mathbf{U}$ can be defined naturally as

$$\frac{dU}{dt} = \lim_{h \to 0} \frac{1}{h} [\mathbf{U}(t + h) - \mathbf{U}(t)]. \tag{96a}$$

It obviously has the components

$$\frac{du_1}{dt}, \frac{du_2}{dt}, \frac{du_3}{dt}. \tag{96b}$$

One easily verifies that this vector differentiation satisfies analogues of the ordinary laws for derivatives:

$$\frac{d}{dt}(\mathbf{U} + \mathbf{V}) = \frac{d}{dt}\mathbf{U} + \frac{d}{dt}\mathbf{V}; \qquad \frac{d}{dt}(\lambda \mathbf{U}) = \frac{d\lambda}{dt}\mathbf{U} + \lambda \frac{d}{dt}\mathbf{U} \tag{97a}$$

$$\frac{d}{dt}(\mathbf{U} \cdot \mathbf{V}) = \mathbf{U} \cdot \frac{d\mathbf{V}}{dt} + \frac{d\mathbf{U}}{dt} \cdot \mathbf{V} \tag{97b}$$

$$\frac{d}{dt}(\mathbf{U} \times \mathbf{V}) = \mathbf{U} \times \frac{d\mathbf{V}}{dt} + \frac{d\mathbf{U}}{dt} \times \mathbf{V}. \tag{97c}$$

We apply these notions to the case where the family of vectors consists of the *position vectors* $\mathbf{X} = \mathbf{X}(t) = \overrightarrow{OP}$ of the points P on a curve in space given in parametric representation:

$$x_1 = \phi_1(t), \; x_2 = \phi_2(t), \; x_3 = \phi_3(t).$$

Then

$$\mathbf{X} = (x_1, x_2, x_3) = (\phi_1(t), \phi_2(t), \phi_3(t)).$$

The vector $d\mathbf{X}/dt$ has the direction of the *tangent* to the curve at the point corresponding to t. For the vector $\Delta\mathbf{X} = \mathbf{X}(t + \Delta t) - \mathbf{X}(t)$ has the direction of the line segment joining the points with parameter values t and $t + \Delta t$. The same holds for the vector $\Delta\mathbf{X}/\Delta t$, when $\Delta t > 0$. As $\Delta t \to 0$ the direction of this chord approaches the direction of the tangent. If instead of t we introduce as parameter the length of arc s of the curve measured from a definite starting point, we can prove that

(98) $$\left|\frac{d\mathbf{X}}{ds}\right|^2 = \frac{d\mathbf{X}}{ds}\cdot\frac{d\mathbf{X}}{ds} = 1.$$

The proof follows exactly the same lines as the corresponding proof for plane curves (see Volume I, p. 354). Thus, $d\mathbf{X}/ds$ is a unit vector. Differentiating both sides of equation (98) with respect to s, using rule (97b), we obtain

(99) $$\frac{d\mathbf{X}}{ds}\cdot\frac{d^2\mathbf{X}}{ds^2} + \frac{d^2\mathbf{X}}{ds^2}\cdot\frac{d\mathbf{X}}{ds} = 2\frac{d\mathbf{X}}{ds}\cdot\frac{d^2\mathbf{X}}{ds^2} = 0.$$

This equation states that the vector

$$\frac{d^2\mathbf{X}}{ds^2} = \left(\frac{d^2x_1}{ds^2}, \frac{d^2x_2}{ds^2}, \frac{d^2x_3}{ds^2}\right)$$

is *perpendicular to the tangent.* This vector we call the *curvature vector* or *principal normal vector,* and its length

(100) $$k = \frac{1}{\rho} = \left|\frac{d^2\mathbf{X}}{ds^2}\right|$$

we call the *curvature* of the curve at the corresponding point. The reciprocal $\rho = 1/k$ of the curvature we call the *radius of curvature,* as before. The point obtained by measuring from the point on the curve a length ρ in the direction of the principal normal vector is called the *center of curvature.*

We shall show that this definition of curvature agrees with the one given for plane curves in Volume I (p. 354). For each s the vector $\mathbf{Y} = d\mathbf{X}/ds$ is of length 1 and has the direction of the tangent. If we think of the vectors $\mathbf{Y}(s + \Delta s)$ and $\mathbf{Y}(s)$ as having the origin as common initial point, then the difference $\Delta\mathbf{Y} = \mathbf{Y}(s + \Delta s) - \mathbf{Y}(s)$ is represented by the vector joining the end points. The angle β between the tangents to the curve at the points with parameters s and $s + \Delta s$ is equal to the angle between the vectors $\mathbf{Y}(s)$ and $\mathbf{Y}(s + \Delta s)$. Then

$$|\Delta\mathbf{Y}| = |\mathbf{Y}(s + \Delta s) - \mathbf{Y}(s)| = 2\sin\frac{\beta}{2},$$

since

$$|\mathbf{Y}(s)| = |\mathbf{Y}(s + \Delta s)| = 1.$$

Using

$$\frac{2\sin\beta/2}{\beta}\to 1 \qquad \text{for} \qquad \beta\to 0,$$

we find that

$$\left|\frac{d^2\mathbf{X}}{ds^2}\right| = \left|\frac{d\mathbf{Y}}{ds}\right| = \lim_{\Delta s\to 0}\left|\frac{\Delta\mathbf{Y}}{\Delta s}\right| = \lim_{\Delta s\to 0}\frac{\beta}{\Delta s}$$

Hence, k is the limit of the ratio of the angle between the tangents at two points of the curve and the length of arc between those points as the points approach each other. But this limit defines curvature for plane curves.[1]

The curvature vector plays an important part in mechanics. We suppose that a particle moving along a curve has the position vector $\mathbf{X}(t)$ at the time t. The velocity of the motion is then given both in magnitude and direction by the vector $d\mathbf{X}/dt$. Similarly, the acceleration is given by the vector $d^2\mathbf{X}/dt^2$. By the chain rule, we have

$$\frac{d\mathbf{X}}{dt} = \frac{ds}{dt}\frac{d\mathbf{X}}{ds}$$

and

$$\frac{d^2\mathbf{X}}{dt^2} = \frac{d^2s}{dt^2}\frac{d\mathbf{X}}{ds} + \left(\frac{ds}{dt}\right)^2\frac{d^2\mathbf{X}}{ds^2}\,. \tag{101}$$

In view of what we know already about the first and second derivatives of the vector $\mathbf{X}$ with respect to s, equation (101) expresses the following facts: the *acceleration vector* of the motion is the sum of two vectors. One of these is directed along the tangent to the curve and its length is equal to d^2s/dt^2, that is, to the acceleration of the point in its path (the rate of change of speed or *tangential acceleration*). The other is directed normal to the path toward the center of curvature, and its length is equal to the square of the speed multiplied by the curvature (the *normal acceleration*). For a particle of unit mass

[1]In the case of space curves, we cannot, as for plane curves, identify β with the increment $\Delta\alpha$ of an angle of inclination α. The reason is that the angle between $\mathbf{Y}(s)$ and $\mathbf{Y}(s+\Delta s)$ is generally not equal to the difference of the angles the vectors $\mathbf{Y}(s)$ and $\mathbf{Y}(s+\Delta s)$ form with some fixed third direction. *Angles between directions in space are not additive*, as in the plane.

the acceleration vector is equal to the force acting on the particle. If no force acts in the direction of the curve (as is the case for a particle constrained to move along a curve subject only to the reaction forces acting normal to the curve), the tangential acceleration vanishes and the total acceleration is normal to the curve and of magnitude equal to the square of the velocity multiplied by the curvature.

Exercies 2.5

1. Verify that the position vector $\overrightarrow{PQ}$ of a point Q with respect to a point P behaves like a vector in a change of coordinates.
2. Derive the following identities.

   (a) $\operatorname{grad}(\alpha\beta) = \alpha \operatorname{grad}\beta + \beta \operatorname{grad}\alpha$

   (b) $\operatorname{div}(\alpha\mathbf{U}) = \mathbf{U}\cdot\operatorname{grad}\alpha + \alpha \operatorname{div}\mathbf{U}$

   (c) $\operatorname{curl}(\alpha\mathbf{U}) = \operatorname{grad}\alpha \times \mathbf{U} + \alpha \operatorname{curl}\mathbf{U}$

   (d) $\operatorname{div}(\mathbf{U}\times\mathbf{V}) = \mathbf{V}\cdot\operatorname{curl}\mathbf{U} - \mathbf{U}\cdot\operatorname{curl}\mathbf{V}$.
3. Let $\mathbf{U}\cdot\nabla$ be the symbol for the operator
$$\mathbf{U}_x\frac{\partial}{\partial x} + \mathbf{U}_y\frac{\partial}{\partial y} + \mathbf{U}_z\frac{\partial}{\partial z}.$$
   Show that

   (a) $\operatorname{grad}(\mathbf{U}\cdot\mathbf{V}) = \mathbf{U}\cdot\nabla\mathbf{V} + \mathbf{V}\cdot\nabla\mathbf{U} + \mathbf{U}\times\operatorname{curl}\mathbf{V} + \mathbf{V}\times\operatorname{curl}\mathbf{U}$

   (b) $\operatorname{curl}(\mathbf{U}\times\mathbf{V}) = \mathbf{U}\operatorname{div}\mathbf{V} - \mathbf{V}\operatorname{div}\mathbf{U} + \mathbf{V}\cdot\nabla\mathbf{U} - \mathbf{U}\cdot\nabla\mathbf{V}$.
4. For the Laplacian operator Δ establish
$$\Delta\mathbf{U} = \operatorname{grad}\operatorname{div}\mathbf{U} - \operatorname{curl}\operatorname{curl}\mathbf{U}$$
5. Find the equation of the so-called osculating plane of a curve $x = f(t)$, $y = g(t)$, $z = h(t)$ at the point t_0, that is, the limit of the planes passing through three points of the curve as these points approach the point with parameter t_0.
6. Show that the curvature vector and the tangent vector both lie in the osculating plane.
7. Let C be a smooth curve with a continuously turning tangent. Let d denote the shortest distance between two points on the curve and l the length of arc between the two points. Prove that $d - l = o(d)$ when d is small.
8. Prove that the curvature of the curve $\mathbf{X} = \mathbf{X}(t)$, t being an arbitrary parameter, is given by
$$k = \frac{\{|\mathbf{X}'|^2\,|\mathbf{X}''|^2 - (\mathbf{X}'\cdot\mathbf{X}'')^2\}^{1/2}}{|\mathbf{X}'|^3}.$$
9. If $\mathbf{X} = \mathbf{X}(s)$ is any parametric representation of a curve, then the vector $d^2\mathbf{X}/dt^2$ with initial point $\mathbf{X}$ lies in the osculating plane at $\mathbf{X}$.
10. If C is a continuously differentiable closed curve and A a point not on C, there is a point B on C that has a shorter distance from A than any other point on C. Prove that the line AB is normal to the curve.

11. A curve is drawn on the cylinder $x^2 + y^2 = a^2$ such that the angle between the z-axis and the tangent at any point P of the curve is equal to the angle between the y-axis and the tangent plane at P to the cylinder. Prove that the coordinates of any point P of the curve can be expressed in terms of a parameter θ by the equations
$$x = a \cos\theta, \quad y = a\sin\theta, \quad z = c \pm a \log \sin\theta,$$
and that the curvature of the curve is $(1/a)\sin\theta\,(1 + \sin^2\theta)^{1/2}$.

12. Find the equation of the osculating plane (cf. Exercise 5) at the point θ of the curve $x = \cos\theta$, $y = \sin\theta$, $z = f(\theta)$. Show that if $f(\theta) = (\cosh A\theta)/A$, each osculating plane touches a sphere whose center is the origin and whose radius is $\sqrt{(1 + 1/A^2)}$.

13. (a) Prove that the equation of the plane passing through the three points t_1, t_2, t_3 on the curve
$$x = \frac{1}{3}at^3, \quad y = \frac{1}{2}bt^2, \quad z = ct$$
is
$$\frac{3x}{a} - 2(t_1 + t_2 + t_3)\frac{y}{b} + (t_2t_3 + t_3t_1 + t_1t_2)\frac{z}{c} - t_1t_2t_3 = 0.$$
(b) Show that the point of intersection of the osculating planes at t_1, t_2, t_3 lies in this plane.

14. Let $\mathbf{X} = \mathbf{X}(s)$ be an arbitrary curve in space, such that the vector $\mathbf{X}(s)$ is three times continuously differentiable (s is the length of arc). Find the center of the sphere of closest contact with the curve at the point s.

15. If $\mathbf{X} = \mathbf{X}(s)$ is a curve on a sphere of unit radius where s is arclength, then
$$|\dddot{\mathbf{X}}|^2 - |\ddot{\mathbf{X}}|^4 = |\dddot{\mathbf{X}}|^2 - (\dot{\mathbf{X}} \cdot \dddot{\mathbf{X}})^2 = (\dot{\mathbf{X}} \cdot [\ddot{\mathbf{X}} \times \dddot{\mathbf{X}}])^2.$$
holds.

16. The limit of the ratio of the angle between the osculating planes at two neighboring points of a curve and of the length of arc between these two points (i.e., the derivative of the unit normal vector with respect to the arc s) is called the *torsion* of the curve. Let $\xi_1(s)$, $\xi_2(s)$ denote the unit vector along the tangent and the curvature vector of the curve $\mathbf{X}(s)$; by $\xi_3(s)$ we mean the unit vector orthogonal to ξ_1 and ξ_2 (the so-called *binormal* vector), which is given by $[\xi_1 \times \xi_2]$.
Prove Frenet's formulae
$$\dot{\xi}_1 = \frac{\xi_2}{\rho},$$
$$\dot{\xi}_2 = -\frac{\xi_1}{\rho} + \frac{\xi_3}{\tau},$$
$$\dot{\xi}_3 = -\frac{\xi_2}{\tau},$$
where $1/\rho = k$ is the curvature and $1/\tau$ the torsion of $x(s)$.

17. Using the vectors ξ_1, ξ_2, ξ_3 of Exercise 16 as coordinate vectors, find expressions for (a) the vector $\dddot{\mathbf{X}}$, (b) the vector from the point $\mathbf{X}$ to the center of the sphere of closest contact at $\mathbf{X}$.

18. Show that a curve of zero torsion is a plane curve.
19. Consider a fixed point A in space and a variable point P whose motion is given as a function of the time. Denoting by $\dot{\mathbf{P}}$ the velocity vector of P and by $\mathbf{a}$ a unit vector in the direction from P to A, show that

$$\frac{d}{dt}|\overrightarrow{PA}| = -\mathbf{a}\cdot\dot{\mathbf{P}}$$

20. (a) Let A, B, C be three fixed noncollinear points and let P be a moving point. Let $\mathbf{a}, \mathbf{b}, \mathbf{c}$ be unit vectors in the directions PA, PB, PC, respectively; express the velocity vector $\dot{\mathbf{P}}$ as a linear combination of these vectors:

$$\dot{\mathbf{P}} = \mathbf{a}u + \mathbf{b}v + \mathbf{c}w.$$

Prove that

$$\dot{\mathbf{a}} = \frac{1}{|\mathrm{A}-\mathrm{P}|}\{[(\mathbf{a}\cdot\mathbf{b})v + (\mathbf{a}\cdot\mathbf{c})w]\,\mathbf{a} - v\mathbf{b} - w\mathbf{c}\}.$$

(b) Prove that the acceleration vector $\ddot{\mathbf{P}}$ of the point P is

$$\ddot{\mathbf{P}} = \alpha\mathbf{a} + \beta\mathbf{b} + \gamma\mathbf{c},$$

where

$$\alpha = \dot{u} + uv\left(\frac{\mathbf{a}\cdot\mathbf{b}}{|A-P|} - \frac{1}{|B-P|}\right) + uw\left(\frac{\mathbf{a}\cdot\mathbf{c}}{|A-P|} - \frac{1}{|C-P|}\right)$$

with similar expressions for β and γ.

21. Prove that if $z = u(x, y)$ represents the surface formed by the tangents of an arbitrary curve, then (a) every osculating plane of the curve is a tangent plane to the surface and (b) $u(x, y)$ satisfies the equation

$$u_{xx}u_{yy} - u_{xy}^2 = 0.$$

CHAPTER 3

Developments and Applications of the Differential Calculus

3.1 Implicit Functions

a. General Remarks

Frequently in analytical geometry the equation of a curve is given not in the form $y = f(x)$ but in the form $F(x, y) = 0$. A straight line may be represented in this way by the equation $ax + by + c = 0$, and an ellipse, by the equation $x^2/a^2 + y^2/b^2 = 1$. To obtain the equation of the curve in the form $y = f(x)$ we must "solve" the equation $F(x, y) = 0$ for y. In Volume I we considered the special problem of finding the inverse of a function $y = f(x)$, that is, the problem of solving the equation $F(x, y) = y - f(x) = 0$ for the variable x.

These examples suggest the importance of methods for solving an equation $F(x, y) = 0$ for x or for y. We shall find such methods even for equations involving functions of more than two variables.

In the simplest cases, such as the foregoing equations for the straight line and ellipse, the solution can readily be found in terms of elementary functions. In other cases, the solution can be approximated as closely as we desire. For many purposes, however, it is preferable not to work with the solved form of the equation or with these approximations but instead to draw conclusions about the solution by directly studying the function $F(x, y)$, in which neither of the variables x, y is given preference over the other.

Not every equation $F(x, y) = 0$ is the implicit representation of a function $y = f(x)$ or $x = \phi(y)$. It is easy to give examples of equations $F(x, y) = 0$ that permit no solution in terms of functions of one variable. Thus, the equation $x^2 + y^2 = 0$ is satisfied by the single pair of values $x = 0$, $y = 0$ only, while the equation $x^2 + y^2 + 1 = 0$ is satisfied by no real values at all. It is therefore necessary to investigate more closely the circumstances under which an equation $F(x, y) = 0$ defines a function $y = f(x)$ and the properties of this function.

Exercises 3.1a

1. Suppose that for some pair of values (a, b), $f(a, b) = 0$. If a is known, give a constructive iterative method for finding b. Under what conditions on f will this method work?

b. Geometrical Interpretation

To clarify the situation we represent the function $F(x, y)$ by the surface $z = F(x, y)$ in three-dimensional space. The solutions of the equation $F(x, y) = 0$ are the same as the simultaneous solutions of the two equations $z = F(x, y)$ and $z = 0$. Geometrically, our problem is to find whether the surface $z = F(x, y)$ intersects the x, y-plane in curves $y = f(x)$ or $x = \phi(y)$. (How *far* such a curve of intersection may extend does not concern us here.)

A first possibility is that the surface and the plane have no point in common. For example the paraboloid $z = F(x, y) = x^2 + y^2 + 1$ lies entirely above the x, y-plane. Here there is no curve of intersection. Obviously, we need consider only cases in which there is at least one point (x_0, y_0) at which $F(x_0, y_0) = 0$; the point (x_0, y_0) constitutes an "initial point" for our solution.

Knowing an initial solution, we have two possibilities: either the tangent plane at the point (x_0, y_0) is horizontal or it is not. If the tangent plane is horizontal, we can readily show by means of examples that it may be impossible to extend a solution $y = f(x)$ or $x = \phi(y)$ from (x_0, y_0). For example, the paraboloid $z = x^2 + y^2$ has the initial solution $x = 0$, $y = 0$, but contains no other point in the x, y-plane. In contrast, the surface $z = xy$ with the initial solution $x = 0, y = 0$ intersects the x, y-plane along the lines $x = 0$ and $y = 0$; but in no neighborhood of the origin can we represent the *whole* intersection by a function $y = f(x)$ or by a function $x = \phi(y)$, (see Figs. 3.1 and 3.2). On the other hand, it is quite possible for the equation $F(x, y) = 0$ to have such a solution even when the tangent plane at the initial solution is horizontal, as in the case $F(x, y) = (y - x)^4 = 0$. In the exceptional case of a horizontal tangent plane, therefore, no definite general statement can be made.

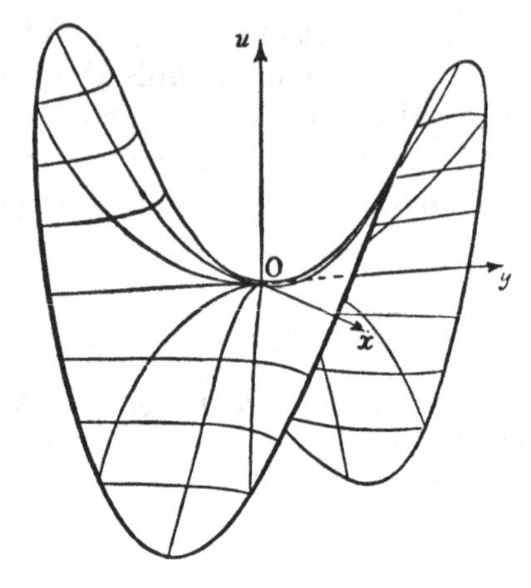

Figure 3.1 The surface $u = xy$.

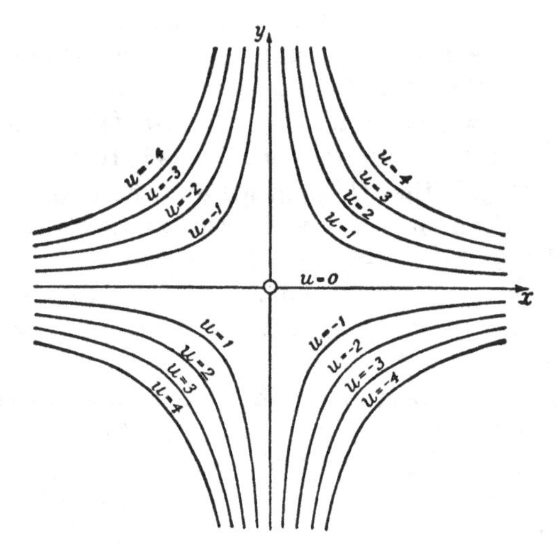

Figure 3.2 Contour lines of $u = xy$.

The remaining possiblity is that the tangent plane at the initial solution is not horizontal. Then, thinking intuitively of the surface $z = F(x, y)$ as approximated by the tangent plane in a neighborhood of the initial solution, we may expect that the surface cannot bend fast enough to avoid cutting the x,y-plane near (x_0, y_0) in a single well-defined curve of intersection and that a portion of the curve near the initial solution can be represented by the equation $y = f(x)$

or $x = \phi(y)$. Analytically, the statement that the tangent plane is not horizontal means that $F_x(x_0, y_0)$ and $F_y(x_0, y_0)$ are not both zero (see p. 47). This is the basis for the discussion in the next subsection.

Exercises 3.1b

1. By examining the surface of $z = f(x, y)$, determine whether the equation $f(x, y) = 0$ can be solved for y as a function of x in a neighborhood of the indicated point (x_0, y_0) for

   (a) $f(x, y) = x^2 - y^2, \quad x_0 = y_0 = 0$

   (b) $f(x, y) = [\log (x + y)]^{1/2}, \quad x_0 = 1.5, \quad y_0 = -.5$

   (c) $f(x, y) = \sin [\pi (x + y)] - 1, \quad x_0 = y_0 = 1/4$

   (d) $f(x, y) = x^2 + y^2 - y, \quad x_0 = y_0 = 0.$

c. *The Implicit Function Theorem*

We now state sufficient conditions for the existence of implicit functions and at the same time give a rule for differentiating them: *Let $F(x, y)$ have continuous derivatives F_x and F_y in a neighborhood of a point (x_0, y_0), where*

$$F(x_0, y_0) = 0, \quad F_y(x_0, y_0) \neq 0. \tag{1}$$

Then centered at the point (x_0, y_0), there is some rectangle

$$x_0 - \alpha \leqq x \leqq x_0 + \alpha, \quad y_0 - \beta \leqq y \leqq y_0 + \beta \tag{2}$$

such that for every x in the interval I given by $x_0 - \alpha \leqq x \leqq x_0 + \alpha$ the equation $F(x, y) = 0$ has exactly one solution $y = f(x)$ lying in the interval $y_0 - \beta \leqq y \leqq y_0 + \beta$. This function f satisfies the initial condition $y_0 = f(x_0)$ and, for every x in I,

$$F(x, f(x)) = 0. \tag{3}$$

$$y_0 - \beta \leqq f(x) \leqq y_0 + \beta \tag{3a}$$

$$F_y(x, f(x)) \neq 0. \tag{3b}$$

Furthermore, f is continuous and has a continuous derivative in I, given by the equation

$$y' = f'(x) = -\frac{F_x}{F_y}. \tag{4}$$

This is a strictly *local* existence theorem for solutions of the equation $F(x, y) = 0$ in the neighborhood of an initial solution (x_0, y_0). It does not indicate how to find such an initial solution or how to decide if the equation $F(x, y) = 0$ is satisfied for any (x, y) at all. These are *global* questions and beyond the scope of the theorem. *Uniqueness* and *regularity* of the solution $y = f(x)$, also, can be guaranteed only locally, that is, when y is restricted to the interval $y_0 - \beta < y < y_0 + \beta$. The need for such restrictions is evident from the simple example of the equation

$$F(x, y) = x^2 + y^2 - 1 = 0.$$

For every x with $-1 < x < 1$ the equation has two different solutions $y = \pm\sqrt{1 - x^2}$. A single-valued solution $y = f(x)$ is obtained by prescribing arbitrarily one of the signs at each x. It is clear that in this way we can find solutions that are discontinuous for every x, choosing, for example, the positive sign for rational x and the negative one for irrational x. Continuous solutions $y = f(x)$ are obtained if we restrict y to a constant sign. This sign can be fixed by choosing for a given x_0 in $-1 < x_0 < 1$ one of the two possible values y_0 for which $x_0^2 + y_0^2 = 1$. A unique continuous solution $y = f(x)$ with $y_0 = f(x_0)$ is obtained then for all x in $-1 < x < 1$ by requiring y to satisfy $x^2 + y^2 = 1$ and to have the same sign as y_0. Geometrically, the graph of f is either the upper or the lower semicircle, whichever contains the point (x_0, y_0). The function f has a continuous derivative

$$y' = -\frac{F_x}{F_y} = -\frac{x}{y} = -\frac{x}{f(x)}$$

for $-1 < x < 1$. With y defined to be zero for $x = \pm 1$, the solution $y = f(x)$ will be continuous in the closed interval $-1 \leqq x \leqq 1$. However, the derivative y' then becomes infinite at the end points of the interval, since $F_y = 0$ there.

We shall prove the general theorem in the next section. We observe here only that once the existence and the differentiability of the function $f(x)$ satisfying (3) have been established, we can find an explicit expression for $f'(x)$ by applying the chain rule [see (18) p. 55] to differentiate $F(x, y)$. This yields

$$F_x + F_y f'(x) = 0,$$

and leads to formula (4) as long as $F_y \neq 0$. Equivalently, if the equation $F(x, y) = 0$ determines y as a function of x, we conclude that

$$dF = F_x\,dx + F_y\,dy = 0$$

and, hence, that

$$dy = \frac{dy}{dx}\,dx = -\frac{F_x}{F_y}\,dx.$$

An implicit function $y = f(x)$ can be differentiated to any given order, provided the function $F(x, y)$ possesses continuous partial derivatives of that same order. For example, if $F(x, y)$ has continuous first and second derivatives in the rectangle (2), the right side of equation (4) is a compound function of x:

$$-\frac{F_x(x, f(x))}{F_y(x, f(x))}.$$

Since, by (3b), the denominator does not vanish and since $f(x)$ already is known to have a continuous first derivative, we conclude from (4) that y' has a continuous derivative; by the chain rule y'' is given by

$$y'' = -\frac{F_yF_{xx} + F_yF_{xy}f' - F_xF_{xy} - F_xF_{yy}f'}{F_y^2}.$$

Substituting the expression (4) for f', we find that

$$\text{(5)} \qquad y'' = -\frac{F_y^2F_{xx} - 2F_xF_yF_{xy} + F_x^2F_{yy}}{F_y^3}.$$

The rules (4) and (5) for finding the derivatives of an implicit function $y = f(x)$ can be used whenever the existence of f in an interval has been established from the general theorem on implicit functions, even in cases where it is impossible to express y explicitly in terms of elementary functions (rational functions, trigonometric functions, etc.). Even if we can solve the equation $F(x, y) = 0$ explicitly for y, it is usually easier to find the derivatives of y from the formulae (4) and (5), without making use of any explicit representation of $y = f(x)$.

Examples

1. The equation of the *lemniscate* (Volume I, p. 102)

$$F(x, y) = (x^2 + y^2)^2 - 2a^2(x^2 - y^2) = 0$$

is not easily solved for y. For $x = 0$, $y = 0$ we obtain $F = 0$, $F_x = 0$, $F_y = 0$. Here our theorem fails, as might be expected from the fact that two different branches of the lemniscate pass through the origin. However, at all points of the curve where $y \neq 0$, our rule applies, and the derivative of the function $y = f(x)$ is given by

$$y' = -\frac{F_x}{F_y} = -\frac{4x(x^2 + y^2) - 4a^2x}{4y(x^2 + y^2) + 4a^2y}.$$

We can obtain important information about the curve from this equation, without using the explicit expression for y. For example, maxima or minima might occur where $y' = 0$, that is, for $x = 0$ or for $x^2 + y^2 = a^2$. From the equation of the lemniscate, $y = 0$ when $x = 0$; but at the origin there is no extreme value (cf. Fig. 1.S.3, Volume I, p. 103). The two equations therefore give the four points $\left(\pm \frac{a}{2}\sqrt{3}, \pm \frac{a}{2}\right)$ as the maxima and minima.

2. The *folium of Descartes* has the equation

$$F(x, y) = x^3 + y^3 - 3axy = 0$$

(cf. Fig 3.3), with awkward explicit solutions. At the origin, where the curve intersects itself, our rule again fails, since at that point $F = F_x = F_y = 0$. For all points at which $y^2 \neq ax$ we have

$$y' = -\frac{F_x}{F_y} = -\frac{x^2 - ay}{y^2 - ax}.$$

Accordingly, there is a zero of the derivative when $x^2 - ay = 0$ or, if we use the equation of the curve, when

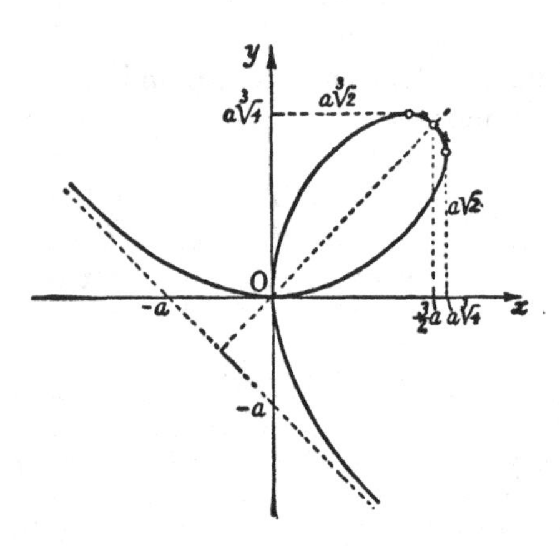

Figure 3.3 Folium of Descartes.

$$x = a\sqrt[3]{2}, \quad y = a\sqrt[3]{4}.$$

Exercises 3.1c

1. Prove that the following equations have unique solutions for y near the points indicated:
   (a) $x^2 + xy + y^2 = 7 \qquad (2, 1)$
   (b) $x \cos xy = 0 \qquad (1, \pi/2)$
   (c) $xy + \log xy = 1 \qquad (1, 1)$
   (d) $x^5 + y^5 + xy = 3 \qquad (1, 1)$.
2. Find the first derivatives of the solutions in Exercise 1 and give their values at the indicated points.
3. Find the second derivatives of the solutions in Exercise 1 and give their values at the indicated points.
4. Which of the implicitly defined functions of Exercise 1 are convex at the indicated points.
5. Find the maximum and minimum values of the function y that satisfies the equation $x^2 + xy + y^2 = 27$.
6. Let $f_y(x, y)$ be continuous on a neighborhood of the point (x_0, y_0). Show that the equation

$$y = y_0 + \int_{x_0}^{x} f(\xi, y)d\xi$$

determines y as a function of x in some interval about $x = x_0$.

d. *Proof of the Implicit Function Theorem*

Existence of the implicit function follows directly from the intermediate value theorem (see Volume I, p. 44). Assume that $F(x, y)$ is defined and has continuous first derivatives in a neighborhood of the point (x_0, y_0), and let

$$F(x_0, y_0) = 0, \quad F_y(x_0, y_0) \neq 0.$$

Without loss of generality we assume that $m = F_y(x_0, y_0) > 0$. Otherwise, we merely replace the function F by $-F$, which leaves the points described by the equation $F(x, y) = 0$ unaltered. Since $F_y(x, y)$ is continuous, we can find a rectangle R with center (x_0, y_0) and so small that R lies completely in the domain of F and $F_y(x, y) > m/2$ throughout R. Let R be the rectangle

$$x_0 - a \leqq x \leqq x_0 + a, \quad y_0 - \beta \leqq y \leqq y_0 + \beta$$

(see Fig. 3.4). Since $F_x(x, y)$ also is continuous, we conclude that F_x

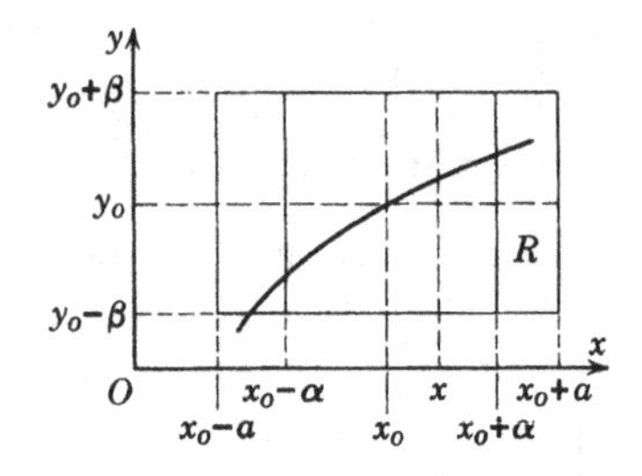

Figure 3.4

is bounded in R. Thus, there exist positive constants m, M such that

$$(6) \qquad F_y(x, y) > \frac{m}{2}, \qquad |F_x(x, y)| \leqq M \qquad \text{for} \qquad (x, y) \text{ in } R.$$

For any fixed x between $x_0 - a$ and $x_0 + a$ the expression $F(x, y)$ is a continuous and monotonically increasing function of y for $y_0 - \beta \leqq y \leqq y_0 + \beta$. If

$$(7) \qquad F(x, y_0 + \beta) > 0, \qquad F(x, y_0 - \beta) < 0,$$

we can be sure that there exists a single value y intermediate between $y_0 - \beta$ and $y_0 + \beta$ at which $F(x, y)$ vanishes. For the given x the equation $F(x, y)$ will then have a single solution $y = f(x)$ for which

$$y_0 - \beta < y < y_0 + \beta.$$

To prove (7), we observe that by the mean value theorem

$$F(x, y_0) = F(x, y_0) - F(x_0, y_0) = F_x(\xi, y_0)(x - x_0).,$$

where ξ is intermediate between x_0 and x. Hence, if α denotes a number between 0 and a, we have

$$|F(x, y_0)| \leqq |F_x(\xi, y_0)|\,|x - x_0| \leqq M\alpha \qquad \text{for} \qquad |x - x_0| \leqq \alpha.$$

Similarly, it follows from $F_y > m/2$ that

$$F(x, y_0 + \beta) = [F(x, y_0 + \beta) - F(x, y_0)] + F(x, y_0) > \frac{1}{2} m\beta - M\alpha,$$

$$F(x, y_0 - \beta) = -[F(x, y_0) - F(x, y_0 - \beta)] + F(x, y_0) < -\frac{1}{2} m\beta + M\alpha.$$

Thus, the inequalities (7) hold for any x in the interval $x_0 - \alpha \leqq x \leqq$

$x_0 + \alpha$ provided we take α so small that $\alpha \leqq a$ and $\alpha < m\beta/2M$.

For any x with $|x - x_0| \leqq \alpha$ this proves existence and uniqueness of a solution $y = f(x)$ of the equation $F(x, y) = 0$ such that $|y - y_0| \leqq \beta$ and $F_y(x, y) > m/2 > 0$. For $x = x_0$ the equation $F(x, y) = 0$ has the solution $y = y_0$ corresponding to our initial point. Since y_0 certainly lies between $y_0 - \beta$ and $y_0 - \beta$, we see that $f(x_0) = y_0$. Continuity and differentiability of $f(x)$ now follow from the mean value theorem for functions of several variables applied to $F(x, y)$ [see (33) p. 67]. Let x and $x + h$ be two values between $x_0 - \alpha$ and $x_0 + \alpha$. Let $y = f(x)$ and $y + k = f(x + h)$ be the corresponding values of f where y and $y + k$ lie between $y_0 - \beta$ and $y_0 + \beta$. Then $F(x, y) = 0$, $F(x + h, y + k) = 0$. It follows that

$$\begin{aligned} 0 &= F(x + h, y + k) - F(x, y) \\ &= F_x(x + \theta h, y + \theta k)\, h + F_y(x + \theta h, y + \theta k) k, \end{aligned}$$

where θ is a suitable intermediate value between 0 and 1.[1] Using $F_y \neq 0$, we can divide by F_y and find that

$$\frac{k}{h} = -\frac{F_x(x + \theta h, y + \theta k)}{F_y(x + \theta h, y + \theta k)}. \tag{8}$$

Since $|F_x| \leqq M$, $|F_y| > m/2$ for all points of our rectangle, we find that the right-hand side is bounded by $2M/m$. Thus

$$|k| \leqq \frac{2M}{m} |h|.$$

Hence, $k = f(x + h) - f(x) \to 0$ for $h \to 0$, which shows that $y = f(x)$ is a continuous function. We conclude from (8) that for fixed x and for $y = f(x)$,

$$\lim_{h \to 0} \frac{f(x + h) - f(x)}{h} = -\lim_{h \to 0} \frac{F_x(x + \theta h, y + \theta k)}{F_x(x + \theta h, y + \theta k)} = -\frac{F_x(x, y)}{F_y(x, y)}.$$

This establishes the differentiability of f and at the same time yields formula (4) for the derivative.

The proof hinges on the assumption $F_y(x_0, y_0) \neq 0$, from which we could conclude that F_y is of constant sign in a sufficiently small

[1]Observe that the mean value theorem can be applied here, since the segment joining any two points of the rectangle $|x - x_0| \leqq \alpha$, $|y - y_0| \leqq \beta$ lies wholly within the rectangle.

neighborhood of (x_0, y_0) and that $F(x, y)$ for fixed x is a monotone function of y.

The proof merely tells us that the function $y = f(x)$ exists. It is a typical example of a pure "existence theorem," in which the practical possibility of calculating the solution is not considered. Of course, we could apply any of the numerical methods discussed in Volume I (pp. 494 ff.) to approximate the solution y of the equation $F(x, y) = 0$ for fixed x.

Exercises 3.1d

1. Give an example of a function $f(x, y)$ such that (a) $f(x, y) = 0$ can be solved for y as a function of x near $x = x_0$, $y = y_0$, and (b) $f_y(x_0, y_0) = 0$.
2. Give an example of an equation $F(x, y) = 0$ that can be solved for y as a function $y = f(x)$ near a point (x_0, y_0), such that f is not differentiable at x_0.
3. Let $\phi(x)$ be defined for all real values of x. Show that the equation $F(x, y) = y^3 - y^2 + (1 + x^2)\, y - \phi(x) = 0$ defines a unique value of y for each value of x.

e. *The Implicit Function Theorem for More Than Two Independent Variables*

The implicit function theorem can be extended to a function of several independent variables as follows:

Let $F(x, y, \ldots, z, u)$ be a continuous function of the independent variables $x, y, \ldots z, u$, with continuous partial derivatives $F_x, F_y, \ldots, F_z, F_u$. Let $(x_0, y_0, \ldots, z_0, u_0)$ be an interior point of the domain of definition of F, for which

$$F(x_0, y_0, \ldots, z_0, u_0) = 0 \quad \text{and} \quad F_u(x_0, y_0, \ldots, z_0, u_0) \neq 0.$$

Then we can mark off an interval $u_0 - \beta \leqq u \leqq u_0 + \beta$ about u_0 and a rectangular region R containing $(x_0, y_0, \ldots, z_0)$ in its interior such that for every $(x, y, \ldots, z)$ in R, the equation $F(x, y, \ldots, z, u) = 0$ is satisfied by exactly one value of u in the interval $u_0 - \beta \leqq u \leqq u_0 + \beta$.[1] *For this value of u, which we denote by $u = f(x, y, \ldots, z)$, the equation*

$$F(x, y, \ldots, z, f(x, y, \ldots, z)) = 0$$

holds identically in R; in addition,

[1]The value β and the rectangular region R are not determined uniquely. The assertion of the theorem is valid if β is any sufficiently small positive number and if we choose R (depending on β) sufficiently small.

$$u_0 = f(x_0, y_0, \ldots, z_0),$$

$$u_0 - \beta < f(x, y, \ldots, z) < u_0 + \beta;\ F_u(x, y, \ldots, z,\ f(x, y, \ldots, z)) \neq 0.$$

The function f is a continuous function of the independent variables x, y, ..., z, and possesses continuous partial derivatives given by the equations

$$F_x + F_u f_x = 0,\ F_y + F_u f_y = 0, \ldots, F_z + F_u f_z = 0. \tag{9a}$$

The proof follows exactly the same lines that were given in the previous section for the solution of the equation $F(x, u) = 0$ and offers no further difficulty.

It is suggestive to combine the differentiation formulae (9a) in the single equation

$$F_x\,dx + F_y\,dy + \cdots + F_z\,dz + F_u\,du = 0. \tag{9b}$$

In words, *if the variables $x, y, \ldots, z, u$, are not independent of one another but are subject to the condition $F(x, y, \ldots, z, u) = 0$, then the linear parts of the increments of these variables are likewise not independent but are connected by the linear equation*

$$dF = F_x\,dx + F_y\,dy + \cdots + F_z\,dz + F_u\,du = 0.$$

If we replace du in (9b) by the expression $u_x dx + u_y dy + \cdots + u_z dz$ and then equate the coefficient of each of the mutually independent differentials $dx, dy, \ldots, dz$ to zero, we retrieve the differentiation formulae (9a).

Incidentally, the concept of implicit function enables us to give a general definition of an *algebraic function.* We say that $u = f(x, y, \ldots)$ is an *algebraic* function of the independent variables $x, y, \ldots$ if u can be defined implicitly by an equation $F(x, y, \ldots, u) = 0$, where F is a polynomial in the arguments $x, y, \ldots, u$; briefly, if u "satisfies an algebraic equation." A function that satisfies no algebraic equation is called *transcendental.*

As an example, we apply our differentiation formulae to the equation of the sphere,

$$F(x, y, u) = x^2 + y^2 + u^2 - 1 = 0.$$

For the partial derivatives, we obtain

$$u_x = -\frac{x}{u}, \quad u_y = -\frac{y}{u},$$

and by further differentiation

$$u_{xx} = -\frac{1}{u} + \frac{x}{u^2} u_x = -\frac{x^2 + u^2}{u^3},$$

$$u_{xy} = \frac{x}{u^2} u_y = -\frac{xy}{u^3},$$

$$u_{yy} = -\frac{1}{u} + \frac{y}{u^2} u_y = -\frac{y^2 + u^2}{u^3}.$$

Exercises 3.1e

1. Show that the equation $x + y + z = \sin xyz$ can be solved for z near (0, 0, 0). Find the partial derivatives of the solution.
2. For each of the following equations examine whether it has a unique solution for z as a function of the remaining variables near the indicated point:

   (a) $\sin x + \cos y + \tan z = 0 \qquad (x = 0,\ y = \frac{\pi}{2},\ z = \pi)$

   (b) $x^2 + 2y^2 + 3z^2 - w = 0 \qquad (x = 1,\ y = 2,\ z = -1,\ w = 8)$

   (c) $1 + x + y = \cosh (x + z) + \sinh (y + z) \qquad (x = y = z = 0)$.
3. Show that $x + y + z + xyz^3 = 0$ defines z implicitly as a function of x and y in a neighborhood of (0, 0, 0). Expand z to fourth order in powers of x and y.

3.2 Curves and Surfaces in Implicit Form

a. *Plane Curves in Implicit Form*

The description of a plane curve by an equation of the form $y = f(x)$ gives asymmetric preference to one of the coordinates. The *tangent* and the *normal* to the curve were found (see Volume I, pp. 344–345) to be given by the respective equations

$$(\eta - y) - (\xi - x)f'(x) = 0 \tag{10a}$$

and

$$(\eta - y)f'(x) + (\xi - x) = 0, \tag{10b}$$

where ξ, η are the "running coordinates" of an arbitrary point on the tangent or normal, and x, y are the coordinates of the point on the curve. The *curvature* of the curve is

(10c) $$k = \frac{f''}{(1 + f'^2)^{3/2}}$$

(see Volume I p. 357). For a point of inflection the condition

(10d) $$f''(x) = 0$$

holds. We shall now obtain the corresponding symmetrical formulae for curves represented implicitly by an equation of the type $F(x, y) = 0$. We do this under the assumption that at the point in question F_x and F_y are not both 0, so that

(11) $$F_x^2 + F_y^2 \neq 0.$$

If we suppose that $F_y \neq 0$, say, we can substitute for $f'(x)$ in (10a, b), its value from (4), p. 221, and at once obtain the equation of the *tangent* in the form

(12a) $$(\xi - x)F_x + (\eta - y)F_y = 0$$

and that of the *normal* in the form

(12b) $$(\xi - x)F_y - (\eta - y)F_x = 0.$$

For $F_y = 0$, $F_x \neq 0$ we obtain the same equations by starting from the solution of the implicit equation $F(x, y) = 0$ in the form $x = g(y)$.

The *direction cosines of the normal* to the curve at the point (x, y)—that is, the direction cosines of the normal to the line with equation (12a) in the ξ, η-plane—are given by

(12c) $$\cos\alpha = \frac{F_x}{\sqrt{F_x^2 + F_y^2}}, \quad \sin\alpha = \frac{F_y}{\sqrt{F_x^2 + F_y^2}}$$

[see (20), p. 135] Similarly, the direction cosines of the tangent to the curve—that is, of the normal to the line (12b)—are

(12d) $$\cos\beta = \frac{-F_y}{\sqrt{F_x^2 + F_y^2}}, \quad \sin\beta = \frac{F_x}{\sqrt{F_x^2 + F_y^2}}.$$

There are actually two directions normal to the curve at a given point, the one with direction cosines (12c) and the opposite one. The normal given by (12c) has the same direction as the vector with components F_x, F_y, the *gradient* of F (see p. 205). We saw on p. 206 that the direction of the gradient vector is the one in which F increases fastest;

thus, at a point of the curve $F(x, y) = 0$ the gradient points into the region $F > 0$ and the same holds for the normal direction determined by the formulae (12c).

Formula (5), p. 223 gave the expression for the second derivative $y'' = f''(x)$ of a function given in explicit form $F(x, y) = 0$. It follows that the necessary condition $f'' = 0$ for the occurrence of a point of inflection can be written as

$$F_y^2 F_{xx} - 2F_x F_y F_{xy} + F_x^2 F_{yy} = 0 \tag{13}$$

for curves given implicitly. In this formula there is no preference for either of the two variables x, y. It is completely symmetric and no longer requires the assumption that $F_y \neq 0$. This symmetric character reflects, of course, the fact that the notion of point of inflection has a geometrical meaning quite independent of any coordinate system.

If we substitute formula (5) for $f''(x)$ into the formula (10c) for the curvature k of the curve, we again obtain an expression[1] symmetric in x and y,

$$k = \frac{F_y^2 F_{xx} - 2F_x F_y F_{xy} + F_x^2 F_{yy}}{(F_x^2 + F_y^2)^{3/2}}. \tag{14a}$$

Introducing the *radius of curvature*

$$\rho = \frac{1}{k}, \tag{14b}$$

we find for the coordinates ξ, η of the *center of curvature*, the point on the inner normal at distance ρ from (x, y) (see Volume I, p. 358),

$$\xi = x - \rho \frac{F_x}{\sqrt{F_x^2 + F_y^2}}, \quad \eta = y - \rho \frac{F_y}{\sqrt{F_x^2 + F_y^2}}. \tag{14c}$$

If instead of the curve $F(x, y) = 0$, we consider the curve

$$F(x, y) = c,$$

where c is a constant, everything in the preceding discussions remains the same. We only have to replace the function $F(x, y)$ by $F(x, y) - c$, which has the same derivatives as the original function. Thus, for

[1]For the sign of the curvature, see Volume I, p. 357. The curvature k defined by formula (14a) is positive if F increases on the "outer" side of the curve, that is, if the tangent to the curve near the point of contact lies in the region $F \geqq 0$.

these curves, the form of the equations of the tangent, normal, and so on are exactly the same as above.

The class of all curves $F(x, y) - c = 0$ that we obtain when we allow c to range through all the values of an interval forms the family of "contour lines," or "level lines," of the function $F(x, y)$; (see p. 14). More generally, *we obtain a one-parameter family of curves* from an equation of the form

$$F(x, y, c) = 0,$$

which for each constant value of the parameter c yields a curve Γ_c in implicit form. For a point (x, y) lying on the curve Γ_c—that is, satisfying the equation $F(x, y, c) = 0$—all the formulae derived previously apply. In particular, the gradient vector $(F_x(x, y, c), F_y(x, y, c))$ is normal to Γ_c at the point (x, y).

As an example, we consider the ellipse

$$F(x, y) = \frac{x^2}{a^2} + \frac{y^2}{b^2} = 1. \tag{15a}$$

By (12a) the equation of the tangent at the point (x, y) is

$$(\xi - x)\frac{x}{a^2} + (\eta - y)\frac{y}{b^2} = 0;$$

hence, from (15a),

$$\frac{\xi x}{a^2} + \frac{\eta y}{b^2} = 1.$$

We find from (14a) that the curvature is

$$k = \frac{a^4 b^4}{(a^4 y^2 + b^4 x^2)^{3/2}}. \tag{15b}$$

If $a > b$, this has its greatest value a/b^2 at the vertices $y = 0$, $x = \pm a$. Its least value b/a^2 occurs at the other vertices $x = 0$, $y = \pm b$.

If two curves $F(x, y) = 0$ and $G(x, y) = 0$ intersect at the point (x, y) the *angle between the curves* is defined as the angle ω formed by their tangents (or normals) at the point of intersection. If we recall that the gradients give the direction of the normals and apply formula (7), p. 128 for the angle between two vectors, we find that

$$\cos \omega = \frac{F_x G_x + F_y G_y}{\sqrt{F_x^2 + F_y^2}\,\sqrt{G_x^2 + G_y^2}}. \tag{16}$$

Here $\cos \omega$ is determined uniquely by the choice of ω as angle between the normals of the two curves in the directions of increasing F and G.

Putting $\omega = \pi/2$ in (16), we obtain the condition for *orthogonality*, that is, for the curves to intersect at right angles at the point (x, y):

$$F_x G_x + F_y G_y = 0. \tag{16a}$$

If the curves *touch*—that is, have a common tangent and normal in the point where they meet—their gradient vectors (F_x, F_y) and (G_x, G_y) must be parallel. This leads to the condition

$$F_x G_y - F_y G_x = 0. \tag{16b}$$

As an example, we consider the family of parabolas

$$F(x, y, c) = y^2 - 2c\left(x + \frac{c}{2}\right) = 0 \tag{17a}$$

(see Fig. 3.9, p. 245), all of which have the origin as focus ("confocal parabolas"). If $c_1 > 0$ and $c_2 < 0$, the two parabolas

$$F(x, y, c_1) = y^2 - 2c_1\left(x + \frac{c_1}{2}\right) = 0$$

and

$$F(x, y, c_2) = y^2 - 2c_2\left(x + \frac{c_2}{2}\right) = 0$$

intersect each other perpendicularly at two points; for at the points of intersection

$$x = -\frac{1}{2}(c_1 + c_2), \quad y^2 = -c_1 c_2,$$

and hence,

$$F_x(x, y, c_1)\, F_x(x, y, c_2) + F_y(x, y, c_1)\, F_y(x, y, c_2)$$
$$= 4(c_1 c_2 + y^2) = 0.$$

By (14a) the curvature of the parabola (17a) is given by

$$k = \frac{c^2}{(c^2 + y^2)^{3/2}}.$$

At the *vertex* $x = -c/2$, $y = 0$, this reduces to

$$k = \frac{1}{|c|}.$$

The center of curvature or center of the *osculating circle* at the vertex has then by (14c) the coordinates

$$\xi = -\frac{c}{2} + |c| \operatorname{sgn} c = \frac{c}{2}, \qquad \eta = 0$$

so that the focus (0, 0) lies halfway between the vertex and the center of curvature.

Exercises 3.2a

1. Find the equations of the tangent and normal for the curves given implicitly by the following relations:

   (a) $x^2 + 2y^2 - xy = 0$

   (b) $e^x \sin y + e^y \cos x = 1$

   (c) $\cosh(x + 1) - \sin y = 0$

   (d) $x^2 + y^2 = y + \sin x$

   (e) $x^3 + y^4 = \cosh y$

   (f) $x^y + y^x = 1$.

2. Calculate the curvature of the curve

   $$\sin x + \cos y = 1$$

   at the origin.

3. Find the curvature of a curve that is given in polar coordinates by the equation $f(r, \theta) = 0$.

4. Prove that the intersections of the curve

   $$(x + y - a)^3 + 27axy = 0$$

   with the line $x + y = a$ are inflections of the curve.

5. Determine a and b so that the conics

   $$4x^2 + 4xy + y^2 - 10x - 10y + 11 = 0$$
   $$(y + bx - 1 - b)^2 - a(by - x + 1 - b) = 0$$

   cut one another orthogonally at the point (1, 1) and have the same curvature at this point.

6. Let K' and K'' be two circles having two points A and B in common. If a circle K is orthogonal to K' and K'', then it is also orthogonal to every circle passing through A and B.

b. Singular Points of Curves

In many of the formulae of the preceding section the expression $F_x^2 + F_y^2$ occurs in the denominator. Accordingly, we may expect something unusual to happen when this quantity vanishes, that is, when $F_x = 0$ and $F_y = 0$ at a point of the curve $F(x,y) = 0$. At such a point the expression $y' = -F_x/F_y$ for the slope of the tangent loses its meaning.

We call a point P of a curve *regular* if in a neighborhood of P either variable x or y can be represented as a continuously differentiable function of the other. In that case, the curve has a tangent at P and is closely approximated by that tangent in a neighborhood of P. If not regular, a point of the curve is called *singular* or a *singularity*.

From the implicit function theorem we know that if $F(x, y)$ has continuous first partial derivatives, then a point of the curve $F(x, y) = 0$ is regular if at that point $F_x^2 + F_y^2 \neq 0$, for if $F_y \neq 0$ at P, we can solve the equation $F(x, y) = 0$ and obtain a unique continuously differentiable solution $y = f(x)$. Similarly, if $F_x \neq 0$ we can solve the equation for x.

An important type of singularity is a *multiple point*, that is, a point through which two or more branches of the curve pass. For example, the origin is a multiple point of the lemniscate (Volume I, p. 102)

$$(x^2 + y^2)^2 - 2a^2(x^2 - y^2) = 0.$$

It is clear that in the neighborhood of a multiple point we cannot express the equation of the curve uniquely in the form $y = f(x)$ or $x = g(y)$.

An example of a singularity that is not a multiple point is furnished by the cubic curve

$$F(x, y) = y^3 - x^2 = 0.$$

(see Fig. 3.5). Here at the origin $F_x = F_y = 0$. Solving for y, we can put the equation of the curve into the form

$$y = f(x) = \sqrt[3]{x^2},$$

where f is continuous but not differentiable at the origin. The curve has a *cusp* at that point.

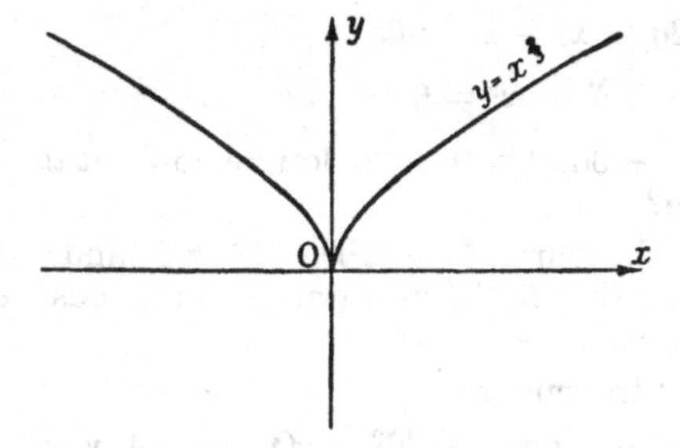

Figure 3.5 The curve $y^3 - x^2 = 0$.

A curve *can be regular* at a point where both F_x and F_y vanish. This is exemplified by

$$F(x, y) = y^3 - x^4 = 0.$$

Here again $F_x = F_y = 0$ at the origin. But solving for y, we find

$$y = f(x) = \sqrt[3]{x^4},$$

where $f(x)$ is continuously differentiable for all x. Thus, the origin is a regular point. Since F is an even function of x, the curve is symmetric with respect to the y-axis. It is convex and touches the x-axis at the origin, like the parabola $y = x^2$. Yet the origin is a somewhat special point for the curve, since there f'' becomes infinite, and there the curve has *infinite curvature*.

The trivial example of the equation

$$F(x, y) = (y - x)^2 = 0$$

representing the straight line $y = x$ shows that no peculiar behavior has to be associated with points of a curve $F(x, y) = 0$ for which $F_x^2 + F_y^2 = 0$. We shall treat singular points more systematically in Appendix 3.

Exercises 3.2b

1. Discuss the singular points of the following curves at the origin:

   (a) $F(x, y) = ax^3 + by^3 - cxy = 0$

   (b) $F(x, y) = (y^2 - 2x^2)^2 - x^5 = 0$

   (c) $F(x, y) = (1 + e^{1/x})y - x = 0$

(d) $F(x, y) = y^2(2a - x) - x^3 = 0$

(e) $F(x, y) = (y - 2x)^2 - x^5 = 0$.

2. The curve $x^3 + y^3 - 3axy = 0$ has a double point at the origin. What are its tangents there?
3. Draw a graph of the curve $(y - x^2)^2 - x^5 = 0$, and show that it has a cusp at the origin. What is the peculiarity of this cusp as compared with the cusp of the curve $x^2 - y^3 = 0$?
4. Show that each of the curves
$$(x \cos \alpha - y \sin \alpha - b)^3 = c(x \sin \alpha + y \cos \alpha)^2,$$
where α is a parameter and b, c constants, has a cusp and that the cusps all lie on a circle.
5. Let (x, y) be a double point of the curve $F(x, y) = 0$. Calculate the angle ϕ between the two tangents at (x, y), assuming that not all the second derivatives of F vanish at (x, y). Find the angle between the tangents at the double point

(a) of the lemniscate,

(b) of the folium of Descartes (cf. p. 224).

6. Find the curvature at the origin of each of the two branches of the curve
$$y(ax + by) = cx^3 + ex^2y + fxy^2 + gy^3.$$

c. *Implicit Representation of Surfaces*

Hitherto, we have usually represented a surface in x, y, z-space by means of a function $z = f(x, y)$. For a given surface in space the preference for the coordinate z implied in this representation may prove inconvenient. It is more natural and more general to represent surfaces in space implicitly by equations of the form $F(x, y, z) = 0$ or $F(x, y, z) =$ constant. For example, it is better to represent a sphere about the origin by the symmetric equation $x^2 + y^2 + z^2 - r^2 = 0$ than by $z = \pm\sqrt{r^2 - x^2 - y^2}$. The explicit representation of the surface appears then as the special implicit representation $F(x, y, z) = z - f(x, y) = 0$.

In order to derive the equation of the tangent plane at a point P of the surface $F(x, y, z) = 0$, we make the assumption that at that point

$$F_x{}^2 + F_y{}^2 + F_z{}^2 \neq 0, \tag{18}$$

that is, that at least one of the partial derivatives is not 0.[1] If, say, $F_z \neq 0$, we can find an explicit equation $z = f(x, y)$ for the surface near P. The tangent plane at P has the equation

[1]Just as for curves, the vanishing of the gradient of F usually corresponds to singular behavior of the surface. We shall not discuss the nature of such singularities.

(19a) $$\zeta - z = (\xi - x)f_x + (\eta - y)f_y$$

in running coordinates ξ, η, ζ (see p. 47). Substituting for the derivatives of f their values $f_x = -F_x/F_z$, $f_y = -F_y/F_z$ in accordance with formulae (9a), p. 229, we obtain the equation of the tangent plane in the form

(19b) $$(\xi - x)F_x + (\eta - y)F_y + (\zeta - z)F_z = 0.$$

The normal to the tangent plane (19b) has the same direction as the gradient vector (F_x, F_y, F_z) (see p. 134). Hence, the direction cosines of the normal are given by the expressions

(19c) $$\cos\alpha = \frac{F_x}{\sqrt{F_x^2 + F_y^2 + F_z^2}}, \quad \cos\beta = \frac{F_y}{\sqrt{F_x^2 + F_y^2 + F_z^2}},$$

$$\cos\gamma = \frac{F_z}{\sqrt{F_x^2 + F_y^2 + F_z^2}}.$$

Here, more precisely, we have taken that normal of the plane that points in the direction of *increasing* F (see p. 206).

If two surfaces $F(x, y, z) = 0$ and $G(x, y, z) = 0$ intersect at a point, the ***angle*** ω ***between the surfaces*** is defined as the angle between their tangent planes or, what is the same thing, the angle between their normals. This is given by

(20a) $$\cos\omega = \frac{F_xG_x + F_yG_y + F_zG_z}{\sqrt{F_x^2 + F_y^2 + F_z^2}\,\sqrt{G_x^2 + G_y^2 + G_z^2}}.$$

In particular, the condition for perpendicularity (orthogonality) is

(20b) $$F_xG_x + F_yG_y + F_zG_z = 0.$$

Instead of a surface given by an equation $F(x, y, z) = 0$, we may consider more generally surfaces given by $F(x, y, z) = c$, where c is a constant. Different values of c yield different *level surfaces* of the function F (see p. 15). At any point (x, y, z) the gradient vector (F_x, F_y, F_z) is normal to the level surface passing through that point. Similarly, equation (19b) gives the tangent plane to the level surface.

As an example, we consider the *sphere*

$$x^2 + y^2 + z^2 = r^2.$$

By (19b), the tangent plane at the point (x, y, z) is

$$(\xi - x)2x + (\eta - y)2y + (\zeta - z)2z = 0$$

or

$$\xi x + \eta y + \zeta z = r^2.$$

The direction cosines of the normal are proportional to x, y, z, that is, the normal coincides with the radius vector drawn from the origin to the point (x, y, z).

For the most general *ellipsoid* with the coordinate axes as principal axes

$$\frac{x^2}{a^2} + \frac{y^2}{b^2} + \frac{z^2}{c^2} = 1$$

the equation of the tangent plane is

$$\frac{\xi x}{a^2} + \frac{\eta y}{b^2} + \frac{\zeta z}{c^2} = 1.$$

Exercises 3.2c

1. Find the tangent plane
   (a) of the surface
   $$x^3 + 2xy^2 - 7z^3 + 3y + 1 = 0$$
   at the point (1, 1, 1);
   (b) of the surface
   $$(x^2 + y^2)^2 + x^2 - y^2 + 7xy + 3x + z^4 - z = 14$$
   at the point (1, 1, 1);
   (c) of the surface
   $$\sin^2 x + \cos (y + z) = \frac{3}{4}$$
   at the point $(\pi/6, \pi/3, 0)$.
   (d) of the surface
   $$1 + x \cos \pi z + y \sin \pi z - z^2 = 0$$
   at the point (0, 0, 1);
   (e) of the surface
   $$\cos x + \cos y + 2 \sin z = 0$$
   at the point $(0, 0, -\pi/2)$;
   (f) of the surface
   $$x^2 + y^2 = z^2 + \sin z$$
   at the point (0, 0, 0).

2. Prove that the three surfaces of the family of surfaces

$$\frac{xy}{z} = u, \quad \sqrt{x^2 + z^2} + \sqrt{y^2 + z^2} = v, \quad \sqrt{x^2 + z^2} - \sqrt{y^2 + z^2} = w$$

that pass through a single point are orthogonal to one another.

3. The points A and B move uniformly with the same velocity, A starting from the origin and moving along the z-axis, B starting from the point $(a, 0, 0)$ and moving parallel to the y-axis. Find the surface generated by the straight lines joining them.

4. Show that the tangent plane at any point of the surface $x^2 + y^2 - z^2 = 1$ meets the surface in two straight lines.

5. If $F(x, y, z) = 1$ is the equation of a surface, F being a homogeneous function of degree h, then the tangent plane at the point (x, y, z) is given by

$$\xi F_x + \eta F_y + \zeta F_z = h.$$

6. Let z be defined as a function of x and y by the equation

$$x^3 + y^3 + z^3 - 3xyz = 0.$$

Express z_x and z_y as functions of x, y, z.

7. Find the angle of intersection of the following pairs of surfaces, at the indicated points:

(a) $2x^4 + 3y^3 - 4z^2 = -4, \quad 1 + x^2 + y^2 = z^2$, at $(0, 0, 1)$

(b) $x^y + y^z = 2, \quad \cosh(x + y - 2) + \sinh(x + z - 1) = 1$, at $(1, 1, 0)$

(c) $x^2 + y^2 = e^z, \quad x^2 + z^2 = e^y$, at $(1, 0, 0)$

(d) $1 + \sinh(x/\sqrt{z}) = \cosh(y/\sqrt{z}), \quad x^2 + y^2 = z^2 - 1$, at $(0, 0, 1)$

(e) $\cos \pi(x^2 + y) + \sin \pi(x^2 + z) = 1, \quad x^3 + y^3 = z^3$ at $(0, 0, 0)$.

3.3 Systems of Functions, Transformations, and Mappings

a. *General Remarks*

The results we have obtained for implicit functions now enable us to consider *systems* of functions, that is, to discuss several functions simultaneously. In this section we shall consider the particularly important case of systems in which the number of functions is the same as the number of independent variables. We begin by investigating the meaning of such systems in the case of two independent variables. If the two functions

$$\xi = \phi(x, y) \quad \text{and} \quad \eta = \psi(x, y) \tag{21a}$$

are both continuously differentiable in a set R of the x, y-plane, the *domain* of the functions, we can interpret this system of functions in two different ways. The first ("active") interpretation is by means of a *mapping* or *transformation*. (The second, as a coordinate transformation, will be discussed on p. 246). To the point P with coordinates (x, y) in the x, y-plane there corresponds the image point Π with coordinates (ξ, η) in the ξ, η-plane.

An example is the *affine* mapping or transformation

$$\xi = ax + by, \quad \eta = cx + dy$$

where a, b, c, d are constants (see p. 148).

Frequently (x, y) and (ξ, η) are interpreted as points of one and the same plane. In this case we speak of *a mapping, or a transformation of the x,y-plane into itself.*

The fundamental problem connected with a mapping is that of its inversion, the question whether and how x and y can in virtue of the equations $\xi = \phi(x, y)$ and $\eta = \psi(x, y)$ be regarded as functions of ξ and η and how to determine properties of these inverse functions.

If for (x, y) varying over the domain R of the mapping the images (ξ, η) vary over a set B in the ξ, η-plane, we call B the *image set* of R or the *range* of the mapping. If two different points of R always correspond to *two different points* of B, then for each point (ξ, η) of B there is a *single* point (x, y) of R for which (ξ, η) is the image. (The point (x, y) is called the *inverse image*, as opposed to the *image*). That is, we can invert the mapping uniquely, determining x and y as functions

$$x = g(\xi, \eta), \quad y = h(\xi, \eta), \tag{21b}$$

which are defined in B. We then say that the mapping (21a) has a *unique inverse* or is a 1–1 mapping, and we call the transformation (21b) the *inverse mapping* or *transformation* of the original one.

If in this mapping the point $P = (x, y)$ describes a curve in the domain R, its image point (ξ, η) usually will likewise describe a curve in the set B, which is called the *image curve* of the first. For example, to the line $x = c$, which is parallel to the y-axis, there corresponds in the ξ, η-plane the curve given in parametric form by the equations

$$\xi = \phi(c, y), \quad \eta = \psi(c, y), \tag{22a}$$

where y is the parameter. Again, to the line $y = k$ there corresponds the curve

$$\xi = \phi(x, k), \quad \eta = \psi(x, k). \tag{22b}$$

If to c and k we assign sequences of equidistant values $c_1, c_2, c_3, \ldots$ and $k_1, k_2, k_3, \ldots$, then the rectangular "coordinate net" consisting of the lines $x =$ constant and $y =$ constant (e.g., the network of lines on ordinary graph paper) gives rise to a corresponding net of curves, the curvilinear net, in the ξ, η-plane (Figs. 3.6 and 3.7). The two families of curves can be written in implicit form. If we represent the inverse mapping by the equations (21b), the equations of the curves are simply

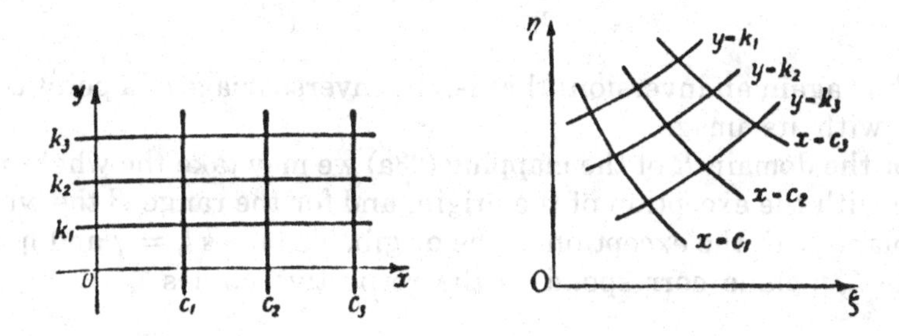

Figure 3.6 and **Figure 3.7** Nets of curves $x =$ constant and $y =$ constant in the x, y-plane and the ξ, η-plane.

(22c) $$g(\xi, \eta) = c \quad \text{and} \quad h(\xi, \eta) = k,$$

respectively. In many situations the curvilinear net furnishes a useful *geometric picture* of the mapping (21a) preferable to the interpretation of the equations as a two-dimensional surface in four-dimensional x, y, ξ, η-space.

In the same way, the two families of lines $\xi = \gamma$ and $\eta = \kappa$ in the ξ, η-plane correspond to the two families of curves

$$\phi(x, y) = \gamma \quad \text{and} \quad \psi(x, y) = \kappa$$

in the x, y-plane.

As an example, we consider the *inversion* (also called *mapping by reciprocal radii* or *reflection with respect to the unit circle*). This transformation is given by the equations

(23a) $$\xi = \frac{x}{x^2 + y^2}, \quad \eta = \frac{y}{x^2 + y^2}$$

To the point $P = (x, y)$ there corresponds the point $\Pi = (\xi, \eta)$ lying on the same ray OP and satisfying the equation

$$\text{(23b)} \qquad \xi^2 + \eta^2 = \frac{1}{x^2 + y^2} \qquad \text{or} \qquad O\Pi = \frac{1}{OP};$$

thus, the length of the position vector $\overrightarrow{OP}$ is the reciprocal of the length of the position vector $\overrightarrow{O\Pi}$. Points inside the unit circle $x^2 + y^2 = 1$ are mapped on points outside the circle and vice versa. From (23b) we find that the *inverse transformation* is..

$$x = \frac{\xi}{\xi^2 + \eta^2}, \qquad y = \frac{\eta}{\xi^2 + \eta^2},$$

which is again an inversion; that is, the inverse image of a point coincides with its image.

For the domain R of the mapping (23a) we may take the whole x, y-plane with the exception of the origin, and for the range B the whole ξ, η-plane with the exception of the origin. The lines $\xi = \gamma$ and $\eta = \kappa$ in the ξ, η-plane correspond to the respective circles

$$x^2 + y^2 - \frac{1}{\gamma} x = 0 \qquad \text{and} \qquad x^2 + y^2 - \frac{1}{\kappa} y = 0$$

in the x, y-plane. In the same way, the rectilinear coordinate net in the x, y-plane corresponds to the two families of circles touching the ξ-axis and η-axis at the origin.

As a further example we consider the mapping

$$\xi = x^2 - y^2, \quad \eta = 2xy.$$

The curves $\xi =$ constant give rise in the x, y-plane to the rectangular hyperbolas $x^2 - y^2 =$ constant, whose asymptotes are the lines $x = y$ and $x = -y$. The lines $\eta =$ constant also correspond to a family of rectangular hyperbolas having the coordinate axes as asymptotes. The hyperbolas of each family cut those of the other family at right angles (Fig. 3.8). The lines parallel to the axes in the x, y-plane correspond to two families of parabolas in the ξ, η-plane, the parabolas $\eta^2 = 4c^2(c^2 - \xi)$ corresponding to the lines $x = c$ and the parabolas $\eta^2 = 4k^2(k^2 + \xi)$ corresponding to the lines $y = k$. All these parabolas have the origin as focus and the ξ-axis as axis; they form a family of confocal and coaxial parabolas (Fig. 3.9).

One-one transformations have an important interpretation and application in the representation of *deformations or motions of continuously distributed substances*, such as fluids. If we think of such a substance as spread out at a given time over a region R and then deformed

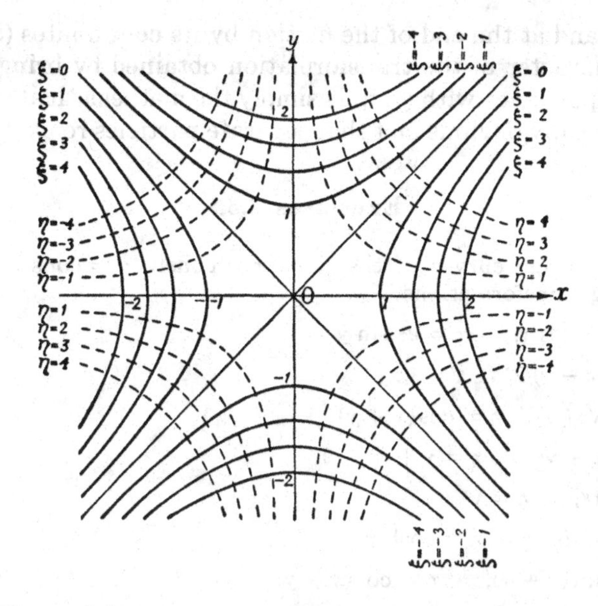

Figure 3.8 Orthogonal families of rectangular hyperbolas.

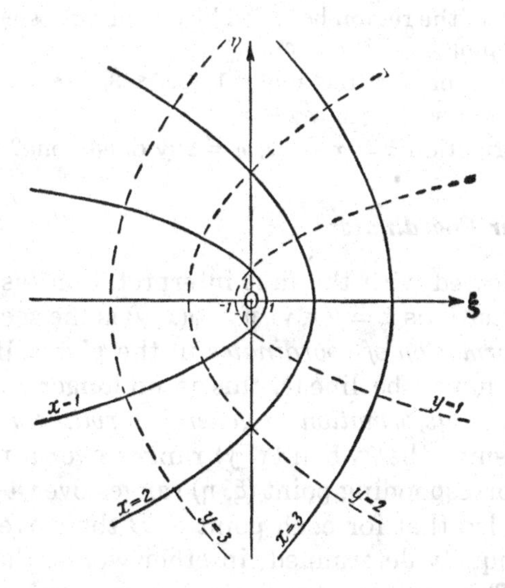

Figure 3.9 Orthogonal families of confocal parabolas.

by a motion, the substance originally spread over R will in general cover a region B different from R. Each particle of the substance can be distinguished at the beginning of the motion by its coordinates

(x, y) in R and at the end of the motion by its coordinates (ξ, η) in B. The 1–1 character of the transformation obtained by bringing (x, y) into correspondence with (ξ, η) is simply the mathematical expression of the physically obvious fact that separate particles remain separate.

Exercises 3.3a

1. Find the image curves of the lines $x =$ const., $y =$ const. under the following transformations:

   (a) $\xi = e^x \cos y, \qquad \eta = e^x \sin y$

   (b) $\xi = (x - y)/2, \qquad \eta = \sqrt{xy}$

   (c) $\xi = \sqrt{x/y}, \qquad \eta = \cos(x + y)$

   (d) $\xi = x + y^2, \qquad \eta = y + x^2 - 1$

   (e) $\xi = x^y, \qquad \eta = y^x$

   (f) $\xi = \sinh x, \qquad \eta = \cosh y$

   (g) $\xi = \sin(x + y), \qquad \eta = \cos(x - y)$

   (h) $\xi = e^{\cos x}, \qquad \eta = e^{\sin y}$.

2. Find the image of the region bounded by the curve $\cosh^2 x + \sinh^2 y = 1$ under the mapping $\xi = e^x$, $\eta = e^y$.
3. Find the image of the rectangle $1 \leqq x \leqq 3$, $4 \leqq y \leqq 16$, under the mapping $\xi = \sqrt{x + y}, \quad \eta = \sqrt{y - x}$.
4. Is the transformation $\xi = x - xy$, $\eta = 2xy$ one-to-one?

b. Curvilinear Coordinates

Closely connected with the first interpretation (as a mapping) of the system of equations $\xi = f(x, y)$, $\eta = \psi(x, y)$ is the second interpretation as a *transformation of coordinates* in the plane. If the functions φ and ψ happen not to be linear, this is no longer an "affine" transformation but a *transformation to general curvilinear coordinates.*

We again assume that when (x, y) ranges over a region R of the x, y-plane the corresponding point (ξ, η) ranges over a region B of the ξ, η-plane and also that for each point of B the corresponding (x, y) in R can be uniquely determined; in other words, that the transformation is 1–1. The inverse transformation we again denote by $x = g(\xi, \eta)$, $y = h(\xi, \eta)$.

By the *coordinates of a point P* in a region R we now mean any number-pair that serves to specify the position of the point P in R uniquely with respect to a given coordinate frame. Rectangular coordinates form the simplest system of coordinates that extend over the whole plane. Another familiar system is the system of polar coordinates in the x, y-plane, introduced by the equations

$$\xi = r = \sqrt{x^2 + y^2}$$

$$\eta = \theta = \arc\tan y/x \qquad (0 \leq \theta < 2\pi).$$

When we are given a system of functions $\xi = \phi(x, y)$, $\eta = \psi(x, y)$ as above, we can in general assign to each point $P(x, y)$ the corresponding values (ξ, η) as new coordinates, for each pair of values (ξ, η) belonging to the region B uniquely determines the pair (x, y), and, thus, uniquely determines the position of the point P in R. The "coordinate lines" ξ = constant and η = constant are then represented in the x, y-plane by two families of curves, which are defined implicitly by the equations $\phi(x,y)$ = constant and $\psi(x,y)$ = constant, respectively. These coordinate curves cover the region R with a coordinate net (usually curved), for which reason the coordinates (ξ, η) are also called *curvilinear coordinates* in R.

We shall once again point out how closely these two interpretations of our system of equations are interrelated. The curves in the ξ, η-plane that in the mapping correspond to straight lines parallel to the axes in the x, y-plane can be directly regarded as the coordinate curves for the curvilinear coordinates $x = g(\xi, \eta)$, $y = h(\xi, \eta)$ in the ξ, η-plane; conversely, the coordinate curves of the curvilinear system $\xi = \phi(x, y)$, $\eta = \psi(x, y)$ in the x, y-plane in the mapping are the images of the straight lines parallel to the axes in the ξ, η-plane. Even in the interpretation of (ξ, η) as curvilinear coordinates in the x,y-plane, we must consider a ξ, η-plane and a region B of that plane in which the point with the coordinates (ξ, η) can vary if we wish to keep the situation clear. The difference is mainly in the point of view.[1] If we are chiefly interested in the region R of the x, y-plane, we regard ξ, η simply as a new means of locating points in the region R, the region B of the ξ, η-plane being then merely subsidiary; while if we are equally interested in the two regions R and B in the x,y-plane and the ξ, η-plane, respectively, it is preferable to regard the system of equations as specifying a correspondence between the two regions, that is, a mapping of one on the other. It is, however, often desirable to keep the two interpretations, mapping, and transformation of coordinates, in mind at the same time.

[1]There is, however, a real difference, in that the equations always define a *mapping*, no matter how many points (x, y) correspond to one point (ξ, η), while they define a *transformation of coordinates* only when the correspondence is 1–1.

If, for example, we introduce polar coordinates (r, θ) and interpret r and θ as rectangular coordinates in an r, θ-plane, the circles $r =$ constant and the lines $\theta =$ constant are mapped on straight lines parallel to the axes in the r, θ-plane. If the region R of the x, y-plane is the circle $x^2 + y^2 \leq 1$, the point (r, θ) of the r, θ-plane will range over a rectangle $0 \leq r \leq 1, \quad 0 \leq \theta \leq 2\pi$, where corresponding points of the sides $\theta = 0$ and $\theta = 2\pi$ are associated with one and the same point of R and the whole side $r = 0$ is the image of the origin $x = 0, y = 0$.

Another example of a curvilinear coordinate system is the system of *parabolic coordinates*. We arrive at these by considering the family of confocal parabolas in the x, y-plane (cf. also p. 234 and Fig. 3.9)

$$y^2 = 2c\left(x + \frac{c}{2}\right),$$

all of which have the origin as focus and the x-axis as axis. Through each point of the plane but the origin there pass two parabolas of the family, one corresponding to a positive parameter value $c = \xi$ and the other to a negative parameter value $c = \eta$. We obtain these two values by solving for c the quadratic equation $y^2 = 2c(x + c/2)$ using the values of x and y corresponding to the point; this gives

$$\xi = -x + \sqrt{x^2 + y^2}, \qquad \eta = -x - \sqrt{x^2 + y^2}.$$

These quantities ξ and η may be introduced as curvilinear coordinates in the x, y-plane, the confocal parabolas then becoming the coordinate curves. These are indicated in Fig. 3.9 if we imagine the symbols (x, y) and (ξ, η) interchanged.

In using parabolic coordinates (ξ, η) we must bear in mind that the *one* pair of values (ξ, η) corresponds to *two* points (x, y) and $(x, -y)$, the two intersections of the corresponding parabolas. Hence, in order to obtain a 1–1 correspondence between the pair (x, y) and the pair (ξ, η), we must restrict ourselves to a half-plane, $y \geq 0$, say. Then every region R in this half-plane is in 1–1 correspondence with a region B of the ξ, η-plane, and the rectangular coordinates (ξ, η) of each point in this region B are exactly the same as the parabolic coordinates of the corresponding point in the region R.

Exercises 3.3b

1. Prove that for $x \neq 1, 0 < y < \pi/2, \xi = (\sin y)/(x - 1), \eta = x \tan y$, define a system of curvilinear coordinates.

2. Find the equation for the circle $x^2 + y^2 = 1$ in terms of the curvilinear coordinates

$$\xi = x^3 + 1, \quad \eta = xy.$$

3. For what points of the x, y-plane can we not use $\xi = xy$ and $\eta = x^2 + y^2$ as curvilinear coordinates?

c. *Extension to More Than Two Independent Variables*

For three or more independent variables the state of affairs is analogous. Thus, a system of three continuously differentiable functions

$$\xi = \phi(x, y, z), \qquad \eta = \psi(x, y, z), \qquad \zeta = \chi(x, y, z),$$

defined in a region R of x, y, z-space, may be regarded as the mapping of the region R on a region B of ξ, η, ζ-space. If this mapping of R on B is 1–1, so that for each image point (ξ, η, ζ) of B the coordinates (x, y, z) of the corresponding point (original point or inverse image) in R can be uniquely calculated by means of functions

$$x = g(\xi, \eta, \zeta), \qquad y = h(\xi, \eta, \zeta), \qquad z = l(\xi, \eta, \zeta),$$

then (ξ, η, ζ) may also be regarded as *general coordinates* of the point P in the region R. The surfaces ξ = constant, η = constant, ζ = constant, or, in other symbols,

$$\phi(x, y, z) = \text{constant}, \quad \psi(x, y, z) = \text{constant}, \quad \chi(x, y, z) = \text{constant},$$

then form a system of three families of surfaces that cover the region R and may be called curvilinear coordinate surfaces.

Just as for two independent variables, we can interpret 1–1 transformations in three dimensions as deformations of a substance spread continuously throughout a region of space.

A very important system of coordinates are the *spherical coordinates*, sometimes called *polar coordinates in space*. These specify the position of a point P in space by three numbers: (1) the distance $r = \sqrt{x^2 + y^2 + z^2}$ from the origin; (2) the geographical longitude ϕ, that is, the angle between the x, z-plane and the plane determined by P and the z-axis; and (3) the polar inclination or complementary latitude θ, that is, the angle between the radius vector OP and the positive z-axis. As we see from Fig. 3.10, the three spherical coordinates r, ϕ, θ are related to the rectangular coordinates by the equations of transformation

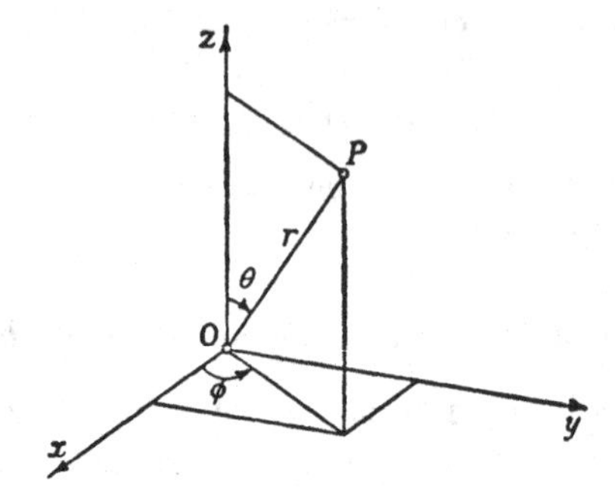

Figure 3.10 Spherical coordinates.

$$x = r \cos \phi \sin \theta,$$

$$y = r \sin \phi \sin \theta,$$

$$z = r \cos \theta,$$

from which we obtain the inverse relations

$$r = \sqrt{x^2 + y^2 + z^2}$$

$$\phi = \text{arc cos} \frac{x}{\sqrt{x^2 + y^2}} = \text{arc sin} \frac{y}{\sqrt{x^2 + y^2}}$$

$$\theta = \text{arc cos} \frac{z}{\sqrt{x^2 + y^2 + z^2}} = \text{arc sin} \frac{\sqrt{x^2 + y^2}}{\sqrt{x^2 + y^2 + z^2}}$$

For polar coordinates in the plane the origin is an exceptional point in that the 1–1 correspondence fails because the angle is indeterminate there. In the same way, for spherical coordinates in space the whole of the z-axis is an exception in that the longitude ϕ is indeterminate there. At the origin itself the polar inclination θ is also indeterminate.

The coordinate surfaces for three-dimensional polar coordinates are as follows; (1) for constant values of r, the concentric spheres about the origin; (2) for constant values of ϕ, the family of half-planes through the z-axis; (3) for constant values of θ, the circular cones with the z-axis as axis and the origin as vertex (Fig. 3.11).

Another coordinate system that is often used is the system of *cylindrical coordinates*. These are obtained by introducing polar coordinates ρ, ϕ in the x, y-plane and retaining z as the third coordinate.

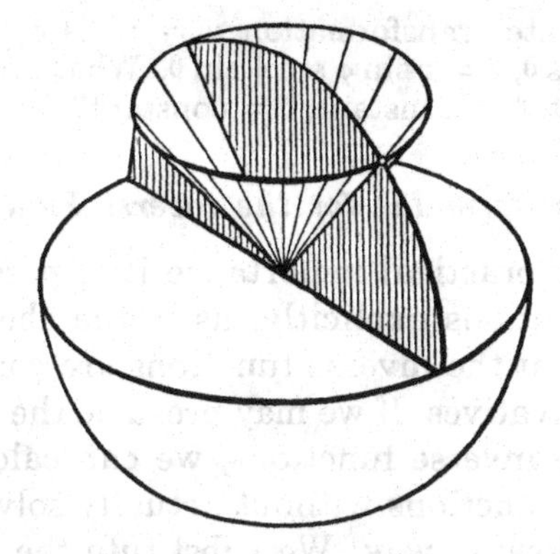

Figure 3.11 Coordinate surfaces for spherical coordinates.

Then the formulae for transformation from rectangular coordinates to cylindrical coordinates are

$$x = \rho \cos \phi,$$

$$y = \rho \sin \phi,$$

$$z = z$$

and the inverse transformation is

$$\rho = \sqrt{x^2 + y^2}$$

$$\phi = \text{arc cos} \frac{x}{\sqrt{x^2 + y^2}} = \text{arc sin} \frac{y}{\sqrt{x^2 + y^2}}$$

$$z = z.$$

The coordinate surfaces ρ = constant are the vertical circular cylinders that intersect the x, y-plane in concentric circles with the origin as center; the surfaces ϕ = constant are the half-planes through the z-axis, and the surfaces z = constant are the planes parallel to the x, y-plane.

Exercises 3.3c

1. Find the inverse of the curvilinear coordinate transformation

$$\xi = \frac{x}{x^2 + y^2 + y^2}, \qquad \eta = \frac{y}{x^2 + y^2 + z^2}, \qquad \zeta = \frac{z}{x^2 + y^2 + z^2},$$

2. Invert the coordinate transformation $w = r\cos\phi$, $x = r\sin\phi\cos\psi$, $y = r\sin\phi\sin\psi\cos\theta$, $z = r\sin\phi\sin\psi\sin\theta$. What are the sets r = constant, ϕ = constant, ψ = constant, θ = constant?

d. Differentiation Formulae for the Inverse Functions

In many cases of practical importance it is possible to solve the given system of equations explicitly, as in the above examples, and thus to recognize that the inverse functions are continuous and possess continuous derivatives. If we may presume the existence and differentiability of the inverse functions, we can calculate the derivatives of the inverse functions without actually solving the equations explictly in the following way: We substitute the inverse functions $x = g(\xi, \eta)$, $y = h(\xi, \eta)$ in the given equations $\xi = \phi(x, y)$, $\eta = \psi(x, y)$. On the right we obtain the compound functions $\phi(g(\xi, \eta), h(\xi, \eta))$ and $\psi(g(\xi, \eta), h(\xi, \eta))$ of ξ and η; but these must be equal to ξ and η, respectively. We now differentiate each of the equations

$$\xi = \phi(g(\xi, \eta), h(\xi, \eta)) \tag{24a}$$

$$\eta = \psi(g(\xi, \eta), h(\xi, \eta))$$

with respect to ξ and to η, regarding ξ and η as independent variables[1] and applying the chain rule to differentiate the compound functions. We then obtain the system of equations

$$1 = \phi_x g_\xi + \phi_y h_\xi, \quad 0 = \phi_x g_\eta + \phi_y h_\eta, \tag{24b}$$

$$0 = \psi_x g_\xi + \psi_y h_\xi, \quad 1 = \psi_x g_\eta + \psi_y h_\eta.$$

Solving these equations, we obtain expressions for the partial derivatives of the inverse functions $x = g(\xi, \eta)$ and $y = h(\xi, \eta)$ with respect to ξ and η, expressed in terms of the derivatives of the original functions $\phi(x, y)$ and $\psi(x, y)$ with respect to x and y, namely,

$$g_\xi = \frac{\psi_y}{D}, \quad g_\eta = -\frac{\phi_y}{D}, \quad h_\xi = -\frac{\psi_x}{D}, \quad h_\eta = \frac{\phi_x}{D}, \tag{24c}$$

or

[1]These equations hold for all values of ξ and η under consideration; as we say, they hold *identically*, in contrast to equations between variables that are satisfied only for *some* of the values of these variables. Such identical equations or *identities*, when differentiated with respect to any of the variables occurring in them, again yield identities as follows immediately from the definition.

(24d) $$x_\xi = \frac{\eta_y}{D}, \qquad x_\eta = -\frac{\xi_y}{D}, \qquad y_\xi = -\frac{\eta_x}{D}, \qquad y_\eta = \frac{\xi_x}{D}.$$

For brevity we have here written

(24e) $$D = \xi_x\eta_y - \xi_y\eta_x = \begin{vmatrix} \dfrac{\partial\xi}{\partial x} & \dfrac{\partial\xi}{\partial y} \\ \dfrac{\partial\eta}{\partial x} & \dfrac{\partial\eta}{\partial y} \end{vmatrix}.$$

This expression D, which we assume is not zero at the point in question, is called the *Jacobian* or *functional determinant* of the functions $\xi = \phi(x, y)$ and $\eta = \psi(x, y)$ with respect to the variables x and y. It plays a major role wherever we consider transformations, as will become apparent in the sequel.

Above, as occasionally elsewhere, we have used the shorter notation $\xi(x, y)$ instead of the more detailed notation $\xi = \phi(x, y)$, which distinguishes between the quantity ξ and its functional expression $\phi(x, y)$. We shall often use similar abbreviations in the future when there is no risk of confusion.

For polar coordinates in the plane expressed in terms of rectangular coordinates,

$$\xi = r = \sqrt{x^2 + y^2} \qquad \text{and} \qquad \eta = \theta = \arctan\frac{y}{x},$$

the partial derivatives are

$$r_x = \frac{x}{\sqrt{x^2 + y^2}} = \frac{x}{r}, \qquad r_y = \frac{y}{\sqrt{x^2 + y^2}} = \frac{y}{r},$$

$$\theta_x = \frac{-y}{x^2 + y^2} = -\frac{y}{r^2}, \qquad \theta_y = \frac{x}{x^2 + y^2} = \frac{x}{r^2}.$$

Hence, the Jacobian has the value

$$D = \frac{x}{r}\frac{x}{r^2} - \frac{y}{r}\left(-\frac{y}{r^2}\right) = \frac{1}{r},$$

and the partial derivatives of the inverse functions (rectangular coordinates expressed in terms of polar coordinates) are, by (24d),

$$x_r = \frac{x}{r}, \qquad x_\theta = -y, \qquad y_r = \frac{y}{r}, \qquad y_\theta = x,$$

as we could have found more easily by direct differentiation of the inverse formulae $x = r\cos\theta$, $y = r\sin\theta$.

The Jacobian occurs so frequently that a special symbol is often used for it[1]:

$$D = \frac{d(\xi, \eta)}{d(x, y)}. \tag{25}$$

The appropriateness of this abbreviation will soon be obvious. From the formulae for the derivatives of the inverse functions (24b), we find that the Jacobian of the functions $x = x(\xi, \eta)$ and $y = y(\xi, \eta)$ with respect to ξ and η is given by the expression

$$\frac{d(x, y)}{d(\xi, \eta)} = x_\xi y_\eta - x_\eta y_\xi = \frac{\xi_x \eta_y - \xi_y \eta_x}{D^2} = \frac{1}{D} = \left(\frac{d(\xi, \eta)}{d(x, y)}\right)^{-1}. \tag{26}$$

That is, *the Jacobian of the inverse system of functions is the reciprocal of the Jacobian of the original system.*[2]

We can also express the second derivatives of the inverse system of functions in terms of the first and second derivatives of the given functions. We have only to differentiate the linear equations (24b) with respect to ξ and to η by means of the chain rule. (We assume, of course, that the given functions possess continuous derivatives of the second order.) We then obtain linear equations from which the required derivatives can readily be calculated.

For example, to calculate the derivatives

$$\frac{\partial^2 x}{\partial \xi^2} = g_{\xi\xi} \quad \text{and} \quad \frac{\partial^2 y}{\partial \xi^2} = h_{\xi\xi}$$

we differentiate the two equations

$$1 = \xi_x x_\xi + \xi_y y_\xi$$
$$0 = \eta_x x_\xi + \eta_y y_\xi$$

once again with respect to ξ and by the chain rule obtain

$$0 = \xi_{xx} x_\xi^2 + 2\xi_{xy} x_\xi y_\xi + \xi_{yy} y_\xi^2 + \xi_x x_{\xi\xi} + \xi_y y_{\xi\xi}, \tag{27a}$$

[1]Often the Jacobian is written with the partial derivative sign as

$$D = \frac{\partial(\xi, \eta)}{\partial(x, y)}.$$

[2]This, of course, is the analogue for the rule for the derivative of the inverse of a function of a single variable (Volume I, p. 207).

$$(27b) \qquad 0 = \eta_{xx}x_\xi^2 + 2\eta_{xy}x_\xi y_\xi + \eta_{yy}y_\xi^2 + \eta_x x_{\xi\xi} + \xi_y y_{\xi\xi}.$$

If we solve this system of linear equations, regarding the quantities $x_{\xi\xi}$ and $y_{\xi\xi}$ as unknowns (the determinant of the system is again D, and therefore, by hypothesis, not zero) and then replace x_ξ and y_ξ by the values already known for them, a brief calculation gives

$$(27c) \qquad x_{\xi\xi} = -\frac{1}{D^3}\begin{vmatrix} \xi_{xx}\eta_y^2 - 2\xi_{xy}\eta_x\eta_y + \xi_{yy}\eta_x^2 & \xi_y \\ \eta_{xx}\eta_y^2 - 2\xi_{xy}\eta_x\eta_y + \eta_{yy}\eta_x^2 & \eta_y \end{vmatrix}$$

and

$$(27d) \qquad y_{\xi\xi} = \frac{1}{D^3}\begin{vmatrix} \xi_{xx}\eta_y^2 - 2\xi_{xy}\eta_x\eta_y + \xi_{yy}\eta_x^2 & \xi_x \\ \eta_{xx}\eta_y^2 - 2\eta_{xy}\eta_x\eta_y + \eta_{yy}\eta_x^2 & \eta_x \end{vmatrix}$$

The third and higher derivatives can be obtained in the same way, by repeated differentiation of the linear system of equations; at each stage we obtain a system of linear equations with the nonvanishing determinant D.

Exercises 3.3d

1. Find the Jacobians of the following transformations:
   (a) $\xi = ax + by, \quad \eta = cx + dy$
   (b) $r = \sqrt{x^2 + y^2}, \quad \theta = \arctan y/x$
   (c) $\xi = x^2, \quad \eta = y^2$
   (d) $\xi = \frac{1}{2}\log(x^2 + y^2), \quad \eta = \arctan\frac{y}{x}$
   (e) $\xi = xy^2, \quad \eta = x^2y$
   (f) $\xi = x^3 - y, \quad \eta = y^3 + x.$
2. For each of the transformations given in Exercise 1, give the points (x, y) lacking neighborhoods where the transformation has an inverse.
3. Find the Jacobian of the transformation $\xi = f(x, y), \eta = g(x, y)$, as well as all partial derivatives of x, y with respect to ξ, η through those of second order, in each of the following cases:
   (a) $\xi = e^x \cos y, \quad \eta = e^x \sin y$
   (b) $\xi = x^2 - y^2, \quad \eta = 2xy$
   (c) $\xi = \tan(x + y), \quad \eta = \cos(x - y), \quad -\pi/2 < x + y < \pi/2$
   (d) $\xi = \sinh x + \cosh y, \quad \eta = -\cosh x + \sinh y$
   (e) $\xi = x^3 + y^3, \quad \eta = xy^2.$

4. A transformation is said to be "conformal" (see p. 288) if the angle between any two curves is preserved
   (a) Prove that the inversion

$$\xi = \frac{x}{x^2 + y^2}, \qquad \eta = \frac{y}{x^2 + y^2}$$

   is a conformal transformation;
   (b) prove that the inverse of any circle is another circle or a straight line;
   (c) find the Jacobian of the inversion.
5. Let K_1, K_2, K_3 be three circles passing through 0 and having distinct pairwise intersections, say P_1, P_2, P_3, at other points. Show that the sum of the angles of the curvilinear triangle $P_1 P_2 P_3$, formed by circular arcs, is π.
6. A transformation of the plane

$$u = \varphi(x, y), \qquad v = \psi(x, y)$$

   is conformal if the functions φ and ψ satisfy the identities

$$\varphi_x = \psi_y, \qquad \varphi_y = -\psi_x.$$

7. Prove that if all the normals of a surface $z = u(x, y)$ meet the z-axis, then the surface is a surface of revolution.
8. The equation

$$\frac{x^2}{a - t} + \frac{y^2}{b - t} = 1 \qquad (a > b)$$

   determines two values of t, depending on x and y:

$$t_1 = \lambda(x, y),$$
$$t_2 = \mu(x, y).$$

   (a) Prove that the curves t_1 = constant and t_2 = constant are ellipses and hyperbolas all having the same foci (confocal conics).
   (b) Prove that the curves t_1 = constant and t_2 = constant are orthogonal.
   (c) t_1 and t_2 may be used as curvilinear coordinates (so-called focal coordinates). Express x and y in terms of these coordinates.
   (d) Express the Jacobian $\partial(t_1, t_2)/\partial(x, y)$ in terms of x and y.
   (e) Find the condition that two curves represented parametrically in the system of focal coordinates by the equations

$$t_1 = f_1(\lambda), \quad t_2 = f_2(\lambda) \qquad \text{and} \qquad t_1 = g_1(\mu), \quad t_2 = g_2(\mu)$$

   are orthogonal to one another.
9. (a) Prove that the equation in t

$$\frac{x^2}{a - t} + \frac{y^2}{b - t} + \frac{z^2}{c - t} = 1 \qquad (a > b > c)$$

   has three distinct real roots t_1, t_2, t_3, which lie respectively in the intervals

$$-\infty < t < c, \quad c < t < b, \qquad b < t < a,$$

provided that the point (x, y, z) does not lie on a coordinate plane.

(b) Prove that the three surfaces $t_1 =$ constant, $t_2 =$ constant, $t_3 =$ constant passing through an arbitrary point are orthogonal to one another.

(c) Express x, y, z in terms of the focal coordinates t_1, t_2, t_3.

10. Prove that the transformation of the x, y-plane given by the equations

$$\xi = \frac{1}{2}\left(x + \frac{x}{x^2 + y^2}\right), \quad \eta = \frac{1}{2}\left(y - \frac{y}{x^2 + y^2}\right)$$

(a) is conformal;

(b) transforms straight lines through the origin and circles with the origin as center in the x, y-plane into confocal conics $t =$ constant given by

$$\frac{\xi^2}{t + 1/2} + \frac{\eta^2}{t - 1/2} = 1.$$

11. For $\xi = f(x,y)$, $\eta = g(x,y)$, and $D = \partial(\xi,\eta)/\partial(x,y) \neq 0$, demonstrate the identities

(a) $$\frac{\partial D}{\partial y} = \frac{\partial(\xi_y, \eta)}{\partial(x, y)} + \frac{\partial(\xi, \eta_y)}{\partial(x, y)},$$

(b) $$D^{-3}\left[\xi_x(\eta_{yy} D - \eta_y D_y) - \xi_y(\eta_{xy} D - \eta_y D_x)\right]$$
$$= D^{-3}\left[\eta_x(\xi_{yy} D - \xi_y D_y) - \eta_y(\xi_{xy} D - \xi_y D_x)\right].$$

e. *Symbolic Product of Mappings*

We begin with some remarks on the composition of transformations. If the transformation

$$\xi = \phi(x, y), \qquad \eta = \psi(x, y) \tag{28a}$$

gives a 1–1 mapping of the points (x, y) of a region R on points (ξ, η) of the region B in the ξ, η-plane and if the equations

$$u = \Phi(\xi, \eta), \qquad v = \Psi(\xi, \eta) \tag{28b}$$

give a 1–1 mapping of the region B on a region R' in the u, v-plane, then a 1–1 mapping of R on R' is generated. This mapping we naturally call the *resultant mapping* or *transformation* and say that it is obtained by composition of the two given mappings and that is represents their *symbolic product*. The resultant transformation is given by the equations

$$u = \Phi(\phi(x, y), \psi(x, y)), \qquad v = \Psi(\phi(x, y), \psi(x, y));$$

from the definition, it follows at once that this mapping is 1–1.

By the rules for differentiating compound functions, we obtain

(29a) $$\frac{\partial u}{\partial x} = \Phi_\xi \phi_x + \Phi_\eta \psi_x, \qquad \frac{\partial u}{\partial y} = \Phi_\xi \phi_y + \Phi_\eta \psi_y,$$

(29b) $$\frac{\partial v}{\partial x} = \Psi_\xi \phi_x + \Psi_\eta \psi_x, \qquad \frac{\partial v}{\partial y} = \Psi_\xi \phi_y + \Psi_\eta \psi_y.$$

In matrix notation (p. 152)

(30) $$\begin{pmatrix} \dfrac{\partial u}{\partial x} & \dfrac{\partial u}{\partial y} \\ \dfrac{\partial v}{\partial x} & \dfrac{\partial v}{\partial y} \end{pmatrix} = \begin{pmatrix} \Phi_\xi & \Phi_\eta \\ \Psi_\xi & \Psi_\eta \end{pmatrix} \begin{pmatrix} \phi_x & \phi_y \\ \psi_x & \psi_y \end{pmatrix}.$$

On comparing this with the law for the multiplication of determinants (cf. p. 172) we find[1] that the Jacobian of u and v with respect to x and y is

(31a) $$\frac{\partial u}{\partial x}\frac{\partial v}{\partial y} - \frac{\partial u}{\partial y}\frac{\partial v}{\partial x} = (\Phi_\xi \Psi_\eta - \Phi_\eta \Psi_\xi)(\phi_x \psi_y - \phi_y \psi_x).$$

In words, *the Jacobian of the symbolic product of two transformations is equal to the product of the Jacobians of the individual transformations,* namely, in the notation (25),

(31b) $$\frac{d(u, v)}{d(x, y)} = \frac{d(u, v)}{d(\xi, \eta)} \frac{d(\xi, \eta)}{d(x, y)}.$$

This equation brings out the appropriateness of our symbol for the Jacobians. *When transformations are combined, the Jacobians behave in the same way as the derivatives behave when functions of one variable are combined.* The Jacobian of the resultant transformation differs from zero, provided the same is true for the individual (or component) transformations.

If, in particular, the second transformation

$$u = \Phi(\xi, \eta), \qquad v = \Psi(\xi, \eta)$$

is the inverse of the first,

$$\xi = \phi(x, y), \qquad \eta = \psi(x, y)$$

[1]The same result can, of course, be obtained by straightforward multiplication.

and if both transformations are differentiable, the resultant transformation will simply be the identical transformation; that is, $u = x$, $v = y$. The Jacobian of this last transformation is obviously 1, so that we again obtain the relation (26).

From this, incidentally, it follows that neither of the two Jacobians can vanish:

$$\frac{d(\xi, \eta)}{d(x, y)} \frac{d(x, y)}{d(\xi, \eta)} = 1.$$

For a pair of continuously differentiable functions $\phi(x, y)$ and $\psi(x, y)$ that has a nonvanishing Jacobian, we can find formulae for the corresponding *mapping of directions* at a point $(x_0, y_0) = P_0$. A curve passing through P_0 can be described parametrically by equations $x = f(t)$, $y = g(t)$, where $f(t_0) = x_0$, $g(t_0) = y_0$. The slope of the curve at P_0 is given by

$$m = \frac{g'(t_0)}{f'(t_0)}.$$

Similarly, the slope of the image curve

$$\xi = \varphi(f(t), g(t)), \qquad \eta = \psi(f(t), g(t))$$

at the point corresponding to P_0 is

$$\mu = \frac{d\eta/dt}{d\xi/dt} = \frac{\psi_x f' + \psi_y g'}{\phi_x f' + \phi_y g'} = \frac{c + dm}{a + bm}, \tag{32}$$

where a, b, c, d are the constants

$$a = \phi_x(x_0, y_0), \; b = \phi_y(x_0, y_0), \; c = \psi_x(x_0, y_0), \; d = \psi_y(x_0, y_0).$$

The relation (32) between the slope m of the original curve at P_0 and the slope μ of the image curve is the same as for the affine mapping

$$\xi = \phi(x_0, y_0) + a(x - x_0) + b(y - y_0),$$

$$\eta = \psi(x_0, y_0) + c(x - x_0) + d(y - y_0).$$

that approximates our mapping near P_0. Since

$$\frac{d\mu}{dm} = \frac{ad - bc}{(a + bm)^2},$$

we find that μ is an increasing function of m for $ad - bc > 0$ and a decreasing function for $ad - bc < 0$.[1]

Increasing slopes correspond to increasing angles of inclination or to counterclockwise rotation of the corresponding directions. Thus, $d\mu/dm > 0$ implies that the counterclockwise sense of rotation is preserved, while it is reversed for $d\mu/dm < 0$. Now, $ad - bc$ is just the Jacobian

$$\frac{d(\xi, \eta)}{d(x, y)} = \begin{vmatrix} \phi_x & \phi_y \\ \psi_x & \psi_y \end{vmatrix}$$

evaluated at the point P_0. It follows that *the mapping* $\xi = \phi(x, y)$, $\eta = \psi(x, y)$ *preserves or reverses orientations near the point* (x_0, y_0) *according to whether the Jacobian at that point is positive or negative.*

Exercises 3.3e

1. For each of the following pairs of transformations find $\partial(u, v)/\partial(x, y)$ first by eliminating ξ and η, then by applying (31b):

(a) $\begin{cases} u = \frac{1}{2}\log(\xi^2 + \eta^2) \\ v = \arc\tan\frac{\eta}{\xi} \end{cases}$ $\qquad$ $\begin{cases} \xi = e^x \cos y \\ \eta = e^x \sin y \end{cases}$

(b) $\begin{cases} u = \xi^2 - \eta^2 \\ v = 2\xi\eta \end{cases}$ $\qquad$ $\begin{cases} \xi = x \cos y \\ \eta = x \sin y \end{cases}$

(c) $\begin{cases} u = e^\xi \cos \eta \\ v = e^\xi \sin \eta \end{cases}$ $\qquad$ $\begin{cases} \xi = x/(x^2 + y^2) \\ \eta = -y/(x^2 + y^2) \end{cases}$

2. In which of the following successive transformations can x, y be defined as continuously differentiable functions of u, v in a neighborhood of the indicated point (u_0, v_0)?

(a) $\xi = e^x \cos y$, $\eta = e^x \sin y$;
$u = \xi^2 - \eta^2$, $v = 2\xi\eta$, $u_0 = 1$, $v_0 = 0$;

(b) $\xi = \cosh x + \sinh y$, $\eta = \sinh x + \cosh y$,
$u = e^{\xi+\eta}$, $v = e^{\xi-\eta}$, $u_0 = v_0 = 1$;

(c) $\xi = x^3 - y^3$, $\eta = x^2 + 2xy^2$;
$u = \xi^5 + \eta$, $v = \eta^5 - \xi$; $u_0 = 1$, $v_0 = 0$.

3. Consider the transformation

$$\begin{cases} u = \varphi(\xi, \eta) \\ v = \psi(\xi, \eta) \end{cases} \qquad \begin{cases} \xi = f(x) \\ \eta = g(y). \end{cases}$$

Show that

[1]More precisely, this holds locally, excluding the directions where m or μ become infinite.

$$\frac{\partial(u, v)}{\partial(x, y)} = f'(x)\, g'(y) \frac{\partial(u, v)}{\partial(\xi, \eta)}.$$

4. If $z = f(x, y)$ and $\xi = \varphi(x, y)$, $\eta = \psi(x, y)$, show that

$$\frac{\partial z}{\partial \xi} = \frac{\partial(z, \eta)}{\partial(x, y)} \Bigg/ \frac{\partial(\xi, \eta)}{\partial(x,y)}$$

and

$$\frac{\partial z}{\partial \eta} = \frac{\partial(\xi, z)}{\partial(x, y)} \Bigg/ \frac{\partial(\xi, \eta)}{\partial(x, y)}$$

provided $\partial(\xi, \eta)/\partial(x, y) \neq 0$.

f. General Theorem on the Inversion of Transformations and of Systems of Implicit Functions. Decomposition into Primitive Mappings

The possibility of inverting a transformation depends on the following general theorem:

Let $\phi(x, y)$ and $\psi(x, y)$ be continuously differentiable functions in a neighborhood of a point (x_0, y_0), for which the Jacobian $D = \phi_x\psi_y - \phi_y\psi_x$ is not zero at (x_0, y_0). Put $u_0 = \phi(x_0, y_0)$, $v_0 = \psi(x_0, y_0)$. Then there exists a neighborhood N of (x_0, y_0) and N' of (u_0, v_0) such that the mapping

(33a) $$u = \phi(x, y), \qquad v = \psi(x, y)$$

has a unique inverse

(33b) $$x = g(u, v), \qquad y = h(u, v)$$

mapping N' into N. The functions g and h satisfy the identities

(33c) $$u = \phi(g(u, v),\ h(u, v)), \qquad v = \psi(g(u, v),\ h(u, v))$$

for (u, v) in N', and the equations

(33d) $$x_0 = g(u_0, v_0), \qquad y_0 = h(u_0, v_0).$$

The inverse functions g, h have continuous derivatives for (u, v) near (u_0, v_0), given by

(33e) $$\frac{\partial x}{\partial u} = \frac{1}{D}\frac{\partial v}{\partial y}, \qquad \frac{\partial x}{\partial v} = -\frac{1}{D}\frac{\partial u}{\partial y}$$

(33f) $$\frac{\partial y}{\partial u} = -\frac{1}{D}\frac{\partial v}{\partial x}, \qquad \frac{\partial y}{\partial v} = \frac{1}{D}\frac{\partial u}{\partial x}.$$

The proof follows from the implicit function theorem on p. 228, which permits one to solve an equation for a single variable. In essence, we invert equations (33a) by solving the first equation for one of the variables x, y and substituting the resulting expression into the second equation, obtaining an equation for the second variable alone.

Since by assumption the Jacobian D does not vanish at the point (x_0, y_0), at least one of the first derivatives of $\phi(x, y)$ differs from zero at that point. Let, say, $\phi_x(x_0, y_0) \neq 0$. We can then solve the equation

$$u = \phi(x, y) \tag{34a}$$

for x. More precisely, we can find positive constants h_1, h_2, h_3 such that for

$$|u - u_0| < h_1, \quad |y - y_0| < h_2 \tag{34b}$$

equation (34a) has a unique solution $x = X(u, y)$ for which $|x - x_0| < h_3$. The function $X(u, y)$ has the domain (34b) and satisfies the equations

$$\phi(X(u, y), y) = u, \quad X(u_0, y_0) = x_0, \tag{34c}$$

and the inequality

$$|X(u, y) - x_0| < h_3. \tag{34d}$$

Moreover, $X(u, y)$ has continuous derivatives, for which, by (34c),

$$\phi_x(X(u, y), y) X_u(u, y) = 1 \tag{34e}$$

$$\phi_x(X(u, y), y) X_y(u, y) + \phi_y(X(u, y), y) = 0. \tag{34f}$$

We assume here that h_2, h_3 are so small that the rectangle

$$|x - x_0| < h_3, \quad |y - y_0| < h_2 \tag{34g}$$

lies in the domain of $\phi(x, y)$, $\psi(x, y)$. Substituting the expression $X(u, y)$ for x into the functions $\psi(x, y)$, we obtain a compound function

$$\psi(X(u, y), y) = \chi(u, y) \tag{34h}$$

with domain (34b). Here, by (34c, f),

$$\chi(u_0, y_0) = \psi(x_0, y_0) = v_0 \tag{34i}$$

(34j) $$\chi_y(u_0, y_0) = \psi_x X_y + \psi_y = -\psi_x \frac{\phi_y}{\phi_x} + \psi_y = \frac{D}{\phi_x} \neq 0;$$

we have $\phi_x \neq 0$ from (34e). It follows that we can find positive constants h_4, h_5, h_6 such that for

(34k) $$|u - u_0| < h_4, \quad |v - v_0| < h_5$$

the equation

(34m) $$\chi(u, y) = v$$

has a unique solution $y = h(u, v)$, for which $|y - y_0| < h_6$. We can assume here that $h_4 \leqq h_1$, $h_6 \leqq h_2$ (see footnote on p. 228).

Finally, we set

(34n) $$X(u, h(u, v)) = g(u, v).$$

The two functions $g(u, v)$, $h(u, v)$ have the domain (34k). By (34c, h) they satisfy the equations

$$\phi(g(u, v), h(u, v)) = \phi(X(u, h(u, v)), h(u, v)) = u$$

$$\psi(g(u, v), h(u, v)) = \psi(X(u, h(u, v)), h(u, v)) = \chi(u, h(u, v)) = v$$

and the inequalities

$$|g(u, v) - x_0| < h_3, \quad |h(u, v) - y_0| < h_6.$$

Formulae (33e, f) for the derivatives of g and h were derived earlier, on p. 253.

To show the uniqueness of the inverse functions, assume that x, y, u, v is any set of values that satisfy the equations (33a) and the inequalities

$$|x - x_0| < h_3, \quad |y - y_0| < h_6, \quad |u - u_0| < h_4, \quad |v - v_0| < h_5.$$

Since (34a, b) hold, we conclude that

(34o) $$x = X(u, y).$$

From (34h) we obtain the equation

$$v = \psi(x, y) = \psi(X(u, y), y) = \chi(u, y),$$

which has the unique solution $y = h(u, v)$. The relation $x = g(u, v)$ then follows from (34n, o). The relations (33d) for g and h follow from the uniqueness of the solution and the assumption that $u_0 = \phi(x_0, y_0)$, $v_0 = \psi(x_0, y_0)$.

We have assumed so far that $\phi_x(x_0, y_0) \neq 0$. If $\phi_x(x_0, y_0) = 0$, but $\phi_y(x_0, y_0) \neq 0$, the inversion of the mapping (33a) proceeds similarly. In this case we solve the first equation of (33a) for y and substitute the resulting function $y = Y(u, x)$ into the second equation, obtaining an equation for x alone.

The inversion of the plane mapping (33a) has been reduced to inversions of mappings in which only one variable is transformed at a time. Generally, we call the transformation (33a) *primitive*, if it leaves one of the coordinates unchanged, that is, if either the function $\phi(x, y)$ is identical with x or the function $\psi(x,y)$ is identical with y. The effect of a primitive transformation of the type $u = \phi(x, y)$, $v = y$ is to move each point in the direction of the x-axis, keeping its ordinate unchanged. After deformation the point has a new abscissa, which depends on both x and y. If the Jacobian ϕ of the primitive mapping is positive, u varies monotonically with x for fixed y.

We shall prove that *we can decompose an arbitrary transformation* (33a) *with nonvanishing Jacobian into primitive transformations in a neighborhood of a point.* This follows readily from our construction of the inverse mapping. If $\phi_x(x_0, y_0) \neq 0$, we represent the mapping (33a) as the symbolic product of the primitive mappings

$$\xi = \phi(x, y), \qquad \eta = y \tag{34p}$$

and

$$u = \xi, \qquad v = \chi(\xi, \eta). \tag{34q}$$

Here the domain R of the first mapping in the x, y-plane shall be a rectangle so small that

$$|x - x_0| < h_3, \quad |y - y_0| < h_2, \quad |\phi(x, y) - u_0| < h_1,$$

while the second mapping has the domain

$$|\xi - u_0| < h_1, \quad |\eta - y_0| < h_2.$$

It follows that the image (ξ, η) of a point (x, y) of R in the mapping (34p), lies in the domain of the mapping (34q) and that

$$x = X(\xi, y).$$

Consequently, also

$$x = X(\phi(x, y), y). \tag{34r}$$

For the mapping compounded from (34p, q) we then have by (34 h, r)

$$u = \phi(x, y)$$

$$v = \chi(\phi(x, y), y) = \psi(X(\phi(x, y), y), y) = \psi(x, y).$$

An analogous decomposition of the mapping (33a) is obtained when $\phi_x(x_0, y_0) = 0$ but $\phi_y(x_0, y_0) \neq 0$. We only have to interchange the roles of the variables x and y.

We cannot expect to resolve a transformation into primitive transformations in one and the same manner throughout the whole open region R. However, since some type of decomposition can be carried out near each point of R, every bounded closed subset of R can be subdivided into a finite number of sets[1] such that in each one of those sets one of the decompositions is possible.

The inversion theorem is a special case of a more general theorem that may be regarded as an extension of the theorem of implicit functions to systems of functions. The theorem of implicit functions (p. 228) applies to the solution of one equation for one of the variables. The general theorem is as follows:

If $\phi(x, y, u, v, \ldots, w)$ and $\psi(x, y, u, v, \ldots, w)$ are continuously differentiable functions of $x, y, u, v, \ldots, w$, and the equations

$$\phi(x, y, u, v, \ldots, w) = 0 \quad \text{and} \quad \psi(x, y, u, v, \ldots, w) = 0$$

are satisfied by a certain set of values $x_0, y_0, u_0, v_0, \ldots, w_0$ and if in addition the Jacobian of ϕ and ψ with respect to x and y differs from zero at that point (that is, $D = \phi_x\psi_y - \phi_y\psi_x \neq 0$), then in the neighborhood of that point the equations $\phi = 0$ and $\psi = 0$ can be solved in one, and only one way for x and y, and this solution gives x and y as continuously differentiable functions of $u, v, \ldots, w$.

The proof of this theorem is similar to that of the inversion theorem above. From the assumption $D \neq 0$ we can conclude that at the point in question some partial derivative does not vanish, say $\phi_x = 0$. By the main theorem of p. 228, if we restrict $x, y, u, v, \ldots, w$ to sufficiently small intervals about $x_0, y_0, u_0, v_0, \ldots, w_0$, respectively, the equation $\phi(x, y, u, v, \ldots, w) = 0$ can be solved in exactly one way for x as a

[1]This follows from the covering theorem, p. 109.

function of the other variables, and this solution $x = X(y, u, v, \ldots, w)$ is a continuously differentiable function of its arguments and has the partial derivative $X_y = -\phi_y/\phi_x$. If we substitute this function $x = X(y, u, v, \ldots, w)$ in $\psi(x, y, u, v, \ldots, w)$, we obtain a function $\psi(x, y, u, v, \ldots, w) = \chi(y, u, v, \ldots, w)$, and

$$\chi_y = -\psi_x \frac{\phi_y}{\phi_x} + \psi_y = \frac{D}{\phi_x}.$$

Hence, in virtue of the assumption that $D \neq 0$, we see that the derivative χ_y is not zero. Thus, if wer estrict $y, u, v, \ldots, w$ to intervals about $y_0, u_0, v_0, \ldots w_0$ contained in the intervals to which they were previously restricted, we can solve the equation $\chi = 0$ in exactly one way for y as a function of $u, y, \ldots, w$, and this solution is continuously differentiable. Substituting this expression for y in the equation $x = X(y, u, v, \ldots, w)$, we find x as a function of $u, v, \ldots, w$. This solution is unique and continuously differentiable, subject to the restriction of $x, y, u, v, \ldots, w$ to sufficiently small intervals about $x_0, y_0, u_0, v_0, \ldots, w_0$, respectively.

Exercises 3.3f

1. Which of the following systems of equations may be solved for x, y as continuously differentiable functions of the remaining variables near the indicated points?

   (a) $e^x \sin u - e^y \cos v + w = 0$
   $x \cosh w - u \sinh y - v^2 = \cosh 1$
   $x = 1, y = 0, u = 0, v = 0, w = 1$

   (b) $u \cos x - v \sin y + w^2 = 1$
   $\cos (x + y) + v = 1,$
   $x = 0, y = \pi/2, u = 1, v = 1, w = 1$

   (c) $x^2 + y^2 + u^2 - v = 0$
   $x^2 - y^2 + 2u - 1 = 0$
   $x = y = u = v = 1$

   (d) $\cos x + t \sin y = 0$
   $\sin x - \cos ty = 0,$
   $x = \pi, y = \pi/2, t = 1.$

g. *Alternate Construction of the Inverse Mapping by the Method of Successive Approximations*

In the preceeding proof the problem of inverting a mapping was reduced to the one-dimensional case and ultimately to the elementary fact that the mappings furnished by continuous monotone functions of a single variable can be inverted. This line of argument has two undesirable features. We are forced to distinguish different cases leading to quite different resolutions (say, for $\phi_x \neq 0$ and $\phi_x = 0$), which do not correspond to any radical change in the character of the original transformation. Moreover, the existence proof is *not constructive;* it does not furnish a practical numerical scheme for inverting mappings. Both of these objectionable features are absent in the method of iteration or of successive approximation that follows the pattern of the numerical methods given in Volume I (p. 502) for the solution of equations for a single unknown quantity. The basic idea is to apply successive corrections to an approximate solution, where the corrections are determined from the *linear equations* best approximating the functional relation in a neighborhood of a point.

We again consider the equations

$$u = \phi(x, y), \quad v = \psi(x, y), \tag{35a}$$

where ϕ and ψ are continuously differentiable functions in an open set R of the x, y-plane. Let (x_0, y_0) be a point of R at which the Jacobian

$$\begin{vmatrix} \phi_x & \phi_y \\ \psi_x & \psi_y \end{vmatrix} \tag{35b}$$

has a value different from zero, and let (u_0, v_0) be the image of (x_0, y_0) in the mapping (35a). We want to show that for (u, v) sufficiently close to (u_0, v_0) there exists a uniquely determined value (x, y) near (x_0, y_0) for which $u = \phi(x, y)$ and $v = \psi(x, y)$.

To obtain the solution we shall use an iteration scheme identical with that for functions of one variable discussed in Volume I (p. 502) in a notation appropriate to the two-dimensional case. We introduce the vectors $\mathbf{U} = (u, v)$, $\mathbf{X} = (x, y)$. We can write the mapping (35a) concisely in the form

$$\mathbf{U} = \mathbf{F}(\mathbf{X}), \tag{35c}$$

where $\mathbf{F}$ is the nonlinear transformation mapping the vector with components x, y onto the vector with components $\phi(x, y)$, $\psi(x, y)$. The differentials dx, dy and du, dv satisfy the linear relations (see p. 49)

$$du = d\phi = \phi_x\, dx + \phi_y\, dy \tag{35d}$$

$$dv = d\psi = \psi_x\, dx + \psi_y\, dy. \tag{35e}$$

If we combine the differentials into vectors $d\mathbf{X} = (dx, dy)$, $d\mathbf{U} = (du, dv)$, we can write[1] the relations (34d, e) as

$$d\mathbf{U} = \mathbf{F}'\, d\mathbf{X}, \tag{35f}$$

where $\mathbf{F}'$ is the square matrix formed from the first derivatives of the mapping functions

$$\mathbf{F}' = \begin{pmatrix} \phi_x & \phi_y \\ \psi_x & \psi_y \end{pmatrix}. \tag{35g}$$

Obviously the matrix $\mathbf{F}'$ *plays the role of the derivative of the vector mapping function* $\mathbf{F}$. The determinant of $\mathbf{F}'$ is just the Jacobian (35b) of the mapping.[2] Generally we shall write $\mathbf{F}' = \mathbf{F}'(\mathbf{X})$ to emphasize the dependence of the matrix $\mathbf{F}'$ on the vector $\mathbf{X} = (x, y)$. For a linear mapping the matrix $\mathbf{F}'$ is constant.

The "size" of the elements of the matrix $\mathbf{F}'$ limits how much the mapping $\mathbf{F}$ can magnify distances. Take two points (x, y) and $(x + h, y + k)$ such that the whole straight line segment joining them lies in the domain of the mapping. By the mean value theorem for functions of several variables (p. 67),

$$\begin{aligned} \phi(x + h, y + k) - \phi(x, y) &= \phi_x h + \phi_y k, \\ \psi(x + h, y + k) - \psi(x, y) &= \psi_x h + \psi_y k, \end{aligned} \tag{36}$$

where the values of the first derivatives are taken at suitable points of the segment joining (x, y) and $(x + h, y + k)$.[3] Let M denote an upper bound for the quantities

$$|\phi_x|, \quad |\phi_y|, \quad |\psi_x|, \quad |\psi_y|$$

taken at all points of the segment joining (x, y) and $(x + h, y + k)$. Then, obviously, the distance of the image points can be estimated by

[1]It is best to interpret (35f) as a relation between three matrices $d\mathbf{U}$, $\mathbf{F}'$, $d\mathbf{X}$, identifying $d\mathbf{X}$ and $d\mathbf{U}$ with matrices with two rows and a single column:

$$d\mathbf{X} = \begin{pmatrix} dx \\ dy \end{pmatrix} \qquad d\mathbf{U} = \begin{pmatrix} du \\ dv \end{pmatrix};$$

see p. 153.

[2]The matrix $\mathbf{F}'$ is often called the *Jacobian matrix* or the *Fréchet derivative of the mapping.*

[3]Generally a different intermediate point has to be used in the first and in the second equation.

$$\text{(36a)} \quad \sqrt{(\phi(x+h, y+k) - \phi(x,y))^2 + (\psi(x+h, y+k) - \psi(x,y))^2}$$
$$\leqq \sqrt{(M|h| + |M|k)^2 + (M|h| + |M|k)^2}$$
$$= \sqrt{2}\, M(|h| + |k|) \leqq 2M\sqrt{h^2 + k^2}.$$

Thus, the distance of the image points is at most $2M$ times that of the original ones. Introducing the vector $\mathbf{Y} = (x + h, y + k)$ we can write (36a) in the form of a Lipschitz condition for the mapping $\mathbf{F}$:

$$\text{(36b)} \qquad |\mathbf{F}(\mathbf{Y}) - \mathbf{F}(\mathbf{X})| \leqq 2M|\mathbf{Y} - \mathbf{X}|,$$

where M is an upper bound for the absolute values of the elements of the matrix $\mathbf{F}'$.[1] In matrix notation equations (36) become

$$\text{(36c)} \qquad \mathbf{F}(\mathbf{Y}) - \mathbf{F}(\mathbf{X}) = \mathbf{H}(\mathbf{X}, \mathbf{Y})(\mathbf{Y} - \mathbf{X})$$

where the matix H satisfies

$$\text{(36d)} \qquad \lim_{Y \to X} \mathbf{H}(\mathbf{X}, \mathbf{Y}) = \mathbf{F}'(\mathbf{X}).$$

We now consider the mapping $\mathbf{U} = \mathbf{F}(\mathbf{X})$ in a neighborhood

$$\text{(37a)} \qquad |\mathbf{X} - \mathbf{X}_0| < \delta$$

of the point $\mathbf{X}_0 = (x_0, y_0)$ in the domain R of $\mathbf{F}$. Let $\mathbf{U}_0 = \mathbf{F}(\mathbf{X}_0) = (u_0, v_0)$. For a fixed $\mathbf{U}$ we write the equation $\mathbf{U} = \mathbf{F}(\mathbf{X})$, which is to be solved for $\mathbf{X}$, in the form

$$\text{(37b)} \qquad \mathbf{X} = \mathbf{G}(\mathbf{X}),$$

where

$$\text{(37c)} \qquad \mathbf{G}(\mathbf{X}) = \mathbf{X} + \mathbf{a}(\mathbf{U} - \mathbf{F}(\mathbf{X}));$$

here $\mathbf{a}$ stands for an appropriately chosen constant nonsingular matrix, which has a reciprocal $\mathbf{a}^{-1}$. Equation (37b) is then equivalent to $\mathbf{a}(\mathbf{U} - \mathbf{F}(\mathbf{X})) = 0$, which by multiplication with $\mathbf{a}^{-1}$ yields

$$\mathbf{a}^{-1}\mathbf{a}(\mathbf{U} - \mathbf{F}(\mathbf{X})) = \mathbf{e}(\mathbf{U} - \mathbf{F}(\mathbf{X})) = \mathbf{U} - \mathbf{F}(\mathbf{X}) = 0,$$

where $\mathbf{e}$ is the unit matrix. Thus, any solution $\mathbf{X}$ of (37b)—that is, any

[1]For mappings $\mathbf{F}$ in n dimensions the factor 2 in (36b) is to be replaced by n.

fixed point of the mapping $\mathbf{G}$—furnishes a solution of $\mathbf{U} = \mathbf{F}(\mathbf{X})$.

We will show that a solution $\mathbf{X}$ of (37b) is given by the limit of the $\mathbf{X}_n$ defined by the recursion formula

$$\mathbf{X}_{n+1} = \mathbf{G}(\mathbf{X}_n) \qquad (n = 0, 1, 2, \ldots), \tag{37d}$$

provided the matrix $\mathbf{G}'(\mathbf{X})$ representing the derivative of the vector mapping $\mathbf{G}$ is of sufficiently small size. More precisely, we require that for all $\mathbf{X}$ in the neighborhood (37a) of $\mathbf{X}_0$ the largest element of the matrix $\mathbf{G}'$ is less than 1/4 in absolute value and that

$$|\mathbf{G}(\mathbf{X}_0) - \mathbf{X}_0| < \frac{1}{2}\delta.$$

First we prove by induction that under the stated assumptions the recursion formula (37d) leads only to vectors satisfying (37a). In this way, one is sure that the $\mathbf{X}_n$ lie in the domain of $\mathbf{G}$, so that the sequence can be continued indefinitely. We find from (36b) with $M = \frac{1}{4}$ that

$$|\mathbf{G}(\mathbf{Y}) - \mathbf{G}(\mathbf{X})| \leqq \frac{1}{2}|\mathbf{Y} - \mathbf{X}| \quad \text{for} \quad |\mathbf{X} - \mathbf{X}_0| < \delta,\ |\mathbf{Y} - \mathbf{X}_0| < \delta. \tag{37e}$$

Now the inequality (37a) is satisfied trivially for $\mathbf{X} = \mathbf{X}_0$. If it holds for $\mathbf{X} = \mathbf{X}_n$, we find for the vector $\mathbf{X}_{n+1}$ defined by (37d) that

$$\begin{aligned}|\mathbf{X}_{n+1} - \mathbf{X}_0| \leqq |\mathbf{X}_{n+1} - \mathbf{X}_1| + |\mathbf{X}_1 - \mathbf{X}_0| &= |\mathbf{G}(\mathbf{X}_n) - \mathbf{G}(\mathbf{X}_0)| \\ &\quad + |\mathbf{G}(\mathbf{X}_0) - \mathbf{X}_0| \leqq \frac{1}{2}|\mathbf{X}_n - \mathbf{X}_0| + \frac{1}{2}\delta < \delta.\end{aligned}$$

This proves that $|\mathbf{X}_n - \mathbf{X}_0| < \delta$ for all n.

In order to see that the $\mathbf{X}_n$ converge, we observe that by (37e)

$$|\mathbf{X}_{n+1} - \mathbf{X}_n| = |\mathbf{G}(\mathbf{X}_n) - \mathbf{G}(\mathbf{X}_{n-1})| \leqq \frac{1}{2}|\mathbf{X}_n - \mathbf{X}_{n-1}|.$$

By the same reasoning

$$|\mathbf{X}_n - \mathbf{X}_{n-1}| \leqq \frac{1}{2}|\mathbf{X}_{n-1} - \mathbf{X}_{n-2}|,$$

$$|\mathbf{X}_{n-1} - \mathbf{X}_{n-2}| \leqq \frac{1}{2}|\mathbf{X}_{n-2} - \mathbf{X}_{n-3}|,$$

and so on. These inequalities together lead to the estimate

$$|\mathbf{X}_{n+1} - \mathbf{X}_n| \leqq \frac{1}{2^n}|\mathbf{X}_1 - \mathbf{X}_0| \leqq \frac{\delta}{2^{n+1}}. \tag{37f}$$

The existence of $\mathbf{X} = \lim_{n\to\infty} \mathbf{X}_n$ follows then by writing $\mathbf{X}$ as sum of an infinite series

$$\mathbf{X} = \mathbf{X}_0 + (\mathbf{X}_1 - \mathbf{X}_0) + (\mathbf{X}_2 - \mathbf{X}_1) + \cdots + (\mathbf{X}_{n+1} - \mathbf{X}_n) + \cdots,$$

whose convergence is established from (37f) by *comparison* (see Volume I, p. 521) with a convergent geometric series. That $\mathbf{X}$ is a solution of (37b) follows immediately from (37d) for $n \to \infty$, using the continutity of $\mathbf{G}(\mathbf{X})$.

By its definition (37c) the function $\mathbf{G}$ depends continuously not only on $\mathbf{X}$ but also on the vector $\mathbf{U}$. The $\mathbf{X}_n$ obtained successively by the recursion formula (37d) then also depend continuously on $\mathbf{U}$.[1] Since the geometric series used in the comparison that establishes the convergence of $\mathbf{X} = \lim_{n\to\infty} \mathbf{X}_n$ does not depend on $\mathbf{U}$, it follows that $\mathbf{X}$ is a *uniform limit of continuous functions of* $\mathbf{U}$ and, hence, is itself a continuous function of $\mathbf{U}$. It is clear, moreover, that $|\mathbf{X} - \mathbf{X}_0| \leqq \delta$, since $|\mathbf{X}_n - \mathbf{X}| < \delta$ for all n. If there existed a second solution $\mathbf{Y}$ with $\mathbf{Y} = \mathbf{G}(\mathbf{Y})$ and $|\mathbf{Y} - \mathbf{X}_0| \leqq \delta$, we would find from (37e) that

$$|\mathbf{Y} - \mathbf{X}| = |\mathbf{G}(\mathbf{Y}) - \mathbf{G}(\mathbf{X})| \leqq \frac{1}{2}|\mathbf{Y} - \mathbf{X}|$$

and, hence, that $|\mathbf{Y} - \mathbf{X}| = 0$ and $\mathbf{Y} = \mathbf{X}$.

In this way, we establish the existence, uniqueness, and continuity of a solution $\mathbf{X}$ of the equation $\mathbf{U} = \mathbf{F}(\mathbf{X})$, for which $|\mathbf{X} - \mathbf{X}_0| \leqq \delta$, *provided* the vector $\mathbf{G}$ defined by (37c) has a derivative $\mathbf{G}'$ with elements less than $\frac{1}{4}$ in absolute value for $|\mathbf{X} - \mathbf{X}_0| \leqq \delta$ and provided

$$|\mathbf{G}(\mathbf{X}_0) - \mathbf{X}_0| < \frac{1}{2}\delta.$$

It is easily seen that these requirements can be satisfied for all $\mathbf{U}$ sufficiently close to $\mathbf{U}_0$ by a suitable choice of the matrix $\mathbf{a}$. By (37c),

$$\mathbf{G}'(\mathbf{X}) = \mathbf{e} - \mathbf{a}\mathbf{F}'(\mathbf{X}),$$

[1]Here we make use of the fact that continuous functions of continuous functions are again continuous.

where $\mathbf{e}$ is the unit matrix. Then, for $\mathbf{X} = \mathbf{X}_0$,

$$\mathbf{G}'(\mathbf{X}_0) = \mathbf{e} - \mathbf{a}\mathbf{F}'(\mathbf{X}_0) = \mathbf{O}$$

if we choose for $\mathbf{a}$ the matrix reciprocal to the matrix $\mathbf{F}'(\mathbf{X}_0)$:

$$\mathbf{a} = (\mathbf{F}'(\mathbf{X}_0))^{-1}.$$

(The existence of this reciprocal follows from our basic assumption that the matrix $\mathbf{F}'(\mathbf{X}_0)$ has a nonvanishing determinant, that is, that the Jacobian of the mapping $\mathbf{F}$ does not vanish at the point $\mathbf{X}_0$). From the assumed continuity of the first derivatives of the mapping $\mathbf{F}$ it follows that $\mathbf{G}'(\mathbf{X})$ depends continuously on $\mathbf{X}$; hence, the elements of $\mathbf{G}'(\mathbf{X})$ are arbitrarily small, for instance, less than $\frac{1}{4}$, for sufficiently small $|\mathbf{X} - \mathbf{X}_0|$, say for

$$|\mathbf{X} - \mathbf{X}_0| \leqq \delta;$$

moreover, by (37c),

$$|\mathbf{G}(\mathbf{X}_0) - \mathbf{X}_0| = |\mathbf{a}(\mathbf{U} - \mathbf{F}(\mathbf{X}_0)| = |\mathbf{a}(\mathbf{U} - \mathbf{U}_0)| < \frac{1}{2}\delta,$$

provided $\mathbf{U}$ lies in a sufficiently small neighborhood of $\mathbf{U}_0$.

This completes the proof for the local existence of a continuous inverse for a continuously differentiable mapping with nonvanishing Jacobian. The existence and continuity of the first derivatives of the inverse mapping follow easily from formulae (36c,d). Let $\mathbf{U} = \mathbf{F}(\mathbf{X})$, where we assume that the Jacobian matrix $\mathbf{F}'(\mathbf{X})$ is non-singular. Then every $\mathbf{V}$ sufficiently close to $\mathbf{U}$ is of the form $\mathbf{V} = \mathbf{F}(\mathbf{Y})$ where $\mathbf{Y}$ tends to $\mathbf{X}$ for $\mathbf{V}$ tending to $\mathbf{U}$. Hence, for $\mathbf{V}$ sufficiently close to $\mathbf{U}$ the matrix $\mathbf{H}(\mathbf{X}, \mathbf{Y})$ also is non-singular. We find then that

$$\begin{aligned} \mathbf{Y} - \mathbf{X} &= (\mathbf{H}(\mathbf{X}, \mathbf{Y}))^{-1}(\mathbf{V} - \mathbf{U}) \\ &= (\mathbf{F}'(\mathbf{X}))^{-1}(\mathbf{V} - \mathbf{U}) + \mathbf{E}(\mathbf{X}, \mathbf{Y})(\mathbf{V} - \mathbf{U}) \end{aligned}$$

where

$$\lim_{V \to U} \mathbf{E}(\mathbf{X}, \mathbf{Y}) = \lim_{Y \to X} \mathbf{E}(\mathbf{X}, \mathbf{Y}) = \mathbf{0}.$$

This relation, however, just expresses that the vector $\mathbf{X}$ satisfying $\mathbf{U} = \mathbf{F}(\mathbf{X})$ is a differentiable function of the vector $\mathbf{U}$, and that the Jacobian matrix of $\mathbf{X}$ with respect to $\mathbf{U}$ is the reciprocal of the matrix

F′(**X**). The same construction of the inverse by *iteration* or *successive approximations* obviously can be applied to mappings in any number of dimensions.

Exercises 3.3g

1. Obtain the iterative approximation (x_2, y_2) for the inverse transformation to

$$u = \frac{1}{2}({}^2x - y^2), \; v = xy$$

by applying (37d) to a neighborhood of $\mathbf{X} = (1, 1)$ or $\mathbf{U} = (0, 1)$.

2. Compare the result of the preceding exercise with the Taylor expansions of x and y to second order in the neighborhood of $u = 1$, $v = 1$.

h. Dependent Functions

If the Jacobian D vanishes at a point (x_0, y_0), no general statement can be made about the possibility of solving the equations (33a) in the neighborhood of that point. Even if inverse functions do happen to exist, they cannot be differentiable, for then the product

$$\frac{d(u, v)}{d(x, y)} \cdot \frac{d(x, y)}{d(u, v)}$$

would vanish, while by p. 259 it must be equal to 1. For example, the equations

$$u = x^3, \qquad v = y$$

can be solved uniquely, in the form

$$x = \sqrt[3]{u}, \qquad y = v,$$

although the Jacobian vanishes at the origin; but the function $\sqrt[3]{u}$ is not differentiable at the origin.

On the other hand, the equations

$$u = x^2 - y^2, \qquad v = 2xy$$

cannot be solved uniquely in the neighborhood of the origin, since the two points (x, y) and $(-x, -y)$ of the x, y-plane both correspond to the same point of the u, v-plane.

If the Jacobian vanishes identically, not merely at the single point (x, y) but at every point in a whole neighborhood of the point (x, y), then the transformation is called *degenerate*. In this case, it can be shown that the functions

$$u = \phi(x, y) \qquad \text{and} \qquad v = \psi(x, y)$$

are dependent, in the sense that one of them is a function of the other one.[1] We first consider the trivial case in which the equations $\phi_x = 0$ and $\phi_y = 0$ hold everywhere, so that the function $\phi(x, y)$ is a constant. We then see that while the point (x, y) ranges over a whole region its image, (u, v) always remains on the line $u =$ constant. That is, a region is mapped only into a line, instead of on a region, so that there is no possibility of a 1–1 mapping of two 2-dimensional regions on one another.

A similar situation arises in the general case in which at least one of the derivatives ϕ_x or ϕ_y does not vanish, but the Jacobian D is still zero. We suppose that at a point (x_0, y_0) of the region under consideration we have $\phi_x \neq 0$. It is then possible to solve the first equation for x in the form $x = X(u, y)$ and to write $v = \psi(X(u, y), y) = \chi(u, y)$, just as on p. 262, for there we made use only of the assumption $\phi_x \neq 0$. In virtue of (34j) and the equation $D = 0$, however, χ_y must be identically 0 in the region where $\phi_x \neq 0$; that is, the quantity $\chi = v$ does not depend on y at all and v is a function of u alone. We conclude, then, that if the Jacobian of the transformation vanishes identically, a region of the x, y-plane is mapped by the transformation on a curve in the u, v-plane instead of on a region, for in a certain interval of values of u only one value of v corresponds to each value of u. Thus, if the Jacobian vanishes identically, the functions are not independent; that is, a relation

$$F(\varphi, \psi) = \psi - \chi(\varphi) = 0$$

exists that is satisfied for all systems of values (x, y) in the region. Conversely, if there exists a curve in the u, v-plane on which the region of the x, y-plane is mapped, then for all points of this region the Jacobian $D = \phi_x\psi_y - \phi_y\psi_x$ must vanish identically, since obviously the mapping cannot be inverted in a full neighborhood of a point.

The exceptional case discussed separately at the begining is obviously included in this general statement. The curve in question is then just the curve $u =$ constant, which is a parallel to the v-axis.

An example of a degenerate transformation is

[1]Vanishing of the Jacobian is also equivalent to *dependence of the vectors* (ϕ_x, ϕ_y) and (ψ_x, ψ_y) formed by the first derivatives of the mapping functions.

$$\xi = x + y, \qquad \eta = (x + y)^2.$$

In this transformation all the points of the x, y-plane are mapped on the points of the parabola $\eta = \xi^2$ in the ξ, η-plane. Inverting the transformation is out of the question, for all the points of the line $x + y = \text{constant}$ are mapped on a single point (ξ, η). As we can easily verify, the value of the Jacobian is 0. The relation between the functions ξ and η, in accordance with the general theorem, is given by the equation

$$F(\xi, \eta) = \xi^2 - \eta = 0.$$

Exercises 3.3h

1. Give an example of a pair of continuously differentiable functions $\xi = f(x, y)$, $\eta = g(x, y)$ that are independent in one region, and not independent in another.
2. Prove that if $\xi = ax + by + c$ and $\eta = \alpha x + \beta y + \gamma$ are dependent, the lines $\xi = 0$ and $\eta = 0$ are parallel.

i. Concluding Remarks

The generalization of the theory to three or more independent variables offers no particular difficulties. The chief difference is that instead of the two-rowed determinant D we have determinants with three or more rows. In the case of transformations with three independent variables

$$\xi = \phi(x, y, z), \qquad \eta = \psi(x, y, z), \qquad \zeta = \chi(x, y, z),$$

$$x = g(\xi, \eta, \zeta), \qquad y = h(\xi, \eta, \zeta), \qquad z = l(\xi, \eta, \zeta),$$

the Jacobian is given by the equation

$$D = \frac{d(\xi, \eta, \zeta)}{d(x, y, z)} = \begin{vmatrix} \phi_x & \psi_x & \chi_x \\ \phi_y & \psi_y & \chi_y \\ \phi_z & \psi_z & \chi_z \end{vmatrix}.$$

In the same way, for transformations

$$\xi_i = \phi_i(x_1, x_2, \ldots, x_n)$$

$$x_i = g_i(\xi_1, \xi_2, \ldots, \xi_n) \qquad (i = 1, 2, \ldots, n)$$

with n independent variables, the Jacobian is

$$\frac{d(\xi_1, \xi_2, \ldots, \xi_n)}{d(x_1, x_2, \ldots, x_n)} = \begin{vmatrix} \dfrac{\partial \phi_1}{\partial x_1}, & \dfrac{\partial \phi_2}{\partial x_1}, & \ldots, & \dfrac{\partial \phi_n}{\partial x_1} \\ \dfrac{\partial \phi_1}{\partial x_2}, & \dfrac{\partial \phi_2}{\partial x_2}, & \ldots, & \dfrac{\partial \phi_n}{\partial x_2} \\ \cdot & \cdot & & \cdot \\ \cdot & \cdot & & \cdot \\ \cdot & \cdot & & \cdot \\ \dfrac{\partial \phi_1}{\partial x_n}, & \dfrac{\partial \phi_2}{\partial x_n}, & \ldots, & \dfrac{\partial \phi_n}{\partial x_n} \end{vmatrix}.$$

For more than two independent variables, it is still true that when transformations are compounded their Jacobians are multiplied together. In symbols,

$$\frac{d(\xi_1, \xi_2, \ldots, \xi_n)}{d(\eta_1, \eta_2, \ldots, \eta_n)} \cdot \frac{d(\eta_1, \eta_2, \ldots, \eta_n)}{d(x_1, x_2, \ldots, x_n)} = \frac{d(\xi_1, \xi_2, \ldots, \xi_n)}{d(x_1, x_2, \ldots, x_n)}$$

In particular, the Jacobian of the inverse transformation is the reciprocal of the Jacobian of the original transformation.

The theorems on the resolution and composition of transformations, on the inversion of a transformation, and on the dependence of transformations remain valid for three and more independent variables. The proofs are similar to those for the case $n = 2$; to avoid unnecessary repetition we omit them. The same holds for the construction of the inverse mapping by the method of iteration.

In the preceding section, we saw that the behavior of a general transformation in many ways resembles that of an affine transformation and that the Jacobian plays the same part as the determinant does in the case of affine transformation. The following remark makes this even clearer. Since the functions $\xi = \phi(x, y)$ and $\eta = \psi(x, y)$ are differentiable in the neighborhood of (x_0, y_0), we can express them in the form

$$\begin{aligned} \xi - \xi_0 &= (x - x_0)\phi_x(x_0, y_0) + (y - y_0)\phi_y(x_0, y_0) \\ &\qquad + \varepsilon \sqrt{(x - x_0)^2 + (y - y_0)^2}, \\ \eta - \eta_0 &= (x - x_0)\psi_x(x_0, y_0) + (y - y_0)\psi_y(x_0, y_0) \\ &\qquad + \delta \sqrt{(x - x_0)^2 + (y - y_0)^2}, \end{aligned}$$

where ε and δ tend to zero with

$$\sqrt{(x-x_0)^2+(y-y_0)^2}.$$

This shows that for sufficiently small values of $|x - x_0|$ and $|y - y_0|$ the transformation can be represented approximately by the affine transformation

$$\xi = \xi_0 + (x - x_0)\phi_x(x_0, y_0) + (y - y_0)\phi_y(x_0, y_0),$$

$$\eta = \eta_0 + (x - x_0)\psi_x(x_0, y_0) + (y - y_0)\psi_y(x_0, y_0),$$

whose determinant is the Jacobian of the original transformation.

Exercises 3.3i

1. Evaluate $\partial(\xi, \eta, \rho)/\partial(x, y, z)$ for each of the following:

 (a) $\xi = e^x \cos y \cos z$
 $\eta = e^x \cos y \sin z$
 $\rho = e^x \sin y$

 (b) $\xi = \cos(x + y) + \cos(y + z)$
 $\eta = \cos(x + y) + \sin(y + z)$
 $\rho = \sin(x + y) + \cos(y + z)$

 (c) $\xi = \cosh x + \log y$
 $\eta = \tanh y - \sinh z$
 $\rho = x - y^z$

 (d) $\xi = x \cos y \sin z$
 $\eta = x \sin y \sin z$
 $\rho = x \cos z$

 (e) $\xi = x \cos y$
 $\eta = x \sin y$
 $\rho = z.$

2. Define dependence of the functions $\xi = f(x, y, z)$, $\eta = g(x, y, z)$, $\rho = h(x, y, z)$, in a region. Generalize the results of Section h to this case.
3. Which of the triples of functions given in Exercise 1 are dependent? Give an equation relating the functions of each such triple.
4. Show that the following three functions are dependent and find a relation connecting them:

$$\xi = x + y + z$$

$$\eta = x^2 + y^2 + z^2$$

$$\zeta = xy + yz + zx.$$

5. Inversion in three dimensions is defined by the formulae

$$\xi = \frac{x}{x^2 + y^2 + z^2}, \quad \eta = \frac{y}{x^2 + y^2 + z^2}, \quad \zeta = \frac{z}{x^2 + y^2 + z^2}.$$

(a) Prove that the angle between any two surfaces is unchanged.

(b) Prove that spheres are transformed either into spheres or into planes.

(c) Find the Jacobian of the transformation.

3.4 Applications

a. *Elements of the Theory of Surfaces*

For surfaces, as for curves, parametric representation is frequently to be preferred to other types of representation. For surfaces, we need two parameters instead of one; we denote them by u and v. A parametric representation may be expressed in the form

$$x = \phi(u, v), \quad y = \psi(u, v), \quad z = \chi(u, v), \tag{39a}$$

where ϕ, ψ, and χ are given functions of the parameters u and v and the point (u, v) ranges over a given region R in the u, v-plane. The corresponding point with the three rectangular coordinates (x, y, z) then ranges over a set in x, y, z-space. Typically, this set is a surface, which can be represented in explicit form $z = f(x, y)$, for we may be able to solve two of our three equations for u and v in terms of the two corresponding rectangular coordinates. If we then substitute the expressions found for u and v in the third equation, we obtain an unsymmetrical representation of the surface $z = f(x, y)$.[1] Hence in order to ensure that the equations really do represent a surface, we have only to assume that the three Jacobians

$$\begin{vmatrix} \psi_u & \psi_v \\ \chi_u & \chi_v \end{vmatrix}, \quad \begin{vmatrix} \chi_u & \chi_v \\ \phi_u & \phi_v \end{vmatrix}, \quad \begin{vmatrix} \phi_u & \phi_v \\ \psi_u & \psi_v \end{vmatrix} \tag{39b}$$

do not all vanish at once; in a single formula, we require that

$$(\phi_u\psi_v - \phi_v\psi_u)^2 + (\psi_u\chi_v - \psi_v\chi_u)^2 + (\chi_u\phi_v - \chi_v\phi_u)^2 > 0. \tag{39c}$$

Then in some neighborhood of each point in space represented by (39a) it is certainly possible to express one of the three coordinates in terms of the other two.

It is advantageous to replace the three equations (39a) in the parametric representation (39a) by a single vector equation

[1]This is actually a special case of the parametric form, as we see by putting $x = u$ and $y = v$.

(40a) $$\mathbf{X} = \Phi(u, v),$$

where $\mathbf{X} = (x, y, z)$ is the *position vector* of a point on the surface, and Φ denotes the vector

$$\Phi(u, v) = (\phi(u, v),\ \psi(u, v),\ \chi(u, v)).$$

At each point with parameters u, v on the surface, we can form the *partial derivatives of the position vector*

(40b) $$\mathbf{X}_u = (\phi_u, \psi_u, \chi_u) \quad \text{and} \quad \mathbf{X}_v = (\phi_v, \psi_v, \chi_v).$$

The total differential of the vector $\mathbf{X}$ is then [cf. formula (15b), p.49]

(40c) $$d\mathbf{X} = (dx, dy, dz) = \mathbf{X}_u\, du + \mathbf{X}_v\, dv.$$

The three determinants (39b) are just the components of the vector product $\mathbf{X}_u \times \mathbf{X}_v$ of the vectors $\mathbf{X}_u$ and $\mathbf{X}_v$(see p. 000). The expression on the left in (39c) represents the square of the length of the vector $\mathbf{X}_u \times \mathbf{X}_v$, so that condition (39c) is equivalent to

(40d) $$\mathbf{X}_u \times \mathbf{X}_v \neq 0.$$

For example, the spherical surface $x^2 + y^2 + z^2 = r^2$ of radius r is represented parametrically by the equations

(40e) $$x = r \cos u \sin v, \qquad y = r \sin u \sin v, \qquad z = r \cos v$$
$$(0 \leqq u < 2\pi, \quad 0 \leqq v \leqq \pi)$$

where $v = \theta$ is the "polar inclination" and $u = \phi$ is the "longitude" of the point on the sphere (cf. p. 250).

This example exhibits one of the advantages of parametric representation. The three coordinates are given explictly as functions of u and v, and these functions are single-valued. If v runs from $\pi/2$ to π, we obtain the lower hemisphere, that is,

$$z = -\sqrt{r^2 - x^2 - y^2},$$

while values of v from 0 to $\pi/2$ give the upper hemisphere. Thus, for the parametric representation it is not necessary, as it is for the representation

$$z = \pm\sqrt{r^2 - x^2 - y^2},$$

to consider two single-valued branches of the function in order to obtain the whole sphere.

We obtain another parametric representation of the sphere by means of *stereographic projection* (see Volume I, p. 21). In order to project the sphere $x^2 + y^2 + z^2 - r^2 = 0$ stereographically from the north pole $(0, 0, r)$ on the equatorial plane $z = 0$, we join each point of the surface to the north pole N by a straight line and call the intersection of this line with the equatorial plane the *stereographic image* of the corresponding point of the sphere (Fig. 3.12) We thus obtain a 1–1 correspondence between the points of the sphere and the points of the plane, except for the north pole N. Using elementary geometry, we readily find that this correspondence is expressed by the formulae

$$\text{(40f)} \quad x = \frac{2r^2u}{u^2 + v^2 + r^2}, \qquad y = \frac{2r^2v}{u^2 + v^2 + r^2}, \qquad z = \frac{(u^2 + v^2 - r^2)r}{u^2 + v^2 + r^2},$$

where (u, v) are the rectangular coordinates of the image-point in the plane. These equations may be regarded as a parametric representation of the sphere, the parameters u and v being rectangular coordinates in the u, v-plane.

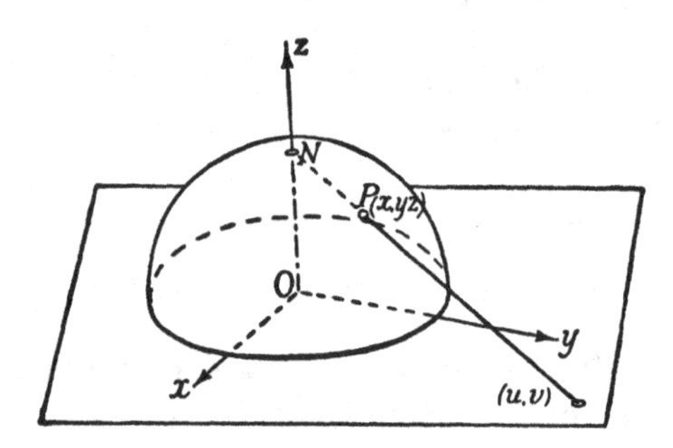

Figure 3.12 Stereographic projection of the sphere

As a further example, we give parametric representations of the surfaces

$$\frac{x^2}{a^2} + \frac{y^2}{b^2} - \frac{z^2}{c^2} = 1 \qquad \text{and} \qquad \frac{x^2}{a^2} + \frac{y^2}{b^2} - \frac{z^2}{c^2} = -1,$$

which are called the *hyperboloid of one sheet* and the *hyperboloid of two sheets* respectively (cf. Figs. 3.13 and 3.14). The hyperboloid of one sheet is represented by

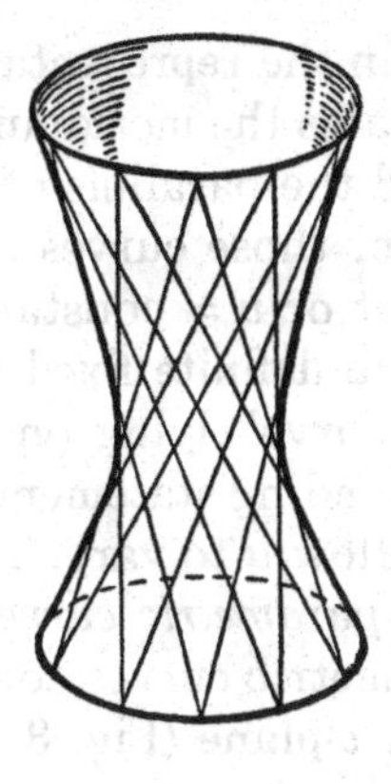

Figure 3.13 Hyperboloid of one sheet.

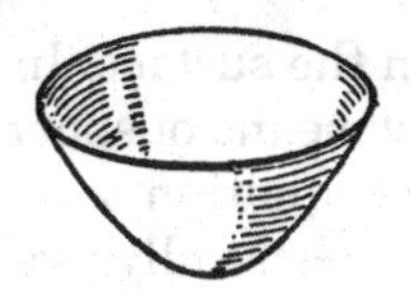

Figure 3.14 Hyperboloid of two sheets.

$$\text{(40g)} \qquad \begin{aligned} x &= a \cos u \cosh v, \\ y &= b \sin u \cosh v, \\ z &= c \sinh v \end{aligned}$$

$$(0 \leqq u < 2\pi, \ -\infty < v < +\infty)$$

and the hyperboloid of two sheets by

$$\text{(40h)} \qquad \begin{aligned} x &= a \cos u \sinh v, \\ y &= b \sin u \sinh v, \\ z &= \pm c \cosh v \end{aligned}$$

$$(0 \leqq u < 2\pi, \ 0 < v < +\infty).$$

In general, we may regard the *parametric representation* of a surface as the *mapping of the region R of the u, v-plane onto the corresponding surface.* To each point of the region R of the u, v-plane there corresponds one point of the surface, and typically the converse is also true.[1]

In the same way, a curve $u = u(t)$, $v = v(t)$ in the u, v-plane corresponds by virtue of the equations

$$x = \phi(u(t), v(t)) = x(t), \ldots$$

[1]This, of course, is not always the case. For example, in the representation (40e) of the sphere by spherical coordinates (p. 279) the poles of the sphere correspond to the whole line segments given by $v = 0$ and $v = \pi$.

to a curve on the surface. In particular, in the representation (40e) of the sphere by means of spherical coordinates the meridians are represented by the equation u = constant and the parallels of latitude by v = constant. Generally, we may consider those curves on a surface that are given by equations u = constant or v = constant. If in our parametric representation we substitute a definite fixed value for u, we obtain a "space curve" or "twisted curve" lying on the surface and having v as parameter, and a corresponding statement holds good if we substitute a fixed value for v and allow u to vary. These curves u = constant and v = constant are the *parametric curves* or *coordinate lines* on the surface. The net of parametric curves corresponds to the net of parallels to the axes in the u, v-plane (Fig. 3.15).

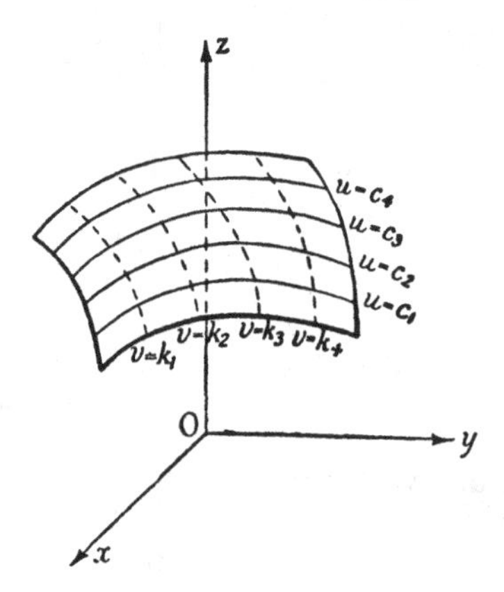

Figure 3.15 Parametric curves u = constant, v = constant.

The tangent to the curve on the surface corresponding to the curve $u = u(t)$, $v = v(t)$ in the u, v-plane has the direction of the vector

$$(41)\quad \mathbf{X}_t = (x_t, y_t, z_t) = \left(x_u \frac{du}{dt} + x_v \frac{dv}{dt},\ y_u \frac{du}{dt} + y_v \frac{dv}{dt},\ z_u \frac{du}{dt} + z_v \frac{dv}{dt}\right)$$

$$= \mathbf{X}_u \frac{du}{dt} + \mathbf{X}_v \frac{dv}{dt}$$

(see p. 212). At a given point of the surface the tangential vectors $\mathbf{X}_t$ of all curves on the surface passing through that point are dependent on the two vectors $\mathbf{X}_u$, $\mathbf{X}_v$, which respectively are tangential to the parametric lines v = constant and u = constant passing through that point. This means that the tangents all lie in the plane through the point *spanned* by the vectors $\mathbf{X}_u$ and $\mathbf{X}_v$, the *tangent plane to the surface* at that point. The *normal* to the surface is perpendicular to all tangential directions, in particular to the vectors $\mathbf{X}_u$ and $\mathbf{X}_v$. It follows (see. p. 182) that the surface normal is parallel to the direction of the vector product

$$\mathbf{X}_u \times \mathbf{X}_v = (y_u z_v - y_v z_u,\ z_u x_v - z_v x_u,\ x_u y_v - x_v y_u). \tag{42}$$

One of the most important tools for investigation of the properties of a given surface is the study of the curves that lie on it. Here we shall only give the expression for s, the length of arc of such a curve. As mentioned on p. 213, (see also Volume I, p. 353)

$$\left(\frac{ds}{dt}\right)^2 = \left(\frac{dx}{dt}\right)^2 + \left(\frac{dy}{dt}\right)^2 + \left(\frac{dz}{dt}\right)^2 = \mathbf{X}_t \cdot \mathbf{X}_t,$$

so that in view of the equations (41) we obtain

$$\begin{aligned} \left(\frac{ds}{dt}\right)^2 &= \left(\mathbf{X}_u \frac{du}{dt} + \mathbf{X}_v \frac{dv}{dt}\right) \cdot \left(\mathbf{X}_u \frac{du}{dt} + \mathbf{X}_v \frac{dv}{dt}\right) \\ &= \left(x_u \frac{du}{dt} + x_v \frac{dv}{dt}\right)^2 + \left(y_u \frac{du}{dt} + y_v \frac{dv}{dt}\right)^2 + \left(z_u \frac{du}{dt} + z_v \frac{dv}{dt}\right)^2 \\ &= E\left(\frac{du}{dt}\right)^2 + 2F\frac{du}{dt}\frac{dv}{dt} + G\left(\frac{dv}{dt}\right)^2. \end{aligned} \tag{43}$$

Here the coefficients E, F, G, the *Gaussian fundamental quantities* of the surface, are given by

$$E = \left(\frac{\partial x}{\partial u}\right)^2 + \left(\frac{\partial y}{\partial u}\right)^2 + \left(\frac{\partial z}{\partial u}\right)^2 = \mathbf{X}_u \cdot \mathbf{X}_u \tag{44a}$$

$$F = \frac{\partial x}{\partial u}\frac{\partial x}{\partial v} + \frac{\partial y}{\partial u}\frac{\partial y}{\partial v} + \frac{\partial z}{\partial u}\frac{\partial z}{\partial v} = \mathbf{X}_u \cdot \mathbf{X}_v \tag{44b}$$

$$G = \left(\frac{\partial x}{\partial v}\right)^2 + \left(\frac{\partial y}{\partial v}\right)^2 + \left(\frac{\partial z}{\partial v}\right)^2 = \mathbf{X}_v \cdot \mathbf{X}_v. \tag{44c}$$

These depend only on the surface itself and its parametric representation and not on the particular choice of the curve on the surface. The expression (43) for the derivative of the length of arc s with respect to the parameter t usually is written symbolically without reference to the parameter used along the curve. One says that the *line element* ds is given by the quadratic differential form ("fundamental form")

$$ds^2 = E\,du^2 + 2F\,du\,dv + G\,dv^2. \tag{45}$$

The length of the cross product $\mathbf{X}_u \times \mathbf{X}_v$ can be expressed in terms of E, F, G since (see p. 182)

$$|\mathbf{X}_u \times \mathbf{X}_v|^2 = |\mathbf{X}_u|^2|\mathbf{X}_v|^2 - (\mathbf{X}_u \cdot \mathbf{X}_v)^2 = EG - F^2. \tag{45a}$$

Our original assumption (39c) or (40d) on the parametric representation can thus be formulated as the condition

$$EG - F^2 > 0 \tag{46}$$

for the fundamental quantities.

The direction cosines for one of the two normals to the surface are the components of the unit vector

$$\frac{1}{|\mathbf{X}_u \times \mathbf{X}_v|}\mathbf{X}_u \times \mathbf{X}_v = \frac{1}{\sqrt{EG - F^2}}\mathbf{X}_u \times \mathbf{X}_v.$$

It follows from (42) that the normal for a surface represented parametrically has the direction cosines

$$\cos\alpha = \frac{y_u z_v - y_v z_u}{\sqrt{EG - F^2}}, \quad \cos\beta = \frac{z_u x_v - z_v x_u}{\sqrt{EG - F^2}}, \quad \cos\gamma = \frac{x_u y_v - x_v y_u}{\sqrt{EG - F^2}}. \tag{47}$$

The tangent to a curve $u = u(t)$, $v = v(t)$ on the surface has the direction of the vector

$$\mathbf{X}_t = \mathbf{X}_u \frac{du}{dt} + \mathbf{X}_v \frac{dv}{dt}.$$

If we now consider a second curve $u = u(\tau)$, $v = v(\tau)$ on the surface referred to a parameter τ, its tangent has the direction of the vector

$$\mathbf{X}_\tau = \mathbf{X}_u \frac{du}{d\tau} + \mathbf{X}_v \frac{dv}{d\tau}.$$

If the two curves pass through the same point on the surface, the cosine of the angle of intersection ω is the same as the cosine of the angle between the vectors $\mathbf{X}_t$ and $\mathbf{X}_\tau$. Hence (see p. 131),

$$\cos\omega = \frac{\mathbf{X}_t \cdot \mathbf{X}_\tau}{|\mathbf{X}_t||\mathbf{X}_\tau|}.$$

Here

$$\mathbf{X}_t \cdot \mathbf{X}_\tau = \left(\mathbf{X}_u \frac{du}{dt} + \mathbf{X}_v \frac{dv}{dt}\right) \cdot \left(\mathbf{X}_u \frac{du}{d\tau} + \mathbf{X}_v \frac{dv}{d\tau}\right)$$

$$= E\frac{du}{dt}\frac{du}{d\tau} + F\left(\frac{du}{dt}\frac{dv}{d\tau} + \frac{du}{d\tau}\frac{dv}{dt}\right) + G\frac{dv}{dt}\frac{dv}{d\tau}.$$

Consequently the cosine of the angle between the two curves on the surface is given by

(48) $\cos\omega$

$$= \frac{E\dfrac{du}{dt}\dfrac{du}{d\tau} + F\left(\dfrac{du}{dt}\dfrac{dv}{d\tau} + \dfrac{du}{d\tau}\dfrac{dv}{dt}\right) + G\dfrac{dv}{dt}\dfrac{dv}{d\tau}}{\sqrt{E\left(\dfrac{du}{dt}\right)^2 + 2F\dfrac{du}{dt}\dfrac{dv}{dt} + G\left(\dfrac{dv}{dt}\right)^2}\sqrt{E\left(\dfrac{du}{d\tau}\right)^2 + 2F\dfrac{du}{d\tau}\dfrac{dv}{d\tau} + G\left(\dfrac{dv}{d\tau}\right)^2}}.$$

The *mapping of one plane region on another* may be regarded as a special case of parametric representation, for if the third of our functions $\chi(u, v)$ in (39a) vanishes for all values of u and v under consideration, our equations merely represent the mapping of a region of the u, v-plane on a region of the x, y-plane; or if we prefer to think in terms of transformations of coordinates, the equations define a system of *curvilinear coordinates* in the u, v-region, and the inverse functions (if they exist) define a curvilinear u, v-system of coordinates in the plane x, y-region. In terms of the curvilinear coordinates (u, v) the line element in the x, y-plane is simply [see (44a, b, c)]

$$ds^2 = E\,du^2 + 2F\,du\,dv + G\,dv^2,$$

where

(49a) $$E = \left(\frac{\partial x}{\partial u}\right)^2 + \left(\frac{\partial y}{\partial u}\right)^2,$$

(49b) $$F = \frac{\partial x}{\partial u}\frac{\partial x}{\partial v} + \frac{\partial y}{\partial u}\frac{\partial y}{\partial v},$$

(49c) $$G = \left(\frac{\partial x}{\partial v}\right)^2 + \left(\frac{\partial y}{\partial v}\right)^2.$$

As a further example of the representation of a surface in parametric form we consider the *anchor ring,* or *torus*. This is obtained by rotating a circle about a line which lies in the plane of the circle and does not intersect it (cf. Fig. 3.16). We take the axis of rotation as the z-axis and choose the y-axis in such a way that it passes through the center of the circle, whose y-coordinate we denote by a. If the radius of the circle is $r < |a|$, we obtain

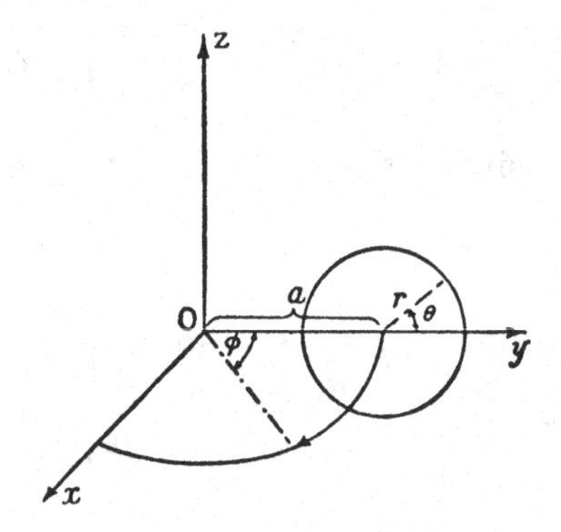

Figure 3.16 Generation of a torus by the rotation of a circle.

$$x = 0, \ y - a = r \cos \theta, \ z = r \sin \theta \ (0 \leqq \theta < 2\pi)$$

as a parametric representation of the circle in the y, z-plane. Now letting the circle rotate about the z-axis, we find that for each point of the circle $x^2 + y^2$ remains constant; that is, $x^2 + y^2 = (a + r \cos \theta)^2$. If ϕ is the angle of rotation about the z-axis, we have

$$x = (a + r \cos \theta) \sin \phi,$$

$$y = (a + r \cos \theta) \cos \phi,$$

$$z = r \sin \theta$$

$$(0 \leqq \phi < 2\pi, \ 0 \leqq \theta < 2\pi)$$

as a parametric representation of the torus in terms of the parameters θ and ϕ. In this representation the torus appears as the image of a square of side 2π in the θ, ϕ-plane, where any pair of boundary points lying on the same line θ = constant or ϕ = constant corresponds to only one point on the surface, and the four corners of the square all correspond to the same point.

For the line element on the anchor ring, we have by (44a, b, c), (45)

$$ds^2 = r^2 \, d\theta^2 + (a + r \cos \theta)^2 d\phi^2.$$

Exercises 3.4a

1. Calculate the line element

   (a) on the sphere

$$x = \cos u \sin v, \qquad y = \sin u \sin v, \qquad z = \cos v;$$

(b) on the hyperboloid

$$x = \cos u \cosh v, \qquad y = \sin u \cosh v, \qquad z = \sinh v;$$

(c) on a surface of revolution given by

$$r = \sqrt{x^2 + y^2} = f(z),$$

using the cylindrical coordinates z and $\theta = \text{arc tan}\,(y/x)$ as coordinates on the surface;

(d) on the quadric $t_3 =$ constant of the family of confocal quadrics given by

$$\frac{x^2}{a - t} + \frac{y^2}{b - t} + \frac{z^2}{c - t} = 1,$$

using t_1 and t_2 as coordinates on the quadric (cf. Exercise 9, p. 256).

2. Find the Gauss fundamental quantities for the catenoid $x = a \cosh (t/a) \cos (\theta/a)$, $y = a \cosh (t/a) \sin (\theta/a)$, $z = t$; show that $E - G = F = 0$.
3. For the surface $x = u \cos v$, $y = u \sin v$, $z = \alpha u + \beta$, α, $\beta =$ constant, show that the images of the lines $u =$ constant, $v =$ constant are orthogonal.
4. What is the fundamental form giving the line element for a surface given by an equation $z = f(x, y)$?
5. Prove that if a new system of curvilinear coordinates r, s is introduced on a surface with parameters u, v by means of the equations

$$u = u(r, s), \qquad v = v(r, s),$$

then

$$E'G' - F'^2 = (EG - F^2)\left\{\frac{d(u, v)}{d(r, s)}\right\}^2,$$

where E', F', G' denote the fundamental quantities taken with respect to r, s and E, F, G those taken with respect to u, v.

6. Let t be a tangent to a surface S at the point P, and consider the sections of S made by all planes containing t. Prove that the centers of curvature of the different sections lie on a circle.
7. If t is a tangent to the surface S at the point P, we call the curvature of the normal plane section through t (i.e., the section through t and the normal) at that point the *curvature k of S in the direction t*. For every tangent at P we take the vector with the direction of t, initial point P, and length $1/\sqrt{k}$. Prove that the final points of these vectors lie on a conic.
8. A curve is given as the intersection of the two surfaces

$$x^2 + y^2 + z^2 = 1$$

$$ax^2 + by^2 + cz^2 = 0$$

Find the equations of

(a) the tangent,

(b) the osculating plane, at any point of the curve.

9. If the coordinates (x, y, z) of a point on a sphere are given by the equations (cf. p. 250)

$$x = a \sin\theta \cos\phi,\ y = a \sin\theta \sin\phi,\ z = a \cos\theta,$$

show that the two curves of the systems $\theta + \phi = \alpha$, $\theta - \phi = \beta$, which pass through any point (θ, ϕ), cut one another at the angle arc cos $\{(1 - \sin^2\theta)/(1 + \sin^2\theta)\}$ (cf. p. 285).

Show that the radius of curvature of either curve is equal to

$$\frac{a(1 + \sin^2\theta)^{3/2}}{(5 + 3\sin^2\theta)^{1/2}}.$$

b. Conformal Transformation in General

A transformation in the plane

$$x = \phi(u, v), \qquad y = \psi(u, v) \tag{50}$$

is called conformal if it maps any two intersecting curves into two others enclosing the same angle as the original ones.

THEOREM. *A necessary and sufficient condition that a continuously differentiable transformation* (50) *should be conformal is that the Cauchy-Riemann equations*

$$\phi_u - \psi_v = 0, \qquad \phi_v + \psi_u = 0 \tag{51a}$$

or

$$\phi_u + \psi_v = 0, \qquad \phi_v - \psi_u = 0 \tag{51b}$$

hold. In the first case the direction of the angles is preserved, in the second case the direction is reversed.[1]

The proof of this follows: If the transformation is conformal, the two orthogonal curves u = constant = u_0, $v = v_0 + t$ and $u = u_0 + \tau$, v = constant = v_0 in the u, v-plane must map into orthogonal curves in the x, y-plane. From the formula (48) for the angle between two curves (p. 285) is follows immediately that

$$0 = F = \phi_u\phi_v + \psi_u\psi_v. \tag{51c}$$

In the same way, the curves corresponding to the lines $u = u_0 + t$, $v = v_0 + t$ and $u = u_0 + \tau$, $v = v_0 - \tau$ must be orthogonal. This gives

[1]This last statement follows directly from the statements on p. 260 concerning the sign of the Jacobian $D = \phi_u\psi_v - \phi_v\psi_u$. In case (51a) holds, we have $D = \phi_u^2 + \phi_v^2 \geqq 0$, in case (51b) $D = -\phi_u^2 - \phi_v^2 \leqq 0$.

(51d) $$0 = E - G = \phi_u^2 + \psi_u^2 - \phi_v^2 - \psi_v^2.$$

Equation (51c) can be written as

$$\phi_u = \lambda\psi_v, \quad \phi_v = -\lambda\psi_u,$$

where λ denotes a constant of proportionality. Introducing this into equation (51d), we immediately get $\lambda^2 = 1$, so that one or the other of our two systems of Cauchy-Riemann equations (51a, b) holds.

That the Cauchy-Riemann equations are a sufficient condition for conformality except at points where all four of the quantities $\phi_u, \phi_v, \psi_u, \psi_v$ are zero is confirmed by the following observations.

Equations (51a) or (51b) yield relations

$$E = G \geqq 0, \qquad F = 0$$

for the fundamental quantities E, F, G, defined by (49a, b, c). By (48) the angle ω between two curves in the x, y-plane is then given by

$$\cos\omega = \frac{\dfrac{du}{dt}\dfrac{du}{d\tau} + \dfrac{dv}{dt}\dfrac{dv}{d\tau}}{\sqrt{\left(\dfrac{du}{dt}\right)^2 + \left(\dfrac{dv}{dt}\right)^2}\sqrt{\left(\dfrac{du}{d\tau}\right)^2 + \left(\dfrac{dv}{d\tau}\right)^2}}.$$

The right side of this equation is just the cosine of the angle between the corresponding curves in the u, v-plane. Thus, the mapping preserves angles between curves, possibly changing their orientation. The only exception is presented by points where $E = F = G = 0$, that is, by points where all first derivatives of both mapping functions vanish.[1]

Exercises 3.4b

1. Investigate the behavior of the mapping $x = u^2 - v^2$, $y = 2uv$. Is it conformal at $u = 2$, $v = 3$? At $u = v = 0$? Why?
2. Where is the mapping $x = \frac{1}{2}\log(u^2 + v^2)$, $y = \arctan v/u$, conformal?
3. Show that if the mappings $(u, v) \to (x, y)$ and $(u, v) \to (\xi, \eta)$ are both conformal, the mapping $(u, v) \to (x\xi - y\eta, x\eta + y\xi)$ is also conformal.
4. (a) Prove that the stereographic projection of the unit sphere on the plane is conformal.
   (b) Prove that circles on the sphere are transformed either into circles or into straight lines in the plane.

[1]There the mapping may actually cease to be conformal.

(c) Prove that in stereographic projection reflection of the spherical surface in the equatorial plane corresponds to an inversion in the u, v-plane.
(d) Find the expression for the line element on the sphere in terms of the parameters u, v.

5. Under what conditions on the Gaussian fundamental coefficients (44) will the mapping from the u, v-plane to the surface $\mathbf{X} = \mathbf{X}(u, v)$ be conformal?
6. Find a conformal mapping of the sphere $x = \cos\theta \sin\phi$, $y = \sin\theta \sin\phi$, $z = \cos\phi$ into the u, v-plane such that $\theta = u$, and $\phi = f(v)$ with $f(0) = \frac{1}{2}\pi$.

3.5 Families of Curves, Families of Surfaces, and Their Envelopes

a. *General Remarks*

On various occasions we have already considered curves or surfaces not as individual configurations but as members of a family of curves or surfaces, such as $f(x, y) = c$, where to each value of c there corresponds a different curve of the family.

For example, the lines parallel to the y-axis in the x, y-plane, that is, the lines $x = c$, form a family of curves. The same is true for the family of concentric circles $x^2 + y^2 = c^2$ about the origin; to each value of c there corresponds a circle of the family, namely, the circle with radius c. Similarly, the rectangular hyperbolas $xy = c$ form a family of curves, sketched in Fig. 3.2. The particular value $c = 0$ corresponds to the degenerate hyperbola consisting of the two coordinate axes. Another example of a family of curves is the set of all the normals to a given curve. If the curve is given in terms of the parameter t by the equations $\xi = \phi(t)$, $\eta = \psi(t)$, we obtain the equation of the family of normals in the form (see Volume I, p. 345)

$$(x - \phi(t))\phi'(t) + (y - \psi(t))\psi'(t) = 0,$$

where t is used instead of c to denote the parameter of the family.

The general concept of a family of curves can be expressed analytically in the following way. Let

$$f(x, y, c)$$

be a continuously differentiable function of the two independent variables x and y and of the parameter c, where the parameter varies in a given interval. (Thus, the parameter is really a third independent variable, which is lettered differently simply because it plays a different part.) Then, if for each value of the parameter c the equation

$$f(x, y, c) = 0 \tag{52a}$$

represents a curve, the aggregate of the curves obtained as c describes its interval is called a *family of curves* depending on the parameter c.

Each curve of such a family may also be represented in parametric form

$$x = \phi(t, c), \qquad y = \psi(t, c), \tag{52b}$$

where c is the parameter distinguishing the different curves of the family and t the parameter along the curve.

For example, the equations

$$x = c \cos t, \qquad y = c \sin t$$

represent the family of concentric circles mentioned above; again the equations

$$x = ct, \qquad y = \frac{1}{t},$$

represent the family of rectangular hyperbolas mentioned above, except for the degenerate hyperbola consisting of the coordinate axes.

Occasionally we are led to consider families of curves that depend on several parameters. For example, the aggregate of all circles $(x - a)^2 + (y - b)^2 = c^2$ in the plane is a family of curves depending on the three parameters a, b, c. If nothing is said to the contrary, we shall always understand a family of curves to be a "one-parameter" family, depending on a single parameter. The other cases we shall distinguish by speaking of two-parameter, three-parameter, or multiparameter families of curves.

Similar statements of course hold for families of surfaces in space. If we are given a continuously differentiable function $f(x, y, z, c)$ and if for each value of the parameter c in a certain definite interval the equation

$$f(x, y, z, c) = 0$$

represents a surface in the space with rectangular coordinates x, y, z, then the aggregate of the surfaces obtained by letting c describe its interval is called a *family of surfaces*, or, more precisely, a *one-para-*

meter family of surfaces with the parameter c. For example, the spheres $x^2 + y^2 + z^2 = c^2$ about the origin form such a family. As with curves, we can also consider families of surfaces depending on several parameters.

Thus, the planes defined by the equation

$$ax + by + \sqrt{1 - a^2 - b^2}\, z + 1 = 0$$

form a two-parameter family depending on the parameters a and b if the parameters a and b range over the region $a^2 + b^2 \leqq 1$. This family of surfaces consists of the class of all planes that are at unit distance from the origin.[1]

Exercises 3.5a

1. Characterize the following families of curves geometrically:

   (a) $\dfrac{x^2}{a^2} + \dfrac{y^2}{b^2} = c^2$, $a, b =$ known constants, $c =$ a parameter

   (b) $x^2 + (y - c)^2 = c^2$, $c =$ parameter

   (c) $x = \cos(c + t)$, $y = \sin(c + t)$, $0 \leqq t \leqq 2\pi$, $c =$ parameter.

2. Describe the one-parameter family of surfaces

$$(x - c)^2 + (y - 1 - c)^2 + (z + \sqrt{2} - 2c)^2 = 1.$$

b. *Envelopes of One-Parameter Families of Curves*

If a family of straight lines consists of the tangents to a plane curve E (e.g., if the family of normals of a curve C is the family of tangents to the evolute E of C; cf. Volume I, p. 424,) we shall say that *the curve E is the envelope of the family of lines.* In the same way, we shall say that the family of circles with radius 1 and center on the x-axis—that is, the family of circles with the equation $(x - c)^2 + y^2 - 1 = 0$—has as its envelope the pair of lines $y = 1$ and $y = -1$, which touch each of the circles (Fig. 3.17). In both examples, we can obtain the point of contact of the envelope and a curve of the family with parameter value c by finding the intersections of the two curves of the family with parameter values c and $c + h$ and then letting h tend to 0. We express this briefly by saying that the envelope is the *locus of the intersections of neighbouring curves.*

For any family of curves *a curve E that at each of its points touches*

[1]Sometimes a one-parametric family of surfaces is referred to as ∞^1 surfaces, a two-parametric family as ∞^2 surfaces, and so on.

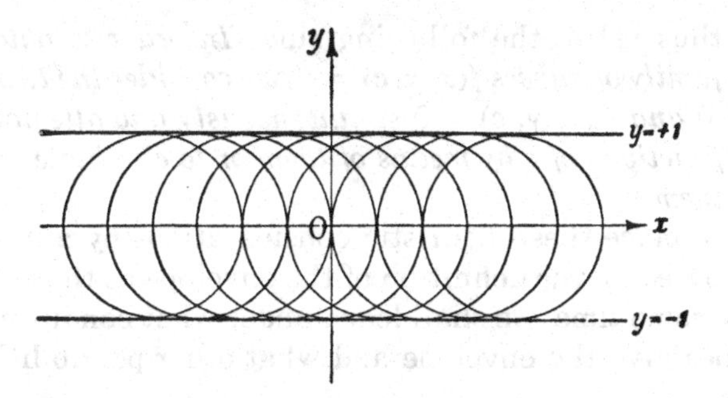

Figure 3.17 Family of circles with envelope.

some one of the curves of the family is called the envelope of the family of curves. The question now arises of finding the envelope E of a given family of curves $f(x, y, c) = 0$. We first make a few plausible remarks in which we assume that an envelope E does exist and that it can be obtained, as in the above cases, as the locus of the intersections of neighboring curves.[1] We then obtain the point of contact of the curve $f(x, y, c) = 0$ with the curve E in the following way: In addition to this curve we consider a neighboring curve $f(x, y, c + h) = 0$, find the intersection of these two curves, and then let h tend to 0. The point of intersection must then approach the point of contact sought. At the point of intersection the equation

$$\frac{f(x, y, c + h) - f(x, y, c)}{h} = 0$$

is true as well as the equations $f(x, y, c + h) = 0$ and $f(x, y, c) = 0$. In the first equation, we pass to the limit $h \to 0$. Since we assume the existence of the partial derivative f_c, this gives the two equations

$$f(x, y, c) = 0, \qquad f_c(x, y, c) = 0 \tag{53}$$

for the point of contact of the curve $f(x, y, c) = 0$ with the envelope. If we can determine x and y as functions of c by means of these equations, we obtain the parametric representation of a curve with the parameter c, and this curve is the envelope. By elimination of the parameter c, the curve can also be represented in the form $g(x, y) = 0$. This equation is called the *discriminant* of the family, and the curve given by the equation $g(x, y) = 0$ is called the *discriminant curve.*

[1]Since this last assumption will be shown by examples to be too restrictive, we shall shortly replace these plausibilities by a more complete discussion.

We are thus led to the following rule: *In order to obtain the envelope of a family of curves $f(x, y, c) = 0$, we consider the two equations $f(x, y, c) = 0$ and $f_c(x, y, c) = 0$ simultaneously and attempt to express x and y as functions of c by means of them or to eliminate the quantity c between them.*

We now replace these heuristic considerations by a more general discussion based on the definition of the envelope as the curve of contact. At the same time, we shall learn under what conditions our rule actually does give the envelope and what other possibilities present themselves.

To begin with, we assume that E is an envelope that can be represented in terms of the parameter c by two continuously differentiable functions

$$x = x(c), \quad y = y(c),$$

where

$$\left(\frac{dx}{dc}\right)^2 + \left(\frac{dy}{dc}\right)^2 \neq 0,$$

and that E at the point with parameter c touches the curve of the family $f(x, y, c) = 0$ with the same value of the parameter c. The equation $f(x, y, c) = 0$ is then satisfied at the point of contact. Consequently, if we substitute the expressions $x(c)$ and $y(c)$ for x and y in this equation, it remains valid for all values of c in the interval. On differentiating with respect to c, we at once obtain

$$f_x \frac{dx}{dc} + f_y \frac{dy}{dc} + f_c = 0.$$

Now the condition of tangency is

$$f_x \frac{dx}{dc} + f_y \frac{dy}{dc} = 0,$$

for the quantities dx/dc and dy/dc are proportional to the direction cosines of the tangent to E and the quantities f_x and f_y are proportional to the direction cosines of the normal to the curve $f(x, y, c) = 0$ of the family, and these directions must be at right angles to one another. It follows that the envelope satisfies the equation $f_c = 0$, and we thus see that equations (53) form a *necessary* condition for the envelope.

In order to find out how far this condition is also *sufficient*, we assume that a curve E represented by two continuously differentiable functions $x = x(c)$ and $y = y(c)$ satisfies the two equations $f(x, y, c) = 0$ and $f_c(x, y, c) = 0$. In $f(x, y, c) = 0$ we again substitute $x(c)$ and $y(c)$ for x and y; this equation then becomes an identity in c. If we differentiate with respect to c and remember that $f_c = 0$, we at once obtain the relation

$$f_x \frac{dx}{dc} + f_y \frac{dy}{dc} = 0,$$

which therefore holds for all points of E. If the two expressions $f_x^2 + f_y^2$ and $(dx/dc)^2 + (dy/dc)^2$ both differ from 0 at a point of E, so that at that point both the curve E and the curve of the family have well-defined tangents, this equation states that the envelope and the curve of the family touch one another. With these additional assumptions our rule is a sufficient condition for the envelope as well as a necessary one. If, however, f_x and f_y both vanish, the curve of the family may have a singular point (cf. p. 236), and we can draw no conclusions about the contact of the curves.

Thus, after we have found the discriminant curve, it is still necessary to make a further investigation in each case, in order to discover whether it is really an envelope or to what extent it fails to be one.

In conclusion, we state the condition for the discriminant curve of a family of curves given in parametric form

$$x = \phi(t, c), \quad y = \psi(t, c),$$

with the curve parameter t. This is

$$\phi_t \psi_c - \phi_c \psi_t = 0.$$

We can readily obtain this condition by passing from the parametric representation of the family to the original expression by elimination of t.

Exercises 3.5b

1. Do the normals to a smooth plane curve always have an envelope?
2. The straight lines

$$y = cx + \psi(c)$$

satisfy the differential equation

$$y = xy' + \psi(y')$$

(Clairaut equation). Obtain a nonparametric equation for the envelope of the family and verify that it, too, must satisfy the differential equation.

c. *Examples*

1. $(x - c)^2 + y^2 = 1$. As we remarked on p. 292, this equation represents the family of circles of unit radius whose centers lie on the x-axis (Fig. 3.17). Geometrically, we see at once that the envelope must consist of the two lines $y = 1$ and $y = -1$. We can verify this by means of our rule; for the two equations $(x - c)^2 + y^2 = 1$ and $-2(x - c) = 0$ immediately give us the envelope in the form $y^2 = 1$.

2. The family of circles of unit radius passing through the origin, whose centers, therefore, must lie on the circle of unit radius about the origin, is given by the equation

$$(x - \cos c)^2 + (y - \sin c)^2 = 1$$

or

$$x^2 + y^2 - 2x \cos c - 2y \sin c = 0.$$

The derivative with respect to c equated to 0 gives $x \sin c - y \cos c = 0$. These two equations are satisfied by the values $x = 0$ and $y = 0$. If, however, $x^2 + y^2 \neq 0$, it readily follows from our equations that $\sin c = y/2$, $\cos c = x/2$, so that on eliminating c we obtain $x^2 + y^2 = 4$. Thus, for the envelope our rule gives us the circle of radius 2 about the origin, as is anticipated by geometrical intuition; but it also gives us the isolated point $x = 0$, $y = 0$.

3. The family of parabolas $(x - c)^2 - 2y = 0$ (cf. Fig. 3.18) also has an envelope, which both by intuition and by our rule is found to be the x-axis.

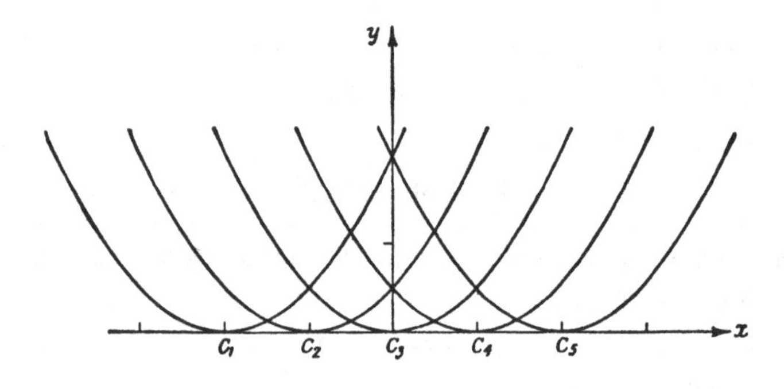

Figure 3.18 Family of parabolas with envelope.

4. We consider the family of circles $(x - 2c)^2 + y^2 - c^2 = 0$ (cf. Fig. 3.19). Differentiation with respect to c gives $2x - 3c = 0$, and by substitution we find that the equation of the envelope is

$$y^2 = \frac{x^2}{3};$$

that is, the envelope consists of the two lines

$$y = \frac{1}{\sqrt{3}}x \quad \text{and} \quad y = -\frac{1}{\sqrt{3}}x.$$

The origin is an exception in that contact does not occur there.

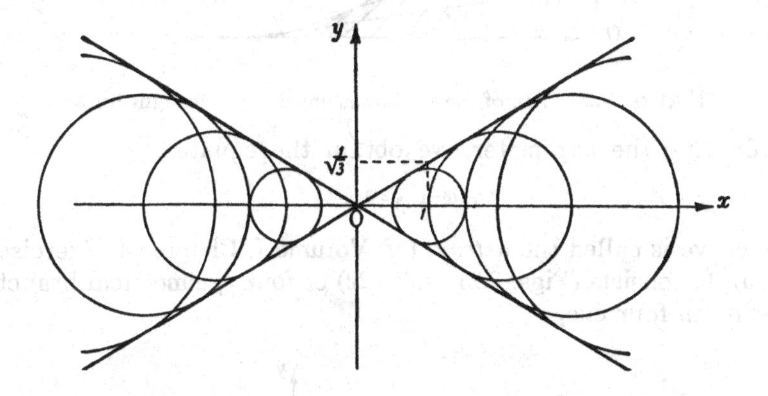

Figure 3.19 The family $(x - 2c)^2 + y^2 - c^2 = 0$.

5. We next consider the family of straight lines on which unit length is cut out by the x- and y-axes. If $\alpha = c$ is the angle indicated in Fig. 3.20, the lines are given by the equation

$$\frac{x}{\cos\alpha} + \frac{y}{\sin\alpha} = 1.$$

The condition for the envelope is

$$\frac{\sin\alpha}{\cos^2\alpha}x - \frac{\cos\alpha}{\sin^2\alpha}y = 0,$$

which, in conjunction with the equation of the lines, gives the envelope in parametric form,

$$x = \cos^3\alpha, \quad y = \sin^3\alpha.$$

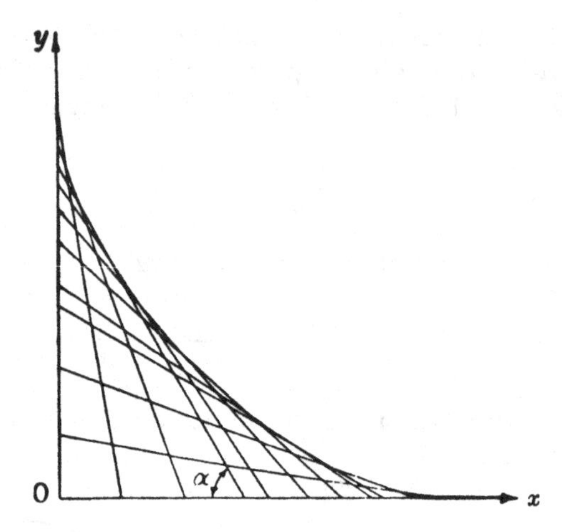

Figure 3.20 Arc of the astroid as envelope of straight lines.

Eliminating the parameter, we obtain the equation

$$x^{2/3} + y^{2/3} = 1.$$

This curve is called the *astroid* (cf. Volume I, Chapter 4, Exercise 1, p. 435). It consists (Figs. 3.21 and 3.22) of four symmetrical branches meeting in four cusps.

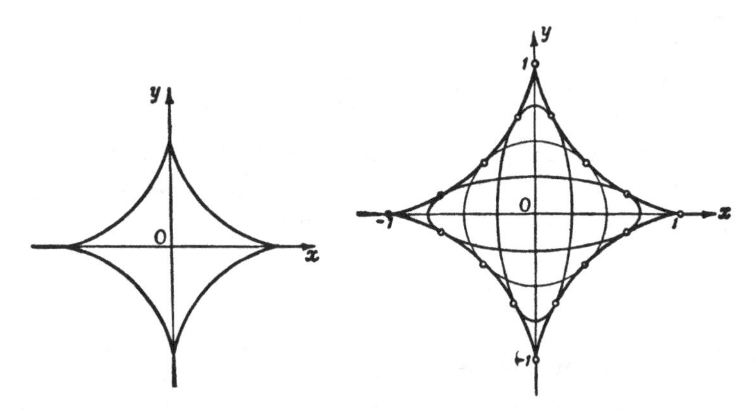

Figure 3.21 Astroid.

Figure 3.22 Astroid as envelope of ellipses.

6. The astroid $x^{2/3} + y^{2/3} = 1$ also appears as the envelope of the family of ellipses

$$\frac{x^2}{c^2} + \frac{y^2}{(1-c)^2} = 1$$

whose semiaxes c and $(1-c)$ have the constant sum 1 (Fig. 3.22).

7. The family of curves $(x-c)^2 - y^3 = 0$ shows that in certain circumstances our process may fail to give an envelope. Here the rule gives the x-axis. But, as Fig. 3.23 shows, this is not an envelope; it is the locus of the cusps of the curves of the family.

8. For the family

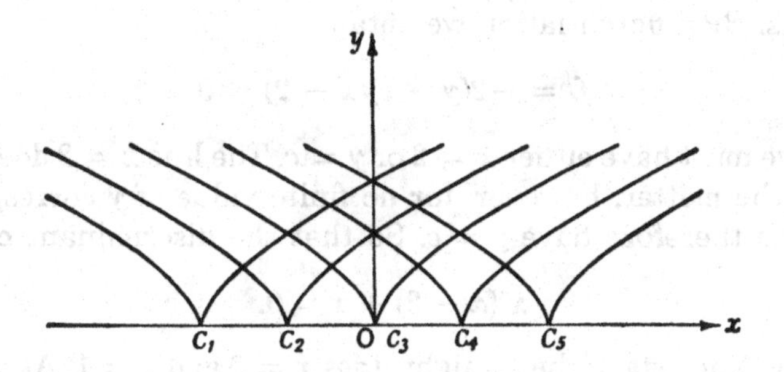

Figure 3.23 The family $(x-c)^2 - y^3 = 0$.

$$(x-c)^3 - y^2 = 0,$$

the discriminant curve is the x-axis (cf. Fig. 3.24). This is again the cusp-locus; but it touches each of the curves, and in this sense must be regarded as the envelope.

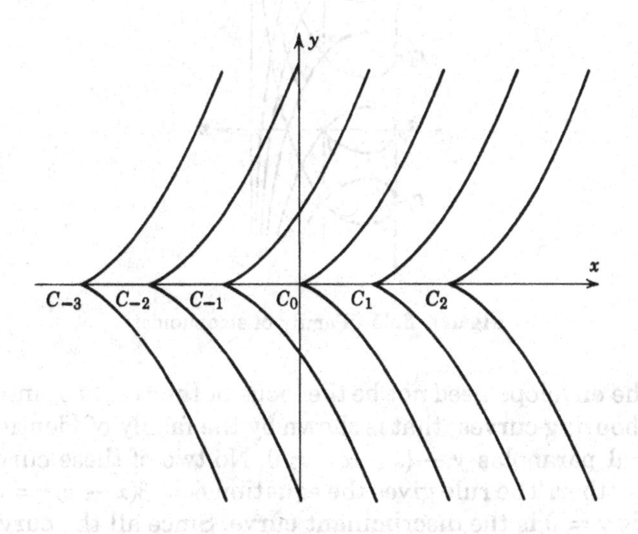

Figure 3.24 The family $(x-c)^3 - y^2 = 0$.

9. The family of *strophoids*

$$[x^2 + (y - c)^2]\,(x - 2) + x = 0$$

(cf. Fig. 3.25) has a discriminant curve consisting of the envelope plus the locus of the double points. The curves of the family are congruent to each other and arise from one another by translation parallel to the y-axis. By differentiation we obtain

$$f_c = -2(y - c)(x - 2) = 0,$$

so that we must have either $x = 2$ or $y = c$. The line $x = 2$ does not enter into the matter, however, for no finite value of y corresponds to $x = 2$. We therefore have $y = c$. So that the discriminant curve is

$$x^2(x - 2) + x = 0.$$

This curve consists of the straight lines $x = 0$ and $x = 1$. As we see in Fig. 3.25, only $x = 0$ is the envelope; the line $x = 1$ passes through the double points of the curves.

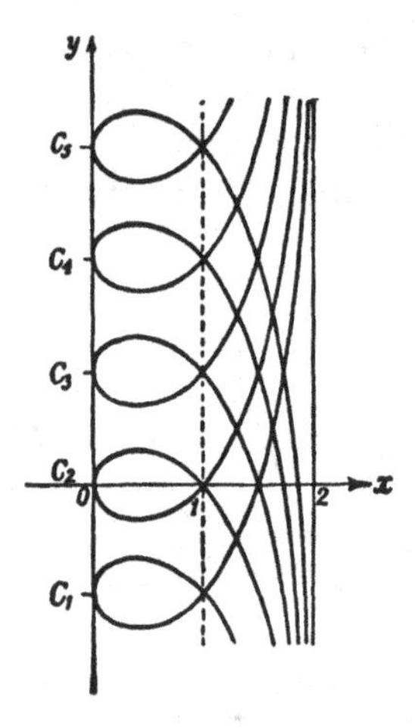

Figure 3.25 Family of strophoids.

10. The envelope need not be the locus of the points of intersection of neighbouring curves; that is shown by the family of identical parallel cubical parabolas $y - (x - c)^3 = 0$. No two of these curves intersect each other. The rule gives the equation $f_c = 3(x - c)^2 = 0$, so that the x-axis $y = 0$ is the discriminant curve. Since all the curves of the family are touched by it, it is also the envelope (Fig. 3.26).

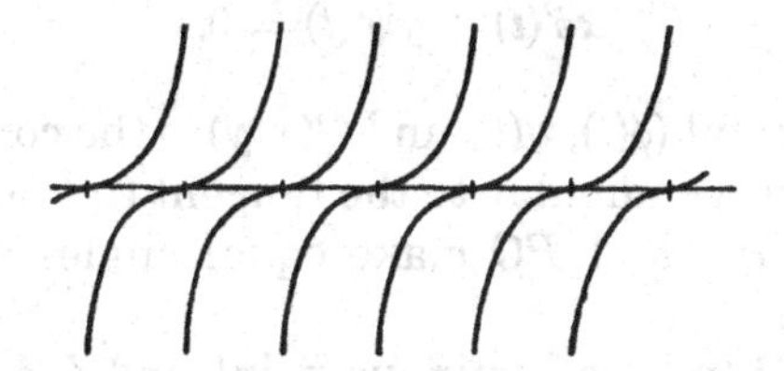

Figure 3.26 Family of cubical parabolas.

11. The notion of the envelope enables us to give a new definition for the *evolute* of a curve C (cf. Volume I, pp. 359, 424 ff.). Let C be given by

$$x = \phi(t), \quad y = \psi(t).$$

We define the evolute E of C as the envelope of the normals of C. Since the normals of C are given by

$$\{x - \phi(t)\}\phi'(t) + \{y - \psi(t)\}\psi'(t) = 0,$$

the envelope is found by differentiating this equation with respect to t:

$$0 = \{x - \phi(t)\}\phi''(t) + \{y - \psi(t)\}\psi''(t) - \phi'^2(t) - \psi'^2(t).$$

From this equation and the preceding one, we obtain the parametric representation of the envelope,

$$x = \phi(t) - \psi'(t)\frac{\phi'^2 + \psi'^2}{\psi''\phi' - \phi''\psi'} = \phi - \frac{\psi'\rho}{\sqrt{\phi'^2 + \psi'^2}},$$

$$y = \psi(t) + \phi'(t)\frac{\phi'^2 + \psi'^2}{\psi''\phi' - \phi''\psi'} = \psi + \frac{\phi'\rho}{\sqrt{\phi'^2 + \psi'^2}},$$

where

$$\rho = \frac{(\phi'^2 + \psi'^2)^{3/2}}{\psi''\phi' - \phi''\psi'}$$

denotes the radius of curvature (cf. Volume I, p. 358). These equations are identical with those given in Volume I (p. 359) for the evolute.

12. Let a curve C be given by $x = \phi(t)$, $y = \psi(t)$. We form the envelope E of the circles having their centers on C and passing through the origin O. Since the circles are given by

$$x^2 + y^2 - 2x\phi(t) - 2y\psi(t) = 0,$$

the equation of E is

$$x\phi'(t) + y\psi'(t) = 0.$$

Hence, if P is the point $(\phi(t), \psi(t))$ and $Q(x, y)$ is the corresponding point of E, then OQ is perpendicular to the tangent to C at P. Since by definition $PQ = PO$, PO and PQ make equal angles with the tangent to C at P.

If we imagine O to be a luminous point and C a reflecting curve, then QP is the reflected ray corresponding to OP. The envelope of the reflected rays is called the *caustic* of C with respect to O. *The caustic is the evolute of E:* the reflected ray PQ is normal to E, since a circle with center P touches E at Q, and the envelope of the normals of E is its evolute, as we saw in the preceding example.

For example, let C be a circle passing through O. Then E is the path described by the point O' of a circle C' congruent to C that rolls on C and starts with O and O' coincident, for during the motion O and O' always occupy symmetrical positions with respect to the common tangent of the two circles. Thus, E will be a special epicycloid, in fact, a cardioid (cf. Volume I, p. 329 ff.). As the evolute of an epicycloid is a similar epicycloid (cf. Volume I, p. 439), the caustic of C with respect to O is in this case a cardioid.

Exercises 3.5c

1. A projectile fired from the origin at initial angle of inclination α and fixed initial speed v travels in a parabolic trajectory given by the equations

$$x = (v \cos \alpha)\, t$$

$$y = (v \sin \alpha)\, t - \frac{1}{2} gt^2,$$

where g is the constant acceleration of gravity.

(a) Find the envelope of the family of trajectories with parameter α.

(b) Show that no point above the envelope can be hit by the projectile.

(c) Show that every point below the envelope can be hit in two ways, that is, that such a point lies on two trajectories.

2. Obtain the envelopes of the following families of curves:

(a) $y = cx + 1/c$.

(b) $y^2 = c(x - c)$

(c) $cx^2 + y^2/c = 1$

(d) $(x - c)^2 + y^2 = a^2c^2/(1 + a^2)$, $a =$ constant.

3. Let C be an arbitrary curve in the plane, and consider the circles of radius p whose centers lie on C. Prove that the envelope of these circles

is formed by the two curves parallel to C at the distance p (cf. the definition of parallel curves, Volume I, p. 291).

4. A family of straight lines in space may be given as the intersection of two planes depending on a parameter t:

$$a(t)x + b(t)y + c(t)z = 1$$
$$d(t)x + e(t)y + f(t)z = 1.$$

Prove that if these straight lines are tangents to some curve, (i.e., possess an envelope), then

$$\begin{vmatrix} a-d & b-e & c-f \\ a' & b' & c' \\ d' & e' & f' \end{vmatrix} = 0.$$

5. If a plane curve C is given by $x = f(t)$, $y = g(t)$, its polar reciprocal C' is defined as the envelope of the family of straight lines

$$\xi f(t) + \eta g(t) = 1,$$

where (ξ, η) are running coordinates.

(a) Prove that C is also the polar reciprocal of C'.

(b) Find the polar reciprocal of the circle $(x - a)^2 + (y - b)^2 = 1$.

(c) Find the polar reciprocal of the ellipse $x^2/a^2 + y^2/b^2 = 1$.

6. A circle of radius a rolls on a fixed straight line, carrying a tangent fixed relatively to the circle. Taking axes at the point of contact where the moving tangent coincides with the fixed line, show that the envelope of the tangent is given by

$$x = a(\theta + \cos\theta \sin\theta - \sin\theta)$$
$$y = a(\cos^2\theta - \cos\theta).$$

7. Find the envelope of a variable circle in a plane which passes through a fixed point O, and whose center describes a given conic with center O.

8. (a) If Γ is a plane curve and O a point in its plane, the locus Γ' of the orthogonal projections of O on a variable tangent of Γ is called the pedal curve of Γ with respect to the point O. Prove that if the point M describes the curve Γ, the pedal curve Γ' is the envelope of the variable circle with the radius vector OM as diameter.

(b) What is the envelope like if Γ is a circle and O a point on its circumference?

9. MM' is a variable chord of an ellipse parallel to the minor axis. Find the envelope of the variable circle with MM' as diameter.

d. *Envelopes of Families of Surfaces*

The remarks made about the envelopes of families of curves apply with but little alteration to families of surfaces also. Given a one-parameter family of surfaces $f(x, y, z, c) = 0$ defined for an interval of parameter values c, we shall say that a surface E is the envelope of the family if it touches each surface of the family along a whole curve and if, further, these curves of contact form a one-parameter family of curves on E that completely cover E.

An example is given by the family of all spheres of unit radius with centers on the z-axis. We see intuitively that the envelope is the cylinder $x^2 + y^2 - 1 = 0$ with unit radius and axis along the z-axis; the family of curves of contact is simply the family of circles parallel to the x, y-plane, with unit radius and center on the z-axis.[1]

As on p. 292, if we assume that the envelope does exist we can find it by the following heuristic method: We first consider surfaces $f(x, y, z, c) = 0$ and $f(x, y, z, c + h) = 0$ corresponding to two different parameter values c and $c + h$. These two equations determine the curve of intersection of the two surfaces (we expressly assume that such a curve of intersection exists). As a consequence of the two equations above, this curve also satisfies the third equation

$$\frac{f(x, y, z, c + h) - f(x, y, z, c)}{h} = 0.$$

If we let h tend to zero, the curve of intersection will approach a definite limiting position, and this limit curve is determined by the two equations

$$f(x, y, z, c) = 0, \quad f_c(x, y, z, c) = 0. \tag{54}$$

This curve is often referred to in a nonrigorous intuitive way as the intersection of neighboring surfaces of the family. It is a function of the parameter c, so that the curves of intersection for all the different values of c form a one-parameter family of curves in space. If we eliminate the quantity c from the two equations above, we obtain an equation that is called the *discriminant*. As on p. 293, we can show that the envelope must satisfy this discriminant equation.

Just as in the case of plane curves, we may readily convince ourselves that a plane touching the discriminant surface also touches the corresponding surface of the family, provided that $f_x^2 + f_y^2 + f_z^2 \neq 0$. Hence, the discriminant surface again gives the envelopes of the family and the loci of the singularities of the surfaces of the family.

As a first example, we consider the family of spheres

[1]The envelope of spheres of constant radius whose centers lie on a given curve are called *tube-surfaces*.

$$x^2 + y^2 + (z - c)^2 - 1 = 0$$

mentioned above. To find the envelope we have the additional equation

$$-2(z - c) = 0.$$

For fixed values of c these two equations obviously represent the circle of unit radius parallel to the x, y-plane at the height $z = c$. If we eliminate the parameter c between the two equations, we obtain the equation of the envelope in the form $x^2 + y^2 - 1 = 0$, which is the equation of the right circular cylinder with unit radius and the z-axis.

For families of surfaces it is also possible to find envelopes of two-parameter families $f(x, y, z, c_1, c_2) = 0$. (For families of curves, however, the concept of envelope has a meaning only for one-parameter families.) For example, we consider the family of all spheres with unit radius and center on the x, y-plane, represented by the equation

$$(x - c_1)^2 + (y - c_2)^2 + z^2 - 1 = 0.$$

Intuition tells us at once that the two planes $z = 1$ and $z = -1$ touch a surface of the family at every point. In general, we shall say that a surface E is the envelope of a two-parameter family of surfaces if at every point P of E the surface E touches a surface of the family in such a way that as P ranges over E, the parameter values c_1, c_2 corresponding to the surface touching E at P range over a region of the c_1,c_2-plane, and in addition different points (c_1, c_2) correspond to different points P of E. A surface of the family then touches the envelope at a *point* and not, as before, along a whole curve.

With assumptions similar to those made in the case of plane curves, we find that the point of contact of a surface of the family with the envelope, if it exists, must satisfy the equations

$$f(x, y, z, c_1, c_2) = 0, \quad f_{c_1}(x, y, z, c_1, c_2) = 0, \quad f_{c_2}(x, y, z, c_1, c_2) = 0.$$

From these three equations we determine the point of contact of a given surface of the family by assigning the corresponding values to the parameters. Conversely, if we eliminate the parameters c_1 and c_2, we obtain an equation that the envelope must satisfy.

For example, the family of spheres with unit radius and center on the x, y-plane is given by the equation

$$f(x, y, z, c_1, c_2) = (x - c_1)^2 + (y - c_2)^2 + z^2 - 1 = 0$$

with the two parameters c_1 and c_2. The rule for forming the envelope gives the two equations

$$f_{c_1} = -2(x - c_1) = 0 \qquad \text{and} \qquad f_{c_2} = -2(y - c_2) = 0.$$

Thus, for the discriminant equation, we have $z^2 - 1 = 0$, and in fact, the two planes $z = 1$ and $z = -1$ are envelopes, as we have already seen intuitively.

Exercises 3.5d

1. What is the envelope of the family of ellipsoids of constant volume (i.e., fixed product of the semiaxes) with common center at O and axes parallel to the coordinate axes?
2. What is the envelope of the family of planes $ax + by + cz = 1$, where $\sqrt{a^2 + b^2 + c^2} = 1$?
3. (a) Find the envelope of the two-parameter family of planes for which

   $$OP + OQ + OR = \text{constant} = 1,$$

   where P, Q, R denote the points of intersection of the planes with the coordinate axes and O the origin.

   (b) Find the envelope of the planes for which

   $$OP^2 + OQ^2 + OR^2 = 1.$$
4. A family of planes is given by

   $$x \cos t + y \sin t + z = t,$$

   where t is a parameter.

   (a) Find the equation of the envelope for the planes in cylindrical coordinates (r, z, θ).

   (b) Prove that the envelope consists of the tangents to a certain curve.
5. Let $z = u(x, y)$ be the equation of a tube-surface, that is, the envelope of a family of spheres of unit radius with their centers on some curve $y = f(x)$ in the x, y-plane. Prove that $u^2 (u_x^2 + u_y^2 + 1) = 1$.
6. Find the envelope of the family of spheres that touch the three spheres

   $$S_1\colon \left(x - \frac{3}{2}\right)^2 + y^2 + z^2 = \frac{9}{4},$$

   $$S_2\colon x^2 + \left(y - \frac{3}{2}\right)^2 + z^2 = \frac{9}{4},$$

   $$S_3\colon x^2 + y^2 + \left(z - \frac{3}{2}\right)^2 = \frac{9}{4}.$$
7. Let Γ be a plane curve and Γ' its pedal curve as described in Exercise 8, p. 303

   (a) Let M be a point describing the curve Γ. What is the envelope of the variable sphere with the radius vector OM as diameter?

(b) What is the envelope of the variable spheres if Γ is a circle and O a point on its circumference?

8. Show that the surface $xyz =$ constant is the envelope of the family of planes that form, with the coordinate planes, a tetrahedron of constant volume (i.e., fixed product of the intercepts).

9. A plane moves so as to touch the parabolas $z = 0$, $y^2 = 4x$ and $y = 0$, $z^2 = 4x$. Show that its envelope consists of two parabolic cylinders.

3.6 Alternating Differential Forms

a. Definition of Alternating Differential Forms

In Chapter 1 (p. 84) we considered the general linear differential form

$$L = A(x, y, z)\,dx + B(x, y, z)\,dy + C(x, y, z)\,dz \tag{55a}$$

in three independent variables. Along any curve Γ with parameter representation $x = \phi(t)$, $y = \psi(t)$, $z = \chi(t)$ the form L determines values

$$\frac{L}{dt} = A\frac{dx}{dt} + B\frac{dy}{dt} + C\frac{dz}{dt} = A\dot{\phi} + B\dot{\psi} + C\dot{\chi}, \tag{55b}$$

which depend on the special parametric representation of Γ. If Γ is referred to a different parameter t, we obtain

$$\begin{aligned}\frac{L}{d\tau} = A\frac{dx}{d\tau} + B\frac{dy}{d\tau} + C\frac{dz}{d\tau} &= \left(A\frac{dx}{dt} + B\frac{dy}{dt} + C\frac{dz}{dt}\right)\frac{dt}{d\tau}\\ &= \frac{L}{dt}\frac{dt}{d\tau}.\end{aligned} \tag{55c}$$

However, the integral

$$\int_\Gamma L = \int \frac{L}{dt}\,dt = \int\left(A\frac{dx}{dt} + B\frac{dy}{dt} + C\frac{dz}{dt}\right)dt$$

depends only on the curve Γ (and its orientation) and not on the particular parametric representation.

Similarly, we can consider a differential form ω which is quadratic in dx, dy, and dz, namely, a linear combination ω of the symbols $dx\,dx$, $dx\,dy$, $dx\,dz$, $dy\,dx$, $dy\,dy$, $dy\,dz$, $dz\,dx$, $dz\,dy$, $dz\,dz$ with coefficients that are functions of x, y, z. Upon any surface S in space with parametric representation $x = \phi(s, t)$, $y = \psi(s, t)$, $z = \chi(s, t)$, the form ω defines values $\omega/ds\,dt$ if we agree that the quotients

$$\frac{dx\,dx}{ds\,dt},\ \frac{dx\,dy}{ds\,dt},\ \frac{dx\,dz}{ds\,dt},\ \ldots$$

are to stand respectively for the Jacobians

$$\frac{d(x, x)}{d(s, t)},\ \frac{d(x, y)}{d(s, t)},\ \frac{d(x, z)}{d(s, t)},\ \ldots.[1]$$

We do not distinguish between two differential forms ω that yield the same values $\omega/ds\,dt$ at each point of the surface. In view of the alternating character of determinants, namely, that.

$$\frac{d(x, x)}{d(s, t)} = 0, \qquad \frac{d(x, y)}{d(s, t)} = -\frac{d(y, x)}{d(s, t)}, \ldots,$$

we see that the terms of ω with $dx\,dx$, $dy\,dy$, $dz\,dz$ make no contributions and that $dy\,dx$, $dz\,dy$, $dx\,dz$ can be replaced respectively by $-dx\,dy, -dy\,dz, -dz\,dx$. Thus the most general quadratic differential form in dx, dy, dz can be written as

$$\text{(56a)} \qquad \omega = a(x, y, z)\,dy\,dz + b(x, y, z)\,dz\,dx + c(x, y, z)\,dx\,dy.$$

The values that ω associates with the points of a surface S referred to parameters s, t are

$$\text{(56b)} \quad \frac{\omega}{ds\,dt} = a(x, y, z)\frac{d(y, z)}{d(s, t)} + b(x, y, z)\frac{d(z, x)}{d(s, t)} + c(x, y, z)\frac{d(x, y)}{d(s, t)}.$$

Giving S different parameters s', t', we obtain from the multiplication law for Jacobians (see p. 258)

$$\text{(56c)} \qquad \begin{aligned} \frac{\omega}{ds'\,dt'} &= a\frac{d(y, z)}{d(s', t')} + b\frac{d(z, x)}{d(s', t')} + c\frac{d(x, y)}{d(s', t')} \\ &= \frac{\omega}{ds\,dt}\frac{d(s, t)}{d(s', t')}. \end{aligned}$$

Later (p. 593), we shall also define the double integral

[1]This convention characterizes *alternating* differential forms. In other contexts, nonalternating quadratic differential forms are encountered as well, such as the one giving the square of the line element in space or on a surface (see p. 283):

$$ds^2 = dx^2 + dy^2 + dz^2 = E\,du^2 + 2F\,du\,dv + G\,dv^2.$$

$$\iint_S \omega$$

and see that it does not depend on the particular parameter representation of the surface S.

In a similar way, we can consider a differential form ω that is cubic in dx, dy, dz. Such a form assigns values $\omega/dr\, ds\, dt$ corresponding to any parametric representation

$$x = \phi(r, s, t), \quad y = \psi(r, s, t), \quad z = \chi(r, s, t),$$

where again we interpret the quotients

$$\frac{dx\, dx\, dx}{dr\ ds\ dt}, \frac{dx\, dy\, dz}{dr\ ds\ dt}, \ldots$$

as the Jacobians

$$\frac{d(x, x, x)}{d(r, s, t)}, \frac{d(x, y, z)}{d(r, s, t)}, \ldots.$$

Since the Jacobians vanish when two of the dependent variables are identical and change signs when two of the dependent variables are interchanged, the cubic differential forms in the three independent variables x, y, z are all of the type

$$\omega = \alpha(x, y, z) dx\, dy\, dz. \tag{56d}$$

Whenever x, y, z are represented as functions of r, s, t, we obtain from ω the value

$$\frac{\omega}{dr\, ds\, dt} = \alpha(x, y, z)\frac{d(x, y, z)}{d(r, s, t)}. \tag{56e}$$

Proceeding in the same manner we could define "alternating" differential forms in dx, dy, dz of degrees 4, 5, ... But all of these are identically 0, since any Jacobians of orders 4, 5, . . . that we could form would have two of the dependent variables identical, and, hence, would vanish.[1]

---

[1]Higher-order forms have, however, a nontrivial meaning in spaces of higher dimensions. In four-dimensional x, y, z, u-space the most general alternating differential forms of order 1, 2, 3, 4 can be written as

(56f) $A\, dx + B\, dy + C\, dz + D\, du$

Exercises 3.6a

1. Find $\omega/du\,dv$ for each of the following:

   (a) $\omega = x\,dy\,dz + y\,dz\,dx + z\,dx\,dy,$

   $x = \cos u \sin v, \quad y = \sin u \sin v, \quad z = \cos v$

   (b) $\omega = (y - z)dy\,dz + (z - x)dz\,dx + (x - y)dx\,dy,$

   $x = au + bv, \quad y = bu + cv, \quad z = cu + av$

   (c) $\omega = dy\,dz + dz\,dx + dx\,dy,$

   $x = u^2 + v^2, \quad y = 2uv, \quad z = u^2 - v^2.$

b. Sums and Products of Differential Forms

Two differential forms of the same order (i.e., either both linear, both quadratic, or both cubic) can be added trivially by adding corresponding coefficients. Thus, for

$$\omega_1 = a_1\,dy\,dz + b_1\,dz\,dx + c_1\,dx\,dy,$$

$$\omega_2 = a_2\,dy\,dz + b_2\,dz\,dx + c_2\,dx\,dy,$$

we define

$$\text{(57a)} \quad \omega_1 + \omega_2 = (a_1 + a_2)dy\,dz + (b_1 + b_2)dz\,dx + (c_1 + c_2)dx\,dy.$$

We can define the product $\omega_1\omega_2$ of any two differential forms ω_1 and ω_2 of the same or of different orders by just substituting for ω_1 and ω_2 their expressions in terms of dx, dy, dz and applying the distributive law of multiplication, taking care, however, to preserve the original order of the differentials in each term.[1] Thus, the product of the two linear forms

$$\omega_1 = A_1\,dx + B_1\,dy + C_1\,dz \qquad \text{and} \qquad \omega_2 = A_2\,dx + B_2\,dy + C_2\,dz$$

would be the quadratic form

(56g) $A\,dx\,dy + B\,dy\,dz + C\,dz\,du + D\,du\,dx + E\,dx\,dz + F\,dy\,du$

(56h) $A\,dy\,dz\,du + B\,dz\,du\,dx + C\,du\,dx\,dy + D\,dx\,dy\,dz$

(56i) $A\,dx\,dy\,dz\,du,$

respectively, with coefficients $A, B, \ldots$, which are functions of x, y, z, u. Forms of order higher than 4 vanish.

[1]The product formed in this way is sometimes denoted by the symbol $\omega_1 \wedge \omega_2$.

$$
\begin{aligned}
\text{(57b)} \quad \omega_1\omega_2 &= (A_1\,dx + B_1\,dy + C_1\,dz)(A_2\,dx + B_2\,dy + C_2\,dz) \\
&= A_1A_2\,dx\,dx + A_1B_2\,dx\,dy + A_1C_2\,dx\,dz + B_1A_2\,dy\,dx \\
&\qquad + B_1B_2\,dy\,dy + B_1C_2\,dy\,dz + C_1A_2\,dz\,dx \\
&\qquad + C_1B_2\,dz\,dy + C_1C_2\,dz\,dz \\
&= (B_1C_2 - C_1B_2)dy\,dz + (C_1A_2 - A_1C_2)dz\,dx \\
&\qquad + (A_1B_2 - B_1A_2)dx\,dy.
\end{aligned}
$$

If we describe the individual forms ω_1 and ω_2 by the "coefficient vectors" $\mathbf{R}_1 = (A_1, B_1, C_1)$ and $\mathbf{R}_2 = (A_2, B_2, C_2)$, then the coefficients of the product $\omega_1\omega_2$ are just the components of the *vector product* $\mathbf{R}_1 \times \mathbf{R}_2$ (see p. 181). Clearly, the product of the forms is not commutative. Here, for example, $\omega_1\omega_2 = -\omega_2\omega_1$.

Multiplying the first-order form

$$\omega_1 = A\,dx + B\,dy + C\,dz$$

with the second-order form

$$\omega_2 = a\,dy\,dz + b\,dz\,dx + c\,dx\,dy,$$

we obtain similarly

$$
\begin{aligned}
\text{(57c)} \quad \omega_1\omega_2 &= (A\,dx + B\,dy + C\,dz)(a\,dy\,dz + b\,dz\,dx + c\,dx\,dy) \\
&= Aa\,dx\,dy\,dz + Ab\,dx\,dz\,dx + Ac\,dx\,dx\,dy \\
&\quad + Ba\,dy\,dy\,dz + Bb\,dy\,dz\,dx + Bc\,dy\,dx\,dy \\
&\quad + Ca\,dz\,dy\,dz + Cb\,dz\,dz\,dx + Cc\,dz\,dx\,dy \\
&= (Aa + Bb + Cc)dx\,dy\,dz.
\end{aligned}
$$

We observe that in this case the coefficient of $\omega_1\omega_2$ is the scalar product of the coefficient vectors (A, B, C) and (a, b, c). Here, incidentally, $\omega_1\,\omega_2 = \omega_2\,\omega_1$.

Forming the product of a first- and a third-order form, of two second-order forms, or of a second- and a third-order form yields forms of order higher than 3, which vanish. For the sake of completeness it is convenient to define differential forms of order 0 as the scalars $\alpha(x, y, z)$. The product of a form α of order 0 with a form ω of any order $k = 0, 1, 2, 3$ is then obtained by multiplying each of the coefficients of ω by the scalar α.

It is easily seen from the definition that products of differential forms are associative. For three linear forms

$$L_i = A_i\,dx + B_i\,dy + C_i\,dz \qquad (i = 1, 2, 3).$$

for example, as is to be proved in Exercise 5,

$$L_1(L_2L_3) = \begin{vmatrix} A_1 & B_1 & C_1 \\ A_2 & B_2 & C_2 \\ A_3 & B_3 & C_3 \end{vmatrix} dx\,dy\,dz. \tag{57d}$$

and for $(L_1\,L_2)\,L_3$ we obtain the same evaluation.

Of course, a greater variety of products of differential forms can be formed when the number of independent variables is greater than 3.

Exercises 3.6b

1. Evaluate the following products:
   (a) $(x\,dx + y\,dy)(x\,dx - y\,dy)$
   (b) $[(x^2 + y^2)dx + 2xy\,dy]\,[2xy\,dx + (x^2 - y^2)dy]$
   (c) $(a\,dx + b\,dy)(a\,dy\,dz + b\,dz\,dx + c\,dx\,dy)$
   (d) $(dx + dy + dz)(dy\,dz - dx\,dy)$.
2. For any form ω of order 1 in x, y, z, show that $\omega^2 = 0$.
3. For first-order forms ω_1, ω_2 in three variables, show that
$$(\omega_1 + \omega_2)(\omega_1 - \omega_2) = 2\omega_2\omega_1.$$
4. Show for first-order forms in three variables that
$$(\omega_1 + \omega_2 + \omega_3 + \omega_4)(\omega_1 - \omega_2 + \omega_3 - \omega_4) = 2(\omega_2 + \omega_4)(\omega_1 + \omega_3).$$
5. Derive (57d).

c. *Exterior Derivatives of Differential Forms*

For a differential form of order 0, that is, for a scalar $\alpha(x, y, z)$ we have by definition

$$d\alpha = \alpha_x\,dx + \alpha_y\,dy + \alpha_z\,dz. \tag{58a}$$

The coefficients of this differential form are just the components of the vector we denoted by grad α on p. 206. More generally, we define the *exterior derivative* $d\omega$ of any differential form ω. For this purpose, we write out ω as a sum of terms where each term is a product of certain of the differentials dx, dy, dz *preceded* by a scalar factor and replace each of the scalar factors by its differential, formed in the ordinary sense. Thus, for a first order form

$$L = A\,dx + B\,dy + C\,dz,$$

we find for dL the second-order differential form

(58b)

$$\begin{aligned} dL &= dA\,dx + dB\,dy + dC\,dz \\ &= (A_x\,dx + A_y\,dy + A_z\,dz)dx \\ &\quad + (B_x\,dx + B_y\,dy + B_z\,dz)dy + (C_x\,dx + C_y\,dy + C_z\,dz)dz \\ &= (C_y - B_z)dy\,dz + (A_z - C_x)dz\,dx + (B_x - A_y)dx\,dy. \end{aligned}$$

If we associate with L the vector $\mathbf{R} = (A, B, C)$, we have the remarkable fact that *the coefficients of dL are just the components of the curl of* $\mathbf{R}$ (see p. 209).

For a second-order form

$$\omega = a\,dy\,dz + b\,dz\,dx + c\,dx\,dy$$

the exterior derivative $d\omega$ is the third-order form

(58c)

$$\begin{aligned} d\omega &= da\,dy\,dz + db\,dz\,dx + dc\,dx\,dy \\ &= (a_x\,dx + a_y\,dy + a_z\,dz)dy\,dz \\ &\quad + (b_x\,dx + b_y\,dy + b_z\,dz)dz\,dx \\ &\quad + (c_x\,dx + c_y\,dy + c_z\,dz)dx\,dy \\ &= (a_x + b_y + c_z)dx\,dy\,dz. \end{aligned}$$

Hence, if the coefficients of ω are combined into the vector $\mathbf{R} = (a, b, c)$, then the coefficient of $d\omega$ is the scalar div $\mathbf{R}$ (see p. 210).

The derivative of a third-order differential form is of fourth order and, hence, vanishes.

An important general rule ("Poincaré lemma") is that *the second exterior derivative of any differential form* ω *vanishes:*

(58d)

$$dd\omega = 0.$$

In three-space this only has to be proved for the cases where ω either is of order 0 or 1. Now if ω is a scalar $\alpha(x, y, z)$, we have by (58a, b)

$$d^2\omega = d(\alpha_x\,dx + \alpha_y\,dy + \alpha_z\,dz) = 0.$$

This is really only a different way of expressing the rule stated on p. 210 that curl (grad α) $= \mathbf{0}$ for any scalar α. Similarly, we find from (58b, c) for the case of a first-order differential form

$$\omega = A\,dx + B\,dy + C\,dz$$

that

$$d^2\omega = d[(C_y - B_z)dy\,dz + (A_z - C_x)dz\,dx + (B_x - A_y)dx\,dy] = 0.$$

This again is nothing else but the rule div (curl **R**) = 0 valid for any vector **R** (see p. 211).

The inverse problem of finding a form τ that has a given form ω as its exterior derivative is basic. We should like to represent a given differential form ω as

$$\omega = d\tau \tag{58e}$$

with a suitable differential form τ. We call ω an *exact,* or *total, differential* when such a representation is possible. Applying rule (59) to the differential τ, we see that *a necessary condition for* ω *to be an exact differential is that* $d\omega = 0$.[1] It turns out that this condition is also sufficient; that is, for $d\omega = 0$ *the equation* (58e) *has a solution* τ, *provided we restrict ourselves to a rectangular neighborhood of a point* (x_0, y_0, z_0) *interior to the domain of definition*[2] *of* ω.

We prove this statement separately for each order of ω. If ω is of order 1, say

$$\omega = A\,dx + B\,dy + C\,dz,$$

then, by (58b), the condition $d\omega = 0$ is equivalent to the relations

$$C_y - B_z = 0, \quad A_z - C_x = 0, \quad B_x - A_y = 0. \tag{58f}$$

But these are just the *integrability conditions* that permit us to represent ω as the total differential of some function f, provided we restrict the point (x, y, z) to a rectangular parallelepiped containing (x_0, y_0, z_0) or, more generally, to a simply connected set (see p. 104).

For ω of order 2,

$$\omega = a\,dy\,dz + b\,dz\,dx + c\,dx\,dy,$$

the condition $d\omega = 0$ by (58c) is equivalent to

$$a_x + b_y + c_z = 0. \tag{58g}$$

[1]Forms ω for which $d\omega = 0$ are called *closed.*

[2]We always assume that the differential forms considered here have coefficients with as many continuous derivatives as are needed for our arguments to hold.

Assume that this condition is satisfied in the rectangular parallelepiped

$$|x - x_0| < r_1, \quad |y - y_0| < r_2, \quad |z - z_0| < r_3.$$

We have to show that $\omega = d\tau$, where τ is of the form

$$\tau = A\,dx + B\,dy + C\,dz.$$

This means functions A, B, C have to be found for which

$$a = C_y - B_z, \quad b = A_z - C_x, \quad c = B_x - A_y.$$

We try to satisfy these equations with the choice $C \equiv 0$. Then A and B have to be of the form

$$A(x, y, z) = \alpha(x, y) + \int_{z_0}^{z} b(x, y, \zeta)\,d\zeta,$$

$$B(x, y, z) = \beta(x, y) - \int_{z_0}^{z} a(x, y, \zeta)\,d\zeta$$

in order to satisfy the first two equations. It follows, using condition (58g), that

$$\frac{\partial}{\partial z}(B_x - A_y) = \frac{\partial}{\partial x}B_z - \frac{\partial}{\partial y}A_z = -a_x - b_y = c_z.$$

Hence $B_x - A_y - c$ does not depend on z. The third equation $c = B_x - A_y$ will be satisfied for all z in question if it holds for $z = z_0$. Hence, we only have to determine the functions $\alpha(x, y)$ and $\beta(x, y)$ in such a way that

$$\beta_x(x, y) - \alpha_y(x, y) = c(x, y, z_0).$$

This is achieved by taking

$$\alpha(x, y) = 0, \quad \beta(x, y) = \int_{x_0}^{x} c(\xi, y, z_0)\,d\xi,$$

for example.

Finally, for a third-order operator

$$\omega = a(x, y, z)\,dx\,dy\,dz$$

the condition $d\omega = 0$ is always satisfied. We want to represent ω in the form $\omega = d\tau$, where τ is a second-order differential form

$$\tau = a\,dy\,dz + b\,dz\,dx + c\,dx\,dy.$$

By (58c) this amounts to finding functions a, b, c for which

$$a_x + b_y + c_z = \alpha.$$

One solution clearly is given by

$$a(x, y, z) = b(x, y, z) = 0, \quad c(x, y, z) = \int_{z_0}^{z} \alpha(x, y, \zeta)d\zeta.$$

This proves our theorem.

Exercises 3.6c

1. Evaluate $d\omega$ for each of the following:
   (a) $\omega = \text{arc tan } y/x$
   (b) $\omega = y\,dx - x\,dy$
   (c) $\omega = f(x, y)\,dx\,dy$
   (d) $\omega = x^2 \cos y \sin z\,dy\,dz - x \sin y \sin z\,dz\,dx + x \cos z\,dx\,dy$
   (e) $\omega = (z^2 - y^2)x\,dy\,dz + (x^2 - z^2)y\,dz\,dx + (y^2 - x^2)z\,dx\,dy.$
2. For first-order forms in three variables, show that
$$d(\omega_1\omega_2) = \omega_1(d\omega_2) + (d\omega_1)\omega_2.$$
3. Show that any product of exact first-order forms in three variables is exact.

d. Exterior Differential Forms in Arbitrary Coordinates

So far, we have always looked at differential forms as linear combinations of alternating products of the differentials dx, dy, dz of the Cartesian coordinates x, y, z in space. We made essential use of this representation of forms in terms of dx, dy, dz in defining the product of two forms and the derivative of a form. The usefulness of alternating differential forms in applications depends on the fact that these forms can be defined and operations on forms can be performed in the same way when three-dimensional[1] euclidean space is referred

[1]The dimension 3 is chosen here only for the sake of definiteness. All these considerations are equally valid for any other number of dimensions.

to any *curvilinear coordinates* u, v, w. More generally, this holds on any noneuclidean three-dimensional space or *manifold*[1] referred to parameters u, v, w, for example, on a three-dimensional "surface" in four-dimensional euclidean space. What is important is that operations on forms can be defined in an *invariant manner,* without reference to a special coordinate system, and that the resulting formulae look the same in every system.

In this context, one thinks of the points P of the three-dimensional space or of a manifold Σ as *geometric* objects that exist independently of any coordinate system. A scalar f is a function of P with real numbers as values (that is, a *mapping* of Σ into the real number axis). There are, however, many ways of describing points P by *curvilinear coordinates,* that is, by triples of numbers (u, v, w), for example, by rectangular coordinates or spherical coordinates in euclidean space. We always assume that any two such coordinate systems, say u, v, w and u', v', w', are related by transformation equations

$$u' = \phi(u, v, w), \quad v' = \psi(u, v, w), \quad w' = \chi(u, v, w),$$

where ϕ, ψ, χ are continuous functions with as many continuous derivatives as required for our operations, and with a Jacobian $\dfrac{d(u', v', w')}{d(u, v, w)}$ that does not vanish.[2] In that case u, v, w can be expressed by similar formulae in terms of u', v', w'. In a given coordinate system u, v, w a scalar $f = f(P)$ becomes a function $f(u, v, w)$ of the coordinates u, v, w of the point P. In different coordinate systems, the functions representing the same scalar are generally quite different.

On the manifold Σ let C be a curve with the parametric representation $P = P(t)$; with every real number t of a certain interval the parametric equation associates a point P of the manifold Σ. Any scalar $f(P)$ defined on Σ yields a function of t along C obtained by forming the composition $f(P(t))$. If this function is differentiable, it makes sense to form the derivative df/dt, which is defined for the given curve and parametric representation of C, independently of any curvilinear coordinate system used for Σ. In a given coordinate system the coordinates u, v, w of a point P themselves are functions $u = u(t)$, $v = v(t)$, $w = w(t)$; and $f(P(t))$ is given by the compound function

[1]Generally we use the term "manifold" to denote a parametrically given set of any number of dimensions $m \leqq n$ in n-dimensional euclidean space.

[2]The particular representation of the transformation involving univalued functions ϕ, ψ, χ needs to be valid only locally, that is, in a sufficiently small neighborhood of some point.

$f(u(t), v(t), w(t))$. Assuming $f(u, v, w)$ and $u(t), v(t), w(t)$ to have continuous derivatives, we find from the chain rule of differentiation that in the particular u, v, w-system df/dt takes the form

$$\frac{df}{dt} = \frac{\partial f}{\partial u}\frac{du}{dt} + \frac{\partial f}{\partial v}\frac{dv}{dt} + \frac{\partial f}{\partial w}\frac{dw}{dt}. \tag{59}$$

A zero-order differential form in Σ is just a scalar f. The general first-order differential form ω is defined as a formal expression of the type

$$\omega = \sum_{i=1}^{N} a_i \, df_i,$$

where $a_1, \ldots, a_N, f_1, \ldots, f_N$ are given scalars. Along any curve C referred to a parameter t, we associate with ω the function of t, denoted by ω/dt, which is defined by

$$\frac{\omega}{dt} = \sum_{i=1}^{N} a_i \frac{df_i}{dt}.$$

Two forms

$$\omega = \sum_{i=1}^{N} a_i \, df_i \qquad \text{and} \qquad \omega' = \sum_{i=1}^{m} b_i \, dg_i$$

are considered equal if

$$\frac{\omega}{dt} = \frac{\omega'}{dt}$$

for any curve C and any parameter t along C.

In a particular u, v, w-coordinate system ω/dt becomes

$$\frac{\omega}{dt} = \sum_{i=1}^{N} a_i \left(\frac{\partial f_i}{\partial u}\frac{du}{dt} + \frac{\partial f_i}{\partial v}\frac{dv}{dt} + \frac{\partial f_i}{\partial w}\frac{dw}{dt} \right) = A\frac{du}{dt} + B\frac{dv}{dt} + C\frac{dw}{dt},$$

where

$$A = \sum_{i=1}^{N} a_i \frac{\partial f_i}{\partial u}, \qquad B = \sum_{i=1}^{N} a_i \frac{\partial f_i}{\partial v}, \qquad C = \sum_{i=1}^{N} a_i \frac{\partial f_i}{\partial w}$$

are scalars defined in Σ. By our definition of equality of first-order differential forms, we can write ω as

$$\omega = A\,du + B\,dv + C\,dw$$

Here the coefficients A, B, C of ω referred to a particular coordinate system u, v, w are determined uniquely, for if we take for the curve C a "coordinate line," say $u = t$, $v =$ constant, $w =$ constant, we find

$$\frac{\omega}{dt} = \frac{\omega}{du} = A,$$

and similarly,

$$\frac{\omega}{dv} = B, \qquad \frac{\omega}{dw} = C.$$

Thus, in any particular coordinate system u, v, w, we can write ω as

$$\omega = \frac{\omega}{du}\,du + \frac{\omega}{dv}\,dv + \frac{\omega}{dw}\,dw, \tag{60}$$

where ω/du really stands for the partial derivative formed along a curve where v and w are constant. This formula can be regarded as an extension of the chain rule (59) from the differential df of any scalar f to a general first-order differential form ω.

We can define now in exactly the same manner a *second-order alternating differential form* ω as a formal expression of the type

$$\omega = \sum_{i=1}^{N} a_i\,df_i\,dg_i, \tag{61a}$$

where $a_1, \ldots, a_N, f_1, \ldots, f_N, g_1, \ldots, g_N$ are scalars defined on $\sum$. On any surface S in $\sum$ referred to parameters s, t, we associate with the form ω the values $\omega/ds\,dt$ defined by

$$\frac{\omega}{ds\,dt} = \sum_{i=1}^{N} a_i \frac{d(f_i, g_i)}{d(s, t)} = \sum_{i=1}^{N} a_i \begin{vmatrix} \dfrac{\partial f_i}{\partial s} & \dfrac{\partial f_i}{\partial t} \\ \dfrac{\partial g_i}{\partial s} & \dfrac{\partial g_i}{\partial t} \end{vmatrix}. \tag{61b}$$

Two forms ω and ω', although represented with the help of different scalars, are considered identical when they determine the same values $\omega/ds\,dt = \omega'/ds\,dt$ on each surface for every parameter representation. Now in any particular coordinate system u, v, w we have for two scalars f, g

$$\begin{vmatrix} f_s & f_t \\ g_s & g_t \end{vmatrix} = \begin{vmatrix} f_u u_s + f_v v_s + f_w w_s & f_u u_t + f_v v_t + f_w w_t \\ g_u u_s + g_v v_s + g_w w_s & g_u u_t + g_v v_t + g_w w_t \end{vmatrix}$$
$$= (f_v g_w - f_w g_v)(v_s w_t - v_t w_s) + (f_w g_u - f_u g_w)(w_s u_t - w_t u_s) + (f_u g_v - f_v g_u)(u_s v_t - u_t v_s);$$

hence,

(61c) $$\frac{\omega}{ds\,dt} = a\frac{d(v, w)}{d(s, t)} + b\frac{d(w, u)}{d(s, t)} + c\frac{d(u, v)}{d(s, t)},$$

where

(61d) $$a = \sum_{i=1}^{N} a_i \frac{d(f_i, g_i)}{d(v, w)}, \qquad b = \sum_{i=1}^{N} a_i \frac{d(f_i, g_i)}{d(w, u)},$$
$$c = \sum_{i=1}^{N} a_i \frac{d(f_i, g_i)}{d(u, v)}.$$

Thus, we can write ω in the u, v, w-system as

(61e) $$\omega = a\,dv\,dw + b\,dw\,du + c\,du\,dv.$$

The coefficients a, b, c in this representation of ω are again determined uniquely; they are given by

$$a = \frac{\omega}{dv\,dw}, \qquad b = \frac{\omega}{dw\,du}, \qquad c = \frac{\omega}{du\,dv},$$

where $a = \omega/dvdw$ is formed with respect to a coordinate surface $v = s$, $w = t$, $u =$ constant, and similarly for b and c. In the u, v, w-system the symbolic expression (61c) for ω becomes

(61f) $$\omega = \frac{\omega}{dv\,dw}dv\,dw + \frac{\omega}{dw\,du}dw\,du + \frac{\omega}{du\,dv}du\,dv,$$

in analogy to the formula (60) for first-order differential forms.[1]

[1]Formulae (61a, b) retain their validity for second-order forms in n-dimensional space referred to parameters $u_1, \ldots, u_n$. Instead of (61c, d, e, f), we have then

(61g) $$\omega = \sum_{\substack{j,k=1,\ldots,n \\ j<k}} A_{jk}\,du_j\,du_k,$$

where

(61h) $$A_{jk} = \sum_i a_i \frac{d(f_i, g_i)}{d(u_j, u_k)} = \frac{\omega}{du_j\,du_k},$$

as is easily verified.

We define the *product LM* of two first-order forms

$$L = \sum_i a_i \, df_i, \quad M = \sum_i b_k \, dg_k \tag{62a}$$

on a surface with parameters s, t, as that second-order form ω, for which

$$\begin{aligned} \frac{\omega}{ds\,dt} &= \frac{L}{ds}\frac{M}{dt} - \frac{L}{dt}\frac{M}{ds} \\ &= \sum_i a_i \frac{\partial f_i}{\partial s} \sum_k b_k \frac{\partial g_k}{\partial t} - \sum_i a_i \frac{\partial f_i}{\partial t} \sum_k b_k \frac{\partial g_k}{\partial s} \\ &= \sum_{i,k} a_i b_k \frac{d(f_i, g_k)}{d(s, t)}.[1] \end{aligned} \tag{62b}$$

Consequently, if L and M are given by (62a), LM can be identified with the second-order form

$$\omega = \sum_{i,k} a_i b_k \, df_i \, dg_k. \tag{62c}$$

However, the definition of $\omega/ds\,dt = LM/ds\,dt$ given by (62b) does not depend on the particular representation of L and M in terms of scalars a_i, f_i, b_k, g_k; hence, formula (62c) must represent the same form $\omega = LM$ for all representations of the factors L, M.

Another way of generating second-order forms from those of first order is by differentiation. Given the first-order form

$$L = \sum_i a_i df_i \tag{63a}$$

we can define dL without reference to any particular coordinate system by the prescription

$$\begin{aligned} \frac{dL}{ds\,dt} &= \frac{\partial}{\partial s}\frac{L}{dt} - \frac{\partial}{\partial t}\frac{L}{ds} \\ &= \frac{\partial}{\partial s}\sum_i a_i \frac{\partial f_i}{\partial t} - \frac{\partial}{\partial t}\sum_i a_i \frac{\partial f_i}{\partial s} \\ &= \sum_i \left(\frac{\partial a_i}{\partial s}\frac{\partial f_i}{\partial t} - \frac{\partial a_i}{\partial t}\frac{\partial f_i}{\partial s}\right) = \sum \frac{d(a_i, f_i)}{d(s, t)}. \end{aligned} \tag{63b}$$

[1]Here M/ds and M/dt denote "partial" differentiation (or derivatives) with t and s, respectively, held constant. (A consistent distinction between ordinary and partial differentiation can hardly be made.)

This is equivalent to the formula

(63c) $$dL = \sum_i da_i\, df_i,$$

and shows that the second-order form dL does not depend on the particular representation (63a) of L in terms of the scalars a_i, f_i. It is the natural generalization of formula (58b) for the special case of the derivative of a form L expressed as $L = A\,dx + B\,dy + C\,dz$.

In the particular case where the first-order form L is a total differential—that is, $L = df$ with a scalar f—we find, of course, from (63c) that $dL = 0$. Hence, for a 0-order operator f, the rule

$$ddf = 0$$

is verified. When L is represented in terms of a particular coordinate system u, v, w in space by the standard form

$$L = A\,du + B\,dv + C\,dw,$$

we find from (61f), (63b)

$$\begin{aligned} dL &= dA\,du + dB\,dv + dC\,dw \\ &= \frac{dL}{dv\,dw}dv\,dw + \frac{dL}{dw\,du}dw\,du + \frac{dL}{duv}du\,dv \\ &= \left(\frac{\partial}{\partial v}\frac{L}{dw} - \frac{\partial}{\partial w}\frac{L}{dv}\right)dv\,dw + \left(\frac{\partial}{\partial w}\frac{L}{du} - \frac{\partial}{\partial u}\frac{L}{dw}\right)dw\,dv \\ &\qquad + \left(\frac{\partial}{\partial u}\frac{L}{dv} - \frac{\partial}{\partial v}\frac{L}{du}\right)du\,dv \\ &= (C_v - B_w)\,dv\,dw + (A_w - C_u)\,dw\,du + (B_u - A_v)\,du\,dv, \end{aligned}$$

in agreement with formula (58b).

If $dL = 0$, we obtain as before that $C_v - B_w = A_w - C_u = B_u - A_v = 0$. It follows that locally there exists a scalar f for which $A = f_u$, $B = f_v$, $C = f_w$ or $L = df$.

Finally, a third-order alternating differential form is defined by a formal expression

(64a) $$\omega = \sum_{i=1}^{N} a_i\, df_i\, dg_i\, dh_i$$

with scalars a_i, f_i, g_i, h_i. In any parameter system r, s, t in space it defines the values

(64b)
$$\frac{\omega}{dr\,ds\,dt} = \sum_{i=1}^{N} a_i \frac{d(f_i, g_i, h_i)}{d(r, s, t)}.$$

With reference to a particular u, v, w-coordinate system, we can write

(64c)
$$\frac{\omega}{dr\,ds\,dt} = \sum_{i=1}^{N} a_i \frac{d(f_i, g_i, h_i)}{d(u, v, w)} \frac{d(u, v, w)}{d(r, s, t)}.$$

This amounts to the identity

(64d)
$$\omega = \alpha\, du\, dv\, dw,$$

where

(64e)
$$\alpha = \sum_{i=1}^{N} a_i \frac{d(f_i, g_i, h_i)}{d(u, v, w)}.$$ [1]

We can define the product $L\omega$ of a first-order form

$$L = \sum_i a_i\, df_i$$

and a second-order form

$$\omega = \sum_k b_k\, dg_k\, dh_k$$

by specifying that

$$\begin{aligned}\frac{L\omega}{dr\,ds\,dt} &= \frac{L}{dr}\frac{\omega}{ds\,dt} + \frac{L}{ds}\frac{\omega}{dt\,dr} + \frac{L}{dt}\frac{\omega}{dr\,ds}\\ &= \sum_{i,k} a_i b_k \left(\frac{\partial f_i}{\partial r}\frac{d(g_k, h_k)}{d(s, t)} + \frac{\partial f_i}{\partial s}\frac{d(g_k, h_k)}{d(t, r)} + \frac{\partial f_i}{\partial t}\frac{d(g_k, h_k)}{d(r, s)}\right)\\ &= \sum_{i,k} a_i b_k \frac{d(f_i, g_k, h_k)}{d(r, s, t)}.\end{aligned}$$

This amounts to the formula

[1] In n-dimensional space referred to parameters $u_1, \ldots, u_n$, we have instead of (64c, d, e) the formula

$$\omega = \sum_{\substack{j,k,m=1,\ldots,n\\ j<k<m}} A_{jkm}\, du_j\, du_k\, du_m,$$

where

$$A_{jkm} = \sum_i a_i \frac{d(f_i, g_i, h_i)}{d(u_j, u_k, u_m)} = \frac{\omega}{du_j\, du_k\, du_m}.$$

(65a) $$L\omega = \sum_{i,k} a_i b_k \, df_i \, dg_k \, dh_k,$$

as could be expected from the formal multiplication of expressions for L and ω. When L and ω are in their standard form

$$L = A\,du + B\,dv + C\,dw, \quad \omega = a\,dv\,dw + b\,dw\,du + c\,du\,dv$$

for a given u, v, w-coordinate system, the product becomes

(65b) $$L\omega = (Aa + Bb + Cc)\,du\,dv\,dw,$$

in accordance with (57c).

The derivative of the second-order form

$$\omega = \sum a_i \, dg_i \, dh_i$$

can be defined independently of special coordinate systems by the rule

$$\begin{aligned}\frac{d\omega}{dr\,ds\,dt} &= \frac{\partial}{\partial r}\frac{\omega}{ds\,dt} + \frac{\partial}{\partial s}\frac{\omega}{dt\,dr} + \frac{\partial}{\partial t}\frac{\omega}{dr\,ds} \\ &= \frac{\partial}{\partial r}\sum_i a_i \frac{d(g_i, h_i)}{d(s, t)} + \frac{\partial}{\partial s}\sum_i a_i \frac{d(g_i, h_i)}{d(t, r)} + \frac{\partial}{\partial t}\sum_i a_i \frac{d(g_i, h_i)}{d(r, s)}.\end{aligned}$$

Thus,

(66a) $$\frac{d\omega}{dr\,ds\,dt} = \sum_i \frac{d(a_i, g_i, h_i)}{d(r, s, t)},$$

as one verifies easily. Hence, our definition of $d\omega$ implies

(66b) $$d\omega = \sum_i da_i \, dg_i \, dh_i.$$

For ω in the standard form

(66c) $$\omega = a\,dv\,dw + b\,dw\,du + c\,du\,dv$$

we obtain

(66d) $$d\omega = (a_u + b_v + c_w)\,du\,dv\,dw.$$

This special representation for $d\omega$ can again be used as on p. 315 to show that a second-order form ω with $d\omega = 0$ is representable locally as $\omega = dL$, where L is a suitable first-order differential form.

Exercises 3.6d

1. In spherical coordinates, $x = \rho \sin\phi \cos\theta$, $y = \rho \sin\phi \sin\theta$, $z = \rho\cos\phi$, choose unit vectors **u**, **v**, **w**, in the direction of the r, ϕ, θ lines, respectively. Show that $d\mathbf{X} = (dx, dy, dz) = \mathbf{u}d\rho + \mathbf{v}\rho d\phi + \mathbf{w}\rho \sin\phi \, d\theta$. Hence, find the expression for $\nabla f(\rho, \phi, \theta)$ in spherical coordinates, where ∇f is defined by $\nabla f \cdot d\mathbf{X} = df$.

3.7 Maxima and Minima

a. *Necessary Conditions*

For functions of several variables, as for functions of a single variable, one of the most important applications of differentiation is the theory of maxima and minima.

We shall begin by considering a function $u = f(x, y)$ of two independent variables x, y. The *domain* of the function shall be a certain set R in the x, y-plane. We can represent f in x, y, z-space by the surface S with equation $z = f(x, y)$. We say that $f(x, y)$ has a maximum[1] at the point (x_0, y_0) of its domain R if $f(x_0, y_0) \geqq f(x, y)$ for all (x, y) in R. Such a maximum corresponds to a highest point of the surface S. We talk of a *strict* maximum if actually $f(x_0, y_0) > f(x, y)$ for all (x, y) in R that are different from (x_0, y_0), so that the greatest value of the function is reached only at the single point (x_0, y_0). Similarly, $f(x, y)$ is said to have a minimum at the point (x_1, y_1) of R if $f(x_1, y_1) \leqq f(x, y)$ for all (x, y) in R, and a *strict minimum* if $f(x_1, y_1) < f(x, y)$ for all $(x, y) \neq (x_1, y_1)$ in R. The basic theorem of p. 112 assures us that *if R is a closed and bounded set and f continuous in R, then there exist points in R where f has its maximum and also points where f has its minimum.*

As an example consider the function $u = x^2 + y^2$ in the closed disc given by $x^2 + y^2 < 1$. The surface S is the portion of the paraboloid of revolution $z = x^2 + y^2$ lying below the plane $z = 1$. Here the maxima of f occur at all the points of the boundary circle $x^2 + y^2 = 1$, whereas f has a *strict minimum* at the origin.

Calculus applies directly to the determination of *relative* maxima or minima, rather than of absolute extrema. A point (x_0, y_0) of the domain R is a *relative maximum* if $f(x_0, y_0) \geqq f(x, y)$ for all points (x, y) of R that lie in a sufficiently small neighborhood of (x_0, y_0). The value $f(x_0, y_0)$ at a relative maximum does not have to be the greatest value of f in all of R but is a maximum of f if we restrict ourselves to

[1]Also called *absolute maximum* in contrast to the *relative maximum* defined below. The terminology used here is exactly the same as for functions of a single variable; see Volume I (pp. 238 ff.).

points sufficiently close to (x_0, y_0). Relative minima are defined analogously. Every absolute maximum (minimum) also is a relative maximum (minimum), but the converse does not hold.

For example, the function $u = (x^2 + y^2)^3 - 3(x^2 + y^2)$, whose domain shall be the open disc $x^2 + y^2 < 4$, has no maximum but does have a relative maximum at the origin. All points on the circle $x^2 + y^2 = 1$ are minimum points. Here the surface S is generated by rotating the curve $z = x^6 - 3x^2$ about the z-axis.

The definitions of absolute or relative minima for functions $u = f(x, y, z, \ldots)$ of more independent variables are entirely similar.

We shall first give *necessary* conditions for the occurrence of a relative maximum or minimum at an *interior* point (x_0, y_0) of the domain R of the function $f(x, y)$. We use the term relative extremum to include both maxima and minima. Let now (x_0, y_0) be an interior point of the domain R of the function $f(x, y)$, and let f have partial derivatives $f_x(x_0, y_0)$, $f_y(x_0, y_0)$ at that point. *For a relative extremum of f to occur at the point (x_0, y_0), it is necessary* that

$$f_x(x_0, y_0) = 0, \qquad f_y(x_0, y_0) = 0. \tag{67a}$$

The conditions (67a) follow at once from the known conditions for functions of a single variable. Put $\phi(x) = f(x, y_0)$. Then $\phi(x)$ is defined for all x sufficiently close to x_0 and has at x_0 the derivative $\phi(x_0) = f_x(x_0, y_0)$. If $f(x_0, y_0) \geqq f(x, y)$ for all (x, y) in R that are sufficiently close to (x_0, y_0), then, in particular, $\phi(x_0) \geqq \phi(x)$ for all x sufficiently close to x_0. It follows (see Volume I, p. 241) that $\phi'(x_0) = 0$; that is, $f_x(x_0, y_0) = 0$. The second necessary condition $f_y(x_0, y_0) = 0$ is derived similarly.

Geometrically, the vanishing of the partial derivatives of $f(x, y)$ at the point (x_0, y_0) means that at the point $(x_0, y_0, f(x_0, y_0))$ the tangent plane to the surface $z = f(x, y)$ is parallel to the x, y-plane. We call (x_0, y_0) a *stationary* or *critical* point of $f(x, y)$ if the first derivatives $f_x(x_0, y_0)$, $f_y(x_0, y_0)$ both exist and vanish. Hence, every relative extremum in the interior of the domain of a differentiable function f is a critical point of f.

The same result applies to functions $f(x, y, z, \ldots)$ of any number of independent variables. Here $(x_0, y_0, z_0, \ldots)$ is a *stationary* or *critical* point of f if all first derivatives $f_x, f_y, \ldots$ at that point exist and satisfy

$$\begin{aligned} &f_x(x_0, y_0, z_0, \ldots) = 0, \quad f_y(x_0, y_0, z_0, \ldots) = 0,\\ &f_z(x_0, y_0, z_0, \ldots) = 0, \ldots . \end{aligned} \tag{67b}$$

The number of conditions is equal to that of independent variables $x, y, z \ldots$. We can combine the conditions into the single requirement that

$$df = f_x\,dx + f_y\,dy + f_z\,dz + \cdots = 0$$

for $(x, y, z, \ldots) = (x_0, y_0, z_0, \ldots)$ and all $dx, dy, dz, \ldots$.

Since the number of equations (67b) is the same as the number of unknowns $x_0, y_0, z_0, \ldots$ one usually expects to find a finite number of critical points, though, of course, that is not always so. Moreover, a critical point need not by any means be a relative extremum.

Consider, for example, the function $u = xy$. Our two equations (67a) at once give the point $x = 0, y = 0$ as the only critical point. In every neighborhood of (0, 0), however, the function may assume either positive or negative values, depending on the quadrant containing (x, y). The function therefore has no relative extremum at this point. The surface representing the function $u = xy$ geometrically is a hyperbolic paraboloid that has neither a highest nor lowest point, but has a *saddle point* at the origin (see Fig. 3.1).

We see that the maximum and minimum points of a differentiable function either lie on the boundary of the domain of the function or are to be looked for among the critical points of the function. To decide whether a critical point actually is a maximum or minimum requires a special investigation. On p. 349 we shall meet conditions that are sufficient to ensure that a critical point be at least a relative extremum.

The *maximum value* M of a function $f(x, y)$ is the greatest of all values assumed by f at the points of its domain R. The maximum points of f are those for which $f(x, y) = M$.[1] Similarly, the *critical or stationary values* of f are those assumed at critical or stationary points.

b. Examples

1. The function

$$u = \sqrt{1 - x^2 - y^2} \qquad (x^2 + y^2 < 1)$$

has the partial derivatives

[1]Sometimes the term "maximum" is used somewhat ambiguously referring either to the maximum value or an argument point (x, y) where f assumes its maximum value.

$$u_x = -\frac{x}{\sqrt{1 - x^2 - y^2}}, \qquad u_y = -\frac{y}{\sqrt{1 - x^2 - y^2}},$$

and these vanish at the origin. Here we have a maximum, for at all other points (x, y) in the neighborhood of the origin the quantity $1 - x^2 - y^2$ under the square root is less than it is at the origin.

2. We wish to construct the triangle for which the product of the sines of the three angles is greatest; that is, we wish to find the maximum of the function

$$f(x, y) = \sin x \sin y \sin (x + y)$$

in the region $0 \leqq x \leqq \pi$, $0 \leqq y \leqq \pi$, $0 \leqq x + y \leqq \pi$. Since f is positive in the interior of this region, its greatest value is positive. On the boundary of the region, where the equality sign holds in at least one of the inequalities defining the region, we have $f(x, y) = 0$, so that the greatest value must lie in the interior.

If we equate the derivatives to 0, we obtain the two equations

$$\cos x \sin y \sin (x + y) + \sin x \sin y \cos (x + y) = 0,$$

$$\sin x \cos y \sin (x + y) + \sin x \sin y \cos (x + y) = 0.$$

Since $0 < x < \pi$, $0 < y < \pi$, $0 < x + y < \pi$, these give $\tan x = \tan y$, or $x = y$. If we substitute this value in the first equation, we obtain the relation $\sin 3x = 0$; hence, $x = \pi/3$, $y = \pi/3$ is the only stationary point, and the required triangle is equilateral.

3. Three points P_1, P_2, P_3, with coordinates (x_1, y_1), (x_2, y_2), and (x_3, y_3), respectively, are the vertices of an acute-angled triangle. We wish to find a fourth point P with coordinates (x, y) such that the sum of its distances from P_1, P_2, and P_3 is the least possible. This sum of distances is a continuous function of x and y, and at some point P inside a large circle enclosing the triangle it has a least value. This point P cannot lie at a vertex of the triangle, for then the foot of the perpendicular from either of the other two vertices to its opposite side would give a smaller sum of distances. Again, P cannot lie on the circumference of the circle, if this is sufficiently far away from the triangle. With the distances r_i defined by

$$r_i = \sqrt{(x - x_i)^2 + (y - y_i)^2}$$

we wish to minimize the function

$$f(x, y) = r_1 + r_2 + r_3,$$

which is differentiable everywhere except at P_1, P_2, and P_3, We know that at the point P the partial derivatives with respect to x and y must vanish. Thus, by differentiating f, we obtain the conditions

$$\frac{x - x_1}{r_1} + \frac{x - x_2}{r_2} + \frac{x - x_3}{r_3} = 0,$$

$$\frac{y - y_1}{r_1} + \frac{y - y_2}{r_2} + \frac{y - y_3}{r_3} = 0$$

for P. According to these equations, the three plane vectors

$$\left(\frac{x_1 - x}{r_1}, \frac{y_1 - y}{r_1}\right), \left(\frac{x_2 - x}{r_2}, \frac{y_2 - y}{r_2}\right), \left(\frac{x_3 - x}{r_3}, \frac{y_3 - y}{r_3}\right)$$

have the vector sum $\mathbf{0}$. Also, these vectors are each of unit length. When given the common initial point P, their end points form an equilateral triangle; that is, each vector is brought into the direction of the next by a rotation through $\frac{2}{3}\pi$ (Fig. 3.27). Since these three vectors have the same directions as the three vectors from P to P_1, P_2, P_3, it follows that each of the three sides of the triangle must subtend the same angle $\frac{2}{3}\pi$ at the point P.

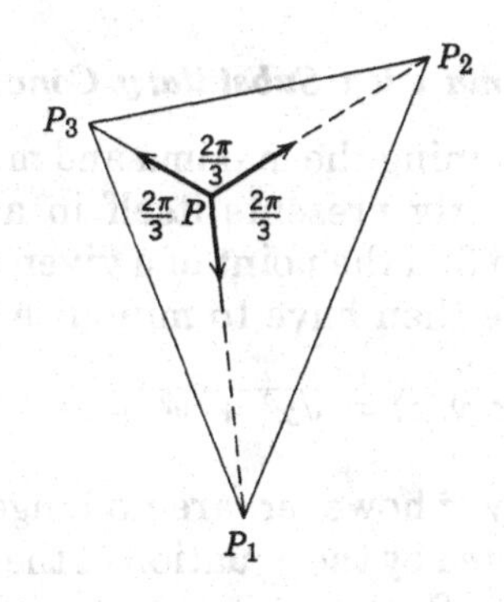

Figure 3.27

Exercises 3.7b

1. Find the stationary points of the following functions and state their nature:

   (a) $f(x, y) = y^2(\sin x - x/2)$

   (b) $f(x, y) = \cos(x + y) + \sin(x - y)$

(c) $f(x, y) = y^x$

(d) $f(x, y) = x/y$

(e) $f(x, y) = ye^{-x^2}$.

2. Determine the maxima and minima of the function

$$(ax^2 + by^2)e^{-x^2-y^2} \qquad (0 < a < b).$$

3. Find the values of x, y which make

$$2x^3 + (x - y)^2 - 6y$$

stationary.

4. The sum of the lengths of the 12 edges of a rectangular block is a; the sum of the areas of the 6 faces is $a^2/25$. Calculate the lengths of the edges when the excess of the volume of the block over that of a cube whose edge is equal to the least edge of the block is greatest.

5. Find the stationary points and state their nature, for the function

$$f(x, y, z) = x^2(y - 1)^2\left(z + \frac{1}{2}\right)^2.$$

6. According to present postal regulations in the United States, a rectangular parcel with side lengths x, y, z inches with $x \leqq y \leqq z$ may be shipped only if $2(x + y) + z \leqq 100$. Find the maximum volume of a shippable parcel under this condition. [Hint. set $z = 100 - 2(x + y)$.]

7. Minimize the sum of the squared distances of a point $\mathbf{X}$ from n given points.

c. *Maxima and Minima with Subsidiary Conditions*

The problem of determining the maxima and minima of functions of several variables frequently presents itself in a different form. For example, we may wish to find the point of a given surface $\phi(x, y, z) = 0$ closest to the origin. We then have to minimize the function

$$f(x, y, z) = \sqrt{x^2 + y^2 + z^2},$$

where the quantities x, y, z however, are no longer three *independent* variables but are connected by the equation of the surface $\phi(x, y, z) = 0$ as a subsidiary condition. Such maxima and minima with subsidiary conditions do not, indeed, represent a fundamentally new problem. Thus in our example we only need solve for one of the variables, say z, as a function of the other two, to reduce the problem to that of determining the stationary values of a function of the two independent variables x, y.

It is, however, more convenient, and also more elegant, to express the conditions for a stationary value in a symmetrical form, in which no preference is given to any one of the variables.

A simple typical case is presented by the problem of *finding the stationary values of a function $f(x, y)$ when the two variables x, y are not mutually independent but are connected by a subsidiary condition*

$$\phi(x, y) = 0.$$

In order to gain geometric insight, we assume first that the subsidiary condition is represented, as in Fig. 3.28, by a curve in the x, y-plane without singularities and that, in addition, the family of curves $f(x, y) = c =$ constant covers a portion of the plane, as in the figure.

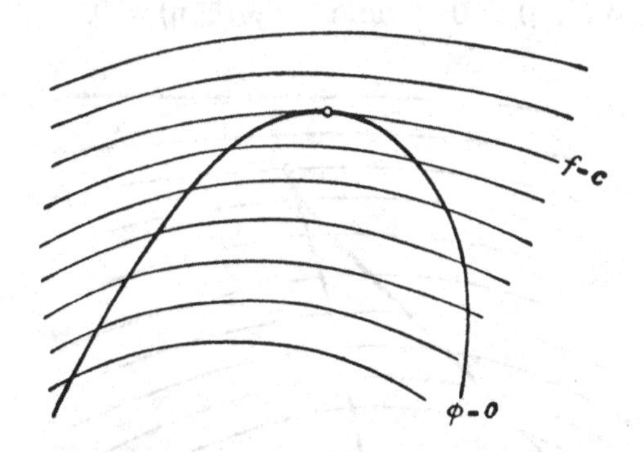

Figure 3.28 Extreme value of f with subsidiary condition $\phi = 0$.

Among the curves of the family that intersect the curve $\phi = 0$, we have to find that one for which the constant c is greatest or least. As we describe the curve $\phi = 0$, we cross the curves $f(x, y) = c$, and in general c changes monotonically; at the point where the sense in which we run through the c-scale is reversed, we may expect an extreme value. From Fig. 3.28 we see that this occurs for the curve of the family that touches the curve $\phi = 0$. The coordinates of the point of contact will be the required values $x = \xi$, $y = \eta$ corresponding to the extreme value of $f(x, y)$. If the two curves $f =$ constant and $\phi = 0$ touch, they have the same tangent. Thus, at the point $x = \xi$, $y = \eta$, the proportional relation

$$f_x : f_y = \phi_x : \phi_y$$

holds; or, if we introduce the constant of proportionality λ, the two equations

$$f_x + \lambda\phi_x = 0$$

$$f_y + \lambda\phi_y = 0$$

are satisfied. These, with the equation

$$\phi(x, y) = 0,$$

serve to determine the coordinates (ξ, η) of the point of contact and also the constant of proportionality λ.

This argument may fail, for example, when the curve $\phi = 0$ has singular point, say a cusp as in Fig. 3.29, at the point (ξ, η) at which it meets a curve $f = c$ with the greatest or least possible c. In this case, however, we have both

$$\phi_x(\xi, \eta) = 0 \qquad \text{and} \qquad \phi_y(\xi, \eta) = 0.$$

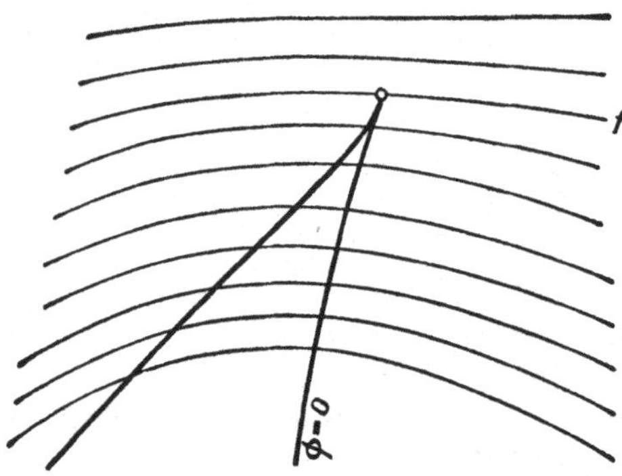

Figure 3.29 Extreme value at a singular point of $\phi = 0$

We are led intuitively to the following rule, which we shall prove in the next subsection:

In order that an extreme value of the function $f(x, y)$ with the subsidiary condition $\phi(x, y) = 0$, may occur at the point $x = \xi$, $y = \eta$, where $\phi_x(\xi, \eta)$ and $\phi_y(\xi, \eta)$ do not both vanish, there must be a constant of proportionality λ such that the two equations

$$\text{(67c)} \qquad f_x(\xi, \eta) + \lambda\phi_x(\xi, \eta) = 0 \qquad \text{and} \qquad f_y(\xi, \eta) + \lambda\phi_y(\xi, \eta) = 0$$

are satisfied together with the equation

$$\text{(67d)} \qquad \phi(\xi, \eta) = 0.$$

This rule is known as *Lagrange's method of undetermined* multipliers, and the factor λ is known as *Lagrange's multiplier.*

We observe that this rule gives as many equations for the determination of the quantities ξ, η, and λ as there are unknowns. We have, therefore, replaced the problem of finding the positions of the extreme values (ξ, η) by a problem in which there is an additional unknown λ but in which we have the advantage of complete symmetry. Lagrange's rule is usually expressed as follows:

To find the extreme values of the function $f(x, y)$ subject to the subsidiary condition $\phi(x, y) = 0$, we add to $f(x, y)$ the product of $\phi(x, y)$ and an unknown factor λ independent of x and y and write down the known necessary conditions,

$$f_x + \lambda\phi_x = 0, \qquad f_y + \lambda\phi_y = 0,$$

for an extreme value of $F = f + \lambda\phi$. In conjunction with the subsidiary condition $\phi = 0$ these serve to determine the coordinates of the extremum and the constant of proportionality.

As an example, we find the extreme values of the function

$$u = xy$$

on the circle with unit radius and center at the origin, that is, with the subsidiary condition

$$x^2 + y^2 - 1 = 0.$$

According to our rule, by differentiating $xy + \lambda(x^2 + y^2 - 1)$ with respect to x and to y, we find that at the stationary points the two equations

$$y + 2\lambda x = 0$$

$$x + 2\lambda y = 0$$

have to be satisfied. In addition we have the subsidiary condition

$$x^2 + y^2 - 1 = 0.$$

On solving, we obtain the four points

$$\xi = \frac{1}{2}\sqrt{2}, \qquad \eta = \frac{1}{2}\sqrt{2},$$

$$\xi = -\frac{1}{2}\sqrt{2}, \qquad \eta = -\frac{1}{2}\sqrt{2},$$

$$\xi = \frac{1}{2}\sqrt{2}, \qquad \eta = -\frac{1}{2}\sqrt{2},$$

$$\xi = -\frac{1}{2}\sqrt{2}, \qquad \eta = \frac{1}{2}\sqrt{2},$$

The first two of these give a maximum value $u = \frac{1}{2}$, and the second two, a minimum value $u = -\frac{1}{2}$, of the function $u = xy$. That the first two do really give the greatest value and the second two the least value of the function u follows from the fact that on the circumference the function must assume a greatest and a least value (cf. p. 325), since the circumference is closed and bounded.

Exercises 3.7c

1. Solve Exercise 6 of Section 3.7b as a problem in maximizing the volume subject to the condition $2(x + y) + z = 100$.
2. Minimize the function $z = x^2y^2$ subject to the condition $x + y = 1$.
3. Maximize the function $z = \cos \pi (x + y)$ subject to the condition $x^2 + y^2 = 1$.
4. In the plane, minimize the sum of the squared distances of a point **X** from n given points subject to the condition that **X** lie on a given line (compare Section 3.7b, Exercise 7).
5. If $C = f(a, b)$ is a true maximum or minimum of $f(x, y)$ subject to the condition $\phi(x, y) = C'$, show that in general $C' = \phi(a, b)$ is a true maximum or minimum of $\phi(x, y)$ subject to the condition $f(x, y) = C$.

d. Proof of the Method of Undetermined Multipliers in the Simplest Case

As we should expect, we arrive at an analytical proof of the method of undetermined multipliers by reducing it to the known case of "free" extreme values. We assume that at an extremum point the two partial derivatives $\phi_x(\xi, \eta)$ and $\phi_y(\xi, \eta)$ do not both vanish; to be specific, we assume that $\phi_y(\xi, \eta) \neq 0$. Then, by the implicit function theorem (p. 221), in a neighborhood of this point the equation $\phi(x, y) = 0$ determines y uniquely as a continuously differentiable function of x, say $y = g(x)$. If we substitute this expression in $f(x, y)$, the function

$$f(x, g(x))$$

must have a free extreme value at the point $x = \xi$. For this the equation

$$f'(x) = f_x + f_y g'(x) = 0$$

must hold at $x = \xi$. In addition, the implicitly defined function

$y = g(x)$ satisfies the relation $\phi_x + \phi_y g'(x) = 0$ identically. If we multiply this equation by $\lambda = -f_y/\phi_y$ and add it to $f_x + f_y g'(x) = 0$, we obtain

$$f_x + \lambda\phi_x = 0,$$

and by the definition of λ, the equation

$$f_y + \lambda\phi_y = 0$$

holds. This establishes the method of undetermined multipliers.

This proof brings out the importance of the assumption that the derivatives ϕ_x and ϕ_y do not both vanish at the point (ξ, η). If both derivatives vanish the rule breaks down, as the following example shows. We wish to make the function

$$f(x, y) = x^2 + y^2$$

a minimum, subject to the condition

$$\phi(x, y) = (x - 1)^3 - y^2 = 0.$$

In Fig. 3.30 the shortest distance from the origin to the curve $(x - 1)^3 - y^2 = 0$ is obviously given by the line joining the origin to the cusp S of the curve (we can easily prove that the unit circle centered at the origin contains no other point of the curve). The coordinates of S—

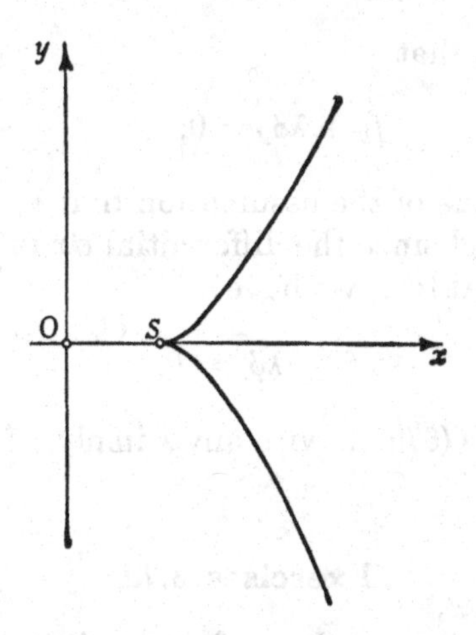

Figure 3.30 The curve $(x - 1)^3 - y^2 = 0$.

that is, $x = 1$ and $y = 0$—satisfy the equations $\phi(x, y) = 0$ and $f_y + \lambda\phi_y = 0$ no matter what value is assigned to λ, but

$$f_x + \lambda\phi_x = 2x + 3\lambda(x-1)^2 = 2 \neq 0.$$

We can state the method of undetermined multipliers in a slightly different way that is particularly convenient for generalization. We have seen that the vanishing of the differential of a function $F(x, y)$ at a given point is a necessary condition for the occurrence of a free extreme value of the function at that point. For the present problem we can similarly make the following statement:

In order for the function $f(x, y)$ to have an extreme value at the point (ξ, η) subject to the subsidiary condition $\phi(x, y) = 0$, the differential df must vanish at that point, where we consider the differentials dx and dy to be not independent but subject to the equation

$$d\phi = \phi_x\, dx + \phi_y\, dy = 0 \tag{67e}$$

deduced from $\phi = 0$. Assume that at the point (ξ, η) the differentials dx and dy satisfy the equation

$$df = f_x(\xi, \eta)\, dx + f_y(\xi, \eta)\, dy = 0 \tag{67f}$$

whenever they satisfy the equation $d\phi = 0$. Multiplying equation (67e) by a number λ and adding to (67f), we obtain

$$(f_x + \lambda\phi_x)\, dx + (f_y + \lambda\phi_y)\, dy = 0.$$

If we determine λ so that

$$f_y + \lambda\phi_y = 0, \tag{67g}$$

as is possible in virtue of the assumption that $\phi_y \neq 0$, it follows that $(f_x + \lambda\phi_x)\, dx = 0$, and since the differential dx in (67e) can be chosen arbitrarily, say, equal to 1, we have

$$f_x + \lambda\phi_x = 0. \tag{67h}$$

Conversely, relations (67g, h) with any λ imply, of course, that $df = 0$ whenever $d\phi = 0$.

Exercises 3.7d

1. Describe the appearance of the surface $z = f(x, y) + \lambda\phi(x, y)$, for λ the Lagrange multiplier and $\phi = 0$ the constraining equation.

e. Generalization of the Method of Undetermined Multipliers

We can extend the method of undetermined multipliers to a greater number of variables and also to a greater number of subsidiary conditions. We shall consider a special case that includes every essential feature. We seek the extreme values of the function

$$u = f(x, y, z, t), \tag{68a}$$

when the four variables x, y, z, t satisfy the two subsidiary conditions

$$\phi(x, y, z, t) = 0, \quad \psi(x, y, z, t) = 0. \tag{68b}$$

We assume that at the point (ξ, η, ζ, τ) the function f takes a value that is an extreme value when compared with the values at all neighboring points satisfying the subsidiary conditions. We require that, in the neighborhood of the point $P = (\xi, \eta, \zeta, \tau)$ two of the variables, say z and t, can be represented as functions of the other two, x and y, by means of the equations (68b). To ensure that such solutions $z = g(x, y)$ and $t = h(x, y)$ can be found, we assume that at the point P the Jacobian

$$\frac{d(\phi, \psi)}{d(z, t)} = \phi_z\psi_t - \phi_t\psi_z \tag{68c}$$

is not zero (cf. p. 265). We now substitute the functions

$$z = g(x, y) \quad \text{and} \quad t = h(x, y)$$

in the function $u = f(x, y, z, t)$, to obtain a function of the two independent variables x and y, and this function must have a free extreme value at the point $x = \xi$, $y = \eta$; that is, its two partial derivatives must vanish at that point. The two equations

$$f_x + f_z\frac{\partial z}{\partial x} + f_t\frac{\partial t}{\partial x} = 0, \tag{69a}$$

$$f_y + f_z\frac{\partial z}{\partial y} + f_t\frac{\partial t}{\partial y} = 0 \tag{69b}$$

must therefore hold. In order to calculate from the subsidiary conditions the four derivatives $\frac{\partial z}{\partial x}, \frac{\partial z}{\partial y}, \frac{\partial t}{\partial x}, \frac{\partial t}{\partial y}$ occurring here, we could write down the two pairs of equations

(69c) $$\phi_x + \phi_z \frac{\partial z}{\partial x} + \phi_t \frac{\partial t}{\partial x} = 0,$$

(69d) $$\psi_x + \psi_z \frac{\partial z}{\partial x} + \psi_t \frac{\partial t}{\partial x} = 0$$

and

(69e) $$\phi_y + \phi_z \frac{\partial z}{\partial y} + \phi_t \frac{\partial t}{\partial y} = 0,$$

(69f) $$\psi_y + \psi_z \frac{\partial z}{\partial y} + \psi_t \frac{\partial t}{\partial y} = 0$$

and solve them for the unknowns $\partial z/\partial x, \ldots, \partial t/\partial y$; this is possible because the Jacobian $d(\phi,\psi)/d(z,t)$ does not vanish. Thus, the problem would be solved.

Instead, we prefer to retain formal symmetry by proceeding as follows. We determine two numbers λ and μ in such a way that the two equations

(70a) $$f_z + \lambda\phi_z + \mu\psi_z = 0,$$

(70b) $$f_t + \lambda\phi_t + \mu\psi_t = 0$$

are satisfied at the point where the extreme value occurs. The determination of these multipliers λ and μ is possible, since we have assumed that the Jacobian $d(\phi,\psi)/d(z,t)$ is not zero. If we multiply the equations (69c, d) by λ and μ, respectively, and add them to the equation (69a), we have

$$f_x + \lambda\phi_x + \mu\psi_x + (f_z + \lambda\phi_z + \mu\psi_z)\frac{\partial z}{\partial x} + (f_t + \lambda\phi_t + \mu\psi_t)\frac{\partial t}{\partial x} = 0.$$

Hence, by the definition (70a, b) of λ and μ,

$$f_x + \lambda\phi_x + \mu\psi_x = 0.$$

Similarly, if we multiply the equations (69e, f) by λ and μ, respectively, and add them to the equation (69b), we obtain the further equation

$$f_y + \lambda\phi_y + \mu\psi_y = 0.$$

We thus arrive at the following result: *If the point (ξ, η, ζ, τ) is an ex-*

tremum of $f(x, y, z, t)$ subject to the subsidiary conditions

(71a) $$\phi(x, y, z, t) = 0,$$

(71b) $$\psi(x, y, z, t) = 0,$$

and if at that point $d(\phi,\psi)/d(z,t)$ is not zero, then two numbers λ and μ exist such that at the point (ξ, η, ζ, τ) the equations

(72a) $$f_x + \lambda\phi_x + \mu\psi_x = 0,$$

(72b) $$f_y + \lambda\phi_y + \mu\psi_y = 0,$$

(72c) $$f_z + \lambda\phi_z + \mu\psi_z = 0,$$

(72d) $$f_t + \lambda\phi_t + \mu\psi_t = 0,$$

and the subsidiary conditions (71a, b) are satisfied.

These last conditions are perfectly symmetrical. Every trace of special emphasis on the two variables x and y has disappeared from them, and we should equally well have obtained (72a, b, c, d) if, instead of assuming that $\partial(\phi, \psi)/\partial(z, t) \neq 0$, we had merely assumed that any one of the Jacobians $\partial(\phi, \psi)/\partial(x, y)$, $\partial(\phi, \psi)/\partial(x, z)$, . . ., $\partial(\phi, \psi)/\partial(z, t)$ did not vanish, so that in the neighborhood of the point in question a certain pair of the quantities x, y, z, t (not necessarily z and t) could be expressed in terms of the other pair. For this symmetry of our equations we have of course paid a price; in addition to the unknowns ξ, η, ζ, τ, we now have λ and μ also. Thus, instead of four unknowns, we now have six, determined by the six equations above.

In exactly the same way, we can state and prove the method of undetermined multipliers for an arbitrary number of variables and an arbitrary number of subsidiary conditions. The general rule is as follows:

If in a function

$$u = f(x_1, x_2, \ldots, x_n)$$

the n variables $x_1, x_2, \ldots, x_n$ are not independent but are connected by the m subsidiary conditions ($m < n$)

$$\phi_1(x_1, x_2, \ldots, x_n) = 0,$$
$$\phi_2(x_1, x_2, \ldots, x_n) = 0,$$
$$\cdots\cdots\cdots\cdots$$
$$\phi_m(x_1, x_2, \ldots, x_n) = 0,$$

then we introduce m multipliers $\lambda_1, \lambda_2, \ldots, \lambda_m$ *and equate the derivatives of the function*

$$F = f + \lambda_1\phi_1 + \lambda_2\phi_2 + \cdots + \lambda_m\phi_m$$

with respect to $x_1, x_2, \ldots, x_n$, *when* $\lambda_1, \lambda_2, \ldots, \lambda_m$ *are constant, to* 0. *The equations*

$$\frac{\partial F}{\partial x_1} = 0, \ldots, \frac{\partial F}{\partial x_n} = 0$$

thus obtained,[1] *together with the m subsidiary conditions*

$$\phi_1 = 0, \ldots, \phi_m = 0,$$

represent a system of $m + n$ *equations for the* $m + n$ *unknown quantities* $x_1, x_2, \ldots, x_n, \lambda_1, \ldots, \lambda_m$. *These equations must be satisfied at any extreme point of f unless every one of the Jacobians of the m functions* $\phi_1, \phi_2, \ldots, \phi_m$ *with respect to m of the variables* $x_1, \ldots, x_n$ *has the value* 0.

We observe that this rule gives us an elegant formal method for determining the points where extreme values occur; however, it merely constitutes a necessary condition. It still remains to investigate the circumstances under which the points that we find by means of the multiplier method actually correspond to a maximum or a minimum of the function. Into this question we shall not enter; its discussion would lead us too far afield. As in the case of free extreme values, when we apply the method of undetermined multipliers we usually know beforehand that an extremum in the interior of the domain of f does exist. If the method determines the point uniquely and the exceptional case (all the Jacobians 0) does not occur anywhere in the region under discussion, then we can be sure that we have really found the point where the extreme value occurs.

Exercises 3.7e

1. Interpret the problem of minimizing $u = f(x, y, z)$ subject to the constraint $\phi(x, y, z) = 0$ geometrically,
2. Give an example of a problem of the form: Extremize $f(x, y, z)$ subject to the constraints $\phi(x, y) = 0$, $\psi(y, z) = 0$. Interpret this geometrically.

f. ***Examples***

1. As a first example we attempt to find the maximum of the function $f(x, y, z) = x^2y^2z^2$ subject to the subsidiary condition $x^2 + y^2$

[1] Which are identical with those for a "free" extremum of the auxiliary function F.

$+ z^2 = c^2$. On the spherical surface $x^2 + y^2 + z^2 = c^2$, the function must assume a greatest value, since the surface is a bounded and closed set. According to the rule, we form the expression

$$F = x^2y^2z^2 + \lambda(x^2 + y^2 + z^2 - c^2)$$

and by differentiation obtain

$$2xy^2z^2 + 2\lambda x = 0,$$

$$2x^2yz^2 + 2\lambda y = 0,$$

$$2x^2y^2z + 2\lambda z = 0.$$

The solutions with $x = 0, y = 0$, or $z = 0$ can be excluded, for at these points the function f takes on its least value, zero. The other solutions of the equation are $x^2 = y^2 = z^2$, $\lambda = -x^4$. Using the subsidiary condition, we obtain the values

$$x = \pm\frac{c}{\sqrt{3}}, \qquad y = \pm\frac{c}{\sqrt{3}}, \qquad z = \pm\frac{c}{\sqrt{3}}$$

for the required coordinates.

At all these points, the function assumes the same value $c^6/27$, which accordingly is the maximum. Hence, any triad of numbers satisfies the relation

$$\sqrt[3]{x^2y^2z^2} \leqq \frac{c^2}{3} = \frac{x^2 + y^2 + z^2}{3},$$

which states that *the geometric mean of three nonnegative numbers* x^2, y^2, z^2 *is never greater than their arithmetic mean.*

One proves similarly for any arbitrary number of positive numbers that the geometric mean never exceeds the arithmetic mean.[1]

2. As a second example we shall seek to find the triangle (with sides x, y, z) with given perimeter $2s$, and the greatest possible area. By the well-known formula of Heron the square of the area is given by

$$f(x, y, z) = s(s - x)(s - y)(s - z).$$

We therefore seek the maximum of this function subject to the subsidiary condition

[1]For another proof, see Volume I, Problem 13, p. 109, or Problem 11, p. 318.

$$\phi = x + y + z - 2s = 0,$$

where x, y, z are restricted by the inequalities

$$x \geqq 0,\ y \geqq 0,\ z \geqq 0,\ x + y \geqq z,\ x + z \geqq y,\ y + z \geqq x.$$

On the boundary of this closed region (i.e., whenever one of these inequalities becomes an equation), we always have $f = 0$. Consequently, the greatest value of f occurs in the interior and is a maximum. We form the function

$$F(x, y, z) = s(s - x)(s - y)(s - z) + \lambda(x + y + z - 2s),$$

and by differentiation obtain the three conditions

$$-s(s - y)(s - z) + \lambda = 0, \qquad -s(s - x)(s - z) + \lambda = 0,$$

$$-s(s - x)(s - y) + \lambda = 0.$$

By equating the three expressions we obtain $x = y = z = 2s/3$; that is, the solution is an equilateral triangle.

3. We next prove the inequality

$$uv \leqq \frac{1}{\alpha} u^\alpha + \frac{1}{\beta} v^\beta \tag{73a}$$

for every $u \geqq 0$, $v \geqq 0$ and every $\alpha > 0$, $\beta > 0$ for which $1/\alpha + 1/\beta = 1$.

The inequality is certainly valid if either u or v vanishes. We may therefore restrict ourselves to values of u and v such that $uv \neq 0$. If the inequality holds for a pair of numbers u, v, it also holds for all numbers $ut^{1/\alpha}$, $vt^{1/\beta}$ where t is an arbitrary positive number. We need therefore consider only values of u, v for which $uv = 1$. Hence, we have to show that the inequality

$$\frac{1}{\alpha} u^\alpha + \frac{1}{\beta} v^\beta \geqq 1$$

holds for all positive numbers u, v such that $uv = 1$.

To do this, we solve the problem of finding the minimum of

$$\frac{1}{\alpha} u^\alpha + \frac{1}{\beta} v^\beta$$

subject to the subsidiary condition $uv = 1$. This minimum obviously exists and occurs at a point (u, v) where $u \neq 0$, $v \neq 0$. Consequently, there exists a multiplier $-\lambda$ for which we have

$$u^{\alpha-1} - \lambda v = 0 \qquad \text{and} \qquad v^{\beta-1} - \lambda u = 0.$$

On multiplication by u and v, respectively, these equations at once yield $u^\alpha = \lambda$, $v^\beta = \lambda$. Taken with $uv = 1$, the last results imply that $u = v = 1$. The minimum value of

$$\frac{1}{\alpha} u^\alpha + \frac{1}{\beta} v^\beta$$

is, therefore, $1/\alpha + 1/\beta = 1$. That is, the statement that

$$\frac{1}{\alpha} u^\alpha + \frac{1}{\beta} v^\beta \geqq 1$$

when $uv = 1$ is proved.

If in the inequality (73a) we replace u and v by

$$u = u_i \Big/ \left(\sum_{i=1}^{n} u_i^\alpha\right)^{1/\alpha} \qquad \text{and} \qquad v = v_i \Big/ \left(\sum_{i=1}^{n} v_i^\beta\right)^{1/\beta},$$

respectively, where $u_1, u_2, \ldots, u_n, v_1, v_2, \ldots, v_n$ are arbitrary non-negative numbers and at least one u and at least one v is not zero and if we sum over $i = 1, \ldots, n$, we obtain *Hölder's inequality*

$$\text{(73b)} \qquad \sum_{i=1}^{n} u_i v_i \leqq \left(\sum_{i=1}^{n} u_i^\alpha\right)^{1/\alpha} \left(\sum_{i=1}^{n} v_i^\beta\right)^{1/\beta}.$$

This holds for any $2n$ numbers u_i, v_i where $u_i \geqq 0$, $v_i \geqq 0$ $(i = 1, 2, \ldots, n)$; not all the u's and not all the v's are zero; and the indices α, β are such that $\alpha > 0$, $\beta > 0$, $1/\alpha + 1/\beta = 1$. The Cauchy-Schwarz inequality is the special case $\alpha = \beta = 2$ of Hölder's inequality.

4. Finally, we seek the point on the closed surface

$$\phi(x, y, z) = 0$$

that is at the least distance from the fixed point (ξ, η, ζ). If the distance is a minimum its square is also a minimum; we accordingly consider the function

$$F(x, y, z) = (x - \xi)^2 + (y - \eta)^2 + (z - \zeta)^2 + \lambda\phi(x, y, z).$$

Differentiation gives the conditions

$$2(x-\xi)+\lambda\phi_x=0,\quad 2(y-\eta)+\lambda\phi_y=0,\quad 2(z-\zeta)+\lambda\phi_z=0,$$

or, in another form,

$$\frac{x-\xi}{\phi_x}=\frac{y-\eta}{\phi_y}=\frac{z-\zeta}{\phi_z}.$$

These equations state that the fixed point (ξ, η, ζ) lies on the normal to the surface at the point of extreme distance (x, y, z). Therefore, in order to travel along the shortest path from a point to a (differentiable) surface, we must travel in a direction normal to the surface. Of course, further discussion is required to decide whether we have found a maximum or a minimum or neither. Consider, for example, a point within a spherical surface. The points of extreme distance lie at the ends of the diameter through the point; the distance to one of these points is a minimum, to the other a maximum.

Exercises 3.7f

1. Find the shortest distance between the plane $Ax + By + Cz = D$ and the point (a, b, c).
2. Find the greatest and least distances of a point on the ellipse $x^2/4 + y^2/1 = 1$ from the straight line $x + y - 4 = 0$.
3. Show that the maximum value of the expression

$$\frac{ax^2+2bxy+cy^2}{ex^2+2fxy+gy^2} \qquad (eg-f^2>0)$$

is equal to the greater of the roots of the equation in λ

$$(ac-b^2)-\lambda(ag-2bf+ec)+\lambda^2(ea-f^2)=0.$$

4. Calculate the maximum values of the following expressions:

(a) $\dfrac{x^2+6xy+3y^2}{x^2-xy+y^2}$

(b) $\dfrac{x^4+2x^3y}{x^4+y^4}$.

5. Find the values of a and b for the ellipse $x^2/a^2 + y^2/b^2 = 1$ of least area containing the circle $(x - 1)^2 + y^2 = 1$ in its interior.
6. Which point of the sphere $x^2 + y^2 + z^2 = 1$ is at the greatest distance from the point $(1, 2, 3)$?
7. Find the point (x, y, z) of the ellipsoid $x^2/a^2 + y^2/b^2 + z^2/c^2 = 1$ for which

(a) $A + B + C$

(b) $\sqrt{A^2 + B^2 + C^2}$,

is a minimum, where A, B, C denote the intercepts that the tangent plane at (x, y, z), where $x > 0, y > 0, z > 0$, makes on the coordinate axes.

8. Find the rectangular parallelepiped of greatest volume inscribed in the ellipsoid $x^2/a^2 + y^2/b^2 + z^2/c^2 = 1$.
9. Find the rectangle of greatest perimeter inscribed in the ellipse $x^2/a^2 + y^2/b^2 = 1$.
10. Find the point of the ellipse $5x^2 - 6xy + 5y^2 = 4$ for which the tangent is at the greatest distance from the origin.
11. Prove that the length l of the greatest axis of the ellipsoid

$$ax^2 + by^2 + cz^2 + 2dxy + 2exz + 2fyz = 1$$

is given by the greatest real root of the equation

$$\begin{vmatrix} a - \frac{1}{l^2} & d & e \\ d & b - \frac{1}{l^2} & f \\ e & f & c - \frac{1}{l^2} \end{vmatrix}$$

12. (a) Maximize $x^a y^b z^c$, where a, b, c are positive constants, subject to the condition $x^k + y^k + z^k = 1$ where x, y, z are nonnegative and $k > 0$.
    (b) From the result of part (a) derive the inequality for any six positive real numbers

$$\left(\frac{u}{a}\right)^a \left(\frac{v}{b}\right)^b \left(\frac{w}{c}\right)^c \leqq \left(\frac{u + v + w}{a + b + c}\right)^{a+b+c}$$

13. Let $P_1P_2P_3P_4$ be a convex quadrilateral. Find the point O for which the sum of the distances from P_1, P_2, P_3, P_4 is a minimum.
14. Find the quadrilateral with given edges a, b, c, d that includes the greatest area.

Appendix

A.1 Sufficient Conditions for Extreme Values

In the theory of maxima and minima in the preceding chapter we contented ourselves with finding necessary conditions for the occurrence of an extreme value. In many cases occurring in actual practice the nature of the "stationary" point thus found can be determined from the special nature of the problem, permitting us to decide whether it is a maximum or a minimum. Yet it is important to have general *sufficient* conditions for the occurrence of relative extrema. Such criteria will be developed here for the typical case of two independent variables.

If we consider a point (x_0, y_0) at which the function is stationary, that is, a point at which both first partial derivatives of the function vanish, an extreme value occurs if and only if the expression

$$f(x_0 + h, y_0 + k) - f(x_0, y_0)$$

has the same sign for all sufficiently small values of h and k. If we expand this expression by Taylor's theorem with the remainder of the third order and use the equations $f_x(x_0, y_0) = 0$ and $f_y(x_0, y_0) = 0$, we obtain

$$f(x_0 + h, y_0 + k) - f(x_0, y_0) = \frac{1}{2}(h^2 f_{xx} + 2hk f_{xy} + k^2 f_{yy}) + \varepsilon\rho^2,$$

where $\rho^2 = h^2 + k^2$ and ε tends to zero with ρ.

This suggests that in a sufficiently small neighborhood of the point (x_0, y_0) the behavior of the functional difference $f(x_0 + h, y_0 + k) - f(x_0, y_0)$ is essentially determined by the expression

$$Q(h, k) = ah^2 + 2bhk + ck^2,$$

where for brevity we have put

$$a = f_{xx}(x_0, y_0), \quad b = f_{xy}(x_0, y_0), \quad c = f_{yy}(x_0, y_0).$$

In order to study the problem of extreme values we must investigate this homogeneous quadratic expression or quadratic form Q in h and k. We assume that the coefficients a, b, c do not all vanish. In the exceptional case where they do all vanish, which we shall not consider, we must begin with a Taylor series extending to terms of higher order.

With regard to the quadratic form Q there are three different possible cases:

1. The form is *definite*. That is, when h and k assume all values, Q assumes values of one sign only and vanishes only for $h = 0$, $k = 0$. We say that the form is *positive definite* or *negative definite* according to whether this sign is positive or negative. For example, the expression $h^2 + k^2$, which we obtain when $a = c = 1$, $b = 0$, is positive definite while the expression $-h^2 + 2hk - 2k^2 = -(h - k)^2 - k^2$ is negative definite.

2. The form is *indefinite*. That is, it can assume values of different sign; for example, the form $Q = 2hk$, which has the value 2 for $h = 1$, $k = 1$ and the value -2 for $h = -1$, $k = 1$.

3. The third possibility is that the form vanishes for values of h, k other than $h = 0$, $k = 0$, but otherwise assumes values of one sign only, for example, the form $(h + k)^2$, which vanishes for all sets of

values h, k such that $h = -k$. Such forms are called *semidefinite.*

The quadratic form $Q = ah^2 + 2bhk + ck^2$ is definite if and only if its *discriminant* $ac - b^2$ satisfies the condition

$$ac - b^2 > 0;$$

it is then positive definite if $a > 0$ (so that $c >$ also); otherwise, it is negative definite.

In order that the form may be indefinite, it is necessary and sufficient that

$$ac - b^2 < 0,$$

while the semi-definite case is characterized by the equation[1]

$$ac - b^2 = 0.$$

We shall now prove the following statements. If the quadratic form $Q(h, k)$ is positive definite, the stationary value assumed for $h = 0$, $k = 0$ is a relative *minimum* (even a *strict* relative minimum). If the form is negative definite, the stationary value is a relative *maximum.* If the form is indefinite, we have neither a maximum nor a minimum; the point is a *saddle point.* Thus, definite character of the form Q is a sufficient condition for an extreme value, while indefinite character of Q excludes the possibility of an extreme value. We shall not consider the semidefinite case, which leads to involved discussions.

In order to prove the first statement, we observe that if Q is a positive definite form, there is a positive number m independent of h and k such that[2]

---

[1]These conditions are easily obtained as follows. Either $a = c = 0$, in which case we must have $b \neq 0$ and the form is, as already remarked, indefinite; the criterion therefore holds for this case; otherwise, we must have, say, $a \neq 0$. We can write

$$ah^2 + 2bhk + ck^2 = a\left[\left(h + \frac{b}{a}k\right)^2 + \frac{ca - b^2}{a^2}k^2\right].$$

This form is obviously definite if $ca - b^2 > 0$, and it then has the same sign as a. It is semidefinite if $ca - b^2 = 0$, for then it vanishes for all values of h, k that satisfy the equation $h/k = -b/a$, but for all other values it has the same sign. It is indefinite if $ca - b^2 < 0$, for it then assumes values of different sign when k vanishes and when $h + (b/a)k$ vanishes.

[2]To see this we consider the quotient $Q(h, k)/(h^2 + k^2)$ as a function of the two quantities $u = h/\sqrt{h^2 + k^2}$ and $v = k/\sqrt{h^2 + k^2}$. Then $u^2 + v^2 = 1$, and the form becomes a continuous function of u and v, which must have a least value $2m$ on the circle $u^2 + v^2 = 1$. This value m obviously satisfies our conditions; it is not zero, for u and v never vanish simultaneously on the circle.

$$Q \geqq 2m(h^2 + k^2) = 2m\rho^2.$$

Therefore,

$$f(x_0 + h, y_0 + k) - f(x_0, y_0) = \frac{1}{2} Q(h, k) + \varepsilon\rho^2 \geqq (m + \varepsilon)\rho^2.$$

If we now choose ρ so small that the number ε is less in absolute value than $\frac{1}{2}m$, we obviously have

$$f(x_0 + h, y_0 + k) - f(x_0, y_0) \geqq \frac{m}{2} \rho^2 > 0.$$

Thus, for this neighborhood of the point (x_0, y_0) the value of the function is everywhere greater than $f(x_0, y_0)$, except of course at (x_0, y_0) itself. In the same way, when the form is negative definite the point is a maximum.

Finally, if the form is indefinite, there is a pair of values (h_1, k_1) for which Q is negative and another pair (h_2, k_2) for which Q is positive. We can therefore find a positive number m such that

$$Q(h_1, k_1) < -2m\rho_1^2,$$

$$Q(h_2, k_2) > 2m\rho_2^2.$$

If we now put $h = th_1$, $k = tk_1$, $\rho^2 = h^2 + k^2$, $(t \neq 0)$—that is, if we consider a point $(x_0 + h, y_0 + k)$ on the line joining (x_0, y_0) to $(x_0 + h_1, y_0 + k_1)$—then from $Q(h, k) = t^2Q(h_1, k_1)$ and $\rho^2 = t^2\rho_1^2$ we have

$$Q(h, k) < -2m\rho^2.$$

Thus, by choice of a sufficiently small t (and corresponding ρ), we can make the expression $f(x_0 + h, y_0 + k) - f(x_0, y_0)$ negative. We need only choose t so small that for $h = th_1$, $k = tk_1$ the absolute value of the quantity ε is less than $\frac{1}{2}m$. For such a set of values we have $f(x_0 + h, y_0 + k) - f(x_0, y_0) < -m\rho^2/2$, so that the value $f(x_0 + h, y_0 + k)$ is less than the stationary value $f(x_0, y_0)$. In the same way, on carrying out the corresponding process for the system $h = th_2$, $k = tk_2$, we find that in an arbitrarily small neighborhood of (x_0, y_0) there are points at which the value of the function is greater than $f(x_0, y_0)$. Thus, we have neither a maximum nor a minimum but, instead, what we call a saddle value.

If $a = b = c = 0$ at the stationary point, so that the quadratic form vanishes identically, and in the semidefinite case, this discussion fails to apply. To obtain sufficient conditions for these cases would lead to involved distinctions.

Thus, we have the following rule for distinguishing maxima and minima:

At a point (x_0, y_0) *where the partial derivatives vanish,*

$$f_x(x_0, y_0) = 0, \quad f_y(x_0, y_0) = 0$$

and the inequality

$$f_{xx}f_{yy} - f_{xy}^2 > 0$$

holds, the function f *has a relative extreme value. This is a relative maximum if* $f_{xx} < 0$ *(and consequently* $f_{yy} < 0$*), and a relative minimum if* $f_{xx} > 0$*. If, on the other hand,*

$$f_{xx}f_{yy} - f_{xy}^2 < 0,$$

the stationary value is neither a maximum nor a minimum. The case

$$f_{xx}f_{yy} - f_{xy}^2 = 0$$

remains undecided.

These conditions have a simple geometrical interpretation. The necessary conditions $f_x = f_y = 0$ state that the tangent plane to the surface $z = f(x, y)$ is horizontal. If we really have an extreme value, then in the neighborhood of the point in question the tangent plane does not intersect the surface. In the case of a saddle point, on the contrary, the plane cuts the surface in a curve that has several branches at the point. This matter will be clearer after the discussion of singular points in section A.3.

As an example we seek the extreme values of the function

$$f(x, y) = x^2 + xy + y^2 + ax + by.$$

If we equate the first derivatives to 0, we obtain the equations

$$2x + y + a = 0, \quad x + 2y + b = 0,$$

which have the solution $x = \frac{1}{3}(b - 2a)$, $y = \frac{1}{3}(a - 2b)$. The expression

$$f_{xx}f_{yy} - f_{xy}^2 = 3$$

is positive, as is $f_{xx} = 2$. The function therefore has a minimum at the point in question.

The function

$$f(x, y) = (y - x^2)^2 + x^5$$

has a stationary point at the origin. There the expression $f_{xx}f_{yy} - f_{xy}^2$ vanishes, and our criterion fails. We readily see, however, that the function has no extreme value there, for in the neighborhood of the origin the function assumes both positive and negative values.

On the other hand, the function

$$f(x, y) = (x - y)^4 + (y - 1)^4$$

has a minimum at the point $x = 1$, $y = 1$, though the expression $f_{xx}f_{yy} - f_{xy}^2$ vanishes there. For

$$f(1 + h, 1 + k) - f(1, 1) = (h - k)^4 + k^4,$$

and this quantity is positive when $\rho \neq 0$.

Exercises A.1

1. Find and characterize the extreme values of the functions:

   (a) $f(x, y) = x^2 - 3xy + y^2$

   (b) $f(x, y) = \cos(x + y) + \sin(x - y) + x^2$

   (c) $f(x, y) = x \cosh y - y^2$.

2. If $\phi(a) = k \neq 0$, $\phi'(a) \neq 0$, and x, y, z satisfy the relation $\phi(x)\phi(y)\phi(z) = k^3$, prove that the function $f(x) + f(y) + f(z)$ has a maximum when $x = y = z = a$, provided that

$$f'(a)\left(\frac{\phi''(a)}{\phi'(a)} - \frac{\phi'(a)}{\phi(a)}\right) > f''(a).$$

3. Let $P_1P_2P_3$ be a plane triangle with all three angles less than 120°. Prove by the criterion of p. 349 or of Exercise 6 below that at the point P interior to $P_1P_2P_3$ such that $\angle P_2PP_3 = \angle P_3PP_1 = \angle P_1PP_2 = 120°$, the sum $PP_1 + PP_2 + PP_3$ is actually a minimum (cf. Example 3, p. 328).

4. Where does the minimum of the sum $PP_1 + PP_2 + PP_3$ occur if in the triangle of Exercise 3 the angle $P_2P_1P_3$ is greater than, or equal to, 120°?

5. (a) Prove that if all the symbols denote positive quantities the stationary value of $lx + my + nz$ subject to condition $x^p + y^p + z^p = c^p$ is $c(l^q + m^q + n^q)^{1/q}$, where $q = p/(p - 1)$.

   (b) Show that the value is a maximum or minimum according to whether $p \gtrless 1$.

6. Generalize the investigation of Section A. 1 to functions of n variables, proving the following results. Let $f(x_1, \ldots, x_n)$ be three times continuously differentiable in the neighborhood of a stationary point $x_1 = x_1^0, \ldots, x_n = x_n^0$, that is, a point where $f_{x_1} = f_{x_2} = f_{x_n} = 0$. Consider the second total differential of f at the point x^0, $d^2f^0 = \sum_{i,k=1}^{n} f_{x_i x_k}{}^0\, dx_i\, dx_k$; this is a quadratic form in the variables $dx_1, \ldots, dx_n$. If this quadratic form is nondegenerate, that is, if

$$D = \begin{vmatrix} f_{x_1x_1}{}^0 & \cdots & f_{x_1x_n}{}^0 \\ \vdots & & \vdots \\ f_{x_nx_1}{}^0 & \cdots & f_{x_nx_n}{}^0 \end{vmatrix} \neq 0,$$

then d^2f^0 may be (1) positive definite, (2) negative definite or (3) indefinite. Prove that these possible cases correspond respectively to the following properties of f at the point x^0: (1) f has a minimum, (2) f has a maximum, (3) f has neither a minimum nor a maximum.

7. To investigate stationary points of $f = f(x_1, \ldots, x_n)$, where the variables satisfy the relations

$$(1) \qquad \phi_1(x_1, \ldots, x_n) = 0, \ldots, \phi_m(x_1, \ldots, x_n) = 0 \qquad (m < n)$$

we may assume that we have found numerical values for the variables and the multipliers λ_μ such that $F = f + \lambda_1\phi_1 + \cdots + \lambda_m\phi_m$ satisfies the equations

$$(2) \qquad \frac{\partial F}{\partial x_1} = 0, \ldots, \frac{\partial F}{\partial x_n} = 0,$$

and such that the Jacobian of $\phi_1, \ldots, \phi_m$ with respect to the variables $x_1, \ldots, x_m$ is not 0. To apply the criterion of Exercise 6 we may proceed as follows: Regarding $x_{m+1}, \ldots, x_n$ as independent variables, by differentiating (1) we can obtain the first and second differentials of $x_1 \ldots, x_m$ as functions of $x_{m+1}, \ldots, x_n$ and finally introduce these values into

$$(3) \qquad d^2f = \sum_{i,k=1}^{n} f_{x_ix_k}\, dx_i\, dx_k + f_{x_1}\, d^2x_1 + \cdots + f_{x_m}\, d^2x_m.$$

Prove the following second rule, not involving the computation of the second differentials $d^2x_1, \ldots, d^2x_m$: Regarding $x_1, \ldots, x_n$ as independent variables, consider

$$d^2F = \sum F_{x_ix_k}\, dx_i\, dx_k = d^2f + \lambda_1 d^2\phi_1 + \cdots + \lambda_m\, d^2\phi_m;$$

compute $dx_1, \ldots, dx_m$ from the equations

$$d\phi_\mu = \phi_{\mu x_1}\, dx_1 + \cdots + \phi_{\mu x_n}\, dx_n = 0 \qquad (\mu = 1, \ldots, m)$$

and introduce these values into d^2F, thus obtaining a quadratic form δ^2F in the variables $dx_{m+1}, \ldots, dx_n$. If this quadratic form is nondegenerate, then f has, respectively, a minimum, a maximum, or neither of these, according to whether δ^2F is positive definite, negative definite, or indefinite.

8. In the problem of finding the maximum of $f = x_1 x_2 \cdots x_n$ subject to the condition $\phi = x_1 + x_2 + \cdots + x_n - a = 0$ $(a > 0)$, the rule of undetermined multipliers gives a stationary value of f at the point $x_1 = x_2 = \cdots = x_n = a/n$. Apply the rule of Exercise 7, instead of the consideration of the absolute maximum, to show that f has a maximum value at this point.
9. Apply the criterion of Exercise 7, to prove that among all triangles of constant perimeter the equilateral triangle has the largest area (cf. p. 341).

A.2 Numbers of Critical Points Related to Indices of a Vector Field

A continuous function $f(x, y)$ defined in a closed and bounded set R certainly has a maximum point and minimum point in R, by our fundamental theorem (see p. 112). If a maximum or minimum point (x_0, y_0) is an interior point of R and if f is a differentiable at (x_0, y_0), then (x_0, y_0) is a critical point of f. In some cases this observation permits us to deduce the existence of at least one critical point of f. For example, if the set R consists of an open, bounded set S and its boundary B and if f is constant on B and differentiable in S, then f has at least one critical point in S. This is just an extension of *Rolle's theorem* (see Volume I. p. 175) to functions of several variables, and it is proved in the same way: The function f has maximum and minimum points. If these all lie on the boundary B where f is constant, then the maximum and minimum value of f coincide; then f is constant in S as well and every point of S is critical. Hence, there is at least one critical point of f in S.

In the case of functions of a single independent variable, more specific information on the number of critical points of a certain type is available. Relative maxima and minima *alternate* (see Volume I, p. 239). Hence, the total numbers of relative maxima and of minima of a function in an interval differ by, at most, 1. This is not true for functions of two variables defined in a set R of the plane. There exists, however, an (intuitively less obvious) relation connecting the total numbers of relative extrema and of saddle points in the interior of R with the values of f on the boundary of R. In order to formulate this relation, we first have to consider the *gradient field* of f and to introduce the notion of *index* of a closed curve with respect to a vector field.

Assume that f is continuous and has continuous first derivatives in the set R of the x, y-plane. Then f determines at each point of R the two quantities

$$u = f_x(x, y), \quad v = f_y(x, y). \tag{74}$$

These can be interpreted as the components of a certain vector, the *gradient* of f. The gradients at the various points of R form a *vector field.* The critical points of R are those where the gradient vanishes. At all other points, the gradient vector has a uniquely determined direction described, for example, by its *direction cosines*

$$\xi = \frac{u}{\sqrt{u^2 + v^2}} \quad \text{and} \quad \eta = \frac{v}{\sqrt{u^2 + v^2}}$$

(see Volume I, p. 383). Clearly, ξ and η are continuous functions of (x, y) at every noncritical point of R. We can put

$$\xi = \cos\theta, \quad \eta = \sin\theta,$$

where, however, the angle θ—the *inclination* of the vector (u, v)—is determined only within whole multiples of 2π. In general, it is not possible to select one definite value for θ that will then vary continuously with (x, y). On the other hand, the differential

$$\begin{aligned} d\theta &= d \arctan\frac{v}{u} = \frac{u\,dv - v\,du}{u^2 + v^2} \\ &= \frac{(uv_x - vu_x)dx + (uv_y - vu_y)dy}{u^2 + v^2} \end{aligned} \tag{75}$$

is defined unambiguously for every noncritical point (x, y) of R.

Now let C be an oriented closed curve that lies in R and does not pass through any critical point of f. We define the *Poincaré index* I_C of C with respect to the vector field as the number

$$I_C = \frac{1}{2\pi}\int_C d\theta = \frac{1}{2\pi}\int_C \frac{u\,dv - v\,du}{u^2 + v^2}. \tag{76}$$

If C is given parametrically by

$$x = \phi(t), \quad y = \psi(t) \qquad (a \leq t \leq b),$$

where ϕ and ψ have the same values at the two end points of the t-interval and where the orientation of C corresponds to the sense of increasing t, then the index of C is given by the integral

$$I_C = \frac{1}{2\pi}\int_a^b \left(\frac{u}{u^2 + v^2}\frac{dv}{dt} - \frac{v}{u^2 + v^2}\frac{du}{dt}\right) dt.$$

Since, after traversing the curve C, we return to the same point (x, y), the values for θ corresponding to $t = a$ and $t = b$ can only differ by a multiple of 2π. Hence, I_C is always an integer. This integer counts the total number of counterclockwise rotations performed by the vector (u, v) as we go around the curve C in the sense indicated by its orientation.[1] Of course, I_C changes sign when we change the orientation of C. As an illustration, consider the function

$$f(x, y) = x^2 + y^2.$$

Here the gradient

$$(u, v) = (2x, 2y)$$

at any point (x, y) has the direction of the radius vector from the origin. Assume we make use of a right-handed coordinate system. For a closed curve C that does not pass through the origin the index,

$$I_C = \frac{1}{2\pi} \int_C \frac{x\,dy - y\,dx}{x^2 + y^2}$$

measures the total number of counterclockwise turns performed by the radius from the origin in going around the curve C. This is exactly the formula for the number of times the curve C winds about the origin derived in Volume I (p. 434).

Generally, at points where u and v do not both vanish, the differential $d\theta$ of equation (75) satisfies the integrability condition

$$\left(\frac{uv_x - vu_x}{u^2 + v^2}\right)_y = \left(\frac{uv_y - vu_y}{u^2 + v^2}\right)_x,$$

which can be verified directly and, of course, only reflects the relation

$$\left[\left(\arc\tan \frac{v}{u}\right)_x\right]_y = \left[\left(\arc\tan \frac{v}{u}\right)_y\right]_x,$$

which holds in spite of the possible multiple-valuedness of the function arc tan (v/u). *It follows from the fundamental theorem on line integrals* (see p. 104 and p. 97) *that* $I_C = 0$ *if* C *is the boundary of a simply connected subset of* R *that contains no critical points of* f.

[1]For the definition of "index" it is not necessary that the vector field be a gradient field.

More generally, consider a multiply connected set R with a number of closed boundary curves $C_1, C_2, \ldots, C_n$. Let the x, y-coordinate system be right-handed, as usual. Assume each C_i is oriented in such a way that we leave R to our left in traversing C_i in the sense corresponding to its orientation. Assume that we can divide R into simply connected sets R_k by suitable auxiliary arcs joining various C_i (cf. Fig. 3.31). Let f have no critical points in R. Then,

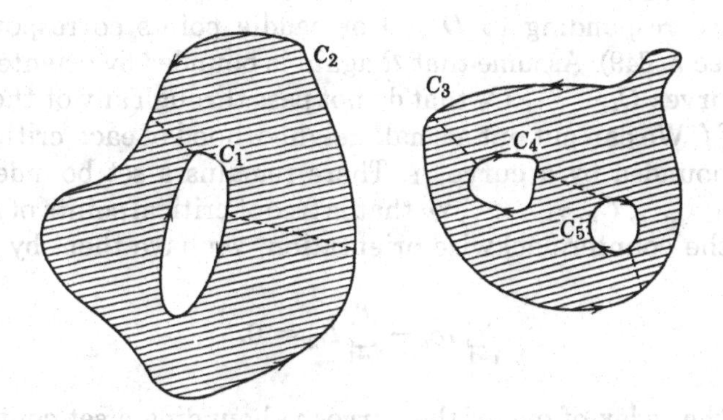

Figure 3.31 Multiply connected region with positively oriented boundary curves C_i divided into simply connected sets.

$$\int d\theta = 0$$

when extended over the boundary of any R_k traversed in the counterclockwise sense. Forming the sum of the integrals over the boundaries of all the R_k, we see that the contributions from the auxiliary arcs cancel out (see p. 94) and we find that

$$0 = \sum_i \int_{C_i} d\theta.$$

This means, however, that

$$\sum_{i=1} I_{C_i} = 0 \tag{77}$$

if the C_i are closed curves forming the boundary of a set R free of critical points of f, and with a sense of orientation leaving R to the left.

As a consequence we obtain the theorem that *there exists at least one critical point in R, whenever the sum of the indices of the boundary curves of R (oriented as explained) is different from zero.*

More precise information on the number of critical points in R is obtained if we assume that f has continuous second derivatives in R, that f has only a finite number of critical points $(x_1, y_1), \ldots, (x_N, y_N)$, and that at each critical point the discriminant

$$D = f_{xx}f_{yy} - f_{xy}^2$$

does not vanish. All critical points are then either relative maxima or minima corresponding to $D > 0$ or saddle points corresponding to $D < 0$ (see p. 349). Assume that R again is bounded by oriented simple closed curves $C_1, \ldots, C_n$ that do not pass through any of the critical points of f. We can cut out a small neighborhood of each critical point (x_k, y_k) bounded by a curve γ_k. There remains a set bounded by the curves $C_1, \ldots, C_n, \gamma_1, \ldots, \gamma_N$ that is free of critical points of f. Giving each γ_k the counterclockwise orientation, we have then, by (77),

$$\sum_{i=1}^{n} I_{C_i} - \sum_{k=1}^{N} I_{\gamma_k} = 0. \tag{78}$$

Now the index of one of the curves γ_k bounding a set containing a single critical point (x_k, y_k) just depends on the *type* of that point, as we shall show.

Let γ_k be a small circle

$$x = x_k + r\cos t, \quad y = y_k + r\sin t$$

of radius r and center at the critical point (x_k, y_k). By Taylor's theorem, we have on γ_k

$$\begin{aligned} u = f_x(x, y) &= (x - x_k)f_{xx}(x_k, y_k) + (y - y_k)f_{xy}(x_k, y_k) + \cdots \\ &= r(a\cos t + b\sin t) + O(r^2) \end{aligned} \tag{79a}$$

$$\begin{aligned} v = f_y(x, y) &= (x - x_k)f_{xy}(x_k, y_k) + (y - y_k)f_{yy}(x_k, y_k) + \cdots \\ &= r(b\cos t + c\sin t) + O(r^2), \end{aligned} \tag{79b}$$

where we put

$$a = f_{xx}(x_k, y_k), \quad b = f_{xy}(x_k, y_k), \quad c = f_{yy}(x_k, y_k).$$

In order to find out how often the vector (u, v) turns in the counterclockwise sense as t varies from $(0, 2\pi)$ we observe that the point in the plane with coordinates (u, v) (that is, the point whose position vector has components u, v) approximately describes the ellipse E with parametric representation

$$u = r(a \cos t + b \sin t), \quad v = r(b \cos t + c \sin t). \tag{80}$$

This ellipse has its center at the origin and has the nonparametric equation

$$(cu - bv)^2 + (av - bu)^2 = r^2(ac - b^2)^2.$$

It is clear that the point (u, v) describes the ellipse E in (80) exactly once as t increases from 0 to 2π, so that the index of γ_k certainly is either $+1$ or -1 depending on the counterclockwise or clockwise sense of E corresponding to increasing t. Now the linear mapping

$$u = r(au + bv), \quad v = r(bu + cv)$$

clearly takes the circle

$$u = \cos t, \quad v = \sin t$$

in the u, v-plane (where increasing t correspond to the counterclockwise sense on the circle) into E. Since sense of curves is preserved or inverted according to the sign of the Jacobian $r^2(ac - b^2)$ of the mapping (see p. 260), we see that

$$\begin{aligned} I_{\gamma_k} &= \operatorname{sgn}(ac - b^2) = \operatorname{sgn}[f_{xx}(x_k, y_k)f_{yy}(x_k, y_k) - f_{xy}{}^2(x_k, y_k)] \\ &= \operatorname{sgn} D(x_k, y_k).^1 \end{aligned}$$

It follows from (78) that

$$\sum_{i=1}^{n} I_{C_i} = \sum_{k=1}^{N} \operatorname{sgn} D(x_k, y_k).$$

As observed earlier $\operatorname{sgn} D(x_k, y_k) = +1$ when the critical point (x_k, y_k) is either a relative maximum or minimum, and $\operatorname{sgn} D(x_k, y_k) = -1$, when

[1]The same result can be obtained analytically by observing that, by formulae(79a, b),

$$\begin{aligned} \lim_{r\to 0} I_{\gamma_k} &= \lim_{r\to 0} \frac{1}{2\pi} \int_{\gamma_k} \frac{u\,dv - v\,du}{u^2 + v^2} \\ &= \frac{1}{2\pi} \int_0^{2\pi} \frac{ac - b^2}{(a \cos t + b \sin t)^2 + (b \cos t + c \sin t)^2}\,dt. \end{aligned}$$

The integral can be evaluated explicitly (see Volume I, p. 294) and has the value $2\pi \operatorname{sgn}(ac - b^2)$.

it is a saddle point. Let M_0, M_1, M_2 denote, respectively, the numbers of minima, saddle points, and maxima in R. Our result becomes the *Poincaré identity.*[1]

$$\sum_{i=1}^{N} I_{C_i} = M_0 - M_1 + M_2. \tag{81}$$

In words, *the excess of the number of relative maxima and minima of f in R over the number of saddle points equals the sum of the indices of the boundary curves of R with respect to the gradient field of f, where each boundary curve is oriented so as to leave R on the left-hand side.*

The result is particularly simple when f is constant along each boundary curve C_i of R. The gradient vector of f then is perpendicular to C (see p. 233) and has the direction of either the exterior or the interior normal of C_i. If no critical point of f lies on C_i and C_i is a smooth closed simple curve the direction of the gradient varies continuously and cannot jump at any point of C_i from that of exterior to that of interior normal or vice verse. It is clear then that the gradient vector turns exactly once along C_i, and in the same sense as the tangent vector of C_i with which the gradient forms a fixed angle. Thus, $I_{C_i} = +1$ when C_i has the counterclockwise sense, and -1 when it has the clockwise one. It is easily seen that with our convention about the orientation of the boundary curves of R a boundary curve C_i has counterclockwise orientation when it forms the "outer" boundary of one of the disconnected pieces making up R and has clockwise orientation if it bounds one of the "holes" in R (see Fig. 3.31). It follows that for f constant on the boundary curves

$$M_0 - M_1 + M_2 = N_0 - N_1, \tag{82}$$

where N_0 is the number of connected components of R and N_1 is the total number of holes in R (the "connectivity" of R).

Take, for example, the case where R is a circular disc. Here $N_0 = 1$, $N_1 = 0$, and thus, for f constant on the boundary,

$$M_0 - M_1 + M_2 = 1.$$

We find here that the *total number of critical points in the interior of R is*

$$M_0 + M_1 + M_2 = 1 + 2M_1$$

[1]The corresponding formulae for functions of more than two independent variables are those of M. Morse.

and, hence, certainly is an odd number. Moreover, if the number $M_0 + M_2$ of relative extrema of f exceeds 1, *then f has at least one saddle point in R.*

For a circular ring R we have

$$N_0 = 1, \quad N_1 = 1,$$

and thus, for f constant on each boundary curve,

$$M_0 - M_1 + M_2 = 0.$$

Take the case where f has the *same* constant value on each of the two boundary curves. Then f is either constant everywhere or assumes its maximum or minimum in the interior of R. If we postulate that f has only critical points with $f_{xx}f_{yy} - f_{xy}^2 \neq 0$ the case of constant f is excluded. It follows then that $M_0 + M_2 > 0$ and, hence, that $M_1 > 0$. Hence, *a function in a circular ring that vanishes everywhere on the boundary has at least one critical point with* $f_{xx}f_{yy} - f_{xy}^2 \leq 0$ *in the interior.*

Exercises A.2

1. Give an example of a continuous function f that has a singularity at the origin of index
   (a) -1;
   (b) -2;
   (c) $-n$, where n is a natural number.
2. Give an example of a function f, not required to be continuous, which has a singularity at the origin of index
   (a) 2;
   (b) n, where n is a natural number.
3. Let the closed convex region R in the x, y-plane be bounded by a closed convex curve C with continuously turning tangent. Let

   $$\xi = f(x, y), \quad \eta = g(x, y)$$

   be a continuously differentiable mapping of R into itself. Prove that the mapping has at least one "fixed point" in R, that is, that there exists a point (x, y) in R such that

   $$x = f(x, y), \quad y = g(x, y).$$

   The analogous fixed point theorem in n dimensions is due to Brouwer. [Hint. Consider the field of vectors with components $u = f(x, y) - x$, $v = g(x, y) - y$.]

A.3 Singular Points of Plane Curves

On p. 236 we saw that a curve $f(x, y) = 0$ in general has a singularity at a point $x = x_0$, $y = y_0$ such that the three equations

$$f(x_0, y_0) = 0, \quad f_x(x_0, y_0) = 0, \quad f_y(x_0, y_0) = 0$$

hold. In order to study these singular points systematically, we assume that in the neighbourhood of (x_0, y_0) the function $f(x, y)$ has continuous derivatives up to the second order and that at that point the second derivatives do not all vanish. By expanding in a Taylor series up to terms of second order, we obtain the equation of the curve in the form

$$\begin{aligned} 2f(x, y) = (x - x_0)^2 f_{xx}(x_0, y_0) &+ 2(x - x_0)(y - y_0) f_{xy}(x_0, y_0) \\ &+ (y - y_0)^2 f_{yy}(x_0, y_0) + \varepsilon\rho^2 = 0, \end{aligned}$$

where we have put $\rho^2 = (x - x_0)^2 + (y - y_0)^2$ and ε tends to 0 with ρ.

Using a parameter t, we can write the equation of the general straight line through the point (x_0, y_0) in the form

$$x - x_0 = at, \quad y - y_0 = bt,$$

where a and b are two arbitrary constants that we may suppose to be so chosen that $a^2 + b^2 = 1$. To determine the point of intersection of this line with the curve $f(x, y) = 0$, we substitute these expressions in the above expansion for $f(x, y)$. For the point of intersection, we thus obtain the equation

$$a^2t^2 f_{xx} + 2abt^2 f_{xy} + b^2t^2 f_{yy} + \varepsilon t^2 = 0.$$

A first solution is $t = 0$, that is, the point (x_0, y_0) itself, as is obvious. However, it is noteworthy that the left-hand side of the equation is divisible by t^2, so that $t = 0$ is a *double root* of the equation. For this reason the singular points are also sometimes called *double points* of the curve. If we remove the factor t^2, we are left with the equation

$$a^2 f_{xx} + 2ab f_{xy} + b^2 f_{yy} + \varepsilon = 0.$$

We now inquire whether it is possible for the line to intersect the curve in another point that tends to (x_0, y_0) as the line tends to some particular limiting position. Such a limiting position of a secant we of course call a tangent. To discuss this, we observe that as a point tends to (x_0, y_0) the quantity t tends to 0, and therefore, ε also tends to 0. If the equation above is still to be satisfied, the expression $a^2 f_{xx} + 2ab f_{xy} + b^2 f_{yy}$ must also tend to 0, that is, for the limiting position of the line, we must have

$$a^2 f_{xx} + 2ab f_{xy} + b^2 f_{yy} = 0.$$

This equation gives us a quadratic condition determining the ratio a/b, which fixes the slope of a tangent.

If the discriminant of the equation is negative, that is, if

$$f_{xx} f_{yy} - f_{xy}^2 < 0,$$

we obtain two distinct real tangents. The curve has a *double point*, or *node*, like that exhibited by the lemniscate $(x^2 + y^2)^2 - (x^2 - y^2) = 0$ at the origin or by the strophoid $(x^2 + y^2)(x - 2a) + a^2 x = 0$ at the point $x_0 = a$, $y_0 = 0$.

If the discriminant vanishes, that is, if

$$f_{xx} f_{yy} - f_{xy}^2 = 0,$$

we obtain two coincident tangents; it is then possible that two branches of the curve touch one another or that the curve has a *cusp*.[1]

Finally, if

$$f_{xx} f_{yy} - f_{xy}^2 > 0,$$

there is no (real) tangent at all. This occurs for example in the case of the so-called *isolated points* of an algebraic curve. These are points at which the equation of the curve is satisfied but in whose neighborhood no other point of the curve lies.

The curve $(x^2 - a^2)^2 + (y^2 - b^2)^2 = a^4 + b^4$ exemplifies this. The values $x = 0$, $y = 0$ satisfy the equation, but for all other values in the region $|x| < a\sqrt{2}$, $|y| < b\sqrt{2}$ the left-hand side is less than the right.

We have omitted the case in which all the derivatives of the second order vanish. This case leads to involved considerations and we shall not investigate it. Through such a point, several branches of the curve may pass, or singularities of other types may occur.

[1]In this case, the curve need not have a singularity at all; for example, $f(x, y) = (x - y)^2$ at the origin.

Finally, we shall briefly mention the connection between these matters and the theory of maxima and minima. Because the first derivatives vanish, the equation of the tangent plane to the surface $z = f(x, y)$ at a stationary point (x_0, y_0) is simply

$$z - f(x_0, y_0) = 0.$$

The equation

$$f(x, y) - f(x_0, y_0) = 0$$

therefore gives us the projection on the x,y-plane of the curve of intersection of the tangent plane with the surface, and we see that the point (x_0, y_0) is a singular point of this curve. If this is an isolated point, in a certain neighborhood the tangent plane has no other point in common with the surface, and the function $f(x, y)$ has a maximum or a minimum at the point (x_0, y_0) (cf. p. 349). If, however, the singular point is a multiple point, the tangent plane cuts the surface in a curve with two branches, and (x_0, y_0) is a saddle point. These remarks lead us precisely to the sufficient conditions that we found earlier in Section A.1.

Exercises A.3

1. Find the singular points of the following curves and discuss their nature:

(a) $(x^2 + y^2)^2 - 2c^2(x^2 - y^2) = 0,\ c \neq 0$

(b) $x^2 + y^2 - 2x^3 - 2y^3 + 2x^2y^2 = 0$

(c) $x^4 + y^4 - 2(x - y)^2 = 0$

(d) $x^5 - x^4 + 2x^2y - y^2 = 0.$

A.4 Singular Points of Surfaces

In a similar way we can discuss a singular point of a surface $f(x, y, z) = 0$, that is, a point for which

$$f = 0, \quad f_x = f_y = f_z = 0.$$

Without loss of generality we may take the point as the origin O. If we write

$$f_{xx} = \alpha, \quad f_{yy} = \beta, \quad f_{zz} = \gamma, \quad f_{xy} = \lambda, \quad f_{yz} = \mu, \quad f_{xz} = \nu$$

for the values at this point, we obtain the equation

$$\alpha x^2 + \beta y^2 + \gamma z^2 + 2\lambda xy + 2\mu yz + 2\nu xz = 0$$

for a point (x, y, z) that lies on a tangent to the surface at O.

This equation represents a quadratic cone touching the surface at the singular point (instead of the tangent plane at an ordinary point of the surface) if we assume that not all of the quantities $\alpha, \beta, \ldots, \nu$ vanish and that the above equation has real solutions other than $x = y = z = 0$.

Exercises A.4

1. Using the results of Exercise 6 of A.1 examine the behavior of a surface in a neighborhood of a singular point.

A.5 Connection Between Euler's and Lagrange's Representations of the Motion of a Fluid

Let (a, b, c) be the coordinates of a particle at the time $t = 0$ in a moving continuum (liquid or gas). The motion can then be represented by the three functions

$$x = x(a, b, c, t),$$
$$y = y(a, b, c, t),$$
$$z = z(a, b, c, t),$$

or in terms of a position vector $\mathbf{X} = \mathbf{X}(a, b, c, t)$. Velocity and acceleration are given by the derivatives with respect to the time t. Thus, the velocity vector is $\dot{\mathbf{X}}$ with components $\dot{x}, \dot{y}, \dot{z}$, and the acceleration vector is $\ddot{\mathbf{X}}$ with components $\ddot{x}, \ddot{y}, \ddot{z}$, all of which appear as functions of the initial position (a, b, c) and the parameter t. For each value of t we have a transformation of the coordinates (a, b, c) belonging to the different points of the moving continuum into the coordinates (x, y, z) at the time t. This is the so-called *Lagrange representation of the motion*. Another representation introduced by Euler is based upon the knowledge of three functions

$$u(x, y, z, t), \quad v(x, y, z, t), \quad w(x, y, z, t)$$

representing the components $\dot{x}, \dot{y}, \dot{z}$ of the velocity $\dot{\mathbf{X}}$ of the motion at the point (x, y, z) at the time t.

In order to pass from the first representation to the second we have to use the first representation to calculate a, b, c as functions of $x, y,$

z, and t and to substitute these expressions in the expressions for $\dot{x}(a, b, c, t)$, $\dot{y}(a, b, c, t)$, $\dot{z}(a, b, c, t)$:

$$u(x, y, z, t) = \dot{x}(a(x, y, z, t),\ b(x, y, z, t),\ c(x, y, z, t), t), \ldots .$$

We then get the components of the acceleration from

$$\dot{x}(a, b, c, t) = u(x(a, b, c, t),\ y(a, b, c, t),\ z(a, b, c, t), t), \ldots .$$

by differentiation with respect to t for fixed a, b, c:

$$\ddot{x} = u_x\dot{x} + u_y\dot{y} + u_z\dot{z} + u_t, \ldots$$

or

$$\ddot{x} = u_x u + u_y v + u_z w + u_t,$$

$$\ddot{y} = v_x u + v_y v + v_z w + v_t,$$

$$\ddot{z} = w_x u + w_y v + w_z w + w_t.$$

In the mechanics of a continuum, the following equation connecting Euler's and Lagrange's representations is fundamental:

$$\operatorname{div} \dot{\mathbf{X}} = u_x + v_y + w_z = \frac{\dot{D}}{D},$$

where

$$D(x, y, z, t) = \frac{d(x, y, z)}{d(a, b, c)}$$

is the Jacobian characterizing the transformation.

The reader may complete the proof of this and the corresponding theorem in two dimensions by using the various rules for the differentiation of implicit functions (see p. 252).

Exercises A.5

1. What is the physical interpretation of the relations $u_t = v_t = w_t = 0$.
2. Interpret the relations

$$\ddot{x} = u_x u + u_y v + u_z w + u_t,$$

$$\ddot{y} = v_x u + v_y v + v_z w + v_t,$$

$$\ddot{z} = w_x u + w_y v + w_z w + w_t$$

physically; rewrite these relations using vector notation.

A.6 Tangential Representation of a Closed Curve and the Isoperimetric Inequality

A family of straight lines with parameter α may be given by

$$x \cos \alpha + y \sin \alpha - p(\alpha) = 0, \cdots \tag{83}$$

where p(α) denotes a function that is twice continuously differentiable and periodic of period 2π (here p represents the distance of the line of the family with normal direction α from the origin). The envelope C of these lines is a closed curve satisfying (83) and the further equation

$$-x \sin \alpha + y \cos \alpha - p'(\alpha) = 0.$$

Hence,

$$\begin{aligned} x &= p \cos \alpha - p' \sin \alpha \\ y &= p \sin \alpha + p' \cos \alpha \end{aligned} \tag{84}$$

is the parametric representation of C (α being the parameter). Formula (83) gives the equation of the tangents of C and is referred to as the *tangential equation*[1] of C, and $p(\alpha)$ as the *support function* of C.

Since

$$x' = -(p + p'') \sin \alpha, \qquad y' = (p + p'')\cos \alpha,$$

we at once have the following expressions for the length L and area A of C:

$$L = \int_0^{2\pi} \sqrt{x'^2 + y'^2}\, d\alpha = \int_0^{2\pi} (p + p'')d\alpha = \int_0^{2\pi} p\, d\alpha$$

$$A = \frac{1}{2}\int_0^{2\pi} (xy' - yx')d\alpha = \frac{1}{2}\int_0^{2\pi} (p + p'')p\, d\alpha = \frac{1}{2}\int_0^{2\pi} (p^2 - p'^2)d\alpha,$$

since $p'(\alpha)$ is also a function of period 2π.[2]

---

[1]The representation of C in the form (84) is valid for any closed convex curve whose curvature is finite and positive, and varies continuously along C.

[2]Since $p(\alpha) + c$ is obviously the support function of the parallel curve at a distance c from C, the formulae for the area and the length of a parallel curve (cf. Volume I, p. 437, Exercise 7, and its solution in A. Blank: Problems in Calculus and Analysis, p. 188) are easily derived from these expressions.

From this we deduce the *isoperimetric inequality*

$$L^2 \geqq 4\pi A,$$

where the equality sign holds for the circle only. This may also be expressed by the statement: *Among all closed curves of given length the circle has the greatest area.*

For the proof we make use of the Fourier expansion of $p(\alpha)$ (Volume I, p. 594),

$$p(\alpha) = \frac{a_0}{2} + \sum_{\nu=1}^{\infty} (a_\nu \cos \nu\alpha + b_\nu \sin \nu\alpha);$$

then

$$p'(\alpha) = \sum_{\nu=1}^{\infty} \nu(b_\nu \cos \nu\alpha - a_\nu \sin \nu\alpha),$$

so that (using the orthogonality relations of Volume I, p. 593) we have

$$L = \pi a_0,$$

$$A = \frac{\pi}{2}\left(\frac{a_0{}^2}{2} - \sum_{\nu=2}^{\infty} (\nu^2 - 1)(a_\nu{}^2 + b_\nu{}^2)\right).$$

Thus,

$$A \leqq \frac{\pi a_0{}^2}{4} = \frac{L^2}{4\pi};$$

in particular, $A = L^2/4\pi$ only if $a_\nu = b_\nu = 0$ for $\nu \geqq 2$; that is, $p(\alpha) = a_0/2 + a_1 \cos \alpha + b_1 \sin \alpha$. The latter equation defines a cirlce, as is easily proved from (84).

Exercises A.6

1. Find the equations of the envelopes, their lengths, and contained areas, for each of the following families of straight lines:

   (a) $(x + 2) \cos \alpha + y \sin \alpha + 2 = 0$

   (b) $x \cos \alpha + y \sin \alpha + \frac{1}{2} \sin 2\alpha = 0$.

2. Compare the formulae for area and length. Can there exist curves of arbitrarily large length enclosing arbitrarily small area?
3. Can every closed curve be represented as the envelope of lines (83)?

CHAPTER 4

Multiple Integrals

Differentiation and operations with derivatives for functions of several variables are directly reducible to their anologues for functions of one variable. Integration and its relation to differentiation are more involved, since the concept of integral can be generalized for functions of several variables in a variety of ways. Thus, for a function $f(x, y, z)$ of three independent variables, we have to consider integrals over surfaces and lines, as well as integrals over regions of space. Nonetheless, all questions of integration will be related to the original concept of the integral of a function of a single independent variable.

For simplicity we shall work mainly in the plane, (i.e., with two independent variables). However, all arguments apply equally well to higher dimensions with mere changes of terminology ("area" by "volume," "square" by "cube," etc.).

4.1 Area in the Plane

a. *Definition of the Jordan Measure of Area*

In Volume I we expressed the area of a region in the x, y-plane by integrals of functions of a single variable. The basic idea (which led us to the notion of *integral* in the first place) was to approximate the region by simpler regions consisting of a finite number of rectangles. For a more systematic development of areas that immediately carries over to volumes in three or more dimensions, it is desirable to give a direct definition that is not tied to the idea of integration of functions of one variable and corresponds more closely to the intuitive notion of the area of a region as the "number of square units" contained in the region. At the same time, this new and more natural definition is more general and avoids all extraneous discussion of the regularity of the boundary, which becomes inevitable whenever we try to reduce areas to single integrals. As usual, we postpone rigorous existence proofs to the Appendix of this chapter. Those proofs only present systematically what should already be more or less obvious to the reader from the informal discussions of ideas and purposes presented in the main text.

In defining areas, we accept the intuitive idea that the area $A(S)$ of a set S should be a nonnegative number attached to S that has the following properties:

1. If S is a square of side k then $A = k^2$.
2. Additivity: *The area of the whole is the sum of the areas of its parts.* More precisely, if S consists of nonoverlapping[1] sets $S_1, \ldots, S_N$ of areas $A(S_1) \ldots, A(S_N)$, respectively, then the area of S is

$$A(S) = A(S_1) + \cdots + A(S_N)$$

On the basis of these simple requirements, we shall be able to assign a value $A(S)$ to most of the two-dimensional sets A encountered in practice although not to all imaginable sets S in the plane.

To arrive at a uniquely determined value $A(S)$ for a bounded set S, we use very special divisions of the plane into squares; it will be shown subsequently that every other way of dividing the plane into squares (or rectangles) will lead to the same area. Congruent squares provide the easiest way of covering the plane without gaps or overlap. We use the grid attached to our coordinate system provided by the lines $x = 0, \pm 1, \pm 2, \pm 3, \ldots$ and $y = 0, \pm 1, \pm 2, \ldots$, which divide the whole plane into *closed* squares of side 1. We denote by $A_0^+(S)$ the number of squares having points in common with S and by $A_0^-(S)$ the number of those completely contained in S. We next divide each square into four equal squares of side $\frac{1}{2}$ and area $\frac{1}{4}$ and denote by $A_1^+(S)$ one-fourth of the number of those subsquares having points with S and by $A_1^-(S)$ one-fourth of the number of those completely contained in S. Since each unit square completely contained in S gives rise to four subsquares completely contained in S we have $A_0^-(S) \leqq A_1^-(S)$, and similarly $A_0^+(S) \geqq A_1^+(S)$. We next divide each square of side $\frac{1}{2}$ further into 4 squares of side $\frac{1}{4}$. One-sixteenth of those squares having points

[1]The sets are *nonoverlapping* if every interior point of one of the sets is exterior to all the other sets. We call the sets *disjoint* if every point of one of the sets belongs to no others.

in common with S and one sixteenth of those contained in S will be denoted, respectively, by $A_2^+(S)$ and $A_2^-(S)$. Proceeding in this fashion, we associate values $A_n^+(S)$ and $A_n^-(S)$ with a division of the plane into squares of side 2^{-n}(see Fig. 4.1). It is clear that the values $A_n^+(S)$ form a

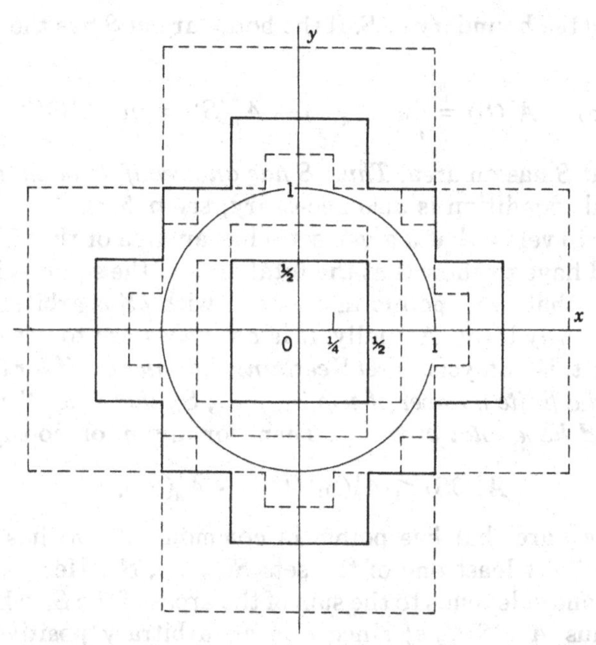

Figure 4.1 Interior and exterior approximations to the area of the unit disk $x^2 + y^2 \leq 1$, for $n = 0, 1, 2$, where $A_0^- = 0$, $A_1^- = 1$, $A_2^- = 2$, $A_2^+ = 4\frac{1}{4}$, $A_1^+ = 6$, $A_0^+ = 12$.

monotone decreasing and bounded sequence that converges toward a value $A^+(S)$, while the $A_n^-(S)$ increase monotonically and converge towards a value $A^-(S)$. The value $A^-(S)$ represents the *inner area*, the closest we can approximate the area of S from below by congruent squares contained in S; the *outer area* $A^+(S)$ represents the best upper bound obtainable by covering S by congruent squares. If both values agree, we say that S is *Jordan-measurable* and call the common value $A^-(S) = A^+(S)$ the *content*, or the *Jordan-measure*, of S. We shall use the simpler term *area* $A(S)$ for the content of S, and shall say "S has an area" instead of using the clumsier phrase "S is Jordan-measurable" to denote the fact that $A^-(S) = A^+(S)$, (which is true for almost all sets occurring in practice).

The difference $A_n^+(S) - A_n^-(S)$ represents the total area of the squares in the nth subdivision that have points in common with S

without lying completely in S. All these squares contain boundary points of S, so that

$$A_n^+(S) - A_n^-(S) \leqq A_n^+(\partial S)$$

where ∂S is the boundary of S. If the boundary of S has the area 0, we find that

$$A^+(S) - A^-(S) = \lim_{n\to\infty} [A_n^+(S) - A_n^-(S)] = \lim_{n\to\infty} A_n^+(\partial S) = 0,$$

that is, that S has an area. *Thus, S has an area if its boundary ∂S has area* 0. (This condition is also necessary; see p. 518).

In order to verify that a given set S has an area or that ∂S has area 0 we would have to show that the total area of the squares in the nth subdivision that have points in common with ∂S is arbitrarily small for n sufficiently large. Actually, it is not necessary to use squares of side 2^{-n} for this analysis. *A set S certainly has an area if for every $\varepsilon > 0$ we can find a finite number of sets $S_1, \ldots, S_N$ that cover the boundary ∂S of S and have total area $< \varepsilon$*. Then, for any n, obviously

$$A_n^+(\partial S) \leqq A_n^+(S_1) + \cdots A_n^+(S_N),$$

since any square that has points in common with ∂S has points in common with at least one of the sets $S_1, \ldots, S_N$. Here, for $n \to \infty$, the right-hand side tends to the sum of the areas of the S_i, which is less than ε; thus $A^+(\partial S) \leqq \varepsilon$; since ε is an arbitrary positive number, we conclude that $A^+(\partial S) = 0$.

This criterion is sufficient to establish that most of the common regions S encountered in analysis have area. In particular, it is sufficient to know that the boundary of S consists of a finite number of arcs each of which has a continuous nonparametric representation $y = f(x)$ or $x = g(y)$ with f or g, respectively, continuous in a finite closed interval. The uniform continuity of continuous functions in bounded closed intervals immediately permits us to show that these arcs can be covered by a finite number of rectangles of arbitrarily small total area.[1]

b. A Set That Does Not Have an Area

An example of a set that does not have an area in our sense (or is not "Jordan-measurable") is the set S of "rational" points in the unit-square, that is, the set of points whose coordinates x, y are both

[1]We leave as an exercise for the reader to prove that a rectangle with sides parallel to the axes has an area (as defined here) equal to the product of two adjacent sides.

rational numbers between 0 and 1. It is evident from the density property of rational and irrational numbers that

$$A_n^+ = 1, \quad A_n^- = 0$$

for all n, so that S has outer area 1 and inner area 0. This agrees with the fact that the boundary ∂S of S consists of the whole closed unit-square and has area 1. If we cover S in any way by a finite number of closed sets $S_1, \ldots, S_N$ with areas $A(S_1), \ldots, A(S_N)$, respectively, then

$$A(S_1) + \cdots + A(S_N) \geqq 1$$

since the S_i necessarily also cover the boundary ∂S of S (see Exercise 6). Paradoxically, however, it is possible to cover S by an *infinite* number of closed sets S_i of arbitrarily small total area. We only have to use the fact that the pairs (x, y) of rational numbers form a denumerable set (see Volume I, p. 98).[1] Thus, the points of S can be arranged into an infinite sequence $(x_1, y_1), (x_2, y_2), (x_3, y_3), \ldots$. Let ε be an arbitrary positive number. Denote for each integer $m > 0$ by S_m a square of area $\varepsilon 2^{-m}$ and center (x_m, y_m). Then the S_m cover the whole set S, while their total area is given by

$$\frac{\varepsilon}{2} + \frac{\varepsilon}{4} + \frac{\varepsilon}{8} + \frac{\varepsilon}{16} + \cdots = \varepsilon.$$

Thus, coverings by ***infinitely many unequal*** squares can lead to a substantial lowering of the upper bound $A^+(S)$ for the "area" of S, reflecting more closely the "rarity" of the rational points among the real ones. One of the starting points in the refined theory of measuring sets, originated by Lebesgue, is to define the outer area of a set as the greatest lower bound of the sum of areas of any ***finite or infinite*** set of squares covering it. For our set S this outer Lebesgue area has the value 0, the same as the inner area of S. Incidentally, for a closed and bounded set S the two definitions of outer area agree, since by the

[1] We can arrange them, for example, in groups, according to the size of the larger of the two denominators; each group has only a finite number of elements:

$$\left(\frac{1}{2}, \frac{1}{2}\right); \; \left(\frac{1}{3}, \frac{1}{3}\right), \; \left(\frac{1}{3}, \frac{1}{2}\right), \; \left(\frac{1}{3}, \frac{2}{3}\right), \; \left(\frac{1}{2}, \frac{1}{3}\right), \; \left(\frac{1}{2}, \frac{2}{3}\right),$$

$$\left(\frac{2}{3}, \frac{1}{3}\right), \; \left(\frac{2}{3}, \frac{1}{2}\right), \; \left(\frac{2}{3}, \frac{2}{3}\right); \; \left(\frac{1}{4}, \frac{1}{4}\right), \; \left(\frac{1}{4}, \frac{1}{3}\right), \ldots.$$

Heine-Borel theorem (cf. p. 109) any infinite covering of S already contains a finite covering.

c. Rules for Operations with Areas

In most cases that interest us we can establish the existence of an area of a set S by verifying that S is bounded by a finite number of arcs with continuous nonparametric representation. For that reason one might be tempted to exclude all other regions with more complicated boundaries from consideration. It turns out however that such a restriction not only results in a loss of generality but actually complicates matters, since we have to make sure that the regions resulting from the operations of set union and intersection again have simple boundaries. The advantage of our general definition of area as *content* is that it is based on the primitive notion of counting of squares; nothing is postulated about the boundary at all beyond the requirement that it can be covered by a finite number of squares of arbitrarily small total area. The boundary of a Jordan-measurable set can be very complicated in detail, consisting perhaps of infinitely many closed curves. These complications will have no effect in the theory of integration, as long as we can show that the total contribution arising from the boundary is negligible.

For work with areas, the operations of dividing a set into subsets and of combining sets into larger ones are basic. The important point is that applying these operations we stay within the class of sets that have areas. We have the fundamental theorem that *the union $S \cup T$ and the intersection $S \cap T$ of two Jordan-measurable sets S and T are again Jordan-measurable.*[1] This follows immediately from the fact that the boundaries of $S \cup T$ and of $S \cap T$ consist of boundary points of S or T and, hence, have again area 0 (see p. 521).

For the important case of two *nonoverlapping* sets S, T—that is, sets such that no interior point of one belongs to the other set or to its boundary—the *law of additivity for areas* holds:

$$A(S \cup T) = A(S) + A(T).$$

More generally, for any finite number of Jordan-measurable sets S_1, $S_2, \ldots, S_N$, no two of which overlap, we have the relation

$$A\left(\bigcup_{i=1}^{N} S_i\right) = \sum_{i=1}^{N} A(S_i). \tag{1}$$

[1]We remind the reader that the union of sets consists of the points belonging to at least one of the sets and the intersection of those points belonging to all.

The proof is trivial on the basis of the inequalities

$$A_n^+\left(\bigcup_{i=1}^{N} S_i\right) \leq \sum_{i=1}^{N} A_n^+(S_i)$$

$$A_n^-\left(\bigcup_{i=1}^{N} S_i\right) \geq \sum_{i=1}^{N} A_n^+(S_i).$$

Here the first inequality follows simply from the fact that any square that has points in common with the union of the S_i must have points in common with at least one of the S_i. The second one follows from the fact that any square contained in one set S_i cannot be contained in any other S_k(since the two are nonoverlapping) but is contained in their union. For $n \to \infty$, we conclude that

$$A^+\left(\bigcup_{i=1}^{N} S_i\right) \leq \sum_{i=1}^{N} A^+(S_i)$$

$$A^-\left(\bigcup_{i=1}^{N} S_i\right) \geq \sum_{i=1}^{N} A^-(S_i).$$

From the assumption that the S_i have areas, that is, that

$$A^+(S_i) = A^-(S_i) = A(S_i),$$

and that the inner area of the union cannot exceed the outer area, the equation (1) follows.

It is now easy to verify that "areas" as defined here can be expressed in terms of integrals in the specific instances considered in Volume I. For example, let the set S consist of the points "below" the graph of a continuous positive function $y = f(x)$ in an interval $a \leq x \leq b$. that is, the set of points (x, y) for which

$$a \leq x \leq b, \quad 0 \leq y \leq f(x).$$

Consider any subdivision of the interval $[a, b]$ into N subintervals of length Δx_i, and let m_i be the minimum and M_i the maximum of $f(x)$ in the ith subinterval. The rectangles with base Δx_i and height m_i are clearly nonoverlapping and their union is contained in S, so that

$$\sum_{i=1}^{N} m_i \,\Delta x_i \leq A(s).$$

Similarly,

$$A(S) \leq \sum_{i=1}^{N} M_i \,\Delta x_i.$$

For continuous f, the lower and upper sums both tend to the integral of f and we arrive at the classical expression

$$A(S) = \int_a^b f(x)\,dx \tag{2}$$

for the area of S.

Exercises 4.1

1. Show that if S and T have area and if S is contained in T, then $A(S) \leqq A(T)$.
2. Under the hypothesis of Exercise 1, show that $T - S$ has area, where $T - S$ is the set of points of T that are not contained in S.
3. Show that if S and T are bounded,
   (a) $A^+(S \cup T) + A^+(S \cap T) \leqq A^+(S) + A^+(T)$
   (b) $A^-(S \cup T) + A^-(S \cap T) \geqq A^-(S) + A^-(T)$
4. Let S and T be any disjoint sets whose union has area. Show that $A^+(S) + A^-(T) = A(S \cup T)$.
5. (a) Show that if a set S has area in one coordinate system, it has area in any other coordinate system obtained by rotation and translation of axes.
   (b) Show that the area of S is the same in both coordinate systems.
6. Let S be covered by a finite collection $S_1, \ldots, S_N$ of closed sets. Show that the collection also covers the boundary ∂S of S.
7. Does the set S of points $(1/p, 1/q)$, where p and q are natural numbers, have an area?

4.2 Double Integrals

a. *The Double Integral as a Volume*

Everything said about areas in the preceding paragraphs carries over immediately to volumes in three or higher dimensions. In defining the volume $V(S)$ of a bounded set S in x, y, z-space, we need only use subdivisions of space into *cubes* of side 2^{-n}. The set S will *have a volume* when its boundary can be covered by a finite number of these cubes of arbitrarily small total volume. This is the case for all bounded sets S whose boundary consists of a finite number of surfaces each of which has a continuous nonparametric representation $z = f(x, y)$ or $y = g(x, z)$ or $x = h(y, z)$ on a closed planar set.

The attempt to represent the volume analytically leads directly to the notion of multiple integral, which has a great variety of applications.

Let R, a Jordan-measurable closed and bounded set in the x, y-plane be the domain of a positive-valued function $z = f(x, y)$. We wish to find the volume "below" the surface $z = f(x, y)$, that is, the volume $V(S)$ of the set S of points (x, y, z) for which

$$(x, y) \in R, \qquad 0 \leqq z \leqq f(x, y).$$

For this purpose, we divide R into nonoverlapping closed Jordan-measurable sets $R_1, \ldots, R_N$. Let m_i be the minimum, and M_i the maximum, of f for (x, y) in R_i. It is easily seen that the cylinder with base R_i and height m_i has the volume $m_iA(R_i)$, where $A(R_i)$ is the area of R_i (Fig. 4.2).[1] These cylinders do not overlap. Similarly, the

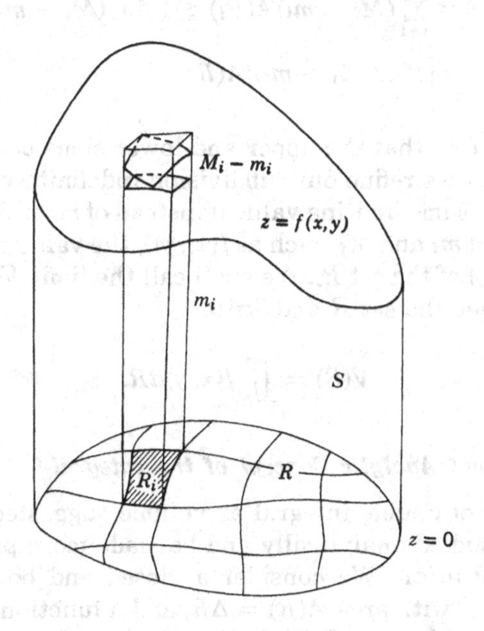

Figure 4.2

cylinders with base R_i and height M_i have volume $M_iA(R_i)$ and do not overlap. It follows that

$$\text{(3a)} \qquad \sum_{i=1}^{N} m_iA(R_i) \leq V(S) \leq \sum_{i=1}^{N} M_iA(R_i)$$

[1]When we divide space into cubes of side 2^{-n}, the cubes having points in common with the cylinder can be arranged into cylindrical "columns" whose cross section is a square having a point in common with R_i and whose height differs by less than 2^{-n} from m_i.

The sums appearing in this inequality we call, respectively, the *lower* and *upper* sums.

We now make our subdivision finer and finer, in the sense that the largest diameter of any R_i occuring in the subdivision tends to zero.[1] The continuous function $f(x, y)$ is *uniformly continuous* in the compact set R, so that the maximum difference $M_i - m_i$ tends to zero with the maximum diameter of the sets R_i in the subdivision. The difference between the upper and lower sums also tends to zero, since

$$\begin{aligned}\sum_{i=1}^{N} M_i A(R_i) - \sum_{i=1}^{N} m_i A(R_i) &= \sum_{i=1}^{N} (M_i - m_i)A(R_i) \leqq [\operatorname*{Max}_i (M_i - m_i)] \sum_{k=1}^{N} A(R_k)\\ &= [\operatorname*{Max}_i (M_i - m_i)]A(R).\end{aligned}$$

It follows from (3a) that the upper and lower sums both converge to the limit $V(S)$ as we refine our subdivision indefinitely. We can obviously obtain the same limiting value if instead of m_i or M_i we take any number between m_i and M_i, such as $f(x_i, y_i)$, the value of the function at a point (x_i, y_i) of the set R_i. We shall call the limit $V(S)$ the double integral of f over the set R and write

$$V(S) = \iint_R f(x, y)dR. \tag{3b}$$

b. The General Analytic Concept of the Integral

The concept of double integral as volume suggested by geometry must now be studied analytically and be made more precise without reference to intuition. We consider a closed and bounded Jordan-measurable set R with area $A(R) = \Delta R$, and a function $f(x, y)$ that is continuous everywhere in R (including the boundary). As before, we subdivide R into N nonoverlapping Jordan-measurable subsets R_1, $R_2, \ldots, R_N$ with areas $\Delta R_1, \ldots, \Delta R_N$. In R_i we choose an arbitrary point (ξ_i, η_i), where the function has the value $f_i = f(\xi_i, \eta_i)$ and we form the sum

$$V_N = \sum_{i=1}^{N} f_i \Delta R_i = \sum_{i=1}^{N} f_i A(R_i).$$

The fundamental existence theorem then states:

[1]The "diameter" of a closed set is the maximum distance of any two points in the set.

If the number N increases beyond all bounds and at the same time the greatest of the diameters of the subregions tends to zero, then V_N tends to a limit V. This limit is independent of the particular nature of the subdivision of the regions R and of the choice of the point (ξ_i, η_i) in R_i. We call the limit V the (double) integral of the function $f(x, y)$ over the region R and denote it by

$$\iint_R f(x, y)dR.$$[1]

COROLLARY. We obtain the same limit if we take the sum only over those subregions R_i that lie entirely in the interior of R, that is, which have no points in common with the boundary of R.[2]

This existence theorem for the integral of a continuous function must be proved in a purely analytical way. The proof, which is very similar to the corresponding proof for one variable, is given in the appendix to this chapter (p. 526).

We now illustrate this concept of an integral by considering some special subdivisions. The simplest case is that in which R is a rectangle $a \leqq x \leqq b$, $c \leqq y \leqq d$ and the subregions R_i are also rectangles (formed by subdividing the x-interval into n equal parts and the y-interval into m equal parts) of lengths

$$h = \frac{b-a}{n} \quad \text{and} \quad k = \frac{d-c}{m}.$$

[1]We can refine this theorem further in a way useful for many purposes. In the subdivision into N subregions it is not necessary to choose a value that is actually assumed by the function $f(x, y)$ at a definite point (ξ_i, η_i) of the corresponding subregion; it is sufficient to choose values that differ from the values of the function $f(\xi_i, \eta_i)$ by quantities that tend uniformly to 0 as the subdivision is made finer. In other words, instead of the values of the function $f(\xi_i, \eta_i)$ we can consider the quantities

$$f_i = f(\xi_i, \eta_i) + \varepsilon_{i,N}$$

where $|\varepsilon_{i,N}| \leqq \varepsilon_N$, $\lim_{N\to\infty} \varepsilon_N = 0$. This theorem is almost trivial, for, since the numbers $\varepsilon_{i,N}$ tend uniformly to 0, the absolute value of the difference between the two sums

$$\sum_1^N f_i \,\Delta R_i \quad \text{and} \quad \sum_1^N (f_i + \varepsilon_{i,N})\Delta R_i$$

is less than $\varepsilon_N \sum \Delta R_i$, and can be made as small as we please if we take the number N sufficiently large. For example, if $f(x, y) = P(x, y)\, Q(x, y)$, we may take $f_i = P_i Q_i$, where P_i and Q_i are the maxima of P and Q in R_i, which are in general not assumed at the same point.

[2]The corollary follows from the fact that not only the boundary ∂R of R but also the set of all points sufficiently close to ∂R can be covered by squares of arbitrarily small total area.

The points of subdivision we call $x_0 = a, x_1, x_2, \ldots, x_n = b$ and $y_0 = c, y_1, y_2, \ldots, y_m = d$. They correspond to parallels to the y-axis and x-axis, respectively. We then have $N = nm$. The subregions are all rectangles with area $A(R_i) = \Delta R_i = hk = \Delta x\, \Delta y$, where $h = \Delta x$, $k = \Delta y$. For the point (ξ_i, η_i) we take any point in the corresponding rectangle R_i, and then form the sum

$$\sum_i f(\xi_i, \eta_i)\Delta x\, \Delta y$$

for all the rectangles of the subdivision.

If we now let n and m simultaneously increase beyond all bounds, the sum tends to the integral of the function f over the rectangle R.

These rectangles can also be characterized by two suffixes μ and ν, corresponding to the coordinates $x = a + \nu h$ and $y = c + \mu k$ of the lower left-hand corner of the rectangle in question. Here ν assumes integral values from 0 to $(n - 1)$ and μ from 0 to $(m - 1)$. With this identification of the rectangles by the suffixes ν and μ, we may appropriately write the sum as a double sum[1]

$$\sum_{\nu=0}^{n-1} \sum_{\mu=0}^{m-1} f(\xi_\nu, \eta_\mu)\Delta x\, \Delta y. \tag{3c}$$

Even when R is not a rectangle, it is often convenient to subdivide the region into rectangular subregions R_i. To do this we superimpose on the plane the rectangular net formed by the lines

$$x = \nu h \qquad (\nu = 0, \pm 1, \pm 2, \ldots)$$

$$y = \mu k \qquad (\mu = 0, \pm 1, \pm 2, \ldots),$$

where h and k are numbers chosen arbitrarily. We now consider all those rectangles of the division that lie entirely within R. These rectangles we call R_i. Of course, they do not completely fill the region; on the contrary, in addition to these rectangles R also contains certain regions R_i adjacent to the boundary that are bounded partly by lines of the net and partly by portions of the boundary of R. By the corollary on p. 377 we can calculate the integral of the function f over the region R by summing over the interior rectangles only and then passing to the limit.

Another type of subdivision frequently applied is the subdivision by a polar coordinate net (Fig. 4.3). We subdivide the entire angle 2π

[1]If we are to write the sum in this way, we must suppose that the points (ξ_i, η_i) are chosen so as to lie in vertical or horizontal straight lines.

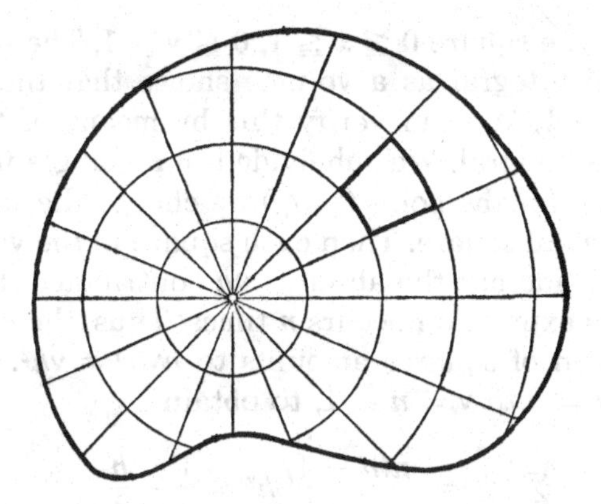

Figure 4.3 Subdivision by polar coordinate nets.

into n parts of magnitude $\Delta\theta = 2\pi/n = h$, and we also choose a second quantity $k = \Delta r$. We now draw the lines $\theta = \nu h (\nu = 0, 1, 2, \ldots, n-1)$ through the origin and also the concentric circles $r_\mu = \mu k$ $(\mu = 1, 2, \ldots)$. Those that lie entirely in the interior of R, we denote by R_i, and their areas, by ΔR_i. We can then regard the integral of the function $f(x, y)$ over the region R as the limit of the sum

$$\sum f(\xi_i, \eta_i)\Delta R_i,$$

where (ξ_i, η_i) is a point chosen arbitrarily in R_i. The sum is taken over all the subregions R_i in the interior of R, and the passage to the limit consists in letting h and k tend simultaneously to zero.

By elementary geometry the area ΔR_i is given by the equation

$$\Delta R_i = \frac{1}{2}(r_{\mu+1}{}^2 - r_\mu{}^2)h = \frac{1}{2}(2\mu + 1)k^2 h,$$

if we assume that R_i lies in the ring bounded by the circles with radii μk and $(\mu + 1)k$.

c. *Examples*

The simplest example is the function $f(x, y) = 1$. Here the limit of the sum is obviously independent of the mode of subdivision and is always equal to the area of the region R. Consequently, the integral of the function $f(x, y) = 1$ over the region is also equal to this area. This might have been expected, for the integral is the volume of the cylinder of unit altitude with the region R as base.

As a further example, we consider the integral of the function

$f(x, y) = x$ over the square $0 \leqq x \leqq 1$, $0 \leqq y \leqq 1$. The intuitive interpretation of the integral as a volume shows that the value of our integral must be $\frac{1}{2}$. We can verify this by means of the analytical definition of the integral. We subdivide the rectangle into squares of side $h = 1/n$, and for the point (ξ_i, η_i) we choose the lower left-hand corner of each small square. Then each square in the vertical column whose left-hand side has the abscissa νh contributes the amount νh^3 to the sum. This expression occurs n times. Thus, the contribution of the whole column of squares amounts to $n\nu h^3 = \nu h^2$. We now form the sum from $\nu = 0$ to $\nu = n - 1$, to obtain

$$\sum_{\nu=0}^{n-1} \nu h^2 = \frac{n(n-1)}{2} h^2 = \frac{1}{2} - \frac{h}{2}.$$

The limit of this expression as $h \to 0$ is $\frac{1}{2}$, as we stated.

In a similar way we can integrate the product xy or, more generally, any function $f(x, y)$ that can be represented as a product of a function of x and a function of y in the form $f(x, y) = \phi(x)\psi(y)$, provided that the region of integration is a rectangle with sides parallel to the axes, say $a \leqq x \leqq b$, $c \leqq y \leqq d$. We use the same division of the rectangle as in (3c), and for the value of the function in each subrectangle we take the value of the function at the lower left-hand corner. The integral is then the limit of the sum

$$hk \sum_{\nu=0}^{n-1} \sum_{\mu=0}^{m-1} \phi(\nu h)\psi(\mu k)$$

which may be written as the product of two sums in the form

$$\sum_{\nu=0}^{n-1} h\phi(\nu h) \sum_{\mu=0}^{m-1} k\psi(\mu k).$$

From the definition of the ordinary integral, as $h \to 0$ and $k \to 0$ these factors tend to the integrals of the corresponding functions over the respective intervals from a to b and from c to d. We thus obtain the general rule that if a function $f(x, y)$ can be represented as a product of two functions $\phi(x)$ and $\psi(y)$, its double integral over a rectangle $a \leqq x \leqq b$, $c \leqq y \leqq d$ can be resolved into the product of two integrals:

$$\iint_R f(x, y)\, dx\, dy = \int_a^b \phi(x)\, dx \cdot \int_c^d \psi(y)\, dy.$$

This rule and the summation rule (cf. (4b), p. 383) yield the integral of any polynomial over a rectangle with sides parallel to the axes.

As a last example, we consider a case in which it is convenient to use a subdivision by the polar coordinate net instead of a subdivision into rectangles. Let the region R be the circle with unit radius and center at the origin, given by $x^2 + y^2 \leq 1$, and let

$$f(x, y) = \sqrt{1 - x^2 - y^2}.$$

The integral of f over R is merely the volume of a hemisphere of unit radius.

We construct the polar coordinate net as before. The subregion lying between the circles with radii $r_\mu = \mu k$ and $r_{\mu+1} = (\mu + 1)k$ and between the lines $\theta = \nu h$ and $\theta = (\nu + 1)h$, where $h = 2\pi/n$ yields the contribution

$$\frac{1}{2}\sqrt{1 - \left(\frac{r_{\mu+1} + r_\mu}{2}\right)^2}\,(r_{\mu+1}^2 - r_\mu^2)h = \sqrt{1 - \rho_\mu^2}\,\rho_\mu kh,$$

where for the value of the function in the subregion R_i we have taken the value that the function assumes on an intermediate circle with the radius $\rho_\mu = (r_{\mu+1} + r_\mu)/2$. All subregions that lie in the same ring give the same contribution, and since there are $n = 2\pi/h$ such regions the contribution of the whole ring is

$$2\pi\sqrt{1 - \rho_\mu^2}\,\rho_\mu k.$$

The integral is therefore the limit of the sum

$$\sum_{\mu=0}^{m-1} 2\pi\sqrt{1 - \rho_\mu^2}\,\rho_\mu k.$$

As we already know, this sum tends to the single integral

$$2\pi\int_0^1 r\sqrt{1 - r^2}\,dr = -\frac{2\pi}{3}\sqrt{(1 - r^2)^3}\,\Big|_0^1 = \frac{2\pi}{3}.$$

We therefore obtain

$$\iint_R \sqrt{1 - x^2 - y^2}\,dR = \frac{2\pi}{3},$$

in agreement with the known formula for the volume of a sphere.

d. Notation. Extensions. Fundamental Rules

The rectangular subdivision of the region R is associated with the symbol for the double integral used since Leibnitz's time. Starting with the symbol

$$\sum_{\nu=0}^{n-1} \sum_{\mu=0}^{m-1} f(\xi_\nu, \eta_\mu) \Delta x \, \Delta y$$

for the sum over the rectangles, we indicate the passage to the limit from the sum to the integral by replacing the double summation sign by a double integral sign and writing the symbol $dx\,dy$ instead of the product of the quantities $\Delta x \, \Delta y$. Accordingly, the double integral is frequently written in the form

$$\iint_R f(x, y) \, dx \, dy$$

instead of the form

$$\iint_R f(x, y) \, dR$$

in which the area ΔR is replaced by the symbol dR. At this stage the symbol $dx\,dy$ merely refers symbolically to the passage to the limit of the above sums of nm terms as $n \to \infty$ and $m \to \infty$.

It is clear that in double integrals, just as in ordinary integrals of a single variable, the notation for the variables of integration is immaterial, so that we could equally well have written

$$\iint_R f(u, v) \, du \, dv \qquad \text{or} \qquad \iint_R f(\xi, \eta) \, d\xi \, d\eta.$$

In introducing the concept of integral, we saw that for a positive function $f(x, y)$ the integral represents the volume under the surface $z = f(x, y)$. In the analytical definition of integral, however, it is quite unnecessary that the function $f(x, y)$ should be positive everywhere; it may be negative, or it may change sign, in which case the surface intersects the region R. Thus, in the general case the integral gives the volume in question with a definite sign, the sign being positive for surfaces or portions of surfaces that lie above the x, y-plane. If the whole surface consists of several such portions, the integral represents the sum of the corresponding volumes taken with their proper signs. In particular, a double integral may vanish, although the function under the integral sign does not vanish everywhere.

For double integrals, as for single integrals, the following fundamental rules hold; their proofs are simple repetitions of those in Volume I (p. 138). If c is a constant, then

$$\iint_R cf(x, y) \, dR = c \iint_R f(x, y) \, dR. \tag{4a}$$

Furthermore, the integral of the sum of two functions is equal to the sum of their two integrals *(linearity of the operation of integration):*

$$\iint_R [f(x, y) + \phi(x, y)]\, dR = \iint_R f(x, y)\, dR + \iint_R \phi(x, y)\, dR. \tag{4b}$$

Finally, if the region R consists of two subregions R' and R'' that have at most portions of the boundary in common, then

$$\iint_R f(x, y)\, dR = \iint_{R'} f(x, y)\, dR + \iint_{R''} f(x, y)\, dR; \tag{4c}$$

that is, *when regions are joined together the corresponding integrals are added* (additivity of integrals).

e. Integral Estimates and the Mean Value Theorem

As for ordinary integrals, there are some very useful estimates for double integrals. Since the proofs are practically the same as those of Volume I (p, 138), we shall be content to merely state the facts.

If $f(x, y) \geqq 0$ in R, then

$$\iint_R f(x, y)\, dR \geqq 0; \tag{5a}$$

similarly, if $f(x, y) \leqq 0$,

$$\iint_R f(x, y)\, dR \leqq 0. \tag{5b}$$

This leads to the following result: *If the inequality*

$$f(x, y) \geqq \phi(x, y) \tag{5c}$$

holds everywhere in R, then

$$\iint_R f(x, y)\, dR \geqq \iint_R \phi(x, y)\, dR. \tag{5d}$$

A direct application of this theorem gives the relations

$$\iint_R f(x, y)\, dR \leqq \iint_R |f(x, y)|\, dR \tag{5e}$$

and

$$\iint_R f(x, y)\, dR \geqq -\iint_R |f(x, y)|\, dR. \tag{5f}$$

We can also combine these two inequalities in a single formula:

$$\left|\iint_R f(x, y)\, dR\right| \leqq \iint_R |f(x, y)|\, dR. \tag{5g}$$

If m is the greatest lower bound and M the least upper bound of the function $f(x, y)$ in R, then

$$m\, \Delta R \leqq \iint_R f(x, y)\, dR \leqq M\, \Delta R, \tag{6}$$

where ΔR is the area of the region R. The integral can then be expressed in the form

$$\iint_R f(x, y)\, dR = \mu\, \Delta R, \tag{7a}$$

where μ lies between m and M. The precise value of μ cannot in general be specified more exactly.[1]

This form of the estimation formula we again call the *mean value theorem of the integral calculus.*

Here again the following generalization holds: If $p(x, y)$ is an arbitrary positive continuous function in R, then

$$\iint_R p(x, y) f(x, y)\, dR = \mu \iint_R p(x, y)\, dR, \tag{7b}$$

where μ denotes a number between the greatest and least values of f that cannot be further specified.

As before, these integral estimates show that *the integral varies continuously with the function.* More precisely, let $f(x, y)$ and $\phi(x, y)$ be two functions that in the whole region R satisfy the inequality

$$|f(x, y) - \phi(x, y)| < \varepsilon,$$

where ε is a fixed positive number. If ΔR is the area of R, then the integrals $\iint_R f(x, y)\, dR$ and $\iint_R \phi(x, y)\, dR$ differ by less than $\varepsilon\, \Delta R$, that is, by less than a number that tends to zero with ε.

In the same way, we see that *the integral of a function* varies continuously with the region. Suppose that two regions R' and R'' are

[1]Just as for integrals of continuous functions of one variable, the value μ is certainly assumed at some point of the set R by the function $f(x, y)$ if R is connected and f is continuous.

obtained from one another by the addition or removal of portions whose total area is less than ε, and let $f(x, y)$ be a function continuous in both regions such that $|f(x, y)| < M$, where M is a fixed number. The two integrals $\iint_{R'} f(x, y)\, dR$ and $\iint_{R''} f(x, y)\, dR$ then differ by less than $M\varepsilon$, that is, by less than a number that tends to zero with ε. The proof of this fact follows at once from formula (4c) of p. 383.

We can therefore calculate the integral over a region R as accurately as we please by taking it over a subregion of R whose total area differs from the area of R by a sufficiently small amount. For example, in the region R, we can construct a polygon whose total area differs by as little as we please from the area of R. In particular, we may suppose this polygon to be bounded by lines parallel to the x- and y-axes alternately, that is, to be pieced together out of rectangles with sides parallel to the axes.

4.3 Integrals over Regions in Three and More Dimensions

Every statement we have made for integrals over regions of the x, y-plane can be extended without further complication or introduction of new ideas to regions in three or more dimensions. For example, to treat the integral over a three-dimensional region R, we need only subdivide R (e.g, by means of a finite number of surfaces with continuous nonparametric representations) into closed nonoverlapping Jordan-measurable subregions $R_1, R_2, \ldots, R_N$ that completely fill R. If $f(x,y,z)$ is a function that is continuous in the closed region R and if (ξ_i, η_i, ζ_i) denotes an arbitrary point in the region R_i, we again form the sum

$$\sum_{i=1}^{N} f(\xi_i, \eta_i, \zeta_i)\Delta R_i,$$

in which ΔR_i denotes the volume of the region R_i. The sum is taken over all the regions R_i or, if it is more convenient, only over those subregions that do not adjoin the boundary of R. If we now let the number of subregions increase beyond all bounds in such a way that the diameter of the largest of them tends to zero, we again find a limit independent of the particular mode of subdivision and of the choice of the intermediate points. This limit we call the integral of $f(x, y, z)$ over the region R, and we denote it by

$$\text{(7c)} \qquad \iiint_R f(x, y, z)\, dR.$$

In particular, if we effect a subdivision of the region into rectangular regions with sides Δx, Δy, Δz, the volumes of the inner regions R_i

will all have the same value $\Delta x\ \Delta y\ \Delta z$. As on p. 382, we indicate the passage to the limit through the notation

$$\iiint_R f(x, y, z)\, dx\, dy\, dz.$$

Apart from the necessary changes in notation, all the facts that we have mentioned for double integrals remain valid for triple integrals.

For regions of more than three dimensions, once we have suitably defined the concept of volume for such regions, the multiple integral can be defined in exactly the same way. If we restrict ourselves to rectangular subregions and define the volume of a rectangular region

$$a_i \leqq x_i \leqq a_i + h_i \qquad (i = 1, 2, \ldots, n)$$

as the product $h_1 h_2 \ldots h_n$, the definition of integral involves nothing new. We denote an integral over the n-dimensional region R by

$$\iint \cdots \int_R f(x_1, x_2, \ldots, x_n)\, dx_1\, dx_2 \cdots dx_n.$$

For more general regions and more general subdivisions we must rely on the abstract definition of volume given in the Appendix.

In what follows, we confine ourselves to integrals in at most three dimensions.

4.4 Space Differentiation. Mass and Density

For functions of one variable, the integrand is the derivative of the integral. This fact represents the fundamental connection between differential and integral calculus. For the multiple integrals of functions of several variables, the same connection exists; but here it is not so fundamental in character.

We consider the multiple integral (domain integral)

$$\iint_B f(x, y)\, dB \qquad \text{or} \qquad \iiint_B f(x, y, z)\, dB$$

of a continuous function of two or three variables over a region B that contains a fixed point P with coordinates (x_0, y_0) or (x_0, y_0, z_0), respectively, and which has the content ΔB. Dividing this integral by the content ΔB, it follows from formula (7a) that the quotient is an intermediate value of the integrand, that is, a number between the greatest and the least values of the integrand in the region. If we let the diameter of the region B about the point P tend to zero, so that the content ΔB also tends to zero, this intermediate value of the function f must tend to its value at the point P. Thus, the passage to the limit yields the respective relations

$$\lim_{\Delta B\to 0}\frac{1}{\Delta B}\iint_B f(x,y)dB = f(x_0, y_0)$$

and

$$\lim_{\Delta B\to 0}\frac{1}{\Delta B}\iiint_B f(x,y,z)dB = f(x_0, y_0, z_0). \tag{8}$$

This limiting process, which parallels the process of differentiation for integrals with one independent variable, we call *space differentiation* of the integral. We see, then, that space *differentiation of a multiple integral gives the integrand.*

We can interpret the relation of integrand to integral in the case of several independent variables, by means of the physical concepts of *density* and *total mass*. We think of a mass of a substance as distributed over a three-dimensioned region R in such a way that an arbitrarily small mass in contained in each sufficiently small subregion. In order to define the specific mass or density at a point P, we first consider a neighborhood B of the point P with content ΔB and divide the total mass in this neighborhood by the content. The quotient we shall call the *mean density* or *average density* in this subregion. If we now let the diameter of B tend to zero, from the average density in the region B we obtain a limit called the *density* at the point P, provided always that such a limit exists independently of the choice of the sequence of regions. If we denote this density by $\mu(x, y, z)$ and assume that it is continuous, we see at once that the process described above yields the same value as the differentiation of the integral

$$\iiint_R \mu(x,y,z)\,dV,$$

taken over the whole region R. This integral taken over the whole region therefore represents the *total mass* of the substance of density μ in the region[1] R.

[1]What we have shown is only that the distribution given by the multiple integral has the same space-derivative as the mass-distribution originally given. It remains to be proved that this implies that the two distributions are actually identical; in other words, that the statement "space differentiation gives the density μ" can be satisfied by only one distribution of mass. The proof, although not difficult, is passed over here. We have to assume that mass is *additive*, that is, that for a region R consisting of two nonoverlapping regions R' and R'', the mass of R is the sum of the masses of R' and R''.

From the physical point of view such a representation of the mass of a substance is naturally an idealization. That this idealization is reasonable, that is, that it approximates to the actual situation with sufficient accuracy, is one of the assumptions of physics.

These ideas, moreover, retain their mathematical significance even when μ is not positive everywhere. Negative densities and masses may also have a physical interpretation, for example, in the study of the distribution of electric charge.

4.5 Reduction of the Multiple Integral to Repeated Single Integrals

The fact that every multiple integral can be reduced to single integrals is of fundamental importance in the evaluation of multiple integrals. It enables us to apply all the methods that we have previously developed for finding indefinite integrals to the evaluation of multiple integrals.

a. *Integrals over a Rectangle*

First we take the region R as a rectangle $a \leqq x \leqq b$, $\alpha \leqq y \leqq \beta$ in the x, y-plane and consider a continuous function $f(x, y)$ in R. We then have the theorem:

To find the double integral of $f(x, y)$ over the region R, We first regard y as constant and integrate $f(x, y)$ with respect to x between the limits a and b. This integral

$$\phi(y) = \int_a^b f(x, y)\, dx$$

is a function of the parameter y, which we integrate between the limits α and β to obtain the double integral. In symbols,

$$\iint_R f(x, y)\, dR = \int_\alpha^\beta \phi(y)\, dy, \qquad \phi(y) = \int_a^b f(x, y)\, dx,$$

or more briefly,

$$\iint_R f(x, y)\, dR = \int_\alpha^\beta dy \int_a^b f(x, y)\, dx. \tag{9a}$$

In order to prove this statement, we return to the definition of the multiple integral (3c). Taking

$$h = \frac{b - a}{m} \quad \text{and} \quad k = \frac{\beta - \alpha}{n},$$

we have

$$\iint_R f(x, y)\, dR = \lim_{\substack{m \to \infty \\ n \to \infty}} \sum_{\nu=1}^{n} \sum_{\mu=1}^{m} f(a + \mu h, \alpha + \nu k)\, hk.$$

Here the limit is to be understood to mean that the sum on the right-hand side differs from the value of the integral by less than an arbitrarily small preassigned positive quantity ε, provided only that the numbers m and n are both larger than a bound N depending only on ε. By introducing the expression[1]

$$\Phi_\nu = \sum_{\mu=1}^{m} f(a + \mu h, \alpha + \nu k)\, h$$

we can write this sum in the form

$$\sum_{\nu=1}^{n} \Phi_\nu k.$$

If we now choose an arbitrary fixed value for ε and for n choose a fixed number greater than N, we know that

$$\left| \iint_R f(x, y)\, dR - k \sum_{\nu=1}^{n} \Phi_\nu \right| < \varepsilon$$

no matter what the number m is, provided only that it is greater than N. If we keep n fixed and let m tend to infinity, the above expression never exceeds ε. According to the definition of the ordinary integral, however, in this limiting process the expression Φ_ν tends to the integral

$$\int_a^b f(x, \alpha + \nu k)\, dx = \phi(\alpha + \nu k),$$

and, therefore, we obtain

$$\left| \iint_R f(x, y)\, dR - k \sum_{\nu=1}^{n} \phi(\alpha + \nu k) \right| \leqq \varepsilon.$$

[1]The root idea of the following proof is simply that of resolving the double limit as m and n increase simultaneously into the two successive single limiting processes: first, $m \to \infty$ when n is fixed, and then, $n \to \infty$.

whatever the value of ε, this inequality holds for all values of n that are greater than a fixed number N depending only on ε. If we now let n tend to ∞ (i.e., let k tend to zero), then by the definition of "integral" and the continuity (see p. 74) of

$$\int_a^b f(x, y)\, dx = \phi(y)$$

we obtain

$$\lim_{n\to\infty} k \sum_{\nu=1}^{n} \phi(\alpha + \nu k) = \int_\alpha^\beta \phi(y)\, dy;$$

whence

$$\left| \iint_R f(x, y)\, dR - \int_\alpha^\beta \phi(y)\, dy \right| \leq \varepsilon.$$

Since ε can be chosen as small as we please and the left-hand side is a fixed number, this inequality can only hold if the left-hand side vanishes, that is, if

$$\iint_R f(x, y)\, dR = \int_\alpha^\beta dy \int_a^b f(x, y)\, dx.$$

This gives the required transformation.

The result permits one to *reduce double integration to two successive single integrations.*

Since the parts played by x and y are interchangeable, no further proof is required to show that the equation

$$\iint_R f(x, y)\, dR = \int_a^b dx \int_\alpha^\beta f(x, y)\, dy \tag{9b}$$

is also true.

b. Change of Order of Integration. Differentiation under the Integral Sign

The two formulae (9a), (9b) yield the relation

$$\int_\alpha^\beta dy \int_a^b f(x, y)\, dx = \int_a^b dx \int_\alpha^\beta f(x, y)\, dy \tag{9c}$$

(already proved in a different way on p. 80) or, in words:

In the repeated integration of a continuous function with constant limits of integration the order of integration can be reversed.

The theorem on the change of order in integration has many applications. In particular, it is frequently used in the explicit calculation of simple definite integrals for which no indefinite integral can be found.

As an example (for further examples see the Appendix), we consider the integral

$$I = \int_0^\infty \frac{e^{-ax} - e^{-bx}}{x}\, dx,$$

which converges for $a > 0$, $b > 0$. We can express I as a repeated integral in the form

$$I = \int_0^\infty dx \int_a^b e^{-xy}\, dy.$$

In this improper repeated integral we cannot at once apply our theorem on change of order. If, however, we write

$$I = \lim_{T\to\infty} \int_0^T dx \int_a^b e^{-xy}\, dy,$$

we obtain by changing the order of integration

$$I = \lim_{T\to\infty} \int_a^b \frac{1 - e^{-Ty}}{y}\, dy = \log\frac{b}{a} - \lim_{T\to\infty} \int_a^b \frac{e^{-Ty}}{y}\, dy.$$

In virtue of the relation

$$\int_a^b \frac{e^{-Ty}}{y}\, dy = \int_{Ta}^{Tb} \frac{e^{-y}}{y}\, dy,$$

the second integral tends to zero as T increases; hence,

$$I = \int_0^\infty \frac{e^{-ax} - e^{-bx}}{x}\, dx = \log\frac{b}{a}. \tag{11a}$$

In a similar way we can prove the following general theorem:

If $f(t)$ is sectionally smooth for $t \geqq 0$ and if the integral

$$\int_1^\infty \frac{f(t)}{t}\, dt$$

exists, then for positive a and b

$$I = \int_0^\infty \frac{f(ax) - f(bx)}{x}\, dx = f(0) \log \frac{b}{a}. \tag{11b}$$

Here we can again express the single integral as the repeated integral

$$I = \int_0^\infty dx \int_b^a f'(xy)\, dy$$

and change the order of integration.

c. Reduction of Double to Single Integrals for More General Regions

By a simple extension of the results already obtained, we can derive analogous results for regions more general than rectangles. We begin by considering a *convex region R*, that is, a region whose boundary curve is not cut by any straight line in more than two points unless the whole straight line between these two points is a part of the boundary (Fig. 4.4). We suppose that the region lies between the *lines of*

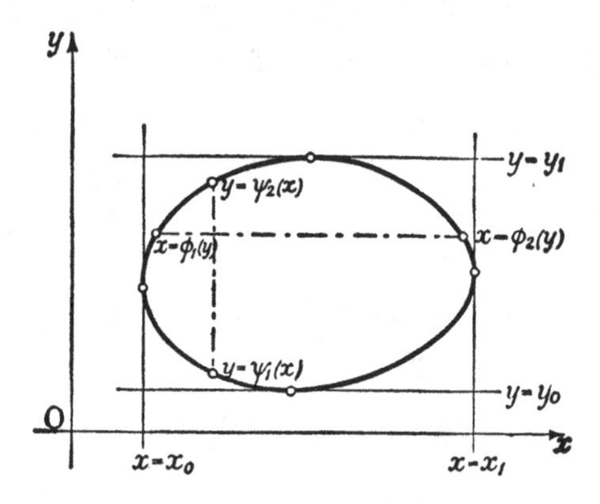

Figure 4.4 General convex region of integration.

support (i.e., lines containing a boundary point of R but not separating any two points of R) $x = x_0$, $x = x_1$ and $y = y_0$, $y = y_1$, respectively. Since the x-coordinate for any point of R lies in the interval $x_0 \leqq x \leqq x_1$ and the y-coordinate in the interval $y_0 \leqq y \leqq y_1$, we consider the integrals

$$\int_{\phi_1(y)}^{\phi_2(y)} f(x, y)\, dx$$

and

$$\int_{\psi_1(x)}^{\psi_2(x)} f(x, y)\, dy,$$

which are taken along the segments in which the lines $y =$ constant and $x =$ constant, respectively, intersect the region. Here $\phi_2(y)$ and $\phi_1(y)$ denote the abscissae of the points in which the boundary of the region is intersected by the line $y =$ constant, and $\psi_2(x)$ and $\psi_1(x)$ the ordinates of the points in which the boundary is intersected by the lines $x =$ constant. The integral

$$\int_{\phi_1(y)}^{\phi_2(y)} f(x, y)\, dx$$

is therefore a function of the parameter y, where the parameter appears both under the integral sign and in the upper and lower limits, and a similar statement holds for the integral

$$\int_{\psi_1(x)}^{\psi_2(x)} f(x, y)\, dy$$

as a function of x. The resolution into repeated integrals is then given by the equations

$$\text{(12)} \qquad \iint_R f(x, y)\, dR = \int_{y_0}^{y_1} dy \int_{\phi_1(y)}^{\phi_2(y)} f(x, y)\, dx$$

$$= \int_{x_0}^{x_1} dx \int_{\psi_1(x)}^{\psi_2(x)} f(x, y)\, dy.$$

To prove this we first choose a sequence of points on the arc $y = \psi_2(x)$, the distance between successive points being less than a positive number δ. We join successive points by paths, each consisting of a horizontal and a vertical line segment lying in R. The lower boundary $y = \psi_1(x)$, we treat similarly, choosing points with the same abscissae as on the upper boundary. We thus obtain a region $\bar{R}$ in R, consisting of a finite number of rectangles, where the boundary of $\bar{R}$ above and below is presented by sectionally constant functions $y = \bar{\psi}_2(x)$ and $y = \bar{\psi}_1(x)$, respectively (cf. Fig. 4.5). By the known theorem for rectangles, we have

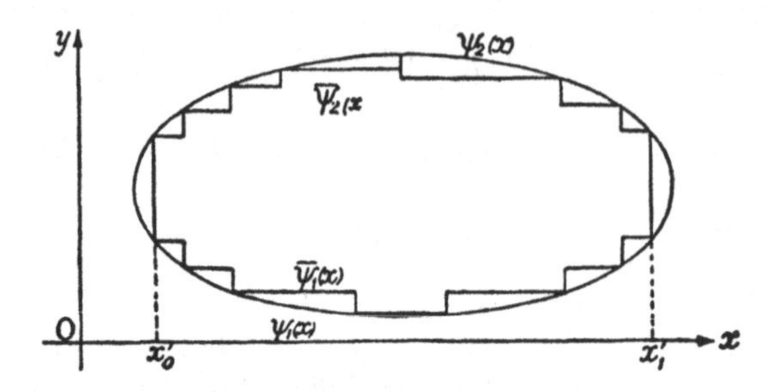

Figure 4.5

$$\iint_{\bar{R}} f(x, y)\, dR = \int_{x_0'}^{x_1'} dx \int_{\bar{\psi}_1(x)}^{\bar{\psi}_2(x)} f(x, y)\, dy.$$

Since $\psi_1(x)$ and $\psi_2(x)$ are uniformly continuous, as $\delta \to 0$, the functions $\bar{\psi}_1(x)$ and $\bar{\psi}_2(x)$ tend uniformly to $\psi_1(x)$ and $\psi_2(x)$, respectively, and so,

$$\lim_{\delta \to 0} \int_{\bar{\psi}_1(x)}^{\bar{\psi}_2(x)} f(x, y)\, dy = \int_{\psi_1(x)}^{\psi_2(x)} f(x, y)\, dy$$

uniformly in x. It follows that

$$\lim_{\delta \to 0} \int_{x_0'}^{x_1'} dx \int_{\bar{\psi}_1(x)}^{\bar{\psi}_2(x)} f(x, y)\, dx = \int_{x_0}^{x_1} dx \int_{\psi_1(x)}^{\psi_2(x)} f(x, y)\, dx.$$

On the other hand, as $\delta \to 0$, the region $\bar{R}$ tends to R. Hence,

$$\lim_{\delta \to 0} \iint_{\bar{R}} f(x, y)\, dR = \iint_R f(x, y)\, dR.$$

Combining the three equations, we have

$$\iint_R f(x, y)\, dR = \int_{x_0}^{x_1} dx \int_{\psi_1(x)}^{\psi_2(x)} f(x, y)\, dy.$$

The other statement can be established in a similar way.

A similar argument is available if we abandon the hypothesis of convexity and consider regions of the form indicated in Fig. 4.6. We assume merely that the boundary curve of the region is intersected by every parallel to the x-axis and by every parallel to the y-axis in a bounded number of points or intervals. By $\int f(x, y)\, dy$, we then mean the sum of the integrals of the function $f(x, y)$ for a fixed x, taken over all the intervals that the line x = constant has in common with the closed region. For nonconvex regions the number of these intervals

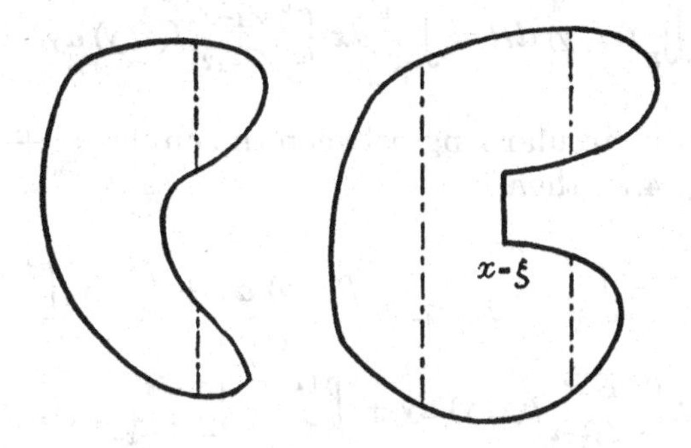

Figure 4.6 Nonconvex regions of integration.

may exceed unity. It may change suddenly at a point $x = \xi$ (as in fig. 4.6, right) in such a way that the expression $\int f(x, y)\,dy$ has a jump-discontinuity at this point. Without essential changes in the proof, however, the resolution of the double integral

$$\iint_R f(x, y)\,dR = \int dx \int f(x, y)\,dy$$

remains valid, the integration with respect to x being taken along the whole interval $x_0 \leqq x \leqq x_1$ over which the region R lies. Naturally, the corresponding resolution

$$\iint_R f(x, y)\,dR = \int dy \int f(x, y)\,dx$$

also holds.

In the example of the circle defined by $x^2 + y^2 \leqq 1$, we have

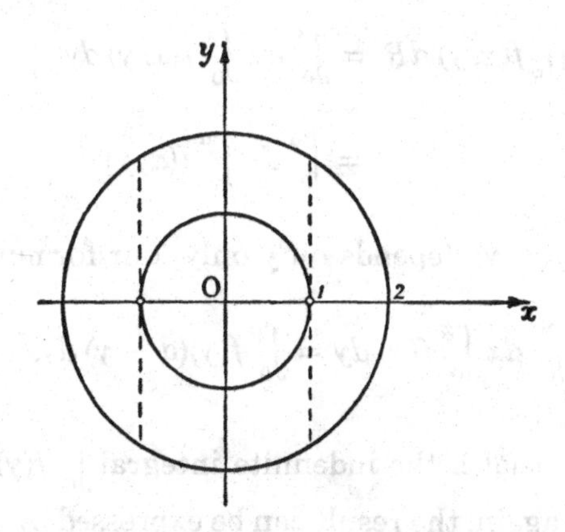

Figure 4.7 Circular ring as region of integration.

$$\iint_R f(x, y)\, dR = \int_{-1}^{+1} dx \int_{-\sqrt{1-x^2}}^{+\sqrt{1-x^2}} f(x, y)\, dy.$$

If the region is a circular ring between the circles $x^2 + y^2 = 1$ and $x^2 + y^2 = 4$ (Fig. 4.7), then

$$\iint_R f(x, y)\, dx\, dy = \int_{-2}^{-1} dx \int_{-\sqrt{4-x^2}}^{+\sqrt{4-x^2}} f(x, y)\, dy + \int_{1}^{2} dx \int_{-\sqrt{4-x^2}}^{+\sqrt{4-x^2}} f(x, y)\, dy$$

$$+ \int_{-1}^{+1} dx \int_{-\sqrt{4-x^2}}^{\sqrt{1-x^2}} f(x, y)\, dy + \int_{-1}^{+1} dx \int_{-\sqrt{1-x^2}}^{+\sqrt{4-x^2}} f(x, y)\, dy.$$

As a final example we take as the region R a triangle (Fig. 4.8) bounded by the lines $x = y$, $y = 0$, and $x = \alpha$ ($\alpha > 0$). Integrating either first with respect to x, or with respect to y, we obtain

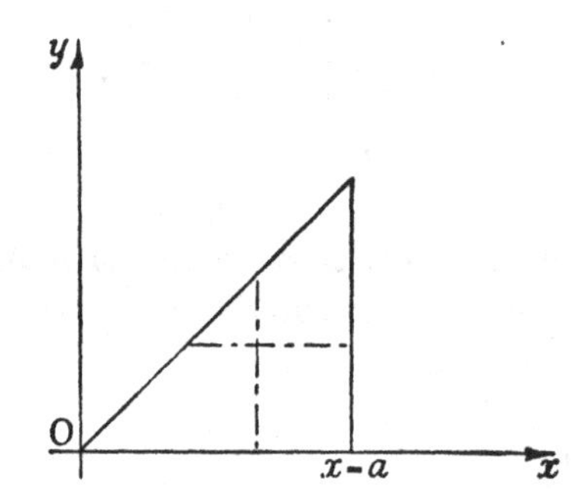

Figure 4.8 Triangle as region of integration.

(13a)
$$\iint_R f(x, y)\, dR = \int_0^{\alpha} dx \int_0^{x} f(x, y)\, dy$$
$$= \int_0^{\alpha} dy \int_y^{\alpha} f(x, y)\, dx.$$

In particular, if $f(x, y)$ depends on y only, our formula gives

(13b)
$$\int_0^{\alpha} dx \int_0^{x} f(y)\, dy = \int_0^{\alpha} f(y)(\alpha - y)\, dy.$$

From this we see that if the indefinite integral $\int_0^x f(y)\, dy$ of a function $f(y)$ is integrated again, the result can be expressed by a single integral (cf. Volume I, p. 320).

d. Extension of the Results to Regions in Several Dimensions

The corresponding theorems in more than two dimensions are so closely analogous to those already given that it is sufficient to state them without proof. If we first consider the rectangular region $x_0 \leqq x \leqq x_1, y_0 \leqq y \leqq y_1, z_0 \leqq z \leqq z_1$, and a function $f(x,y,z)$ continuous in this region, we can reduce the triple integral

$$V = \iiint_R f(x, y, z)\, dR$$

in several ways to single integrals or double integrals. Thus,

$$\iiint_R f(x, y, z)\, dR = \int_{z_0}^{z_1} dz \iint_B f(x, y, z)\, dx\, dy. \tag{14a}$$

Here

$$\iint_B f(x, y, z)\, dx\, dy$$

is the double integral of the function taken over the rectangle B described by $x_0 \leqq x \leqq x_1$, $y_0 \leqq y \leqq y_1$, z being kept constant as a parameter during this integration so that the double integral is a function of the parameter z. Either of the remaining coordinates x and y can be singled out in the same way.

Moreover, the triple integral V can also be represented as a repeated integral in the form of a succession of three single integrations. In this representation we first consider the expression

$$\int_{z_0}^{z_1} f(x, y, z)\, dz,$$

x and y being fixed, and then consider

$$\int_{y_1}^{y_1} dy \int_{z_0}^{z_1} f(x, y, z)\, dz,$$

x being fixed. We finally obtain

$$V = \int_{x_0}^{x_1} dx \int_{y_0}^{y_1} dy \int_{z_0}^{z_1} f(x, y, z)\, dz. \tag{14b}$$

In this repeated integral we could equally well have integrated first with respect to x, then with respect to y, and finally with respect to z and we could have made any other change in the order of integration, since the repeated integral is always equal to the triple integral. We therefore have the following theorem:

A repeated integral of a continuous function throughout a closed rectangular region is independent of the order of integration.

The way in which the resolution is to be performed for nonrectangular regions in three dimensions scarcely requires special mention.[1] We content ourselves with writing down the resolution for a spherical region $x^2 + y^2 + z^2 \leq 1$:

$$(15) \quad \iiint_R f(x, y, z)\, dx\, dy\, dz = \int_{-1}^{+1} dx \int_{-\sqrt{1-x^2}}^{+\sqrt{1-x^2}} dy \int_{-\sqrt{1-x^2-y^2}}^{+\sqrt{1-x^2-y^2}} f(x, y, z)\, dz.$$

4.6 Transformation of Multiple Integrals

a. *Transformation of Integrals in the Plane*

The introduction of a new variable of integration is one of the chief methods for transforming and simplifying single integrals. The introduction of new variables is also extremely important for multiple integrals. In spite of their reduction to single integrals, the explicit evaluation of multiple integrals is generally more difficult than for one independent variable and integration in terms of elementary functions is less likely. Yet often we can evaluate such integrals by introducing new variables in place of the original ones under the integral sign. Quite apart from the question of the explicit evaluation of double integrals, the transformation theory is important for the complete mastery of the concept of integral that it gives us.

The important special transformation to polar coordinates has already been indicated on p. 378. Here we shall proceed at once to general transformations. First, we consider the case of a double integral

$$\iint_R f(x, y)\, dR = \iint f(x, y)\, dx\, dy,$$

taken over a region R of the x, y-plane. Let the equations

$$x = \phi(u, v), \qquad y = \psi(u, v)$$

give a 1–1 mapping of the region R onto the closed region R' of the u, v-plane. We assume that in the region R the functions ϕ and ψ have continuous partial derivatives of the first order and that their Jacobian

$$D = \begin{vmatrix} \phi_u & \phi_v \\ \psi_u & \psi_v \end{vmatrix} = \phi_u \psi_v - \psi_u \phi_v$$

[1]For a general proof, see the Appendix, p. 531.

never vanishes in R. More precisely, we made the assumption, that the system of functions $x = \phi(u, v)$, $y = \psi(u, v)$ possesses a unique inverse $u = g(x, y)$, $v = h(x, y)$ (p. 261). Moreover, the two families of curves u = constant and v = constant form a net over the region R.

Heuristic considerations readily suggest how the integral $\iint_R f(x, y)dR$ can be expressed as an integral with respect to u and v. We naturally think of calculating the double integral $\iint f(x, y)dR$ by abandoning the rectangular subdivision of the region R and instead using a subdivision into subregions R_i by means of curves of the net u = constant or v = constant. We therefore consider the values $u = \nu h$ and $v = \mu k$, where $h = \Delta u$ and $k = \Delta v$ are given numbers and ν and μ take all integer values such that the lines $u = \nu h$ and $v = \mu k$ intersect R' (so that their images are curves in R). These curves define a number of meshes, and for the subregions R_i we choose those meshes that lie in the interior of R (Figs. 4.9 and 4.10). We now have to find the area of such a mesh.

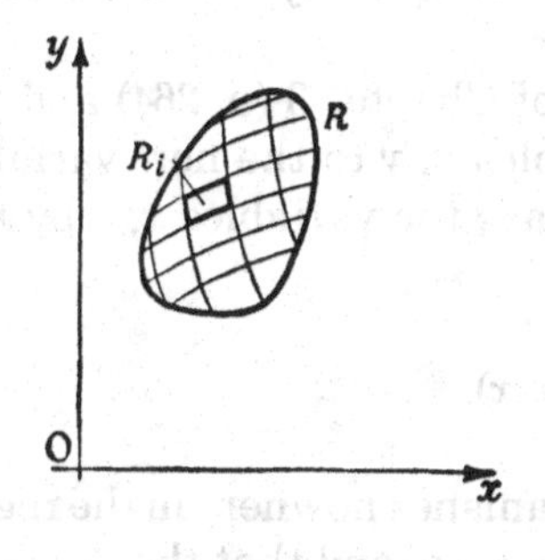

Figure 4.9

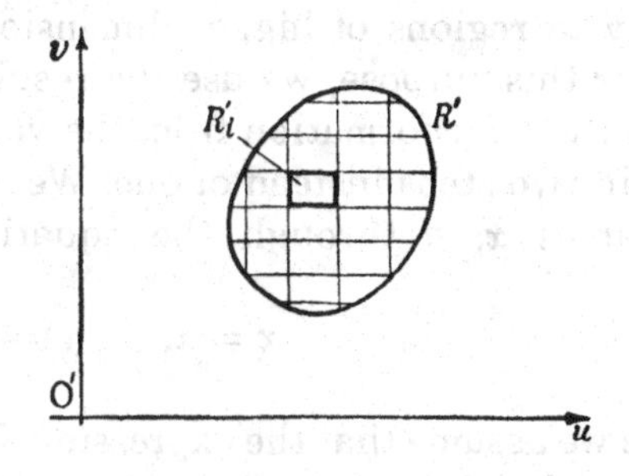

Figure 4.10

If the mesh, instead of being bounded by curves, were a parallelogram with vertices corresponding to the values (u_ν, v_μ), $(u_\nu + h, v_\mu)$, $(u_\nu, v_\mu + k)$, and $(u_\nu + h, v_\mu + k)$, then by a formula of analytical geometry (cf. Chapter 2, p. 180) the area of the mesh would be the absolute value of the determinant

$$\begin{vmatrix} \phi(u_\nu + h, v_\mu) - \phi(u_\nu, v_\mu) & \phi(u_\nu, v_\mu + k) - \phi(u_\nu, v_\mu) \\ \psi(u_\nu + h, v_\mu) - \psi(u_\nu, v_\mu) & \psi(u_\nu, v_\mu + k) - \psi(u_\nu, v_\mu) \end{vmatrix},$$

which is approximately equal to

$$\begin{vmatrix} \phi_u(u_\nu, v_\mu) & \phi_v(u_\nu, v_\mu) \\ \psi_u(u_\nu, v_\mu) & \psi_v(u_\nu, v_\mu) \end{vmatrix} hk = hkD.$$

On multiplying this expression by the value of the function f in the corresponding mesh, summing over all the regions R_i lying entirely within R, and then passing to the limit as $h \to 0$ and $k \to 0$, we obtain the expression

$$\iint_R f(\phi(u, v), \psi(u, v))\,|D|\; du\, dv$$

for the integral transformed to the new variables.

This discussion is incomplete, however, since we have not shown that it is permissible to replace the curvilinear meshes by parallelograms or to replace the area of such a parallelogram by the expression $|\phi_u\psi_v - \psi_u\,\phi_v|\,hk$; that is, we have not shown that the error introduced in this way vanishes in the limit as $h \to 0$ and $k \to 0$. Instead of completing the proof by making the proper estimates (which will be done in the Appendix), we prefer to prove the transformation formula in a somewhat different way, one that can subsequently be extended directly to regions of higher dimensions.

For this purpose, we use the results of Chapter 3 (p. 264) and perform the transformation from the variables x, y to the new variables u, v in two steps instead of one. We replace the variables x, y by new variables x, v through the equations

$$x = x, \qquad y = \Phi(v, x).$$

Here we assume that the expression Φ_v vanishes nowhere in the region R, say, that Φ_v is everywhere greater than zero, and that the whole region R can be mapped in a 1–1 way on the region B of the x, v-plane. We then map this region B in a 1–1 way on the region R' of the u, v-plane by means of a second transformation

$$x = \Psi(u, v), \qquad v = v,$$

where we further assume that the expression Ψ_u is positive throughout the region B. We now effect the transformation of the integral $\iint_R f(x,y)\,dx\,dy$ in two steps. We start with a subdivision of the region B into rectangular subregions of sides $\Delta x = h$ and $\Delta v = k$ bounded by the lines $x = \text{constant} = x_\nu$ and $v = \text{constant} = v_\mu$ in the x, v-plane. This subdivision of B corresponds to a subdivision of the region R into subregions R_i, each subregion being bounded by two parallel lines $x = x_\nu$ and $x = x_\nu + h$ and by arcs of the two curves $y = \Phi(v_\mu, x)$ and $y = \Phi(v_\mu + k, x)$ (Figs. 4.11 and 4.12). By the elementary inter-

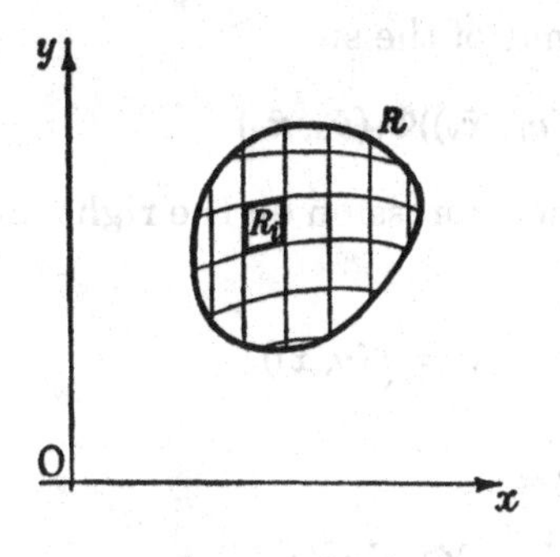

Figure 4.11

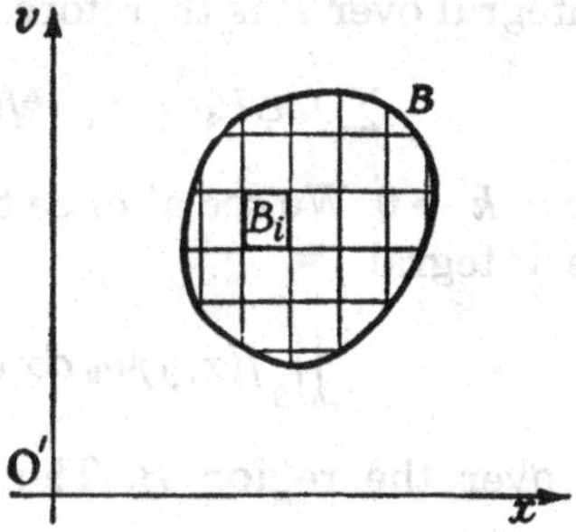

v
B
B_i
O′
x

Figure 4.12

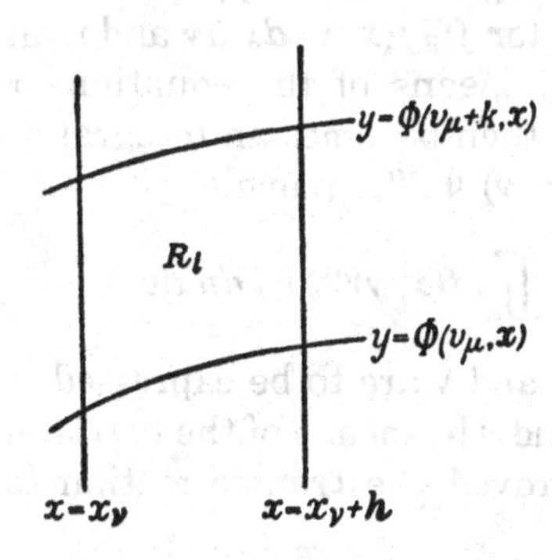

Figure 4.13

pretation of the single integral, the area of the subregion (Fig. 4.13) is

$$\Delta R_i = \int_{x_\nu}^{x_\nu+h} [\Phi(v_\mu + k, x) - \Phi(v_\mu, x)]\, dx.$$

By the mean value theorem of the integral calculus, this can be written in the form

$$\Delta R_i = h[\Phi(v_\mu + k, \bar{x}_\nu) - \Phi(v_\mu, \bar{x}_\nu)],$$

where $\bar{x}_\nu$ is a number between x_ν and $x_\nu + h$. By the mean value theorem of the differential calculus, this finally becomes

$$\Delta R_i = hk\Phi_v(\bar{v}_\mu, \bar{x}_\nu),$$

in which $\bar{v}_\mu$ denotes a value between v_μ and $v_\mu + k$, so that $(\bar{v}_\mu, \bar{x}_\mu)$ are the coordinates of a point of the subregion in B under consideration.

The integral over R is therefore the limit of the sum

$$\sum f_i \, \Delta R_i = \sum hkf(\bar{x}_\nu, \Phi(\bar{v}_\mu, \bar{x}_\nu))\Phi_v(\bar{v}_\mu, \bar{x}_\nu)$$

as $h \to 0$, $k \to 0$. We see at once that the expression on the right tends to the integral

$$\iint_B f(x, y)\Phi_v \, dx \, dv \qquad (y = \Phi(v, x))$$

taken over the region B. Therefore,

$$\iint_R f(x, y) \, dx \, dy = \iint_B f(x, y)\Phi_v \, dx \, dv.$$

To the integral on the right we now apply exactly the same argument as that just employed for $\iint_R f(x, y) \, dx \, dy$ and transform the region B into the region R' by means of the equations $x = \Psi(u, v)$, $v = v$.

The integral over B then becomes an integral over R' with an integrand of the form $f(x, y) \, \Phi_v \Psi_u$, namely,

$$\iint_{R'} f(x, y)\Phi_v\Psi_u \, du \, dv.$$

Here the quantities x and y are to be expressed in terms of the independent variables u and v by means of the two transformations above. We have therefore proved the transformation formula

$$\text{(16a)} \qquad \iint_R f(x, y) \, dx \, dy = \iint_{R'} f(x, y)\Phi_v\Psi_u \, du \, dv.$$

By introducing the direct transformation $x = \phi(u, v)$, $y = \psi(u, v)$ the formula can at once be put in the form stated previously. For

$$\frac{d(x, y)}{d(x, v)} = \Phi_v \qquad \text{and} \qquad \frac{d(x, v)}{d(u, v)} = \Psi_u,$$

and so, by Chapter 3 (p. 258), we have

$$D = \frac{d(x, y)}{d(u, v)} = \Phi_v\Psi_u.$$

We have therefore established the transformation formula whenever the transformation $x = \phi(u, v)$, $y = \psi(u, v)$ can be resolved into a succession of two primitive transformations of the forms[1] $x = x$, $y = \Phi(v, x)$ and $v = v$, $x = \Psi(u, v)$.

[1]We have assumed above that the two derivatives Φ_v and Φ_u are positive, but we easily see that this is not a serious restriction. If it is not satisfied, we merely have to replace $\Phi_v\Psi_u$ by its absolute value in formula (16a).

In Chapter 3 (p. 265), however, we saw that for $D \neq 0$ we can subdivide a closed region R into a finite number of regions in each of which such a resolution is possible, except perhaps that it may be necessary to interchange u and v, but this does not affect the value of the integral. We thus arrive at the following general result:

If the transformation $x = \phi(u, v)$, $y = \psi(u, v)$ represents a continuous 1–1 mapping of the closed Jordan-measurable region R of the x, y-plane on a region R' of the u, v-plane, and if the functions ϕ and ψ have continuous first derivatives and their Jacobian

$$\frac{d(x, y)}{d(u, v)} = \phi_u \psi_v - \psi_u \phi_v$$

is everywhere different from zero, then

$$\iint_R f(x, y)\, dx\, dy = \iint_{R'} f(\phi(u, v), \psi(u, v)) \left| \frac{d(x, y)}{d(u, v)} \right| du\, dv. \tag{16b}$$

For completeness we add that the transformation formula remains valid if the determinant $d(x, y)/d(u, v)$ vanishes without reversing its sign at a finite number of isolated points of the region, for then we have only to cut these points out of R by enclosing them in small circles of radius ρ. The proof is valid for the residual region. If we then let ρ tend to zero, the transformation formula continues to hold for the region R by virtue of the continuity of all the functions involved. This fact permits us to introduce polar coordinates with the origin in the interior of the region; for the Jacobian, being equal to r, vanishes at the origin.

In Chapter 5 we shall return to transformations of integrals and assign a role to the sign of the Jacobian in connection with integrals over *oriented* manifolds. A different method of proving the transformation formula will be given in the Appendix.

b. Regions of More than Two Dimensions

We can, of course, proceed in the same way with regions in space of three or more dimensions and obtain the following general result:

If a closed Jordan-measurable region R of $x, y, z, \ldots$-space is mapped on a region R' of $u, v, w, \ldots$-space by a 1–1 transformation whose Jacobian

$$\frac{d(x, y, z, \ldots)}{d(u, v, w, \ldots)}$$

is everywhere different from zero, then the transformation formula

$$\iint \cdots \int_R f(x, y, z, \ldots)\, dx\, dy\, dx \ldots \tag{17}$$

$$= \iint \cdots \int_{R'} f(x, y, z, \ldots) \left| \frac{d(x, y, z, \ldots)}{d(u, v, w, \ldots)} \right| du\, dv\, dw \ldots$$

holds.

As a special application, we can obtain the *transformation formulas for polar and spherical coordinates.* For polar coordinates in the plane, we write r and θ instead of u and v, and at once obtain $\frac{\partial(x, y)}{\partial(r, \theta)} = r$ (cf. p. 253). For the spherical coordinates in space, defined by the equations

$$x = r \cos \phi \sin \theta, \qquad y = r \sin \phi \sin \theta, \qquad z = r \cos \theta,$$

in which ϕ ranges from 0 to 2π, θ from 0 to π, and r from 0 to $+\infty$, we identify u, v, w with r, θ, ϕ; for the Jacobian we then obtain

$$\frac{d(x, y, z)}{d(r, \theta, \phi)} = \begin{vmatrix} \cos \phi \sin \theta & r \cos \phi \cos \theta & -r \sin \phi \sin \theta \\ \sin \phi \sin \theta & r \sin \phi \cos \theta & r \cos \phi \sin \theta \\ \cos \theta & -r \sin \theta & 0 \end{vmatrix} = r^2 \sin \theta.$$

(The value $r^2 \sin \theta$ is easily obtained by expanding in terms of the minors of the third column.) The transformation to spherical coordinates in space is therefore given by the formula

$$\iiint_R f(x, y, z)\, dx\, dy\, dz = \iiint_{R'} f(x, y, z) r^2 \sin \theta\, dr\, d\theta\, d\phi.$$

As in the corresponding case in the plane, we can also arrive at the transformation formula without using the general theory. We have only to start with a subdivision of space given by the spheres $r =$ constant, the cones $\theta =$ constant, and the planes $\phi =$ constant. The details of this elementary method can be left to the reader.

For spherical coordinates our assumptions are not satisfied when $r = 0$ or $\theta = 0, \pi$ since the Jacobian then vanishes. As in the case of the plane, we can easily convince ourselves that the transformation formula nonetheless remains valid.

Exercises 4.6

1. Perform the following integrations:

   (a) $\int_0^a \int_0^b xy(x^2 - y^2)\,dy\,dx$

   (b) $\int_0^\pi \int_0^\pi \cos(x + y)\,dy\,dx$

   (c) $\int_0^e \int_0^2 \frac{1}{xy}\,dy\,dx$

   (d) $\int_0^a \int_0^b xe^{xy}\,dy\,dx$

   (e) $\int_0^1 \int_0^{\sqrt{1-x^2}} y^2\,dy\,dx.$

   (f) $\int_0^2 \int_0^{2-x} y\,dy\,dx$

2. $\iint x^2y^2\,dx\,dy$ over the circle $x^2 + y^2 \leqq 1$.

3. $\iint \frac{x^3 + y^3 - 3xy(x^2 + y^2)}{(x^2 + y^2)^{3/2}}\,dx\,dy$ over the circle $x^2 + y^2 \leqq 1$.

4. Find the volume between the x, y-plane and the paraboloid $z = 2 - x^2 - y^2$.

5. Evaluate the integral

$$\iint \frac{dx\,dy}{(1 + x^2 + y^2)^2}$$

   taken

   (a) over one loop of the lemniscate $(x^2 + y^2)^2 - (x^2 - y^2) = 0$,

   (b) over the triangle with vertices $(0, 0)$, $(2, 0)$, $(1, \sqrt{3})$.

6. Evaluate the integral

$$\iiint |xyz|\,dx\,dy\,dz$$

   taken throughout the ellipsoid $x^2/a^2 + y^2/b^2 + z^2/c^2 \leqq 1$.

7. Find the volume common to the two cylinders $x^2 + z^2 < 1$ and $y^2 + z^2 < 1$.

8. By integration, find the volume of the smaller of the two portions into which a sphere of radius r is cut by a plane whose perpendicular distance from the center is $h(<r)$.

9. $\iiint (x^2 + y^2 + z^2)\,xyz\,dx\,dy\,dz$ throughout the sphere $x^2 + y^2 + z^2 \leqq r^2$.

10. $\iiint z\,dx\,dy\,dz$ throughout the region defined by the inequalities $x^2 + y^2 \leqq z^2$, $x^2 + y^2 + z^2 \leqq 1$.

11. $\iiint (x+y+z)\,x^2y^2z^2\,dx\,dy\,dz$ throughout the region $x+y+z \leqq 1$, $x \geqq 0$, $y \geqq 0$, $z \geqq 0$.
12. $\displaystyle\iiint \frac{dx\,dy\,dz}{x^2+y^2+(z-2)^2}$ throughout the sphere $x^2+y^2+z^2 \leqq 1$.
13. $\displaystyle\iiint \frac{dx\,dy\,dz}{x^2+y^2+(z-\frac{1}{2})^2}$ throughout the sphere $x^2+y^2+z^2 \leqq 1$.
14. $\displaystyle\iint \frac{dx\,dy}{\sqrt{x^2+y^2}}$ over the square $|x| \leqq 1$, $|y| \leqq 1$.
15. Prove that if $f(x, y)$ is a continuous function on a domain D in the x, y-plane and if for every region R contained in that domain $\int_R f(x, y)\,dx\,dy = 0$, then $f(x, y)$ is identically 0.
16. Prove that

$$\iint_R e^{-(x^2+y^2)}\,dx\,dy = ae^{-a^2}\int_0^\infty \frac{e^{-u^2}}{a^2+u^2}\,du$$

where R denotes the half-plane $x \geqq a > 0$, by applying the transformation

$$x^2+y^2 = u^2+a^2, \qquad y = vx.$$

17. Prove that

$$\left|\iint (u_x{}^2 + u_y{}^2)\,dx\,dy\right|$$

is invariant on inversion.
18. Evaluate the integral

$$I = \iiint \cos(x\xi + y\eta + z\zeta)\,d\xi\,d\eta\,d\zeta$$

taken throughout the sphere $\xi^2+\eta^2+\zeta^2 \leqq 1$.
19. In the integral

$$I = \int_2^4 dx \int_{4/x}^{(20-4x)/(8-x)} (y-4)\,dy$$

change the order of integration and evaluate the integral.

4.7 Improper Multiple Integrals

In the case of functions of one variable, we found it necessary to extend the concept of integral to other functions that are not continuous in the interval of integration. In particular, we considered the integrals of functions with jump-discontinuities and of functions with infinite values; we also considered integrals over infinite intervals of integration. The corresponding extensions of the concept of integral for functions of several variables will now be discussed.

The notion of "integral", as defined on p. 377 (we call it the *Riemann integral*), is not tied to continuity of the integrand $f(x, y)$. As long as f is bounded in the region of integration R, we can always form the upper and lower sums corresponding to a division of R into Jordan-measurable sets R_i. We call f *integrable* (more precisely *Riemann-integrable*) if these upper and lower sums approach the same limit as the division of R is refined indefinitely. This is essentially the procedure we shall follow in the exposition given in the Appendix to this chapter.[1] Strictly speaking the integral of any integrable function is *proper*, even if the function happens to be discontinuous.

In this section, however, we take only the existence of integrals of continuous functions for granted and try by limiting processes to extend the notion of integral and to prove its existence for wider classes of functions. We leave open the question whether *improper* integrals defined in this way are really identical with proper Riemann integrals obtained directly from upper and lower sums of subdivisions of R.[2]

a. Improper Integrals of Functions over Bounded Sets

The functions we aim to integrate are, in most cases, continuous in a certain region R except at isolated points or along certain curves, where the functions are not defined or are unbounded, or where their continuity is doubtful. In all cases that interest us the set of points of exceptional behavior for the function has area 0 (the word "area" is used here exclusively in the sense of Jordan-measure or content).[3] We may then cut away from R a set s of small area containing the exceptional points, integrate f over the remainder, and take the limit of the integrals of f over $R - s$ as the area of s tends to 0. If this limit exists, it defines the "improper" integral of f over R. Since we do not want the limit to depend on the particular way in which we approximate the set R, we shall confine ourselves to the simplest situation (corresponding to "absolute convergence" in contrast to "conditional convergence" in infinite series) where not only f but also $|f|$, has an improper integral.

Let the region of integration R be bounded and have an area. Assume that we can find a "monotone" sequence of closed subregions R_n(i.e.,

[1]We there use only subdivisions into squares in defining the integral. But this restriction can be shown to be inessential.

[2]This actually always is the case when f is bounded and is continuous except possibly on a set of points of content 0, provided R is bounded and Jordan-measurable.

[3]More refined notions, like the Lebesgue integral, are needed to integrate some functions whose points of discontinuity form a set of positive Jordan measure.

$R_n \subset R_{n+1} \subset R$) in each of which $f(x, y)$ is defined and continuous. Assume moreover that the areas $A(R_n)$ of the sets R_n approach the area $A(R)$ and that the integrals

$$\iint_{R_n} |f(x, y)|\, dx\, dy \tag{19a}$$

are bounded independently of n. Then

$$I = \lim_{n\to\infty} \iint_{R_n} f(x, y)\, dx\, dy \tag{19b}$$

exists. This limit will be shown to be independent of the particular approximating sequence R_n, and will be used to define the improper integral

$$I = \iint_R f(x, y)\, dx\, dy. \tag{19c}$$

Before proving this theorem, we illustrate the ideas by some typical examples.

The function

$$f(x, y) = \log \sqrt{x^2 + y^2}$$

becomes infinite at the origin of the x, y-plane. Therefore, in order to calculate the integral of f over a region R containing the origin, for example, over the circle $x^2 + y^2 \leq 1$, we must cut out the origin by surrounding it with a region s whose area tends to 0. We must then investigate the convergence of the integral taken over the residual region $R - s$. We take for s the circular disk s_n of radius $1/n$. Let R_n be the region obtained from R by cutting out s_n Let, in turn, R be contained in a circle of radius ρ about the origin. Transforming to polar coordinates, we have

$$\iint_{R_n} |f|\, dx\, dy = \iint_{R_n} |f|\, r\, dr\, d\theta \leq \int_{1/n}^{\rho} dr \int_0^{2\pi} d\theta\, r|\log r|$$

$$= 2\pi \int_{1/n}^{\rho} r|\log r|\, dr.$$

The transformation thus yields a new integrand $r|\log r|$ that is bounded and even continuous if defined as 0 for $r = 0$. Hence, uniformly for all n,

$$\iint_{R_n} |f|\,dx\,dy \leqq 2\pi \int_0^\rho r\,|\log r|\,dr.$$

The existence of the improper integral

$$\iint_R \log\sqrt{x^2+y^2}\,dx\,dy = \lim_{n\to\infty}\iint_{R_n} \log\sqrt{x^2+y^2}\,dx\,dy$$

follows. For example, if R is the unit disk we find

(20a)
$$\iint_{x^2+y^2<1} \log\sqrt{x^2+y^2}\,dx\,dy = \int_0^1 dr \int_0^{2\pi} d\theta\, r\log r$$

$$= 2\pi \int_0^1 r\log r\,dr$$

$$= 2\pi\left(\frac{1}{2}r^2\log r - \frac{1}{4}r^2\right)_0^1$$

$$= -\frac{\pi}{2}.$$

As a further example, we consider the integral

(20b)
$$\iint_R \frac{dx\,dy}{\sqrt{(x^2+y^2)^\alpha}}$$

taken over the same region. Here we obtain immediately

$$\iint_{R_n} |f|\,dx\,dy \leq \int_{1/n}^\rho dr \int_0^{2\pi} d\theta\,|f|\,r\,dr\,d\theta$$

$$= 2\pi \int_{1/n}^\rho r^{1-\alpha}\,dr.$$

From Volume I (p. 305) we know that the integral $\int_0^\rho r^{1-\alpha}\,dr$ is convergent if and only if $\alpha < 2$. We therefore conclude that the double integral (20b) likewise is convergent if and only if $\alpha < 2$. This remark can readily be extended into a *sufficient* (but by no means necessary) criterion for the convergence of improper double integrals, which is applicable in many special cases.

If the function $f(x, y)$ is continuous in the region R everywhere except at one point, which we take as the origin, and if there exists a fixed bound M and a positive number $\alpha < 2$ such that

(21a)
$$|f(x,y)| < \frac{M}{\sqrt{(x^2+y^2)^\alpha}}$$

everywhere in R for $(x, y) \neq (0, 0)$, then the integral

$$\iint_R f(x, y)\, dx\, dy \tag{21b}$$

converges.

We can treat the triple integral

$$\iiint_R \frac{dx\, dy\, dz}{\sqrt{(x^2 + y^2 + z^2)^\alpha}}$$

in a similar way. If R contains the origin, we introduce spherical coordinates and obtain

$$\iiint_R r^{2-\alpha} \sin\theta\, dr\, d\phi\, d\theta.$$

A discussion similar to the preceding one shows us that convergence occurs when $\alpha < 3$. Again, more generally, we see that

$$\iiint_R f(x, y, z)\, dx\, dy\, dz \tag{22a}$$

converges if $f(x, y, z)$ is continuous in R except at the origin provided that there exists a bound M and a constant $\alpha < 3$ for which

$$|f(x, y, z)| \leqq \frac{M}{\sqrt{(x^2 + y^2 + z^2)^\alpha}}. \tag{22b}$$

In consequence, for an everywhere continuous function $g(x, y, z)$, the improper integral

$$\iiint_R \frac{g(x, y, z)}{\sqrt{(x^2 + y^2 + z^2)^\alpha}}\, dx\, dy\, dz \tag{22c}$$

exists, if $\alpha < 3$. Improper integrals can also exist for integrands that are infinite along whole curves, not only at single points. In the simplest case, the integrand is infinite on a portion of a straight line, say a segment of the y-axis. In this case, if the relation

$$|f(x, y)| < \frac{M}{|x|^\alpha} \tag{23}$$

is valid everywhere in R for $x \neq 0$, where M is a fixed bound and $\alpha < 1$, then again the improper integral of f over R exists. For the proof, we only have to cut out from R a strip about the y-axis and let the width of the strip tend to 0.

Integrals like

$$\iint_R \frac{dx\,dy}{x^3},$$

violating our restriction on the exponent α, may sometimes still be defined in a "conditional" sense, in which the value depends on the precise manner of approximation to R. Here, for example, the integral can be defined as the limit of integrals over the regions obtained by cutting out of R a strip *symmetric* to the y-axis. Other approximations may lead to different values for the integral or even to divergence.

b. Proof of the General Convergence Theorem for Improper Integrals

We consider the set R of area $A(R)$ and a sequence of closed subsets R_n whose areas $A(R_n)$ tend to $A(R)$ for $n \to \infty$. Here the R_n shall expand monotonically inside R:

$$R_1 \subset R_2 \subset R_3 \subset \cdots \subset R. \tag{24a}$$

The function $f(x, y)$ is assumed to be continuous in each R_n. Moreover, there shall exist a constant μ such that

$$\iint_{R_n} |f(x, y)|\, dx\, dy \leq \mu \tag{24b}$$

for all n.

Because of (24a) the integrals

$$\iint_{R_n} |f|\, dx\, dy$$

obviously form a monotone increasing bounded sequence and thus have a limit for $n \to \infty$. By the Cauchy convergence test, for every $\varepsilon > 0$ we can find an $N = N(\varepsilon)$ such that, for $m > n > N(\varepsilon)$,

$$\iint_{R_m} |f|\, dx\, dy - \iint_{R_n} |f|\, dx\, dy = \iint_{R_m - R_n} |f|\, dx\, dy < \varepsilon. \tag{24c}$$

Let

$$I_n = \iint_{R_n} f(x, y)\, dx\, dy.$$

Clearly the I also satisfy the Cauchy test, since, by (5g),

$$\left|\iint_{R_m} f\,dx\,dy - \iint_{R_n} f\,dx\,dy\right| = \left|\iint_{R_m - R_n} f\,dx\,dy\right|$$
$$\leqq \iint_{R_m - R_n} |f|\,dx\,dy < \varepsilon$$

for $m > n > N(\varepsilon)$. It follows that

$$I = \lim_{n\to\infty} \iint_{R_n} f(x, y)\,dx\,dy$$

exists.

It remains to be shown that the value I does not depend on the particular approximating sequence R_n used. Let S be any closed Jordan-measurable subset of R in which f is continuous. Let M be an upper bound for $|f|$ in S. Then, by the mean value theorem of integral calculus (see p. 384),[1]

$$\left|\iint_S f\,dx\,dy - \iint_{S\cap R_n} f\,dx\,dy\right| = \left|\iint_{S-R_n} f\,dx\,dy\right|$$
$$\leqq \iint_{S-R_n} |f|\,dx\,\,dy \leqq MA(S - R_n) \leqq MA(R - R_n)$$
$$= M[A(R) - A(R_n)].$$

It follows from our assumption $\lim_{n\to\infty} A(R_n) = A(R)$ that

$$\iint_S f\,dx\,dy = \lim_{n\to\infty} \iint_{S\cap R_n} f\,dx\,dy \tag{24d}$$

Applying this relation to $|f|$ instead of f, and using (24b), we find

$$\iint_S |f|\,dx\,dy = \lim_{n\to\infty} \iint_{S\cap R_n} |f|\,dx\,dy \tag{24e}$$
$$\leqq \lim_{n\to\infty} \iint_{R_n} |f|\,dx\,dy \leqq \mu.$$

Thus, the estimate (24b) has been extended to more general subsets S of R.

We can also extend (24c). We have, using (24d).

[1]We remind the reader that $S \cap R_n$ stands for the set of points common to S and R_n and $S - R_n$ for the set of points that belong to S but not to R_n (see p. 116):

$$S - R_n = S - S \cap R_n$$

We write again $A(S - R_n)$ for the area of the set $S - R_n$.

(42f) $$\left|\iint_S f\,dx\,dy - \iint_{S\cap R_n} f\,dx\,dy\right|$$
$$= \lim_{m\to\infty}\left|\iint_{S\cap R_m} f\,dx\,dy - \iint_{S\cap R_n} f\,dx\,dy\right|$$
$$= \lim_{m\to\infty}\left|\iint_{S\cap(R_m-R_n)} f\,dx\,dy\right| \leqq \lim_{m\to\infty}\iint_{R_m-R_n}|f|\,dx\,dy$$
$$= \lim_{m\to\infty}\left(\iint_{R_m}|f|\,dx\,dy - \iint_{R_n}|f|\,dx\,dy\right) < \varepsilon$$

for $n > N(\varepsilon)$. Here N does not depend on the particular set S.

Let now $S_1, S_2, \ldots$ be a sequence of closed subsets of R in which f is continuous and for which

(24g) $$S_1 \subset S_2 \subset S_3 \subset \cdots \subset R$$

and

(24h) $$\lim_{m\to\infty} A(S_m) = A(R).$$

Since by (24e)

$$\iint_{S_m}|f|\,dx\,dy \leqq \mu,$$

we know that

$$J = \lim_{m\to\infty}\iint_{S_m} f\,dx\,dy$$

exists. Then

$$\left|J - \iint_{S_m} f\,dx\,dy\right| < \varepsilon$$

for all sufficiently large m. It follows from (24f) that

$$\left|J - \iint_{S_m\cap R_n} f\,dx\,dy\right| < 2\varepsilon$$

for all m, n that are both sufficiently large. Interchanging the roles of the S_m and R_n, we also have

$$\left|I - \iint_{S_m\cap R_n} f\,dx\,dy\right| < 2\varepsilon$$

for all sufficiently large m, n. Hence, $|J - I| < 4\varepsilon$ for any positive number ε, and thus, $I = J$, which was to be proved.

c. Integrals over Unbounded Regions

A different type of improper integral arises when the integrand f is continuous but the region of integration extends to infinity. Again, we do not try to analyze the most general situation but formulate a convergence criterion applicable to most cases occuring in practice. It is sufficient to treat the case of two independent variables.

We consider an unbounded set R in which the function f is continuous. We *exhaust* R by a monotone sequence of subsets

$$R_1 \subset R_2 \subset R_3 \subset \cdots \subset R$$

each of which is closed, bounded, and Jordan-measurable. Instead of the previous condition $\lim_{n\to\infty} A(R_n) = A(R)$, which might make no sense for unbounded R, we require that every closed and bounded subset of R is contained in at least one of the sets R_m. (If, for example, R is the whole plane, we can choose for the R_n the circular disks of radius n with center at the origin.) If the limit

$$\lim_{n\to\infty} \iint_{R_n} f(x, y)\, dx\, dy$$

exists and is independent of the particular choice of the sequence of subsets R_n, we call it the integral of f over R and denote it by

$$\iint_R f\, dx\, dy.$$

We then have the following *sufficient* condition for existence of the integral:

The improper integral of f over the unbounded set R exists if for one particular sequence R_n (of the type described) the integrals of $|f|$ over R_n are bounded uniformly in n, say if

$$\iint_{R_n} |f|\, dx\, dy \leq \mu$$

for all n.

The proof of this convergence criterion uses the same arguments as the one for improper integrals over bounded sets, and should be carried out as an exercise by the reader.

We illustrate the theorem with the integral

$$\iint_R e^{-x^2-y^2}\, dx\, dy,$$

where the region of integration is the whole x, y-plane. We choose for the sequence R_n of subregions the circular disks of radius n with center at the origin that obviously satisfy all our requirements. Here, transforming to polar coordinates:

$$\iint_{R_n} e^{-x^2-y^2}\,dx\,dy = \iint_{x^2+y^2 \leqq n^2} e^{-x^2-y^2}\,dx\,dy$$

$$= \int_0^n dr \int_0^{2\pi} d\theta\, re^{-r^2}\,dr = 2\pi \int_0^n re^{-r^2}\,dr$$

$$= -\pi e^{-r^2}\Big|_0^n = \pi(1 - e^{-n^2}).$$

This proves the boundedness of the integrals over R_n and, hence, the existence of the integral over R. For $n \to \infty$ we find for the value of our improper integral

$$\iint_R e^{-x^2-y^2}\,dx\,dy = \lim_{n\to\infty} \pi(1 - e^{-n^2}) = \pi.$$

On the other hand, we must obtain the same limit by using instead of the R_n the sequence S_m of squares

$$-m \leqq x \leqq +m, \qquad -m \leqq y \leqq +m.$$

Here we can make use of the fact that the integrand is a product of a function of x and of a function of y (see p. 380) and find

$$\iint_{S_m} e^{-x^2-y^2}\,dx\,dy = \iint_{S_m} e^{-x^2} \cdot e^{-y^2}\,dx\,dy$$

$$= \left(\int_{-m}^m e^{-x^2}\,dx\right)\left(\int_{-m}^m e^{-y^2}\,dy\right) = \left(\int_{-m}^m e^{-x^2}\,dx\right)^2.$$

It follows that

$$\lim_{m\to\infty} \iint_{R_m} e^{-x^2-y^2}\,dx\,dy = \left(\int_{\infty}^{\infty} e^{-x^2}\,dx\right)^2.$$

Since the R_n and S_m must yield the same value for the integral over R, we find that

(25a) $$\int_{-\infty}^{\infty} e^{-x^2}\,dx = \sqrt{\pi}.$$

By using the theory of improper double integrals we have thus evaluated an improper single integral that is of great importance in analysis. This value is difficult to find directly since the *indefinite* integral of e^{-x^2} cannot be expressed in terms of elementary functions.

We can make use of this result to evaluate the *gamma function* (see Volume I, p. 308)

$$\Gamma(n) = \int_0^\infty e^{-t} t^{n-1}\, dt \tag{25b}$$

for the argument $n = \frac{1}{2}$. The substitution $t = x^2$ yields

$$\begin{aligned}\Gamma\left(\frac{1}{2}\right) &= \int_0^\infty \frac{e^{-t}}{\sqrt{t}}\, dt = 2\int_0^\infty e^{-x^2}\, dx \\ &= \int_{-\infty}^\infty e^{-x^2}\, dx = \sqrt{\pi}.\end{aligned} \tag{25c}$$

We can formulate useful convergence tests for improper integrals over unbounded regions by comparison with powers of $\sqrt{x^2 + y^2}$. These are analogous to the test found on p. 409 for functions that are unbounded near the origin. We find that the improper integral of a continuous function $f(x, y)$ over an unbounded region R exists if f everywhere in R satisfies an inequality

$$|f(x, y)| \leq \frac{M}{\sqrt{(x^2 + y^2)^\alpha}}, \tag{26}$$

where M and α are fixed constants and $\alpha > 2$.[1]

Exercises 4.7

1. (a) By transforming to polar coordinates, show that the value of the integral

$$K = \int_0^{a \sin\beta} \left\{\int_{y \cot\beta}^{\sqrt{a^2-y^2}} \log(x^2 + y^2)\, dx\right\} dy \qquad \left(0 < \beta < \frac{\pi}{2}\right)$$

is $a^2\beta(\log a - \frac{1}{2})$.

[1]Behavior at infinity and at the origin are "complementary" in the sense that f is integrable near the origin if (26a) holds for a value $\alpha < 2$. Thus, the *improper integral*

$$\iint \frac{dx\, dy}{\sqrt{(x^2 + y^2)^\alpha}}$$

extended over the whole plane exists for no value of α.

(b) Change the order of integration in the original integral.

2. Integrate

(a) $\iint \frac{1}{(x^2+y^2+1)^2}\,dx\,dy$ over the x, y-plane,

(b) $\iiint \frac{1}{(x^2+y^2+z^2+1)^2}\,dx\,dy\,dz$ over x, y, z-space.

3. Show that the order of integration in

$$I = \int_0^1 \left\{\int_0^1 \frac{y-x}{(x+y)^3}\,dx\right\} dy$$

cannot be reversed.

4.8 Geometrical Applications

a. *Elementary Calculation of Volumes*

The concept of volume forms the starting-point of our definition of "integral." Here we use multiple integrals in order to calculate the volumes of several solids.

For example, in order to calculate the volume of the *ellipsoid of revolution*

$$\frac{x^2+y^2}{a^2} + \frac{z^2}{b^2} = 1$$

we write the equation in the form

$$z = \pm \frac{b}{a}\sqrt{a^2 - x^2 - y^2}.$$

The volume of the half of the ellipsoid above the x, y-plane is therefore given by the double integral [see (3b)],

$$\frac{V}{2} = \frac{b}{a}\iint \sqrt{a^2 - x^2 - y^2}\,dx\,dy$$

taken over the circle $x^2 + y^2 \leqq a^2$. If we transform to polar coordinates, the double integral becomes

$$\iint r\sqrt{a^2 - r^2}\,dr\,d\theta,$$

whence, on resolution into single integrals

$$\frac{V}{2} = \frac{b}{a}\int_0^{2\pi} d\theta \int_0^a r\ \sqrt{a^2 + r^2}\,dr = 2\pi\frac{b}{a}\int_0^a r\sqrt{a^2 - r^2}\,dr,$$

which gives the required value,

$$V = \frac{4}{3}\pi a^2 b.$$

To calculate the volume of the general ellipsoid

$$\text{(27a)}\qquad \frac{x^2}{a^2} + \frac{y^2}{b^2} + \frac{z^2}{c^2} = 1$$

we make the transformation

$$x = a\rho\cos\theta, \qquad y = b\rho\sin\theta, \qquad \frac{d(x, y)}{d(\rho, \theta)} = ab\rho$$

and for half the volume obtain

$$\frac{V}{2} = c\iint_R \sqrt{1 - \frac{x^2}{a^2} - \frac{y^2}{b^2}}\,dx\,dy = abc\iint_{R'} \rho\sqrt{1 - \rho^2}\,d\rho\,d\theta.$$

Here the region R' is the rectangle $0 \leqq \rho \leqq 1$, $0 \leqq \theta \leqq 2\pi$. Thus,

$$\frac{V}{2} = abc\int_0^{2\pi} d\theta \int_0^1 \rho\sqrt{1 - \rho^2}\,d\rho = \frac{2}{3}\pi abc$$

or

$$\text{(27b)}\qquad V = \frac{4}{3}\pi abc.$$

Finally, we shall calculate the volume of the pyramid enclosed by the three coordinate planes and the plane $ax + by + cz - 1 = 0$, where we assume that a, b, and c are positive. For the volume we obtain

$$V = \frac{1}{c}\iint_R (1 - ax - by)\,dx\,dy,$$

where the region of integration is the triangle $0 \leqq x \leqq 1/a, 0 \leqq y \leqq (1 - ax)/b$ in the x, y-plane. Therefore,

$$V = \frac{1}{c}\int_0^{1/a} dx \int_0^{(1-ax)/b} (1 - ax - by)dy.$$

Integration with respect to y gives

$$(1 - ax)y - \frac{b}{2}y^2 \Big|_0^{(1-ax)/b} = \frac{(1 - ax)^2}{2b},$$

and if we integrate again by means of the substitution $1 - ax = t$, we obtain

$$V = \frac{1}{2bc}\int_0^{1/a} (1 - ax)^2 dx = -\frac{1}{6abc}(1 - ax)^3 \Big|_0^{1/a} = \frac{1}{6abc}.$$

This result agrees, of course, with the rule of elementary geometry that the volume of a pyramid is one-third of the product of base and altitude.

In order to calculate the volume of a more complicated solid we can subdivide the solid into pieces whose volumes can be expressed directly by double integrals. Later, however (in particular in the next chapter), we shall obtain expressions for the volume bounded by a closed surface that do not involve this subdivision.

b. *General Remarks on the Calculation of Volumes. Solids of Revolution. Volumes in Spherical Coordinates*

Just as we can express the area of a plane region R by the double integral

$$\iint_R dR = \iint_R dx\, dy,$$

we may also express the volume of a three-dimensional region R by the integral

$$V = \iiint_R dx\, dy\, dz$$

over the region R. In fact this point of view exactly corresponds to our definition of integral (cf. Appendix, p. 517) and expresses the geometrical fact that we can find the volume of a region by cutting space into identical cubes, finding the total volume of the cubes contained entirely in R, and then letting the diameter of the cubes tend to zero. The resolution of this integral for V into an integral $\int dz \iint dx\, dy$

[see (14a), p. 397] expresses *Cavalieri's principle,* known to us from elementary geometry, according to which the volume of a solid is determined if we know the area of every plane cross section that is perpendicular to a definite line, say the z-axis. The general expression given above for the volume of a three-dimensional region enables us at once to find various formulae for calculating volumes. For this purpose, it often is useful to introduce new independent variables into the integral instead of x, y, z.

The most important examples are given by *spherical* coordinates and by *cyclindrical* coordinates. Let us calculate, for example, *the volume of a solid of revolution* obtained by rotating a curve $x = \phi(z)$ about the z-axis. We assume that the curve does not cross the z-axis and that the solid of revolution is bounded above and below by planes $z =$ constant. The solid is therefore defined by inequalities of the form $a \leqq z \leqq b$ and $0 \leqq \sqrt{x^2 + y^2} \leqq \phi(z)$. Its volume is given by the integral above. In terms of the cylindrical coordinates

$$z, \quad \rho = \sqrt{x^2 + y^2}, \qquad \theta = \operatorname{arc\,cos} \frac{x}{\rho} = \operatorname{arc\,sin} \frac{y}{\rho}$$

the expression for the volume becomes

$$V = \iiint_R dx\, dy\, dz = \int_a^b dz \int_0^{2\pi} d\theta \int_0^{\phi(z)} \rho\, d\rho.$$

If we perform the single integrations, we at once obtain

$$V = \pi \int_a^b \phi(z)^2\, dz. \tag{28a}$$

We can also give a more intuitive derivation of this formula (see Volume I, p. 374). We cut the solid of revolution into small slices

$$z_\nu \leqq z \leqq z_{\nu+1}$$

by planes perpendicular to the z-axis, and we denote by m_ν the minimum and by M_ν the maximum of the distance $\phi(z)$ from the axis in this slice. The volume of the slice lies then between the volumes of two cylinders with altitude

$$\Delta z = z_{\nu+1} - z_\nu$$

and radii m_ν and M_ν, respectively. Hence,

$$\sum m_v{}^2\pi\,\Delta z \leq V \leq \sum M_v{}^2\pi\,\Delta z.$$

By the definition of the ordinary integral, therefore,

$$V = \pi \int_a^b \phi(z)^2\, dz.$$

If the region R contains the origin O of a spherical coordinate system (r, θ, ϕ) and if the surface is given by an equation

$$r = f(\theta, \phi)$$

where the function $f(\theta, \phi)$ is single-valued, it is frequently advantageous to use these spherical coordinates instead of (x, y, z) in calculating the volume. If we substitute the value of the Jacobian

$$\frac{d(x, y, z)}{d(r, \theta, \phi)} = r^2 \sin\theta$$

(as calculated on p. 000) in the transformation formula, we at once obtain the expression

$$V = \iiint_R r^2 \sin\theta\, dr\, d\theta\, d\phi = \int_0^{2\pi} d\phi \int_0^{\pi} \sin\theta\, d\theta \int_0^{f(\theta,\phi)} r^2\, dr$$

for the volume. Integration with respect to r gives

$$V = \frac{1}{3}\int_0^{2\pi} d\phi \int_0^{\pi} f^3(\theta, \phi)\sin\theta\, d\theta. \tag{28b}$$

In the special case of the sphere, for which $f(\theta, \phi) = \mathrm{R}$ is constant, this at once yields the volume $(4/3)\pi R^3$.

c. *Area of a Curved Surface*

We expressed the length of a curve by an ordinary integral (Volume I, p. 349). We now wish to find an analogous expression for the area of a curved surface by means of a double integral. We defined the length of a curve as the limiting value of the length of an inscribed polygon when the lengths of the individual sides tend to zero. This suggests that we define the area of a surface analogously as follows: In the curved surface we inscribe a polyhedron formed of plane triangles, determine the area of the polyhedron, make the inscribed net of triangles finer by letting the length of the longest side tend to zero, and seek to find the limiting value of the area of the polyhedron.

This limiting value would then be called the area of the curved surface. It turns out, however, that such a definition of area would have no precise meaning, for in general this process does not yield a definite limiting value. This phenomenon may be explained in the following way: a polygon inscribed in a smooth curve always has the property (expressed by the mean value theorem of the differential calculus) that the direction of the individual side of the polygon approaches the direction of the curve as closely as we please if the subdivision is fine enough. With curved surfaces the situation is quite different. The sides of a polyhedron inscribed in a curved surface may be inclined to the tangent plane to the surface at a neighboring point as steeply as we please, even if the polyhedral faces have arbitrarily small diameters. The area of such a polyhedron, therefore, cannot by any means be regarded as an approximation to the area of the curved surface. In the Appendix we shall consider an example of this state of affairs in detail (pp. 540).

In the definition of the length of a smooth curve, however, we can, instead of using an *inscribed* polygon, equally well use a *circumscribed* one, that is, a polygon of which every side touches the curve. The definition of the length of a curve as the limit of the length of a circumscribed polygon can easily be extended to curved surfaces, if first modified as follows: we obtain the length of a curve $y = f(x)$ that has a continuous derivative $f'(x)$ and lies between the abscissae a and b by subdividing the interval between a and b at the points $x_0, x_1, \ldots,$ x_n into n equal or different parts, choosing an arbitrary point ξ_ν in the νth subinterval, constructing the tangent to the curve at this point, and measuring the length l_ν of the portion of this tangent lying in the strip $x_\nu \leqq x \leqq x_{\nu+1}$ (Fig. 4.14). If we let n increase beyond all

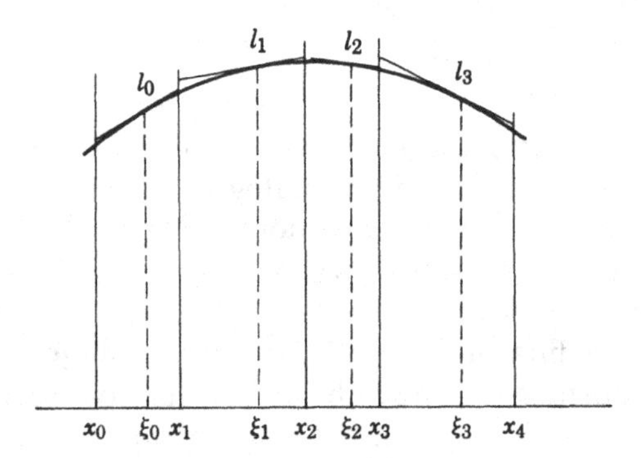

Figure 4.14

bounds and at the same time let the length of the longest subinterval tend to 0, the sum

$$\sum_{\nu=0}^{n-1} l_\nu$$

then tends to the length of the curve, that is, to the integral

$$\int_a^b \sqrt{1 + f'(x)^2}\, dx.$$

This statement follows from the fact that

$$l_\nu = (x_{\nu+1} - x_\nu)\sqrt{1 + f'(\xi_\nu)^2}.$$

We now define the area of a curved surface similarly. We begin by considering a surface represented by a function $z = f(x,y)$ with continuous derivatives on a region R of the x, y-plane. We subdivide R into n subregions $R_1, R_2, \ldots, R_n$ with the areas $\Delta R_1, \ldots, \Delta R_n$, and in these subregions we choose points $(\xi_1, \eta_1), \ldots, (\xi_n, \eta_n)$. At the point of the surface with the coordinates ξ_ν, η_ν and $\zeta_\nu = f(\xi_\nu, \eta_\nu)$ we construct the tangent plane and find the area of the portion of this plane lying above the region R_ν (Fig. 4.15). If α_ν is the angle that the tangent plane

$$z - \zeta_\nu = f_x(\xi_\nu, \eta_\nu)(x - \xi_\nu) + f_y(\xi_\nu, \eta_\nu)(y - \eta_\nu)$$

makes with the x, y-plane and if $\Delta\tau_\nu$ is the area of the portion τ_ν of the

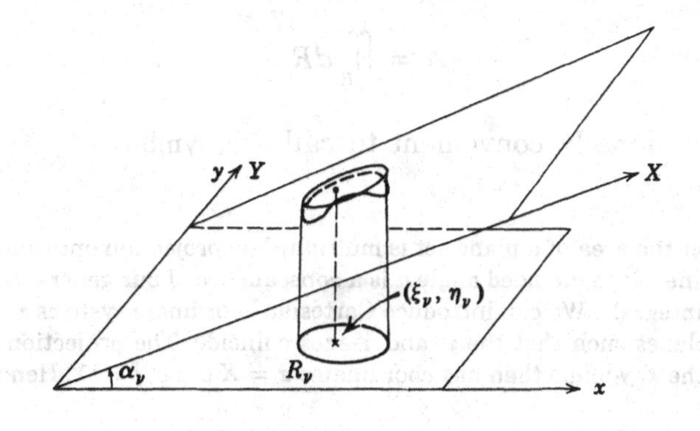

Figure 4.15

tangent plane above R_ν, then the region R_ν is the projection of τ_ν on the x, y-plane,[1] so that

$$\Delta R_\nu = \Delta\tau_\nu \cos\alpha_\nu.$$

Again (cf. Chapter 3, p. 239),

$$\cos\alpha_\nu = \frac{1}{\sqrt{1 + f_x^2(\xi_\nu, \eta_\nu) + f_y^2(\xi_\nu, \eta_\nu)}},$$

and therefore,

$$\Delta\tau_\nu = \sqrt{1 + f_x^2(\xi_\nu, \eta_\nu) + f_y^2(\xi_\nu, \eta_\nu)} \cdot \Delta R_\nu.$$

We form the sum of all these areas

$$\sum_{\nu=1}^{n} \Delta\tau_\nu$$

and let n increase beyond all bounds, at the same time letting the diameter of the largest subdivision tend to zero. According to our definition of "integral" this sum will have the limit

$$A = \iint_R \sqrt{1 + f_x^2 + f_y^2}\, dR. \tag{29a}$$

This integral, which is independent of the mode of subdivision of the region R, we now use to *define the area of the given surface.* If the surface happens to be a plane surface, this definition agrees with the preceding; for example, if $z = f(x, y) = 0$, we have

$$A = \iint_R dR.$$

It is occasionally convenient to call the symbol

[1]The fact that the area of a plane set is multiplied on projection onto another plane with the cosine of the included angle α is a consequence of our general substitution formula for integrals. We can introduce Cartesian coordinate systems x, y and X, Y in the two planes such that the y- and Y-axes coincide. The projection of a point (X, Y) onto the x, y-plane then has coordinates $x = X\cos\alpha$, $y = Y$. Hence, the projected area is

$$\iint dx\,dy = \iint \frac{d(x, y)}{d(X, Y)}\, dX\,dY = \iint dX\,dY\cos\alpha.$$

$$d\sigma = \sqrt{1 + f_x^2 + f_y^2}\, dR = \sqrt{1 + f_x^2 + f_y^2}\, dx\, dy$$

the element of area of the surface $z = f(x, y)$. The area integral can then be written symbolically in the form

$$\iint_R d\sigma.$$

We arrive at another form of the expression for the area if we think of the surface as given by an equation $\phi(x, y, z) = 0$ instead of $z = f(x, y)$. If we assume that $\phi_z \neq 0$, on the surface the equations

$$\frac{\partial z}{\partial x} = -\frac{\phi_x}{\phi_z}, \quad \frac{\partial z}{\partial y} = -\frac{\phi_y}{\phi_z}$$

at once give the expression

$$\iint_R \sqrt{\phi_x^2 + \phi_y^2 + \phi_z^2} \left|\frac{1}{\phi_z}\right| dx\, dy \tag{29b}$$

for the area, where the region R is again the projection of the surface on the x, y-plane.

Let us apply the area formula to the area of a spherical surface. The equation

$$z = \sqrt{R^2 - x^2 - y^2}$$

represents a hemisphere of radius R. We have

$$\frac{\partial z}{\partial x} = -\frac{x}{\sqrt{R^2 - x^2 - y^2}}, \quad \frac{\partial z}{\partial y} = -\frac{y}{\sqrt{R^2 - x^2 - y^2}}.$$

The area of the full sphere is therefore given by the integral

$$A = 2R \iint \frac{dx\, dy}{\sqrt{R^2 - x^2 - y^2}},$$

where the region of integration is the circle of radius R lying in the x, y-plane and having the origin as its center. Introducing polar coordinates and resolving the integral into single integrals we obtain

$$A = 2R \int_0^{2\pi} d\theta \int_0^R \frac{r\, dr}{\sqrt{R^2 - r^2}} = 4\pi R \int_0^R \frac{r\, dr}{\sqrt{R^2 - r^2}}.$$

The ordinary integral on the right can easily be evaluated by means of the substitution $R^2 - r^2 = u$; we have

$$A = -4\pi R\sqrt{R^2 - r^2}\Big|_0^R = 4\pi R^2,$$

in agreement with the result of Archimedes.

In the definition of "area", we have hitherto singled out the coordinate z. If the surface had been given by an equation of the form $x = x(y, z)$ or $y = y(x, z)$, however, we could have represented the area similarly by the integrals

$$\iint \sqrt{1 + x_y^2 + x_z^2}\, dy\, dz \quad \text{or} \quad \iint \sqrt{1 + y_x^2 + y_z^2}\, dz\, dx$$

or, if the surface were given implicitly, by

$$\iint \sqrt{\phi_x^2 + \phi_y^2 + \phi_z^2}\left|\frac{1}{\phi_y}\right| dz\, dx \tag{29c}$$

or

$$\iint \sqrt{\phi_x^2 + \phi_y^2 + \phi_z^2}\left|\frac{1}{\phi_x}\right| dy\, dz. \tag{29d}$$

That all these expressions do actually define the same area can be verified directly. To this end, we apply the transformation

$$x = x(y, z),$$
$$y = y$$

to the integral

$$\iint \frac{\sqrt{\phi_x^2 + \phi_y^2 + \phi_z^2}}{|\phi_z|}\, dx\, dy.$$

Here $x = x(y, z)$ is found by solving the equation $\phi(x, y, z) = 0$ for x. The Jacobian is

$$\frac{d(x, y)}{d(y, z)} = \frac{\phi_z}{\phi_x},$$

and therefore,

$$\iint_R \frac{\sqrt{\phi_x^2 + \phi_y^2 + \phi_z^2}}{|\phi_z|}\, dx\, dy = \iint_{R'} \frac{\sqrt{\phi_x^2 + \phi_y^2 + \phi_z^2}}{|\phi_x|}\, dy\, dz.$$

The integral on the right is to be taken over the projection R' of the surface on the y, z-plane.

If we wish to get rid of any special assumption about the position of the surface relative to the coordinate system, we must represent the surface in the parametric form

$$x = \phi(u, v), \qquad y = \psi(u, v), \qquad z = \chi(u, v)$$

and express the area of the surface as an integral over the parameter domain R. A definite region R of the u, v-plane then corresponds to the surface. In order to introduce the parameters u and v in (29a), we first consider a portion of the surface near a point at which the Jacobian

$$\frac{d(x, y)}{d(u, v)} = D$$

is different from zero. For this portion we can solve for u and v as functions of x and y and obtain (see p. 261)

$$u_x = \frac{\psi_v}{D}, \qquad v_x = -\frac{\psi_u}{D},$$

$$u_y = -\frac{\phi_v}{D}, \qquad v_y = \frac{\phi_u}{D}.$$

for their partial derivatives. Through the equations

$$\frac{\partial z}{\partial x} = \frac{\partial z}{\partial u} u_x + \frac{\partial z}{\partial v} v_x \qquad \text{and} \qquad \frac{\partial z}{\partial y} = \frac{\partial z}{\partial u} u_y + \frac{\partial z}{\partial v} v_y$$

we obtain the expression

$$\sqrt{1 + \left(\frac{\partial z}{\partial x}\right)^2 + \left(\frac{\partial z}{\partial y}\right)^2}$$

$$= \frac{1}{D} \sqrt{(\phi_u \psi_v - \psi_u \phi_v)^2 + (\psi_u \chi_v - \chi_u \psi_v)^2 + (\chi_u \phi_v - \phi_u \chi_v)^2}.$$

If we now introduce u and v as new independent variables and apply the rules for the transformation of double integrals (16b), p. 403 we find that the area A' of the portion of the surface corresponding to a parameter region R' is

$$A' = \iint_{R'} \sqrt{(\phi_u \psi_v - \psi_u \phi_v)^2 + (\psi_u \chi_v - \chi_u \psi_v)^2 + (\chi_u \phi_v - \phi_u \chi_v)^2}\, du\, dv.$$

In this expression no distinction appears between the coordinates x, y, and z. Since we arrive at the same integral expression for the area no matter which one of the special nonparametric representations we start with, it follows that all these expressions are equal and represent the area.

So far we have only considered a portion of the surface on which one particular Jacobian does not vanish. We reach the same result, however, no matter which of the three Jacobians does not vanish. If then we suppose that at each point of the surface at least *one* of the Jacobians is not zero, we can subdivide the whole surface into portions like the above and thus find that the same integral still gives the area A of the whole surface:

(30a)

$$A = \iint_R \sqrt{(\phi_u\psi_v - \psi_u\phi_v)^2 + (\psi_u\chi_v - \chi_u\psi_v)^2 + (\chi_u\phi_v - \phi_u\chi_v)^2}\, du\, dv.$$

The expression for the area of a surface in parametric representation can be put in another noteworthy form if we make use of the coefficients of the line element (cf. Chapter 3, p. 283)

$$ds^2 = E\, du^2 + 2F\, du\, dv + G\, dv^2,$$

that is, of the expressions

$$E = \phi_u^2 + \psi_u^2 + \chi_u^2,$$

$$F = \phi_u\phi_v + \psi_u\psi_v + \chi_u\chi_v,$$

$$G = \phi_v^2 + \psi_v^2 + \chi_v^2.$$

A simple calculation shows that (see p. 284)

$$\text{(30b)} \quad EG - F^2 = (\phi_u\psi_v - \psi_u\phi_v)^2 + (\psi_u\chi_v - \chi_u\psi_v)^2 + (\chi_u\phi_v - \phi_u\chi_v)^2.$$

Thus, for the area we obtain the expression

$$\text{(30c)} \qquad A = \iint \sqrt{EG - F^2}\, du\, dv,$$

and for the element of area

$$\text{(30d)} \qquad d\sigma = \sqrt{EG - F^2}\, du\, dv.$$

As an example, we again consider the area of a sphere with radius R, which we now represent parametrically by the equations

$$x = R\cos u \sin v,$$

$$y = R\sin u \sin v,$$

$$z = R\cos v,$$

where u and v range over the region $0 \leqq u \leqq 2\pi$ and $0 \leqq v \leqq \pi$. A simple calculation shows that here

$$d\sigma = R^2 \sin v\, du\, dv, \tag{30e}$$

which once more gives us the expression

$$R^2 \int_0^{2\pi} du \int_0^{\pi} \sin v\, dv = 4\pi R^2$$

for the area.

More generally, we can apply formula (30d) to the *surface of revolution* formed by rotating the curve $z = \phi(x)$ about the z-axis. If we refer the surface to polar coordinates (u, v) in the x, y-plane as parameters, we obtain

$$x = u\cos v, \qquad y = u\sin v, \qquad z = \phi(\sqrt{x^2 + y^2}) = \phi(u).$$

Then,

$$E = 1 + \phi'^2(u), \qquad F = 0, \qquad G = u^2,$$

and the area is given in the form

$$\int_0^{2\pi} dv \int_{u_0}^{u_1} u\sqrt{1 + \phi'^2(u)}\, du = 2\pi \int_{u_0}^{u_1} u\sqrt{1 + \phi'^2(u)}\, du. \tag{31a}$$

If instead of u we introduce the length of arc s of the meridian curve $z = \phi(u)$ as parameter, we obtain the *area of the surface of revolution* in the form

$$2\pi \int_{s_0}^{s_1} u\, ds, \tag{31b}$$

where u is the distance from the axis of the point on the rotating curve corresponding to s (Guildin's rule; cf. Volume I, p. 374).

We apply this rule to calculate the surface area of the torus (cf. Chapter 3, p. 286) obtained by rotating the circle $(x - a)^2 + z^2 = r^2$ about the z-axis. If we introduce the length of arc s of the circle as a parameter, we have $u = a + r\cos(s/r)$, and the area is therefore

$$2\pi \int_0^{2\pi r} u\, ds = 2\pi \int_0^{2\pi r} \left(a + r\cos\frac{s}{r}\right) ds = 2\pi a \cdot 2\pi r.$$

The area of a torus is therefore equal to the product of the circumference of the generating circle and the length of the path described by the center of the circle.

Exercises 4.8

1. Calculate the volume of the solid defined by

$$\frac{\{\sqrt{x^2 + y^2} - 1\}^2}{a^2} + \frac{z^2}{b^2} \leqq 1 \qquad (a < 1).$$

2. Find the volume cut off from the paraboloid $(x^2/a^2) + (y^2/b^2) = z$ by the plane $z = h$.
3. Find the volume cut off from the ellipsoid $(x^2/a^2) + (y^2/b^2) + (z^2/c^2) = 1$ by the plane $lx + my + nz = p$.
4. (a) Show that if any closed curve $\theta = f(\phi)$ is drawn on the surface $r^2 = a^2 \cos 2\theta$ (r, θ, ϕ being spherical coordinates in space), the area of the surface so enclosed is equal to the area enclosed by the projection of the curve on the sphere $r = a$, the origin of coordinates being the vertex of projection.
   (b) Express the area by a simple integral.
   (c) Find the area of the whole surface.
5. Find the volume and surface area of the solid generated by rotating the triangle ABC about the side AB.
6. Find the surface area of the paraboloid $z = x^2 + y^2$ intercepted between the cylinders $x^2 + y^2 = a$ and $x^2 + y^2 = b$, where $a = \frac{1}{4}[(2m - 1)^2 - 1]$ and $b = \frac{1}{4}[(2n - 1)^2 + 1]$, m and n being natural numbers with $n > m$.
7. Find the surface area of the section cut out of the cylinder $x^2 + z^2 = a^2$ by the cylinder $x^2 + y^2 = b^2$, where $0 < b \leqq a$ and $z \geqq 0$.
8. Show that the area Σ of the right conoid

$$x = r\cos\theta, \qquad y = r\sin\theta, \qquad z = f(\theta),$$

included between two planes through the axis of z and the cylinder with generating lines parallel to this axis and cross section $r = f'(\theta)$, and the area of its orthogonal projection on $z = 0$ are in the ratio $[\sqrt{2} + \log(1 + \sqrt{2})]:1$.

4.9 Physical Applications

In Section 4.4 (p. 386) we have already seen how the concept of *mass* is connected with that of a multiple integral. Here we shall study some of the other concepts of mechanics. We begin with a detailed study of moment and of moment of inertia.

a. *Moments and Center of Mass*

The moment with respect to the x,y-plane of a particle with mass m is defined as the product mz of the mass and the z-coordinate. Similarly, the moment with respect to the y,z-plane is mx and that with respect to the z,x-plane is my. The moments of several particles combine additively; that is, the three *moments of a system of particles* with masses $m_1, m_2, \ldots, m_n$ and coordinates $(x_1, y_1, z_1), \ldots, (x_n, y_n, z_n)$ are given by the expressions

$$T_x = \sum_{\nu=1}^{n} m_\nu x_\nu, \qquad T_y = \sum_{\nu=1}^{n} m_\nu y_\nu, \qquad T_z = \sum_{\nu=1}^{n} m_\nu z_\nu. \tag{32a}$$

If we deal with a mass distributed with continuous density $\mu = \mu(x, y, z)$ through a region in space or over a surface or curve, we define the moment of the mass-distribution by a limiting process, as in Volume I (p. 373) and thus express the moments by integrals. For example, given a distribution in space we subdivide the region R into n subregions, imagine the total mass of each subregion concentrated at any one of its points, and then form the moment of the system of these n particles. We see at once that as $n \to \infty$ and the greatest diameter of the subregions tends at the same time to zero, the sums tend to the limits

$$T_x = \iiint_R \mu x \, dx \, dy \, dz, \quad T_y = \iiint_R \mu y \, dx \, dy \, dz, \tag{32b}$$

$$T_z = \iiint_R \mu z \, dx \, dy \, dz,$$

which we call the *moments of the volume-distribution.*

Similarly, if the mass is distributed over a surface S given by the equations $x = \phi(u, v)$, $y = \psi(u, v)$, $z = \chi(u, v)$ with density $\mu(u, v)$, we define the *moments of the surface distribution* by the expressions

$$T_x = \iint_S \mu x \, d\sigma = \iint_R \mu x \sqrt{EG - F^2} \, du \, dv,$$

$$(32c) \qquad T_y = \iint_S \mu y \, d\sigma = \iint_R \mu y \sqrt{EG - F^2} \, du \, dv,$$

$$T_z = \iint_S \mu z \, d\sigma = \iint_R \mu z \sqrt{EG - F^2} \, du \, dv.$$

Finally, the *moments of a curve* $x(s)$, $y(s)$, $z(s)$ in space with mass density $\mu(s)$ are defined by the expressions

$$(32d) \qquad T_x = \int_{s_0}^{s_1} \mu x \, ds, \qquad T_y = \int_{s_0}^{s_1} \mu y \, ds, \qquad T_z = \int_{s_0}^{s_1} \mu z \, ds,$$

where s denotes the length of arc.

The *center of mass* of a mass of total amount M distributed through a region R is defined as the point with coordinates

$$(32e) \qquad \xi = \frac{T_x}{M}, \qquad \eta = \frac{T_y}{M}, \qquad \zeta = \frac{T_z}{M}.$$

For a distribution in space, the coordinates of the center of mass are therefore given by the expressions

$$\xi = \frac{1}{M} \iiint_R \mu x \, dx \, dy \, dz, \ldots, \qquad \text{where} \qquad M = \iiint_R \mu \, dx \, dy \, dz.$$

If the mass-distribution is *homogeneous*, $\mu(x, y, z) =$ constant, the center of mass of the region is called its *centroid*.[1]

As our first example, we consider the homogeneous hemispherical region H with mass density 1:

$$x^2 + y^2 + z^2 \leqq 1,$$
$$z \geqq 0.$$

The two moments

$$T_x = \iiint_H x \, dx \, dy \, dz,$$

$$T_y = \iiint_H y \, dx \, dy \, dz$$

are 0, since the respective integrations with respect to x or y give the value 0. For the third,

[1]The centroid is clearly independent of the choice of the constant positive value of the mass density. Thus, it may be thought of as a geometrical concept associated only with the shape of the region R, not dependent on the mass-distribution.

$$T_z = \iiint_H z\,dx\,dy\,dz,$$

we introduce cylindrical coordinates (r, z, θ) by means of the equations

$$z = z, \qquad x = r\cos\theta, \qquad y = r\sin\theta$$

and obtain

$$\begin{aligned} T_z &= \int_0^1 z\,dz \int_0^{\sqrt{1-z^2}} r\,dr \int_0^{2\pi} d\theta = 2\pi \int_0^1 \frac{1-z^2}{2} z\,dz \\ &= \pi\left(\frac{z^2}{2} - \frac{z^4}{4}\right)\bigg|_0^1 = \frac{\pi}{4}. \end{aligned}$$

Since the total mass is $2\pi/3$, the coordinates of the center of mass are $x = 0$, $y = 0$, $z = 3/8$.

Next, we calculate the center of mass of a hemispherical surface of unit radius over which a mass of unit density is uniformly distributed. For the parametric representation

$$x = \cos u \sin v, \qquad y = \sin u \sin v, \qquad z = \cos v$$

we calculate the surface element from formula (30e) on p. 429 and find that

$$d\sigma = \sqrt{EG - F^2}\,du\,dv = \sin v\,du\,dv. \tag{32g}$$

Accordingly, we obtain

$$T_x = \int_0^{\pi/2} \sin^2 v\,dv \int_0^{2\pi} \cos u\,du = 0,$$

$$T_y = \int_0^{\pi/2} \sin^2 v\,dv \int_0^{2\pi} \sin u\,du = 0,$$

$$T_z = \int_0^{\pi/2} \sin v \cos v\,dv \int_0^{2\pi} du = 2\pi\,\frac{\sin^2 v}{2}\bigg|_0^{\pi/2} = \pi$$

for the three moments. Since the total mass is obviously 2π, we see that the center of mass lies at the point with coordinates $x = 0$, $y = 0$, $z = \frac{1}{2}$.

b. *Moment of Inertia*

The generalization of the concept of moment of inertia to a continuous mass-distribution is equally obvious. *The moment of inertia* *of a particle with respect to the x-axis is the product of its mass and of* $\rho^2 = y^2 + z^2$, that is, of the square of the distance of the point from the x-axis. In the same way, we define the moment of inertia about the x-axis of a mass distributed with density $\mu(x, y, z)$ through a region R by the expression

$$\iiint_R \mu(y^2 + z^2)\, dx\, dy\, dz. \tag{33a}$$

The moments of inertia about the other axes are represented by similar expressions. Occasionally, the *moment of inertia with respect to a point,* say the origin, is defined by the expression

$$\iiint_R \mu(x^2 + y^2 + z^2)\, dx\, dy\, dz, \tag{33b}$$

and the *moment of inertia with respect to a plane,* say the y, z-plane, by

$$\iiint_R \mu x^2\, dx\, dy\, dz. \tag{33c}$$

Similarly, the moment of inertia, with respect to the x-axis, of a surface distribution is given by

$$\iint_S \mu(y^2 + z^2)\, d\sigma, \tag{33d}$$

where $\mu(u, v)$ is a continuous function of two parameters u and v.

The moment of inertia of a mass distributed with density $\mu(x, y, z)$ through a region R, with respect to an axis parallel to the x-axis and passing through the point (ξ, η, ζ), is given by the expression

$$\iiint_R \mu[(y - \eta)^2 + (z - \zeta)^2]\, dx\, dy\, dz. \tag{33e}$$

If in particular we let (ξ, η, ζ) be the center of mass and recall the relations (32e) for the coordinates of the center of mass, we at once obtain the equation

$$\begin{aligned}\iiint_R \mu(y^2 + z^2)\, dx\, dy\, dz &= \iiint_R \mu[(y - \eta)^2 + (z - \zeta)^2]\, dx\, dy\, dz \\ &\quad + (\eta^2 + \zeta^2) \iiint_R \mu\, dx\, dy\, dz.\end{aligned} \tag{33f}$$

Since any arbitrary axis of rotation of a body can be chosen as the x-axis, the meaning of this equation can be expressed as follows:

The moment of inertia of a rigid body with respect to an arbitrary axis of rotation is equal to the moment of inertia of the body about a parallel axis through its center of mass plus the product of the total mass and the square of the distance between the center of mass and the axis of rotation (Huygens's theorem).

The physical meaning of the moment of inertia for regions in several dimensions is exactly the same as that already stated in Volume I, p. 375:

The kinetic energy of a body rotating uniformly about an axis is equal to half the product of the square of the angular velocity and the moment of inertia.

We calculate the moment of inertia for some simple cases.

For the sphere V with center at the origin, unit radius and unit density, we see by symmetry that the moment of inertia with respect to any axis through the origin is

$$\begin{aligned} I &= \iiint_V (x^2 + y^2)\, dx\, dy\, dz \\ &= \iiint_V (x^2 + z^2)\, dx\, dy\, dz \\ &= \iiint_V (y^2 + z^2)\, dx\, dy\, dz. \end{aligned}$$

If we add the three integrals, we obtain

$$3I = \iiint_V 2(x^2 + y^2 + z^2)\, dx\, dy\, dz.$$

In spherical coordinates,

$$I = \frac{2}{3}\int_0^1 r^4\, dr \int_0^\pi \sin v\, dv \int_0^{2\pi} du = \frac{2}{3}\cdot\frac{1}{5}\cdot 2\cdot 2\pi = \frac{8\pi}{15}.$$

For a beam with edges a, b, c parallel to the x-axis, the y-axis, and the z-axis, respectively, with unit density and center of mass at the origin, we find that the moment of inertia with respect to the x, y-plane is

$$\int_{-a/2}^{a/2} dx \int_{-b/2}^{b/2} dy \int_{-c/2}^{c/2} z^2\, dz = ab\frac{c^3}{12}.$$

c. The Compound Pendulum

The notion of moment of inertia finds an application in the mathematical treatment of the compound pendulum, that is, of a rigid body which oscillates about a fixed horizontal axis under the influence of gravity.

We consider a plane through G, the center of mass of the rigid body, perpendicular to the axis of rotation; let this plane cut the axis in the point O (Fig. 4.16). The motion of the body is given as a function of time

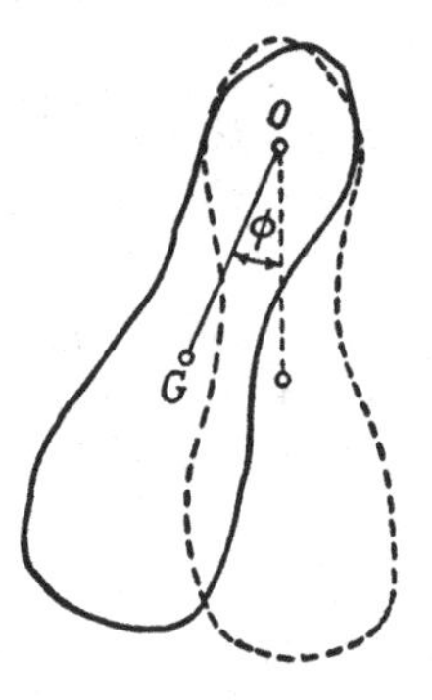

Figure 4.16

by the angle $\phi = \phi(t)$ that OG makes at time t with the downward vertical line through O. In order to determine the function ϕ and also the period of oscillation of the pendulum, we assume a knowledge of certain physical facts (see p. 658). We make use of the law of conservation of energy, which states that during the motion of the body the sum of its kinetic and potential energies remains constant. Here V, the potential energy of the body, is the product Mgh, where M is the total mass, g the gravitational acceleration, and h the height of the center of mass above an arbitrary horizontal line (e.g., above the horizontal line through the lowest position reached by the center of mass during the motion). If we denote by OG, the distance of the center of mass from the axis, by s, then $V = Mgs\,(1 - \cos\phi)$. By p. 435 the kinetic energy is given by $T = \frac{1}{2} I\dot{\phi}^2$, where I is the moment of inertia of the body with respect to the axis of rotation and we have written $\dot{\phi}$ for $d\phi/dt$. The law of conservation of energy therefore gives the equation

$$\frac{1}{2} I\dot{\phi}^2 - Mgs\cos\phi = \text{constant} \tag{34a}$$

If we introduce the constant $l = I/Ms$, this is exactly the same as the equation previously found[1] (Volume I, pp. 408, 410) for the simple pendulum; l is accordingly known as the *length of the equivalent* simple pendulum.

We can now apply the formulas obtained for the simple pendulum (Volume I, p. 410) directly. The *period of oscillation* is given by the formula

$$T = 2\sqrt{\frac{l}{2g}} \int_{-\phi_0}^{\phi_0} \frac{d\phi}{\sqrt{\cos\phi - \cos\phi_0}},$$

where ϕ_0 corresponds to the greatest displacement of the center of mass; for small angles this is approximately

$$T = 2\pi\sqrt{\frac{l}{g}} = 2\pi\sqrt{\frac{I}{Mgs}}.$$

The formula for the simple pendulum is of course included in this as a special case, for if the whole mass M is concentrated at the center of mass, then $I = Ms^2$, so that $l = s$.

Investigating further, we recall that I, the moment of inertia about the axis of rotation, is connected with I_0, the moment of inertia about a parallel axis through the center of mass, by the relation (cf. 33f)

$$I = I_0 + Ms^2.$$

Hence,

$$l = s + \frac{I_0}{Ms},$$

or if we introduce the constant $a = I_0/M$,

$$l = s + \frac{a}{s}.$$

We see at once that in a compound pendulum l always exceeds s, so that the period of a compound pendulum is always greater than

[1]In the notation used here the motion of the point mass in the simple pendulum is described by $x = l\sin\phi$, $y = -l\cos\phi$ and its speed by $l.\dot{\phi}$. Here ϕ, by Volume I, p. 408, satisfies the differential equation

$$\frac{1}{2}(l\dot{\phi})^2 - gl\cos\phi = \text{constant}.$$

that of the simple pendulum obtained by concentrating the mass M at the center of mass. Moreover, the period is the same for all parallel axes at the same distance s from the center of mass, for the length of the equivalent simple pendulum depends only on the two quantities s and $a = I_0/M$ and therefore remains the same, provided neither the direction of the axis of rotation nor its distance from the center of mass is altered.

The formula

$$T = 2\pi \sqrt{\frac{s + a/s}{g}}$$

shows that the period T increases beyond all bounds as s tends to 0 or to infinity. It must therefore have a minimum for some value s_0. By differentiating we obtain

$$s_0 = \sqrt{a} = \sqrt{\frac{I_0}{M}}.$$

A pendulum whose axis is at a distance $s_0 = \sqrt{I_0/M}$ from the center of mass will be relatively insensitive to small displacements of the axis, for in this case dT/ds vanishes, so that first-order changes in s produce only second-order changes in T. This fact has been applied by Professor M. Schuler of Göttingen in the construction of very accurate clocks.

d. *Potential of Attracting Masses*

We have seen in Chapter 2 (p. 208) that Newton's law of gravitation gives the force that a fixed particle Q with coordinates (ξ, η, ζ) and mass m exerts on a second particle P with coordinates (x, y, z) and unit mass, apart from the gravitational constant γ, as

$$m \operatorname{grad} \frac{1}{r},$$

where

$$r = \sqrt{(x - \xi)^2 + (y - \eta)^2 + (z - \zeta)^2}$$

is the distance between the points P and Q. The direction of the force is along the line joining the two particles, and its magnitude is inversely proportional to the square of the distance. Here the *gradient* of a function $f(x, y, z)$ is the vector with components

$$\frac{\partial f}{\partial x}, \frac{\partial f}{\partial y}, \frac{\partial f}{\partial z}.$$

Hence, in our case the force has the components

$$\frac{m(\xi - x)}{r^3}, \quad \frac{m(\eta - y)}{r^3}, \quad \frac{m(\zeta - z)}{r^3}.$$

If we now consider the force exerted on P by a number of points Q_1, $Q_2, \ldots, Q_n$ with respective masses $m_1, m_2, \ldots, m_n$, we can express the total force as the gradient of the quantity

$$\frac{m_1}{r_1} + \frac{m_2}{r_2} + \cdots + \frac{m_n}{r_n},$$

where r_ν denotes the distance of the point Q_ν from the point P. If a force can be expressed as a gradient of a function, it is customary to call this function the *potential of the force*;[1] we accordingly define the *gravitational potential* of the system of particles $Q_1, Q_2, \ldots, Q_n$ at the point P as the expression

$$\sum_{\nu=1}^{n} \frac{m_\nu}{\sqrt{(x - \xi_\nu)^2 + (y - \eta_\nu)^2 + (z - \zeta_\nu)^2}}.$$

We now suppose that instead of being concentrated at a finite number of points the gravitating masses are distributed with continuous density μ over a portion R of space or a surface S or a curve C. Then the potential of this mass-distribution at a point with coordinates (x, y, z) outside the system of masses is defined as

(35a) $$\iiint_R \frac{\mu(\xi, \eta, \zeta)}{r}\, d\xi\, d\eta\, d\zeta,$$

or

(35b) $$\iint_S \frac{\mu}{r}\, d\sigma,$$

or

(35c) $$\int_{s_0}^{s_1} \frac{\mu}{r}\, ds.$$

[1]Often the *negative* of this function, which has the meaning of potential energy, is called the *potential of the forces*.

In the first case, the integration is taken throughout the region R with rectangular coordinates (ξ, η, ζ); in the second case, over the surface S with the element of surface $d\sigma$; and in the third case, along the curve with length of arc s. In all three formulae, r denotes the distance of the point P from the point (ξ, η, ζ) of the region of integration and μ the mass density at the point (ξ, η, ζ). In each case the force of attraction is found by forming the first derivatives of the potential with respect to x, y, z. Working with the potential rather than with the force has the advantage that only one integral instead of three has to be evaluated. The three force components are then obtained as derivatives of the potential.

For example, the potential at the point P with coordinates (x, y, z) due to a sphere K with uniform density 1, with unit radius and with center at the origin, is the integral

$$\iiint_K \frac{d\xi\, d\eta\, d\zeta}{\sqrt{(x-\xi)^2 + (y-\eta)^2 + (z-\zeta)^2}}$$

$$= \int_{-1}^{+1} d\xi \int_{-\sqrt{1-\xi^2}}^{+\sqrt{1-\xi^2}} d\eta \int_{-\sqrt{1-\xi^2-\eta^2}}^{+\sqrt{1-\xi^2-\eta^2}} \frac{1}{r}\, d\zeta.$$

In all the expressions (35a, b, c) the coordinates (x, y, z) of the point P appear not as variables of integration but as parameters, and the potentials are functions of these parameters.

To obtain the components of the force from the potential we have to differentiate the integral with respect to the parameters. The rules for differentiation with respect to a parameter extend directly to multiple integrals, and by p. 74, the differentiation can be performed under the integral sign, provided that the point P does not belong to the region of integration, that is, provided that we are certain that there is no point of the closed region of integration for which the distance r has the value 0. Thus, for example, we find that the *components of the gravitational force* on a unit mass due to a mass distributed with unit density through a region R in space are given by the expressions

$$F_1 = -\iiint_R \frac{x-\xi}{r^3}\, d\xi\, d\eta\, d\zeta, \tag{36}$$

$$F_2 = -\iiint_R \frac{y-\eta}{r^3}\, d\xi\, d\eta\, d\zeta,$$

$$F_3 = -\iiint_R \frac{z-\zeta}{r^3}\, d\xi\, d\eta\, d\zeta.$$

Finally, we point out that the expressions for the potential and its first derivatives continue to have a meaning if the point P lies in the interior of the region of integration. The integrals are then improper integrals, and as is easily shown, their convergence follows from the criteria of Section 4.7

As an illustration, we calculate the potential at an internal point and at an external point due to a spherical surface S with radius a and unit density. If we take the center of the sphere as the origin and let the x-axis pass through the point P (inside or outside the sphere), the point P will have the coordinates $(x, 0, 0)$, and the potential will be

$$U = \iint \frac{d\sigma}{\sqrt{(x-\xi)^2 + \eta^2 + \zeta^2}}.$$

If we introduce spherical coordinates on the sphere through the equations

$$\xi = a \cos\theta,$$
$$\eta = a \sin\theta \cos\phi,$$
$$\zeta = a \sin\theta \sin\phi,$$

then [see (30e), p. 429]

$$U = \int_0^{\pi} \frac{a^2 \sin\theta}{\sqrt{(x - a\cos\theta)^2 + a^2\sin^2\theta}}\, d\theta \int_0^{2\pi} d\phi$$
$$= 2\pi \int_0^{\pi} \frac{a^2 \sin\theta}{\sqrt{x^2 + a^2 - 2ax\cos\theta}}\, d\theta.$$

We put $x^2 + a^2 - 2ax\cos\theta = r^2$, so that $ax \sin\theta\, d\theta = r\, dr$, and (provided that $x \neq 0$) the integral then becomes

$$U = \frac{2\pi a}{x} \int_{|x-a|}^{|x+a|} \frac{r\, dr}{r} = \frac{2\pi a}{x}(|x+a| - |x-a|).$$

For $|x| > a$ we therefore have

$$U = \frac{4\pi a^2}{|x|},$$

and for $|x| < a$,

$$U = 4\pi a.$$

Hence, the potential at an external point is the same as if the whole mass $4\pi a^2$ were concentrated at the center of the sphere. On the other hand, throughout the interior the potential is constant. At the surface of the sphere the potential is continuous; the expression for U is still defined (as an improper integral) and has the value $4\pi a$. The component of force F_x in the x-direction, however, has a jump of amount -4π at the surface of the sphere, for if $|x| > a$, we have

$$F_x = -\frac{4\pi a^2}{x^2}\operatorname{sgn} x,$$

while $F_x = 0$ if $|x| < a$.

The potential of a solid sphere of unit density is found from that of a spherical surface by integrating with respect to a. This gives the value

$$\frac{4\pi a^3}{3|x|}$$

for the potential at an external point. This again is the same as if the total mass $(4/3)\pi a^3$ were concentrated at the center. By differentiation with respect to x we find for a point on the positive x-axis that

$$F_x = -\frac{4\pi a^3}{x^2}.$$

This is Newton's result that the attraction exerted by a solid sphere of constant density on an external point is the same as if the mass of the sphere were concentrated at its center (Volume I, p. 413).

Exercises 4.9

1. (a) Find the position of the centroid of a solid right circular cone.
   (b) What is the position of the centroid of the curved surface of the cone?
2. Find the position of the centroid of the portion of the paraboloid $z^2 + y^2 = px$ cut off by the plane $x = x_0$, where $x_0 < 0$.
3. Find the centroid of the tetrahedron bounded by the three coordinate planes and the plane $x/a + y/b + z/c = 1$.
4. (a) Find the centroid of the hemispherical shell $a^2 \leqq x^2 + y^2 + z^2 \leqq b^2$, $z \geqq 0$.
   (b) Show that the centroid of the hemispherical lamina $x^2 + y^2 + z^2 = a^2$ is the limiting position of the centroid in part (a) as b approaches a.

5. Find the moment of inertia about the z-axis of the homogeneous rectangular parallelopiped of mass m with $0 \leqq x \leqq a, 0 \leqq y \leqq b, 0 \leqq z \leqq c$.
6. Calculate the moment of inertia of the homogeneous solid enclosed between the two cylinders

$$x^2 + y^2 = R \qquad \text{and} \qquad x^2 + y^2 = R' \qquad\qquad (R > R')$$

and the two planes $z = h$ and $z = -h$, with respect to
(a) the z-axis,
(b) the x-axis.
7. Find the mass and moment of inertia about a diameter of a sphere whose density decreases linearly with distance from the center from a value μ_0 at the center to the value μ_1, at the surface.
8. Find the moment of inertia of the ellipsoid $x^2/a^2 + y^2/b^2 + z^2/c^2 \leqq 1$ with respect to
(a) the z-axis,
(b) an arbitrary axis through the origin, given by

$$x{:}y{:}z = \alpha{:}\beta{:}\gamma \qquad\qquad (\alpha^2 + \beta^2 + \gamma^2 = 1).$$

9. If A, B, C denote the moments of inertia of an arbitrary solid of positive density with respect to the x-, y-, and z-axis, then the "triangle inequalities"

$$A + B > C, \quad A + C > B, \quad B + C > A$$

are satisfied.
10. Let O be an arbitrary point and S an arbitrary body. On every ray from O we take the point at the distance $1/\sqrt{I}$ from O, where I denotes the moment of inertia of S with respect to the straight line coinciding with the ray. Prove that the points so constructed form an ellipsoid (the so-called *momental ellipsoid*).
11. Find the momental ellipsoid of the ellipsoid $x^2/a^2 + y^2/b^2 + z^2/c^2 \leqq 1$ at the point (ξ, η, ζ).
12. Find the coordinates of the center of mass of the surface of the sphere $x^2 + y^2 + z^2 = 1$, the density being given by

$$\mu = \frac{1}{\sqrt{(x-1)^2 + y^2 + z^2}}\,.$$

13. Find the x-coordinate of the center of mass of the octant of the ellipsoid

$$x^2/a^2 + y^2/b^2 + z^2/c^2 \leqq 1 \qquad\qquad (x \geqq 0,\ y \geqq 0,\ z \geqq 0).$$

14. A system of masses S consists of two parts S_1 and S_2; I_1, I_2, I are the respective moments of inertia of S_1, S_2, S about three parallel axes passing through the respective centers of mass. Prove that

$$I = I_1 + I_2 + \frac{m_1 m_2}{m_1 + m_2}\, d^2,$$

where m_1 and m_2 are the masses of S_1 and S_2 and d the distance between the axes passing through their centers of mass.

15. Find the envelopes of the planes with respect to which the ellipsoid $(x^2/a^2) + (y^2/b^2) + (z^2/c^2) \leqq 1$ has the same moment of inertia h.
16. Calculate the potential of the homogeneous ellipsoid of revolution
$$\frac{x^2 + y^2}{a^2} + \frac{z^2}{b^2} \leqq 1 \qquad (b > a)$$
at its center.
17. Calculate the potential of a solid of revolution
$$r = \sqrt{x^2 + y^2} \leqq f(z) \qquad (a \leqq z \leqq b)$$
at the origin.
18. Show that at sufficiently great distances the potential of a solid S is approximated by the potential of a particle of the same total mass located at its center of gravity with an error less than some constant divided by the square of the distance.
19. Assuming that the earth is a sphere of radius R for which the density at a distance r from the center is of the form
$$\rho = A - Br^2$$
and the density at the surface is $2\frac{1}{2}$ times the density of water, while the mean density is $5\frac{1}{2}$ times that of water, show that the attraction at an internal point is equal to
$$\frac{1}{11} g \frac{r}{R} \left(20 - 9 \frac{r^2}{R^2}\right),$$
where g is the value of gravity at the surface.
20. A hemisphere of radius a and of uniform density ρ is placed with its center at the origin, so as to lie entirely on the positive side of the x, y-plane. Show that its potential at the point $(0, 0, z)$ is
$$\frac{2\pi\rho}{3z}\left[(a^2 + z^2)^{3/2} - a^3 + \frac{3}{2}a^2 z\right] - \frac{4}{3}\pi\rho z^2 \qquad \text{if} \qquad 0 < z < a$$
and
$$\frac{2\pi\rho}{z}\left[(a^2 + z^2)^{3/2} + a^3 - \frac{3}{2}a^2 z\right] - \frac{2}{3}\pi\rho z^2 \qquad \text{if} \qquad z > a.$$
21. Let (x_1, y_1), (x_2, y_2), (x_3, y_3) be the vertices of a triangle of area A (the order of the suffixes giving the positive orientation). Prove that the moment of inertia of the triangle with respect to the x-axis is given by
$$\frac{A}{6}(y_1^2 + y_2^2 + y_3^2 + y_1 y_2 + y_2 y_3 + y_3 y_1).$$
22. Prove that the attraction at either pole of a uniform spheroid with density ρ and semiaxes a, a, c is equal to

$$2\pi\rho \int_0^{2c} r(1 - \cos\theta)\, dr,$$

where

$$r = 2a^2c\cos\theta/(a^2\cos^2\theta + c^2\sin^2\theta).$$

23. It is known experimentally that a charged conducting spherical lamina (on such a surface the charge distributes itself uniformly) exerts zero force on a point charge inside the sphere. Assuming that point charges repel or attract each other with a force dependent only on the distance between them, prove that this experiment implies Coulomb's law—namely, that point charges attract or repel each other with a force proportional to the inverse square of their separation. This result is the converse of the theorem that the force of gravity of a homogeneous spherical lamina vanishes in its interior.

4.10 Multiple Integrals in Curvilinear Coordinates

a. Resolution of Multiple Integrals

If the region R of the x, y-plane is covered by a family of curves $\phi(x, y) =$ constant, so that each point of R lies on one, and only one, curve of the family, we can take the quantity $\phi(x, y) = \xi$ as a new independent variable; that is, we can take the curves C_ξ represented by $\phi(x, y) =$ constant $= \xi$ as one of the two families of curves in a coordinate grid.

For the second independent variable we can choose the quantity $\eta = y$, provided that we restrict ourselves to a region R in which each pair of curves $\phi(x, y) =$ constant and $y =$ constant intersect in one point.

If we introduce these new variables, a double integral $\iint_R f(x, y)\, dx\, dy$ is transformed as follows [cf. (16b), p. 403]:

$$\iint f(x, y)\, dx\, dy = \iint \frac{f(x, y)}{|\phi_x|}\, d\xi\, d\eta.$$

Keeping ξ constant and integrating the right-hand side with respect to η, the integral with respect to η can be written in the form

$$\int \frac{f(x, y)}{\sqrt{\phi_x^2 + \phi_y^2}} \frac{\sqrt{\phi_x^2 + \phi_y^2}}{|\phi_x|}\, d\eta.$$

Since on C_ξ

$$\frac{ds}{d\eta} = \sqrt{1 + \left(\frac{dx}{dy}\right)^2} = \frac{\sqrt{\phi_x^2 + \phi_y^2}}{|\phi_x|}$$

this integral may be regarded as an integral along the curve $\phi(x, y) = \xi$, the length of arc s being the variable of integration. Thus, we obtain the resolution

$$\iint f(x, y)\, dx\, dy = \int d\xi \int_{C_\xi} \frac{f(x, y)}{\sqrt{\phi_x^2 + \phi_y^2}}\, ds \tag{37a}$$

for our double integral.

The intuitive meaning of this resolution is very easily recognized if we suppose that corresponding to the curves C_ξ there is a family of orthogonal curves (the so-called *orthogonal trajectories*) that intersect each separate curve $\phi = \text{constant} = \xi$ at right angles, in the direction of the vector grad ϕ. If σ is the length of arc on an orthogonal curve represented by the functions $x(\sigma)$ and $y(\sigma)$, then

$$\frac{dx}{d\sigma} = \frac{\phi_x}{\sqrt{\phi_x^2 + \phi_y^2}}, \quad \frac{dy}{d\sigma} = \frac{\phi_y}{\sqrt{\phi_x^2 + \phi_y^2}}.$$

Since

$$\frac{d\xi}{d\sigma} = \phi_x \frac{dx}{d\sigma} + \phi_y \frac{dy}{d\sigma},$$

we obtain

$$\frac{d\xi}{d\sigma} = \sqrt{\phi_x^2 + \phi_y^2} = \sqrt{(\text{grad}\,\phi)^2}. \tag{37b}$$

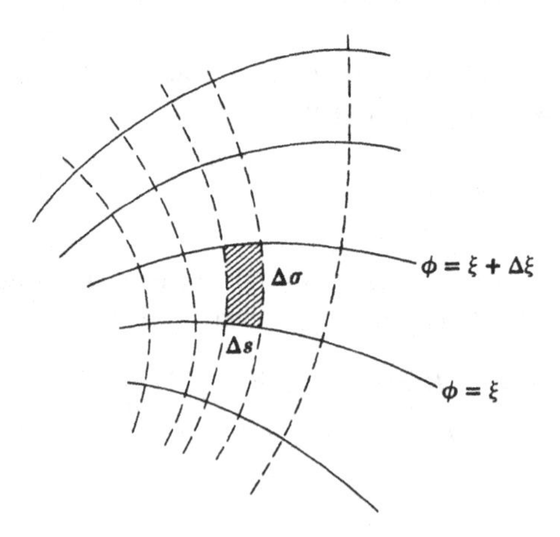

Figure 4.17

We now consider the mesh bounded by two curves $\phi(x, y) = \xi$, $\phi(x, y) = \xi + \Delta\xi$, and two orthogonal curves that cut off a portion of length Δs from $\phi(x, y) = \xi$ (Fig. 4.17). The area of this mesh is given approximately by the product $\Delta s\, \Delta\sigma$, and this in turn is approximately equal to

$$\frac{\Delta s\, \Delta\xi}{\sqrt{\phi_x^2 + \phi_y^2}}.$$

This leads to a new interpretation of the identity (37a);

Instead of calculating a double integral by subdividing the region into "infinitesimal rectangles" with sides parallel to the coordinate axes, we may use the subdivision into infinitesimal curvilinear rectangles determined by the curves $\phi(x, y)$ = constant and their orthogonal trajectories.

A similar resolution can be effected in three-dimensional space. If the region R is covered by a family of surfaces S_ξ given by an equation $\phi(x, y, z)$ = constant $= \xi$ in such a way that through every point there passes one, and only one, surface, then we can take the quantity $\xi = \phi(x, y, z)$ as a variable of integration. In this way we resolve a triple integral

$$\iiint_R f(x, y, z)\, dx\, dy\, dz$$
$$= \int d\xi \iint \frac{f(x, y, z)}{\sqrt{\phi_x^2 + \phi_y^2 + \phi_z^2}} \frac{\sqrt{\phi_x^2 + \phi_y^2 + \phi_z^2}}{|\phi_x|}\, dy\, dz$$

into an integral

$$\iint_{S_\xi} \frac{f(x, y, z)}{\sqrt{\phi_x^2 + \phi_y^2 + \phi_z^2}}\, dS$$

over the surface $\phi = \xi$ with element of area

$$dS = \frac{\sqrt{\phi_x^2 + \phi_y^2 + \phi_z^2}}{|\phi_x|}\, dy\, dz$$

[see (29d), p. 426] and a subsequent integration with respect to ξ:

$$\text{(37c)} \qquad \iiint (fx, y, z)\, dx\, dy\, dz = \int d\xi \iint_{S_\xi} \frac{f(x, y, z)}{\sqrt{\phi_x^2 + \phi_y^2 + \phi_z^2}}\, dS.$$

This formula again permits a geometric interpretation if we introduce the two-parametric family of curves orthogonal at each point to a surface ξ = constant and use, in addition to the S_ξ, coordinate surfaces consisting of those curves.

b. *Application to Areas Swept Out by Moving Curves and Volumes Swept Out by Moving Surfaces. Guldin's Formula. The Polar Planimeter*

The quantity

$$\frac{d\sigma}{d\xi} = \frac{1}{\sqrt{\phi_x^2 + \phi_y^2}}$$

appearing in formulae (37a, b) can be interpreted kinematically if we identify the parameter ξ with the time t. The equation $\phi(x, y) =$ constant $= t$ represents then the position C_t of a moving curve at the time t. The quantity $\Delta\sigma$, which measures distances along the curves orthogonal to the curves C_t, can be thought of as the *normal distance* between the curves C_t and $C_{t+\Delta t}$. Accordingly,

$$c = \frac{d\sigma}{dt} = \frac{1}{\sqrt{\phi_x^2 + \phi_y^2}} \tag{38a}$$

is the *normal velocity* of the moving curve C_t at the time t. This velocity is different at different points of C_t. Similarly, the normal velocity of the moving surface S_t in space with equation $\phi(x, y, z) =$ constant $= t$ is

$$c = \frac{1}{\sqrt{\phi_x^2 + \phi_y^2 + \phi_z^2}}. \tag{38b}$$

In physics, such moving surfaces occur as *wave fronts* (e.g. for electromagnetic waves propagating in a medium).

The normal velocity c of a moving surface S_t (and similarly of a moving curve C_t in the plane) has a particularly simple meaning if S_t consists of individual moving particles. If the position of one of these particles is described by the three functions $x = x(t)$, $y = y(t)$, $z = z(t)$ and if the particle at all times stays on the moving surface, the equation

$$\phi(x(t), y(t), z(t)) = t$$

must hold for all t. Differentiating with respect to t we find the equation

$$1 = \phi_x \frac{dx}{dt} + \phi_y \frac{dy}{dt} + \phi_z \frac{dz}{dt}.$$

If we divide this equation by the absolute gradient of ϕ we obtain the relation

$$c = \pm \left(\xi \frac{dx}{dt} + \eta \frac{dy}{dt} + \zeta \frac{dz}{dt}\right), \tag{38c}$$

where c is the normal velocity defined by (38b), ξ, η, ζ are the direction cosines of one of the normals of S_t, and the positive or negative sign applies according to the normals pointing in the direction of increasing or decreasing t, respectively. If we introduce the unit-normal vector

$$\mathbf{n} = (\xi, \eta, \zeta)$$

and the velocity vector of the particle

$$\mathbf{v} = \left(\frac{dx}{dt}, \frac{dy}{dt}, \frac{dz}{dt}\right)$$

we can represent c by the scalar product

$$c = \pm \mathbf{v} \cdot \mathbf{n} \tag{38d}$$

In words, *the component normal to the surface S_t of the velocity of a particle moving with the surface equals $\pm c$ where c is the normal velocity of S_t. The positive sign* holds when $\mathbf{n}$ is the "forward" normal of S_t, that is, the normal on the side of the surface facing the points to be swept over in the immediate future.

Formula (37c) for $f = 1$ yields an expression for the volume V of the region swept over by a moving surface S_t with normal velocity c:

$$V = \iiint dx\, dy\, dz = \int dt \iint_{S_t} c\, dS. \tag{39a}$$

Similarly, we find for the area A of a region in the plane swept over by a moving curve C_t the expression

$$A = \int dt \int_{C_t} c\, ds. \tag{39b}$$

We apply these results to the case of an area swept over by a straight line segment C_t moving in the plane (Fig. 4.18). The segment can be represented by an equation of the form

$$\xi(t)x + \eta(t)y = p(t), \tag{40a}$$

where (ξ, η) is the unit normal and p the (signed) distance of C_t from the origin. The center of C_t (which is the same as its *centroid*) is at the point [see (32e), p. 432]

$$X(t) = \frac{\int_{C_t} x\, ds}{\int_{C_t} ds}, \quad Y(t) = \frac{\int_{C_t} y\, ds}{\int_C ds}. \tag{40b}$$

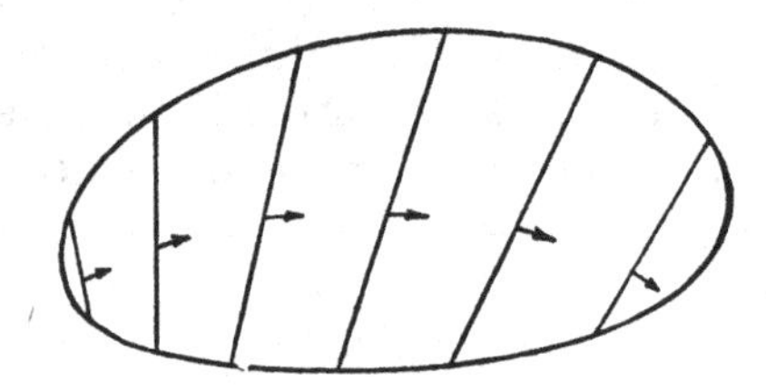

Figure 4.18

Integration of (40a) with respect to s over the segment C_t furnishes the relation

$$\xi(t)X(t) + \eta(t)Y(t) = p(t), \tag{40c}$$

which merely states that the center of C_t lies on C_t. If C_t is thought to consist of individual moving particles the normal component of the velocity of these particles is found from (40a), (38c) to be

$$\mathbf{n} \cdot \mathbf{v} = \xi \frac{dx}{dt} + \eta \frac{dy}{dt} = \frac{dp}{dt} - \frac{d\xi}{dt} x - \frac{d\eta}{dt} y.$$

Hence by (40b), (40c)

$$\begin{aligned} \pm \int_{C_t} c\, ds &= \int_{C_t} \mathbf{n} \cdot \mathbf{v}\, ds = \left(\frac{dp}{dt} - \frac{d\xi}{dt} X - \frac{d\eta}{dt} Y\right) \int_{C_t} ds \\ &= \left(\xi \frac{dX}{dt} + \eta \frac{dY}{dt}\right) \int_{C_t} ds = \mathbf{w} \cdot \mathbf{n} L \end{aligned}$$ [1]

[1]The same formula can also be derived using the expression (38a) for c if one calculates the first derivatives of the function $t = \phi(x, y)$ with respect to x and y from the implicit equation (40a) for the function t.

where

$$\mathbf{w} = \left(\frac{dX}{dt}, \frac{dY}{dt}\right)$$

is the velocity vector of the center (X, Y) of the segment C_t, and

$$L = L(t) = \int_{C_t} ds,$$

the length of C_t. It follows from (39b) that the area swept over by the moving segment C_t is

(41a) $$A = \int \pm L\mathbf{w} \cdot \mathbf{n}\, dt.$$

In the same way, one finds that the volume swept out by a moving plane region S_t of area $A(t)$ and unit normal $\mathbf{n}$ is

(41b) $$V = \int \pm A\mathbf{w} \cdot \mathbf{n}\, dt,$$

where $\mathbf{w}$ is the velocity of the centroid (X, Y, Z) of S_t. In these formulas the positive sign is taken when $\mathbf{n}$ is the "forward normal" of S_t, the one that points in the direction of motion.

Of special interest is the case of formula (41b) in which the centroid (X, Y, Z) of S_t moves along a curve which at every moment is perpendicular to the plane of S_t. In that case, the normal component of velocity of the centroid coincides with the speed of motion of the centroid along its path:

$$\pm \mathbf{w} \cdot \mathbf{n} = \frac{d\sigma}{dt},$$

where σ is the length of arc along the path of the centroid. It follows then that

(42a) $$V = \int A \frac{d\sigma}{dt}\, dt = \int A\, d\sigma.$$

If, moreover, all the plane regions S_t have the same area A, we find that

(42b) $$V = A \int d\sigma,$$

or that the volume swept out by the S_t is equal to their area A multiplied by the length of the path described by their centroids. A particular case is obviously Guldin's rule for the volume of a solid of revolution swept out by rotation of a plane region R about an axis in that plane. The volume is equal to the area A of R multiplied by the length of the path described by the centroid of R during the revolution (see Volume I, p. 374).

Returning to formula (41a) we see that the integral

$$\int L\mathbf{w} \cdot \mathbf{n}\, dt \tag{43a}$$

represents the signed area swept out by the segments C_t, the sign depending on whether the normal $\mathbf{n}$ points in the direction of motion or in the opposite one. The same holds for an integral

$$\int A\mathbf{w} \cdot \mathbf{n}\, dt \tag{43b}$$

associated with volumes swept out by a moving plane area.

These observations allow us to extend our results to cases in which the segment or plane area does not always move in the same sense or covers part of the plane (or space) more than once. The integrals given above will then express the algebraic sum of the areas (or volumes) of the parts of the region described, each taken with the appropriate sign.

As an example, let a segment of constant length move so as to have its end points always on two fixed curves Γ and Γ' in a plane, as in Fig. 4.19. From the arrows showing the positive direction of the normal, we can determine the sign with which each area appears in the integral, and we find that the integral gives the difference between the areas enclosed by Γ and Γ'. If Γ' contains zero area, as when it de-

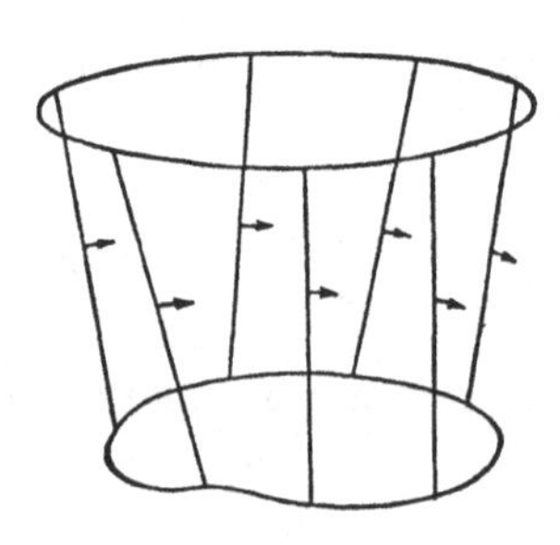

Figure 4.19

generates into a single segment of a curve multiply described, the integral gives the area enclosed by Γ.

This principle is used in the construction of the well-known polar planimeter *(Amsler's planimeter)*. This is a mechanical apparatus for measuring plane areas. It consists of a rigid rod at the center of which is a measuring wheel that can roll on the drawing-paper. The plane of the wheel is perpendicular to the rod. When the instrument is to be used to measure the area enclosed by a curve Γ drawn on the paper, one end of the rod is moved round the curve, while the other is hinged to a rigid arm whose other end pivots about a fixed point O, the pole, exterior to Γ. The hinged end of the rod therefore describes (multiply) an arc of a circle, that is, a closed curve containing zero area. It follows that here the expression (43a) furnishes the area enclosed by Γ. But the integrand $L\mathbf{w}\cdot\mathbf{n}$ is proportional to the angular speed with which the measuring wheel turns, provided that the circumference of the wheel moves on the paper as the rod moves, in which case the position of the wheel is only affected by the motion normal to the rod. The total angle by which the wheel has turned is then proportional to the area enclosed by Γ.

In the instrument as usually constructed the wheel is not exactly at the center of the rod, but this only alters the factor of proportionality in the result, and the factor can be determined directly by a calibration of the instrument.

4.11 Volumes and Surface Areas in Any Number of Dimensions

a. *Surface Areas and Surface Integrals in More than Three Dimensions*

In n-dimensional space described by n coordinates $x_1, \ldots, x_n$ an $(n-1)$-dimensional surface (*hypersurface* or *manifold*) is defined by an implicit equation

$$\phi(x_1, x_2, \ldots, x_n) = \text{constant}, \tag{44a}$$

where at each point of the surface at least one of the first derivatives of ϕ does not vanish. We suppose that a portion S of this surface corresponds to a certain region B in $x_1x_2 \cdots x_{n-1}$-space where $\partial\phi/\partial x_n \neq 0$ and x_n can be calculated from equation (44a) as a function of the other coordinates.

We now define the $(n-1)$-measure of this portion of surface as the integral

(44b) $$A = \iint_B \cdots \int \frac{\sqrt{\phi_{x_1}^2 + \phi_{x_2}^2 + \cdots + \phi_{x_n}^2}}{|\phi_{x_n}|}\, dx_1\, dx_2 \cdots dx_{n-1}.$$

This definition is a formal generalization of formula (29b), p. 425 for areas of surfaces in three-space and can be based on similar intuitive arguments. When there is no danger of confusion, we shall also refer to A simply as "area" even in the case of a hypersurface in n-dimensional space. A more systematic discussion of surfaces, surface areas, and surface integrals will be given in the next chapter. For the moment, we observe only that the quantity A defined by (44b) is independent of the choice of the coordinate x_n for which we solve equation (44a). This may be proved in the same way as was done in the three-dimensional case on p. 426.

More generally, we define the *integral of a function* $f(x_1, \ldots, x_n)$ *over this* $(n-1)$*-dimensional surface as*

(44c) $$\iint_S \cdots \int f(x_1, \ldots, x_n)\, d\sigma$$

$$= \iint_B \cdots \int f(x_1, \ldots, x_n) \frac{\sqrt{\phi_{x_1}^2 + \cdots + \phi_{x_n}^2}}{|\phi_{x_n}|}\, dx_1\, dx_2 \cdots dx_{n-1},$$

where, as before, we suppose that x_n is expressed in terms of $x_1, \ldots, x_{n-1}$ by means of equation (44a). We again find that the value of the expression (44c) is independent of the choice of the variable x_n.

As for two or three dimensions, a multiple *volume integral* over an n-dimensional region R

(45a) $$\iint_R \cdots \int f(x_1, \ldots, x_n)\, dx_1, \ldots, dx_n$$

can be resolved into surface integrals [see formulas (37a, c)]. We assume that the region R is covered by a family of hypersurfaces S_ξ

(45b) $$\phi(x_1, \ldots, x_n) = \text{constant} = \xi$$

in such a way that through each point of R there passes one, and only one, surface. If we replace $x_1, \ldots, x_{n-1}, x_n$ by new independent variables

$$x_1, \ldots, x_{n-1}, \xi = \phi(x_1, \ldots, x_n),$$

the multiple integral (45a) becomes by the rule for transformation of integrals (p. 404)

$$\int d\xi \int \cdots \int \frac{f(x_1, \ldots, x_n)}{|\phi_{x_n}|} dx_1 \cdots dx_{n-1}.$$

Using formula (44c), we obtain the formula

(45c)
$$\iint_R \cdots \int f(x_1, \ldots, x_n) dx_1 \cdots dx_n$$
$$= \int d\xi \int_{S_\xi} \cdots \int \frac{f(x_1, \ldots, x_n)}{\sqrt{\phi_{x_1}^2 + \cdots + \phi_{x_n}^2}} d\sigma,$$

where

(45d)
$$d\sigma = \frac{\sqrt{\phi_{x_1}^2 + \cdots + \phi_{x_n}^2}}{|\phi_{x_n}|} dx_1 \cdots dx_{n-1}$$

is the element of area of the surface S_ξ.

b. *Area and Volume of the n-Dimensional Sphere*

As an application of the formula (45c) for reduction of volume to surface integrals, we shall calculate the area and volume of a sphere of radius R in n-dimensional space, that is, the area of the hypersurface with equation

(46a)
$$x_1^2 + \cdots + x_n^2 = R^2,$$

and the volume of the ball

(46b)
$$x_1^2 + \cdots + x_n^2 \leq R^2.$$

We first derive a general formula that reduces the space integral of a function with spherical symmetry to a single integral. We say the function f of the variables $x_1, \ldots, x_n$ has *spherical symmetry* if

$$f = f(r),$$

where

(46c)
$$r = \sqrt{x_1^2 + \cdots + x_n^2},$$

that is, if f is constant on spheres with centers at the origin. The sphere S_r of radius r about the origin is given by the equation

(46d)
$$\phi(x_1, \ldots, x_n) = \sqrt{x_1^2 + \cdots + x_n^2} = \text{constant} = r.$$

Here

$$\phi_{x_i} = \frac{1}{r} x_i; \quad \sqrt{\phi_{x_1}^2 + \cdots + \phi_{x_n}^2} = 1. \tag{46e}$$

From (45c) we then obtain the volume integral of the function $f(r)$ over the ball (46b), namely,

$$\begin{aligned} \iint \cdots \int f(r)\, dx_1 \cdots dx_n &= \int_0^R f(r)\, dr \int_{S_r} \cdots \int d\sigma \\ &= \int_0^R f(r)\ \Omega_n(r) dr, \end{aligned} \tag{46f}$$

where $\Omega_n(r)$ is the area of the sphere S_r. Here, by (44b), (46e) the area of the hemisphere

$$\phi = \sqrt{x_1^2 + \cdots + x_n^2} = r \qquad (x_n \geqq 0)$$

is

$$\frac{1}{2} \Omega_n(r) = r \int_{B_r} \cdots \int \frac{dx_1 \cdots dx_{n-1}}{x_n}, \tag{47a}$$

where the integration is extended over the $(n-1)$-dimensional ball B_r given by

$$x_1^2 + \cdots + x_{n-1}^2 \leqq r^2,$$

and where

$$x_n = \sqrt{r^2 - x_1^2 - \cdots - x_{n-1}^2}.$$

Replacing $x_1, \ldots, x_{n-1}$ in B_r by the new variables

$$\xi_i = \frac{1}{r} x_i \qquad (i = 1, \ldots, n-1)$$

and putting

$$\xi_n = \frac{1}{r} x_n = \sqrt{1 - \xi_1^2 - \cdots - \xi_{n-1}^2},$$

we obtain from (47a) that

$$\Omega_n(r) = 2r^{n-1} \int \cdots \int \frac{d\xi_1 \cdots d\xi_{n-1}}{\xi_n}, \tag{47b}$$

where the integration is over the unit ball in $n-1$ dimensions

$$\xi_1^2 + \cdots + \xi_{n-1}^2 \leq 1.$$

Formula (47b) can be written as

$$\Omega_n(r) = \omega_n \, r^{n-1}, \tag{47c}$$

where

$$\omega_n = 2\iint\cdots\int \frac{d\xi_1 \cdots d\xi_{n-1}}{\xi_n} = \Omega_n(1)$$

is the area of the unit sphere S_1 in n dimensions. It expresses the intuitively plausible fact that *areas of spheres in n dimensions are proportional to the* $(n-1)$*-st power of their radius.* Formula (46f) for the space integral over the ball (46b) of a function with spherical symmetry now takes the form

$$\iint\cdots\int f(r)\,dx_1 \cdots dx_n = \omega_n \int_0^R f(r) r^{n-1}\,dr. \tag{48a}$$

We can calculate ω_n conveniently from this formula. We choose for $f(r)$ a function for which the integral on the right converges absolutely for $R \to \infty$ and can be evaluated explicitly. The improper integral of $f(r)$ as a function of $x_1, \ldots, x_n$ over the whole space then also converges. We choose for f the function[1]

$$f(r) = \exp(-r^2) = \exp(-x_1^2 - \cdots - x_n^2).$$

The integral of f over the whole space is the limit of integrals over cubes C_a with center at the origin and sides of length $2a$ parallel to the axes. Here

$$\begin{aligned}
&\iint_{C_a}\cdots\int f(r)\,dx_1 \cdots dx_n \\
&= \int_{-a}^{a} dx_1 \int_{-a}^{a} dx_2 \cdots \int_{-a}^{a} dx_n \exp(-x_1^2)\exp(-x_2^2)\cdots\exp(-x_n^2) \\
&= \left(\int_{-a}^{a} e^{-x^2}\,dx\right)^n.
\end{aligned}$$

[1] One conveniently writes $\exp(z)$ for the exponential function e^z in cases where the exponent z is a more complicated expression.

Thus, for $a \to \infty$, we obtain from (48a) the identity

$$\left(\int_{-\infty}^{+\infty} e^{-x^2}\,dx\right)^n = \omega_n \int_0^\infty e^{-r^2} r^{n-1}\,dr. \tag{48b}$$

For the special case $n = 2$, this formula already has been derived by a similar argument on p. 415 and led to the result [see (25a)] that

$$\Gamma\left(\frac{1}{2}\right) = \int_{-\infty}^{\infty} e^{-x^2}\,dx = \sqrt{\pi}. \tag{48c}$$

On the other hand, the substitution $r^2 = s$ shows that

$$\int_0^\infty e^{-r^2} r^{n-1}\,dr = \frac{1}{2}\int_0^\infty e^{-s} s^{(n-2)/2}\,ds = \frac{1}{2}\Gamma\left(\frac{n}{2}\right). \tag{48d}$$

Here $\Gamma(\mu)$ denotes the gamma function defined by

$$\Gamma(\mu) = \int_0^\infty e^{-s} s^{\mu-1}\,ds \qquad (\mu > 0)$$

in Volume I (p. 308).[1] Hence, (48b) leads to the value

$$\omega_n = \frac{2\sqrt{\pi^n}}{\Gamma\left(\frac{n}{2}\right)} \tag{48e}$$

for the surface area of the unit sphere in n dimensions. The value of $\Gamma(n/2)$ for integers n is easily determined from the recursion formula

$$\Gamma(\mu) = (\mu - 1)\,\Gamma(\mu - 1), \tag{48f}$$

which follows directly by integration by parts from the definition of the gamma function (see Volume I, p. 308). Hence, for even n

$$\Gamma\left(\frac{n}{2}\right) = \frac{n-2}{2}\,\frac{n-4}{2}\cdots\frac{2}{2}\Gamma(1) = \left(\frac{n}{2} - 1\right)! \tag{48g}$$

while for odd n, using (48c),

$$\Gamma\left(\frac{n}{2}\right) = \frac{n-2}{2}\,\frac{n-4}{2}\cdots\frac{1}{2}\Gamma\left(\frac{1}{2}\right) = \frac{(n-2)(n-4)\cdots 3\cdot 1}{2^{(n-1)/2}}\sqrt{\pi}. \tag{48h}$$

In this way we obtain from (48e) successively the values

[1]See also pp. 497 of the present volume.

$$\omega_2 = 2\pi, \quad \omega_3 = 4\pi, \quad \omega_4 = 2\pi^2, \quad \omega_5 = \frac{8}{3}\pi^2, \ldots .$$

In order to find the *volume of the n-dimensional ball* $V_n(R)$ of radius R, we put $f = 1$ in formula (48a) and find that

$$V_n(R) = \iint \cdots \int dx_1 \cdots dx_n = \omega_n \int_0^R r^{n-1}\, dr = v_n R^n, \tag{49a}$$

where

$$v_n = \frac{1}{n}\,\omega_n = \frac{\sqrt{\pi}^n}{\Gamma\left(\frac{n+2}{2}\right)} \tag{49b}$$

is the volume of the n-dimensional unit ball. Thus,

$$v_1 = 2, \; v_2 = \pi, \; v_3 = \frac{4}{3}\pi, \; v_4 = \frac{1}{2}\pi^2, \; v_5 = \frac{8}{15}\pi^2, \ldots . \tag{49c}$$

c. *Generalizations. Parametric Representations*

In n-dimensional space we can consider an r-dimensional set for any $r \leqq n$ and seek to define its area. For this purpose a parametric representation is advantageous. Let the r-dimensional set be given by the equations

$$\begin{aligned} x_1 &= \phi_1(u_1, \ldots, u_r) \\ &\cdots\cdots\cdots\cdots \\ x_n &= \phi_n(u_1, \ldots, u_r), \end{aligned}$$

where the functions ϕ_ν possess continuous derivatives in a region B of the variables $(u_1, \ldots, u_r)$. As the variables $u_1, \ldots, u_r$ range over this region, the point $(x_1, \ldots, x_n)$ describes an r-dimensional surface.

From the rectangular matrix (see p. 147),

$$\begin{pmatrix} \frac{\partial x_1}{\partial u_1} & \frac{\partial x_2}{\partial u_1} & \cdots & \frac{\partial x_n}{\partial u_1} \\ \frac{\partial x_1}{\partial u_2} & \frac{\partial x_2}{\partial u_2} & \cdots & \frac{\partial x_n}{\partial u_2} \\ \vdots & \vdots & & \vdots \\ \frac{\partial x_1}{\partial u_r} & \frac{\partial x_2}{\partial u_r} & \cdots & \frac{\partial x_n}{\partial u_r} \end{pmatrix}$$

we now form all possible r-rowed determinants D_i, where $i = 1, 2, \ldots, k = \binom{n}{r}$, the first of which, for example, is the determinant

$$D_1 = \begin{vmatrix} \frac{\partial x_1}{\partial u_1} & \frac{\partial x_2}{\partial u_1} & \cdots & \frac{\partial x_r}{\partial u_1} \\ \frac{\partial x_1}{\partial u_2} & \frac{\partial x_2}{\partial u_2} & \cdots & \frac{\partial x_r}{\partial u_2} \\ \cdot & \cdot & & \cdot \\ \cdot & \cdot & & \cdot \\ \cdot & \cdot & & \cdot \\ \frac{\partial x_1}{\partial u_r} & \frac{\partial x_2}{\partial u_r} & \cdots & \frac{\partial x_r}{\partial u_r} \end{vmatrix}$$

The area of the r-dimensional surface is then given by the integral

$$\text{(50a)} \qquad \int \cdots \int \sqrt{D_1^2 + D_2^2 + \cdots + D_k^2}\, du_1 \cdots du_r \quad ; \quad k = \binom{n}{r}.$$

By means of the theorem on the transformation of multiple integrals (p. 404) and simple calculations with determinants (which we shall omit here), we can prove that the area defined by this expression is not changed if we replace $u_1, \ldots, u_r$ by other parameters. We see also that for $r = 1$ this reduces to the usual formula for the length of arc, and for $r = 2$ in a space of three dimensions it becomes formula (30a), p. 428 for the area.

We prove formula (50a) when $r = n - 1$, where n is arbitrary; that is, we shall prove the following theorem:

If a portion of an $(n - 1)$-dimensional hypersurface in n-dimensional space can be represented parametrically by the equations

$$x_i = \psi_i(u_1, \ldots, u_{n-1}) \qquad (i = 1, \ldots, n),$$

then its area is given by

$$\text{(50b)} \qquad A = \int \cdots \int \sqrt{D_1^2 + \cdots + D_n^2}\, du_1 \cdots du_{n-1},$$

where D_i is the Jacobian of $(n - 1)$ rows given by

$$D_i = \frac{d(x_1, \ldots, x_{i-1}, x_{i+1}, \ldots, x_n)}{d(u_1, \ldots, u_{n-1})}$$
$$= 1 \Big/ \frac{d(u_1, \ldots, u_{n-1})}{d(x_1, \ldots, x_{i-1}, x_{i+1}, \ldots, x_n)}.$$

Here, as always, we assume the existence and continuity of all the derivatives involved.

Without loss of generality we may assume that $\phi_{x_n} \neq 0$. Then, by (44b), A is given by

$$A = \int \cdots \int \frac{|\operatorname{grad} \phi|}{|\phi_{x_n}|} \, dx_1 \cdots dx_{n-1}.$$

We have only to show that

$$\frac{1}{|\phi_{x_n}|} |\operatorname{grad} \phi| \, dx_1 \cdots dx_{n-1} = \sqrt{\sum_i D_i^2} \, du_1 \cdots du_{n-1},$$

or

$$|\operatorname{grad} \phi|^2 = \phi_{x_n}{}^2 \left(\sum_i D_i^2\right) \frac{d(u_1, \ldots, u_{n-1})}{d(x_1, \ldots, x_{n-1})} = \frac{\phi_{x_n}{}^2}{D_n{}^2} \sum_i D_i^2.$$

Now, from the properties of Jacobians,

$$\begin{aligned} \frac{D_i}{D_n} &= \frac{d(x_1, \ldots, x_{i-1}, x_{i+1}, \ldots, x_n)/d(u_1, \ldots, u_{n-1})}{d(x_1, \ldots, x_{n-1})/d(u_1, \ldots, u_{n-1})} \\ &= \frac{d(x_1, \ldots, x_{i-1}, x_{i+1}, \ldots, x_n)}{d(x_1, \ldots, x_{n-1})}. \end{aligned}$$

This last Jacobian corresponds to the introduction of $(x_1, \ldots, x_{i-1}, x_{i+1}, \ldots, x_n)$ instead of $(x_1, \ldots, x_{n-1})$ as independent variables. But as the partial derivatives $\frac{\partial x_n}{\partial x_i}$ are obtained from the equations

$$\phi_{x_n} \frac{\partial x_n}{\partial x_i} + \phi_{x_i} = 0 \qquad (i = 1, \ldots, n-1),$$

we have $D_i/D_n = \pm \phi_{x_i}/\phi_{x_n}$. Hence,

$$\frac{D_i^2}{D_n{}^2} = \frac{\phi_{x_i}{}^2}{\phi_{x_n}{}^2},$$

which proves the formula (50b) for A.

It may be mentioned here that the expression $\sum_i D_i^2$ may be represented as a determinant of $(n-1)$ rows,

(50c) $$W = \sum_{i=1}^{n} D_i^2 = \Gamma(\mathbf{X}_{u_1}, \ldots, \mathbf{X}_{u_{n-1}})$$

$$= \begin{vmatrix} \mathbf{X}_{u_1} \cdot \mathbf{X}_{u_1} & \mathbf{X}_{u_1} \cdot \mathbf{X}_{u_2} & \ldots & \mathbf{X}_{u_1} \cdot \mathbf{X}_{u_{n-1}} \\ \cdots & \cdots & \cdots & \cdots \\ \mathbf{X}_{u_{n-1}} \cdot \mathbf{X}_{u_1} & \cdots & \cdots & \mathbf{X}_{u_{n-1}} \cdot \mathbf{X}_{u_{n-1}} \end{vmatrix}$$

("Gram determinant"; see p. 194), so that

(50d) $$A = \int \cdots \int \sqrt{W}\, du_1 \cdots du_{n-1}.$$

Here, the elements of the determinant are the inner products of the vectors

$$\mathbf{X}_{u_i} = \left(\frac{\partial x_1}{\partial u_i}, \ldots, \frac{\partial x_n}{\partial u_i}\right) \quad \text{and} \quad \mathbf{X}_{u_k} = \left(\frac{\partial x_1}{\partial u_k}, \ldots, \frac{\partial x_n}{\partial u_k}\right),$$

namely, the expressions

(50e) $$\mathbf{X}_{u_i} \cdot \mathbf{X}_{u_k} = \sum_{i=1}^{n} \frac{\partial x_j}{\partial u_i} \frac{\partial x_j}{\partial u_k}.$$

Exercises 4.11

1. Calculate the volume of the n-dimensional ellipsoid

$$\frac{x_1^2}{a_1^2} + \cdots + \frac{x_n^2}{a_n^2} \leqq 1.$$

2. Express the integral I of a function of x_1, depending on x_1 alone, over the unit sphere $x_1^2 + \cdots + x_n^2 = 1$ in n-dimensional space, as a single integral.
3. An *n-simplex* is the intersection in n-dimensional space of $n+1$ half-spaces in general position; that is, any n of the bounding hyperplanes of the half-spaces meet in exactly one point, a *vertex* of the simplex: For example, a triangle in the plane or a tetrahedron in three-dimensional space. Find the volume of the n-simplex bounded by the hyperplanes $x_k \geqq 0$ for $k = 1, 2, \ldots, n$ and

$$\frac{x_1}{a_1} + \frac{x_2}{a_2} + \cdots + \frac{x_n}{a_n} \leqq 1.$$

4.12 Improper Single Integrals as Functions of a Parameter

a. *Uniform Convergence. Continuous Dependence on the Parameter*

Improper integrals frequently appear as functions of a parameter. For example, the integral of the general power

(51a) $$\int_0^1 y^x\,dy = \frac{1}{x+1}$$

is an improper integral for x in the interval $-1 < x < 0$.

We have seen (p. 74) that an integral over a finite interval is continuous when regarded as a function of a parameter, provided that the integrand is continuous. In the case of an infinite interval, however, the situation is not so simple. Let us consider, for example, the integral

(51b) $$F(x) = \int_0^\infty \frac{\sin xy}{y}\,dy.$$

According to whether $x > 0$ or $x < 0$, this is transformed by the substitution $xy = z$ into

$$\int_0^\infty \frac{\sin z}{z}\,dz \quad \text{or} \quad \int_0^{-\infty} \frac{\sin z}{z}\,dz = -\int_0^\infty \frac{\sin z}{z}\,dz.$$

The integral

$$\int_0^\infty \frac{\sin z}{z}\,dz$$

converges, as we have seen in Volume I (p. 310), and in fact has the value $\pi/2$ (Volume I, p. 589). Thus, although the function $(\sin xy)/y$, regarded as a function of x and y, is continuous everywhere and its integral converges for every value of x, the function $F(x)$ is discontinuous:

(51b) $$\int_0^\infty \frac{\sin xy}{y}\,dy = \begin{cases} \dfrac{\pi}{2} & \text{for} \quad x > 0 \\ 0 & \text{for} \quad x = 0 \\ -\dfrac{\pi}{2} & \text{for} \quad x < 0. \end{cases}$$

In itself, this fact is not at all surprising, for it is analogous to the situation of nonuniform convergence for infinite series (Volume I, p. 533), and we must remember that the process of integration is a generalized summation. We can be sure that an infinite series of continuous functions represents a continuous function only if the convergence is *uniform*. Here, in the case of improper integrals depending on a parameter, we must again introduce the concept of uniform convergence.

We say that *the integral*

$$F(x) = \int_0^\infty f(x, y)dy \tag{52a}$$

converges uniformly (in x) in the interval $a \leqq x \leqq b$, provided that the "remainder" of the integral can be made arbitrarily small simultaneously for all values of x in the interval under consideration, or, more precisely, provided that for a given positive number ε, there is a positive number $A = A(\varepsilon)$ that does not depend on x and is such that whenever $B \geqq A$

$$\left|\int_B^\infty f(x, y)dy\right| < \varepsilon. \tag{52b}$$

As a useful test we mention that *the integral*

$$\int_0^\infty f(x, y)dy$$

converges uniformly (and absolutely) if for sufficiently large y, say $y > y_0$, the relation

$$\left|f(x, y)\right| < \frac{M}{y^\alpha} \tag{52c}$$

holds, where M is a positive constant and $\alpha > 1$. For, in this case,

$$\left|\int_B^\infty f(x, y)dy\right| < M\int_B^\infty \frac{dy}{y^\alpha} = M\frac{1}{(\alpha - 1)B^{\alpha-1}} \leqq M\frac{1}{(\alpha - 1)A^{\alpha-1}};$$

the last bound can be made as small as we please by choosing A sufficiently large, and it is independent of x. This is a straightforward analogue of the test for the uniform convergence of series given in Volume I (p. 535).

We readily see that a *uniformly convergent integral of a continuous function is itself a continuous function,* for if we choose A so that

$$\left|\int_A^\infty f(x, y)dy\right| < \varepsilon$$

for all values of x in the interval under consideration, then, from (52a),

$$\left|F(x + h) - F(x)\right| < \left|\int_0^A \{f(x + h, y) - f(x, y)\}\, dy\right| + 2\varepsilon.$$

By virtue of the uniform continuity of the function $f(x, y)$ in a bounded set, we can choose h so small that the finite integral on the right is less than ε, which proves the continuity of the integral.

A similar result holds when the region of integration is finite, but the integrand has a point of infinite discontinuity. Suppose, for example, that the function $f(x, y)$ tends to infinity as $y \to \alpha$. We then say that the *convergent integral*

$$F(x) = \int_{\alpha}^{\beta} f(x, y)dy \tag{53a}$$

converges uniformly in $a \leqq x \leqq b$ if for every positive number ε we can find a number k independent of x such that

$$\left| \int_{\alpha}^{\alpha+h} f(x, y)dy \right| < \varepsilon, \tag{53b}$$

provided $h \leqq k$.

The condition in the neighborhood of the point $y = \alpha$

$$\left| f(x, y) \right| < \frac{M}{(y - \alpha)^{\nu}} \qquad (\nu < 1) \tag{53c}$$

is sufficient for uniform convergence. As before, uniform convergence for a continuous integrand implies that the integral is a continuous function.

If the convergence is uniform in an interval $a \leqq x \leqq b$, the improper integral $F(x)$ is continuous. We can then integrate $F(x)$ over this finite interval and thus form the corresponding improper repeated integral

$$\int_a^b dx \int_0^{\infty} f(x, y)dy$$

for an infinite interval of integration in y, and

$$\int_a^b dx \int_{\alpha}^{\beta} f(x, y)dy$$

for an infinite discontinuity.

Instead of the finite interval $a \leqq x \leqq b$, we can of course also consider an infinite interval of integration for x. But then the repeated integral need not converge. For example, the integral

$$F(x) = \int_0^{\infty} \frac{dy}{x^2 + y^2} = \frac{\pi}{2x}$$

converges uniformly for $x \geqq 1$, but

$$\int_1^\infty F(x)dx$$

does not exist.

b. Integration and Differentiation of Improper Integrals with Respect to a Parameter

It is not true in general that improper integrals may be differentiated or integrated under the sign of integration with respect to a parameter. In other words, limit operations with respect to a parameter and integration cannot generally be executed in reverse order (cf. the example on p. 473).

In order to determine whether the order of integration in improper repeated integrals is reversible, we can often use the following test (or else make a special investigation along the lines of its proof):

If the improper integral

$$F(x) = \int_0^\infty f(x, y)dy \tag{54a}$$

converges uniformly in the interval $\alpha \leqq x \leqq \beta$, *then*

$$\int_\alpha^\beta dx \int_0^\infty f(x, y)dy = \int_0^\infty dy \int_\alpha^\beta f(x, y)dx. \tag{54b}$$

To prove this we put

$$\int_0^\infty f(x, y)dy = \int_0^A f(x, y)dy + R_A(x).$$

By hypothesis, $|R_A(x)| < \varepsilon(A)$, where $\varepsilon(A)$ depends only on A, not on x, and tends to zero as $A \to \infty$. The theorem on p. 80 on interchanging the order of integration yields

$$\begin{aligned}\int_\alpha^\beta dx \int_0^\infty f(x, y)dy &= \int_\alpha^\beta dx \int_0^A f(x, y)dy + \int_\alpha^\beta R_A(x)dx \\ &= \int_0^A dy \int_\alpha^\beta f(x, y)dx + \int_\alpha^\beta R_A(x)dx,\end{aligned}$$

whence by the mean value theorem of the integral calculus

$$\left|\int_\alpha^\beta dx \int_0^\infty f(x, y)dy - \int_0^A dy \int_\alpha^\beta f(x, y)dx\right| \leqq \varepsilon(A)|\beta - \alpha|.$$

If we now let A tend to infinity, we obtain the formula (54b).

If the interval of integration with respect to a parameter is infinite also, the change of order is not always possible, even though the convergence may be uniform. It can, however, be performed if the corresponding improper double integral exists (cf. Chapter 4, pp. 408 ff.). Thus,

$$\int_0^\infty dx \int_0^\infty f(x, y)dy = \int_0^\infty dy \int_0^\infty f(x, y)dx \tag{54c}$$

if the double integral $\iint |f(x, y)|\, dx\, dy$ over the whole first quadrant exists.

Formula (54c) holds since the improper double integral is independent of the mode of approximation to the region of integration. In the one case, we approximate the integral by means of infinite strips parallel to the x-axis, and in the other, by strips parallel to the y-axis.

A similar result also holds if the interval of integration is finite, but the integrand is discontinuous along a finite number of straight lines $y =$ constant or on a finite number of more general curves in the region of integration. The corresponding theorem is as follows:

If the function $f(x, y)$ is discontinuous only along a finite number of straight lines $y = a_1, y = a_2, \ldots, y = a_r$ and if the integral

$$\int_a^b f(x, y)dy$$

converges uniformly in x in the interval $\alpha \leqq x \leqq \beta$, then in this interval it represents a continuous function of x, and

$$\int_\alpha^\beta dx \int_a^b f(x, y)dy = \int_a^b dy \int_\alpha^\beta f(x, y)dx. \tag{54d}$$

That is, under these hypotheses the order of integration can be changed. The proof of the theorem is analogous to the one for formula (54b) given above.

It is equally easy to extend the rules for differentiation with respect to a parameter. The following theorem holds:

If the function $f(x, y)$ has a sectionally continuous derivative with respect to x in the interval $\alpha \leqq x \leqq \beta$ and the two integrals

$$F(x) = \int_0^\infty f(x, y)dy \quad \text{and} \quad \int_0^\infty f_x(x, y)dy \tag{55a}$$

converge uniformly, then

$$F'(x) = \int_0^\infty f_x(x, y)dy. \tag{55b}$$

That is, under these hypotheses, the order of the processes of integration and of differentiation with respect to a parameter can be reversed, for, if we put

$$G(x) = \int_0^\infty f_x(x, y)dy,$$

then (54b) yields

$$\int_\alpha^\xi G(x)dx = \int_\alpha^\xi dx \int_0^\infty f_x(x, y)dy = \int_0^\infty dy \int_\alpha^\xi f_x(x, y)dx.$$

The integrand on the right has the value

$$\int_\alpha^\xi f_x(x, y)dx = f(\xi, y) - f(\alpha, y);$$

therefore,

$$\int_\alpha^\xi G(x)dx = F(\xi) - F(\alpha);$$

hence, if we differentiate and then replace ξ by x, we obtain

$$\frac{dF(x)}{dx} = G(x) = \int_0^\infty f_x(x, y)dy,$$

as was to be proved.

We can similarly extend the rule for differentiation when one of the limits depends on the parameter x (see Chapter 1, p. 77), for we can write

$$\int_{\phi(x)}^\infty f(x, y)dy = \int_{\phi(x)}^a f(x, y)dy + \int_a^\infty f(x, y)dy,$$

where a is any fixed value in the interval of integration. Then we can apply rules previously proved to each of the two terms on the right.

As before our rules of differentiation also hold for improper integrals with finite intervals of integration.

c. Examples

1. We consider the integral

$$\int_0^\infty e^{-xy}\,dy = \frac{1}{x} \qquad (x > 0).$$

If $x \geqq 1$, this integral converges uniformly, since for positive values of A

$$\int_A^\infty e^{-xy}\,dy \leqq \int_A^\infty e^{-y}\,dy = e^{-A},$$

where the final bound no longer depends on x and can be made as small as we please if we choose A sufficiently large. The same is true of the integrals of the partial derivatives of the function with respect to x. By repeated differentiation, we thus obtain

$$\int_0^\infty ye^{-xy}\,dy = \frac{1}{x^2}, \quad \int_0^\infty y^2e^{-xy}\,dy = \frac{2}{x^3}, \ldots, \int_0^\infty y^n e^{-xy}\,dy = \frac{n!}{x^{n+1}}.$$

In particular, for $x = 1$, we have

$$\Gamma(n + 1) = \int_0^\infty y^n e^{-y}\,dy = n!$$

This formula was established differently in Volume I (p. 308).

2. Further, let us consider the integral

$$\int_0^\infty \frac{dy}{x^2 + y^2} = \frac{\pi}{2}\frac{1}{x}.$$

Again it is easy to convince ourselves that if $x \leqq \alpha$, where α is any positive number, all the assumptions required for differentiation under the integral sign are satisfied. By repeated differentiation we therefore obtain the sequence of formulas

$$\int_0^\infty \frac{dy}{(x^2 + y^2)^2} = \frac{\pi}{2}\cdot\frac{1}{2}\cdot\frac{1}{x^3}, \quad \int_0^\infty \frac{dy}{(x^2 + y^2)^3} = \frac{\pi}{2}\cdot\frac{1\cdot 3}{2\cdot 4}\cdot\frac{1}{x^5}, \ldots,$$

$$\int_0^\infty \frac{dy}{(x^2 + y^2)^n} = \frac{\pi}{2}\cdot\frac{1\cdot 3\cdots(2n-3)}{2\cdot 4\cdots(2n-2)}\cdot\frac{1}{x^{2n-1}}.$$

From these formulas we can get another derivation of Wallis's product for π (cf. Volume I, p. 281). For this we put $x = \sqrt{n}$ to obtain

$$\int_0^\infty \frac{dy}{(1+y^2/n)^n} = \frac{\pi}{2} \cdot \frac{1\cdot 3 \cdots (2n-3)}{2\cdot 4 \cdots (2n-2)} \sqrt{n}.$$

As n increases, the left side converges to the integral

$$\int_0^\infty e^{-y^2}\,dy = \frac{1}{2}\sqrt{\pi}.$$

To prove this, we estimate the difference

$$\int_0^\infty e^{-y^2}\,dy - \int_0^\infty \frac{dy}{(1+y^2/n)^n}.$$

This difference satisfies the inequality

$$\left| \int_0^\infty e^{-y^2}\,dy - \int_0^\infty \frac{dy}{(1+y^2/n)^n} \right|$$
$$\leqq \int_0^T \left| e^{-y^2} - \frac{1}{(1+y^2/n)^n} \right| dy + \int_T^\infty e^{-y^2}\,dy + \int_T^\infty \frac{dy}{(1+y^2/n)^n}$$
$$\leqq \int_0^T \left| e^{-y^2} - \frac{1}{(1+y^2/n)^n} \right| dy + \int_T^\infty e^{-y^2}\,dy + \frac{1}{T},$$

since $(1+y^2/n)^n > y^2$. But if we choose T so large that

$$\int_T^\infty e^{-y^2}\,dy + \frac{1}{T} < \frac{\varepsilon}{2}$$

and then choose n so large that

$$\int_0^T \left| e^{-y^2} - \frac{1}{(1+y^2/n)^n} \right| dy < \frac{\varepsilon}{2},$$

as is possible in virtue of the uniform convergence of the limit

$$\lim_{n\to\infty} (1+y^2/n)^{-n} = e^{-y^2}$$

(Volume I, p. 152), it follows at once that

$$\left| \int_0^\infty \left(e^{-y^2} - \frac{1}{(1+y^2/n)^n} \right) dy \right| < \varepsilon.$$

With the value of the integral of e^{-y^2} from (25a), p. 415, this establishes the relation

$$\lim_{n\to\infty} \frac{1\cdot 3\cdots(2n-3)}{2\cdot 4\cdots(2n-2)}\sqrt{n} = \frac{1}{\sqrt{\pi}}, \tag{56}$$

which is equivalent to formula (80) in Volume I (p. 282).

3. With a view to calculating the integral

$$\int_0^\infty \frac{\sin y}{y}\, dy,$$

we shall discuss the function

$$F(x) = \int_0^\infty e^{-xy}\frac{\sin y}{y}\, dy.$$

This integral converges uniformly if $x \geqq 0$, while the integral

$$\int_0^\infty e^{-xy}\sin y\, dy$$

converges uniformly if $x \geqq \delta > 0$, where δ is an arbitrarily small positive number. Both these statements will be proved below. Therefore, $F(x)$ is continuous if $x \geqq 0$; and if $x \geqq \delta$, we have

$$F'(x) = -\int_0^\infty e^{-xy}\sin y\, dy.$$

Integrating by parts twice, we easily evaluate this last integral (see Volume I, p. 277):

$$F'(x) = -\frac{1}{1+x^2}.$$

We integrate this to obtain

$$F(x) = -\operatorname{arc\,tan} x + C,$$

where C is a constant.[1] By virtue of the relation

$$\left|\int_0^\infty e^{-xy}\frac{\sin y}{y}\, dy\right| \leqq \int_0^\infty e^{-xy}\, dy = \left.\frac{e^{-xy}}{x}\right|_\infty^0 = \frac{1}{x},$$

[1]Here arc tan x denotes the principal branch of that function, as defined in Volume I (p. 214).

which holds if $x \geqq \delta$, we see that $\lim_{x\to\infty} F(x) = 0$. Since $\lim_{x\to\infty} \text{arc tan } x = \pi/2$, C must be $\pi/2$, and we obtain

$$F(x) = \frac{\pi}{2} - \text{arc tan } x.$$

Since $F(x)$ is continuous for $x \geqq 0$,

$$\lim_{x\to 0} F(x) = F(0) = \int_0^\infty \frac{\sin y}{y}\, dy,$$

which gives the required formula

$$\int_0^\infty \frac{\sin y}{y}\, dy = \frac{\pi}{2} \tag{57}$$

(cf. Volume I, p. 589).

We prove that

$$\int_0^\infty e^{-xy} \frac{\sin y}{y}\, dy$$

converges uniformly if $x \geqq 0$. If A is an arbitrary number and $k\pi$ is the least multiple of π that exceeds A, we can write the "remainder" of the integral in the form

$$\int_A^\infty e^{-xy} \frac{\sin y}{y}\, dy = \int_A^{k\pi} e^{-xy} \frac{\sin y}{y}\, dy + \sum_{\nu=k}^{\infty} \int_{\nu\pi}^{(\nu+1)\pi} e^{-xy} \frac{\sin y}{y}\, dy.$$

The terms of the series on the right have alternating signs and their absolute values tend monotonically to 0. By Leibnitz's test (Volume I, p. 514), therefore, the series converges and the absolute value of its sum is less than that of its first term. Hence, we have the inequality

$$\left|\int_A^\infty e^{-xy} \frac{\sin y}{y}\, dy\right| < \int_A^{(k+1)\pi} e^{-xy} \frac{|\sin y|}{y}\, dy < \int_A^{(k+1)\pi} \frac{1}{A}\, dy < \frac{2\pi}{A},$$

in which the right side is independent of x and can be made as small as we please. This establishes the uniformity of convergence.

The uniform convergence of

$$\int_0^\infty e^{-xy} \sin y\, dy$$

for $x \geqq \delta > 0$ follows at once from the relation

$$\int_A^\infty \left| e^{-xy} \sin y \right| dy \leq \int_A^\infty e^{-xy}\, dy = \frac{e^{-Ax}}{x} \leq \frac{e^{-A\delta}}{\delta}.$$

4. On p. 466 we learned that *uniform convergence* of the integrals is a sufficient condition for reversibility of the order of integration. Mere *convergence* is not sufficient, as the following example shows:

If we put $f(x, y) = (2 - xy)\, xye^{-xy}$, then, since

$$f(x, y) = \frac{\partial}{\partial y}(xy^2e^{-xy}),$$

the integral

$$\int_0^\infty f(x, y) dy$$

exists for every x in the interval $0 \leq x \leq 1$; in fact, for every such value of x, it has the value 0. Therefore,

$$\int_0^1 dx \int_0^\infty f(x, y) dy = 0.$$

On the other hand, since

$$f(x, y) = \frac{\partial}{\partial x}(x^2ye^{-xy})$$

for every $y \geq 0$, we have

$$\int_0^1 f(x, y) dx = ye^{-y},$$

and, therefore,

$$\int_0^\infty dy \int_0^1 f(x, y) dx = \int_0^\infty ye^{-y}\, dy = \int_0^\infty e^{-y}\, dy = 1.$$

Hence,

$$\int_0^1 dx \int_0^\infty f(x, y) dy \neq \int_0^\infty dy \int_0^1 f(x, y) dx.$$

d. Evaluation of Fresnel's Integrals

Fresnel's integrals

(58a) $$F_1 = \int_{-\infty}^{+\infty} \sin(\tau^2)\, d\tau, \qquad F_2 = \int_{-\infty}^{+\infty} \cos(\tau^2)\, d\tau,$$

are important in optics. In order to evaluate them, we apply the substitution $\tau^2 = t$, obtaining

$$F_1 = \int_0^\infty \frac{\sin t}{\sqrt{t}}\, dt, \quad F_2 = \int_0^\infty \frac{\cos t}{\sqrt{t}}\, dt.$$

Here, we put

$$\frac{1}{\sqrt{t}} = \frac{2}{\sqrt{\pi}} \int_0^\infty e^{-x^2 t}\, dx$$

(this follows from the substitution $x = \tau/\sqrt{t}$) and reverse the order of integration, as is permissible by our rules. (we first restrict the integration with respect to t to a finite interval $0 < a < t < b$, and then let $a \to 0$, $b \to \infty$).

$$F_1 = \frac{2}{\sqrt{\pi}} \int_0^\infty dx \int_0^\infty e^{-x^2 t} \sin t\, dt, \quad F_2 = \frac{2}{\sqrt{\pi}} \int_0^\infty dx \int_0^\infty e^{-x^2 t} \cos t\, dt.$$

Using integration by parts to evaluate the inner integrals, we reduce F_1 and F_2 to the elementary rational integrals

$$F_1 = \frac{2}{\sqrt{\pi}} \int_0^\infty \frac{1}{1 + x^4}\, dx, \quad F_2 = \frac{2}{\sqrt{\pi}} \int_0^\infty \frac{x^2}{1 + x^4}\, dx.$$

The integrals may be evaluated from the formulae given in Volume I (cf. Volume I, p. 290); the second integral can be reduced to the first by means of the substitution $x' = \frac{1}{x}$; both have the value $\frac{\pi}{2\sqrt{2}}$. Consequently,

(58b) $$F_1 = F_2 = \sqrt{\frac{\pi}{2}}.$$

Exercises 4.12

1. Evaluate $\int_0^\infty x^n e^{-x^2}\, dx$.

2. Evaluate

$$F(y) = \int_0^1 x^{y-1}(y \log x + 1) dx.$$

3. Let $f(x, y)$ be twice continuously differentiable and let $u(x, y, z)$ be defined as follows:

$$u(x, y, z) = \int_0^{2\pi} f(x + z\cos\phi, y + z\sin\phi)d\phi.$$

Prove that

$$z(u_{xx} + u_{yy} - u_{zz}) - u_z = 0.$$

4. If $f(x)$ is twice continuously differentiable and

$$u(x, t) = \frac{1}{t^{p-2}} \int_{-t}^{+t} f(x + y)(t^2 - y^2)^{(p-3)/2}\, dy \qquad (p > 1),$$

prove that

$$u_{xx} = \frac{p-1}{t} u_t + u_{tt}.$$

5. How must a, b, c be chosen in order that

$$\int_{-\infty}^{\infty}\int_{-\infty}^{\infty} \exp[-(ax^2 + 2bxy + cy^2)]dx\, dy = 1?$$

6. Evaluate

(a) $\displaystyle\int_{-\infty}^{+\infty}\int_{-\infty}^{+\infty} \exp[-(ax^2 + 2bxy + cy^2)](Ax^2 + 2Bxy + Cy^2)dx\, dy,$

(b) $\displaystyle\int_{-\infty}^{+\infty}\int_{-\infty}^{+\infty} \exp[-(ax^2 + 2bxy + cy^2)](ax^2 + 2bxy + cy^2)dxdy,$

where $a > 0$, $ac - b^2 > 0$.

7. The Bessel function $J_0(x)$ may be defined by

$$J_0(x) = \frac{1}{\pi}\int_{-1}^{+1} \frac{\cos xt}{\sqrt{1 - t^2}}\, dt.$$

Prove that

$$J_0'' + \frac{1}{x} J_0' + J_0 = 0.$$

8. For any nonnegative integral index n the Bessel function $J_n(x)$ may be defined by

$$J_n(x) = \frac{x^n}{1 \cdot 3 \cdot 5 \cdots (2n-1)\pi} \int_{-1}^{+1} (\cos xt)(1 - t^2)^{n-(1/2)}\, dt.$$

Prove that

(a) $J_n'' + \dfrac{1}{x} J_n' + \left(1 - \dfrac{n^2}{x^2}\right) J_n = 0 \qquad (n \geq 0),$

(b) $J_{n+1} = J_{n-1} - 2J_n' \qquad (n \geq 1)$

and

$J_1 = -J_0'.$

9. Evaluate the following integrals:

(a) $K(a) = \int_0^\infty e^{-ax^2} \cos x \, dx$

(b) $\int_0^\infty \frac{e^{-bx} - e^{-ax}}{x} \cos x \, dx$

(c) $I(a) = \int_0^\infty \exp(-x^2 - a^2/x^2) \, dx$

(d) $\int_0^\infty \frac{\sin(ax) J_0(bx)}{x} \, dx$

where J_0 denotes the Bessel function defined in Exercise 7.

10. Prove that

$$\int_0^{n\pi} \frac{\sin^2 ax}{x} \, dx$$

is of the order of $\log n$ when n is large and that

$$\int_0^\infty \frac{\sin^2 ax - \sin^2 bx}{x} \, dx = \frac{1}{2} \log \frac{a}{b}.$$

11. Replace the statement "The integral $\int_0^\infty f(x, y) \, dy$ is not uniformly convergent" by an equivalent statement not involving any form of the words "uniformly convergent".

4.13 The Fourier Integral

a. *Introduction*

The theory given in Section 4.12 is illustrated by *Fourier's integral theorem* (see Volume I, p. 615), which is fundamental in analysis and mathematical physics, We recall that Fourier series represent a sectionally smooth, but otherwise arbitrary, periodic function in terms of trigonometric functions. Fourier's integral gives a corresponding trigonometrical representation of a nonperiodic function $f(x)$ that is defined in the infinite interval $-\infty < x < +\infty$ and has its behavior at infinity restricted in a suitable way to ensure convergence.

We make the following assumptions about the function $f(x)$:

1. In any finite interval $f(x)$ is defined, continuous, and has a continuous first derivative $f'(x)$, except possibly for a finite number of points.

2. Near each exceptional point $f'(x)$ is bounded. At an exceptional point, $f(x)$ takes as its value the arithmetic mean of the limits on the right and left:

(59a) $$f(x) = \frac{1}{2}[f(x+0) + f(x-0)].$$[1]

3. The integral

(59b) $$\int_{-\infty}^{\infty} |f(x)|\,dx = C$$

is convergent.

Then Fourier's integral theorem states:

(60) $$f(x) = \frac{1}{\pi}\int_0^{\infty} d\tau \int_{-\infty}^{\infty} f(t)\cos\tau(t-x)dt.$$

Using the identity

$$\cos\tau(t-x) = \frac{1}{2}(e^{i\tau t - i\tau x} + e^{-i\tau t + i\tau x})$$

and putting

(61a) $$g(\tau) = \frac{1}{\sqrt{2\pi}}\int_{-\infty}^{+\infty} f(t)e^{-i\tau t}\,dt,$$

we can write formula (60) in the form

$$\begin{aligned} f(x) &= \frac{1}{\sqrt{2\pi}}\int_0^{\infty} [e^{i\tau x}g(\tau) + e^{-i\tau x}g(-\tau)]\,d\tau \\ &= \lim_{A\to\infty}\frac{1}{\sqrt{2\pi}}\int_0^{A} [e^{i\tau x}g(\tau) + e^{-i\tau x}g(-\tau)]\,d\tau \\ &= \lim_{A\to\infty}\frac{1}{\sqrt{2\pi}}\int_{-A}^{A} g(\tau)e^{ix\tau}\,d\tau. \end{aligned}$$

Hence, Fourier's theorem becomes

(61b) $$f(x) = \frac{1}{\sqrt{2\pi}}\int_{-\infty}^{\infty} g(\tau)e^{ix\tau}\,d\tau.$$

---

[1]For an exceptional x we do not require that $f'(x)$ be defined. However, the boundedness of f' near an exceptional x implies that the limits $f(x-0)$ and $f(x+0)$, from the left and right, exist.

In the complex form, (61a) associates with a function $f(x)$ another function $g(\tau)$, the *Fourier transform* of f. Fourier's theorem, as given by formula (61b), expresses f in terms of g in a quite symmetric fashion; as a matter of fact, it just states that $f(-x)$ is the Fourier transform of $g(\tau)$. The relation between f and g is reciprocal except for the sign of the exponent and the fact that according to our derivation from (60) the improper integral in (61b) is to be taken in the *restricted sense*

$$\int_{-\infty}^{\infty} = \lim_{A\to\infty} \int_{-A}^{A}.$$

In formula (61a) for g, however, the integral is absolutely convergent by assumption (59b), and the upper and lower limits can tend independently to $+\infty$ and $-\infty$, respectively. The two formulas (61a, b) are reciprocal equations, each yielding the one function in terms of the other.

The Fourier transform $g(\tau)$ of a real-valued function $f(x)$ generally takes complex values. From (61a) we obtain the complex conjugate equation for a real f,

$$\overline{g(\tau)} = \frac{1}{\sqrt{2\pi}} \int_{-\infty}^{+\infty} f(t) e^{i\tau t}\, dt = g(-\tau). \tag{62}$$

When $f(x)$ is an *even* function of x, however, the Fourier transform g is even, too, and is real for real f. Indeed, combining the contributions of t and $-t$ in the integral (61a), we obtain

$$g(\tau) = \frac{2}{\sqrt{2\pi}} \int_0^{\infty} f(t) \cos(\tau t)\, dt, \tag{63a}$$

which implies that $g(\tau) = g(-\tau)$. Formula (61b) can then be written in the form

$$\begin{aligned} f(x) &= \frac{2}{\sqrt{2\pi}} \int_0^{\infty} g(\tau) \cos(\tau x)\, d\tau \\ &= \frac{2}{\pi} \int_0^{\infty} \cos(\tau x) d\tau \int_0^{\infty} f(t) \cos(\tau t) dt. \end{aligned} \tag{63b}$$

Similarly, for an *odd* function $f(x)$,

$$g(\tau) = \frac{-2i}{\sqrt{2\pi}} \int_0^{\infty} f(t) \sin(\tau t)\, dt. \tag{64a}$$

In (64a), g is an odd function with values that are pure imaginary for real f. The reciprocal formula becomes

(64b)
$$f(x) = \frac{2i}{\sqrt{2\pi}} \int_0^\infty g(\tau) \sin(\tau\pi) d\tau$$
$$= \frac{2}{\pi} \int_0^\infty \sin(\tau x) d\tau \int_0^\infty f(t) \sin(\tau t) dt.$$

We illustrate Fourier's integral theorem by examples and then proceed to its proof.

b. *Examples*

1. Let $f(x)$ be the step function defined by $f(x) = 1$ when $x^2 < 1$, $f(x) = 0$ when $x^2 > 1$. By formula (63a) the Fourier transform of f is the function

$$g(\tau) = \frac{2}{\sqrt{2\pi}} \int_0^1 \cos(\tau t) dt = \frac{2}{\sqrt{2\pi}} \frac{\sin\tau}{\tau}.$$

Hence, by (63b),

(65a)
$$f(x) = \frac{2}{\pi} \int_0^\infty \frac{\cos(\tau x)\sin\tau}{\tau} d\tau = \begin{cases} 1 & \text{for} \quad |x| < 1 \\ \frac{1}{2} & \text{for} \quad x = \pm 1 \\ 0 & \text{for} \quad |x| > 1. \end{cases}$$

This integral appears in mathematical literature under the name of *Dirichlet's discontinuous factor*. It shows that an integral can be a discontinuous function of a parameter x although the integrand is continuous in x. Of course, this phenomenon can occur only because the integral is improper.

2. Let $f(x) = e^{-kx}$ for $x > 0$, where k is a positive real number. Defining f as an even function for all x, we find its Fourier transform:

$$g(\tau) = \frac{2}{\sqrt{2\pi}} \int_0^\infty \cos(\tau t)\, e^{-kt}\, dt = \sqrt{\frac{2}{\pi}} \frac{k}{k^2 + \tau^2}$$

[see formula (64), p. 277, of Volume I for the evaluation of the integral]. By (63b) this leads to the equation

(65b)
$$f(x) = \frac{2}{\pi} \int_0^\infty \frac{k\cos(\tau x)}{k^2 + \tau^2} d\tau = e^{-k|x|}.$$

On the other hand, continuing e^{-kx} as an *odd function* of x for negative x, we obtain the Fourier transform

$$g(\tau) = \frac{-2i}{\sqrt{2\pi}} \int_0^\infty \sin(\tau t)\, e^{-kt}\, dt = -i\sqrt{\frac{2}{\pi}} \frac{\tau}{k^2 + \tau^2}$$

and the formula

$$\text{(65c)} \qquad f(x) = \frac{2}{\pi} \int_0^\infty \frac{\tau \sin(\tau x)}{k^2 + \tau^2}\, d\tau = \begin{cases} e^{-kx} & \text{for} \quad x > 0 \\ 0 & \text{for} \quad x = 0 \\ -e^{kx} & \text{for} \quad x < 0. \end{cases}$$

3. The function $f(x) = e^{-x^2/2}$ gives an interesting illustration of our reciprocal formulas. The Fourier transform is

$$g(\tau) = \frac{2}{\sqrt{2\pi}} \int_0^\infty e^{-x^2/2} \cos(x\tau)\, dx.$$

We are handicapped in evaluating g by the fact that no explicit expression for the indefinite integral is available. Curiously enough, g can be found by solving a differential equation. On differentiating the expression for g and integrating by parts, we obtain

$$\begin{aligned} g'(\tau) &= -\frac{2}{\sqrt{2\pi}} \int_0^\infty (xe^{-x^2/2}) \sin(x\tau)\, dx \\ &= \frac{2}{\sqrt{2\pi}} \left[e^{-x^2/2} \sin(x\tau) \Big|_0^\infty - \tau \int_0^\infty e^{-x^2/2} \cos(x\tau)\, dx \right] \\ &= -\tau g(\tau). \end{aligned}$$

It follows that

$$\frac{d}{d\tau} [g(\tau) e^{\tau^2/2}] = (g\tau + g') e^{\tau^2/2} = 0$$

or that

$$g(\tau) e^{\tau^2/2} = \text{constant} = c.$$

Hence, g is of the form

$$g(\tau) = ce^{-\tau^2/2}.$$

Thus, the Fourier transform g of the function $f = e^{-x^2/2}$ has the form

$$g(\tau) = ce^{-\tau^2/2}$$

with a certain constant c. Since [see (25a) p. 415]

$$c = g(0) = \sqrt{\frac{2}{\pi}} \int_0^\infty e^{-x^2/2}\, dx = \frac{2}{\sqrt{\pi}} \int_0^\infty e^{-y^2}\, dy = 1,$$

we find that the *Fourier transform* of $f = e^{-x^2/2}$ *is the same function:*

$$g(\tau) = \frac{2}{\sqrt{2\pi}} \int_0^\infty e^{-x^2/2} \cos(x\tau)\, dx = e^{-\tau^2/2}. \tag{66a}$$

c. *Proof of Fourier's Integral Theorem*

The proof (like the corresponding one for Fourier series in Volume I) is based on a simple lemma ("Riemann-Lebesgue lemma"):

If $\phi(t)$ is bounded and continuous in the open interval $a < t < b$, we have

$$\lim_{A\to\infty} \int_a^b \phi(t) \sin At\, dt = 0. \tag{67}$$

For the proof of the lemma, we assume that $|\phi(t)| < M$ for $a < t < b$. Let ε be a prescribed positive number. Let α and β be chosen so that

$$a < \alpha < a + \frac{\varepsilon}{M}, \quad b - \frac{\varepsilon}{M} < \beta < b, \quad \alpha < \beta.$$

Then,

$$\left| \int_a^b \phi(t) \sin At\, dt \right| \leq \left| \int_\alpha^\beta \phi(t) \sin At\, dt \right| + 2\varepsilon.$$

In the closed interval $\alpha \leqq t \leqq \beta$, the function $\phi(t)$ is uniformly continuous and we can find a δ such that

$$|\phi(t') - \phi(t)| < \frac{\varepsilon}{b - a} \quad \text{for} \quad |t' - t| < \delta.$$

Now, replacing t by $t + \pi/A$ in the integral we have

$$\begin{aligned}\int_{\alpha}^{\beta} \phi(t) \sin At\, dt &= -\int_{\alpha-\pi/A}^{\beta-\pi/A} \phi\left(t + \frac{\pi}{A}\right) \sin At\, dt \\ &= -\int_{\alpha}^{\beta} \phi(t) \sin At\, dt \\ &\quad - \int_{\alpha}^{\beta-\pi/A} \left[\phi\left(t + \frac{\pi}{A}\right) - \phi(t)\right] \sin At\, dt \\ &\quad + \int_{\beta-\pi/A}^{\beta} \phi(t) \sin At\, dt \\ &\quad - \int_{\alpha-\pi/A}^{\alpha} \phi\left(t + \frac{\pi}{A}\right) \sin At\, dt.\end{aligned}$$

Hence, if A is so large that $\pi/A < \delta$ and $2M\pi/A < \varepsilon$, we find that

$$\left| 2\int_{\alpha}^{\beta} \phi(t) \sin At\, dt \right| \leqq \frac{\beta - \alpha - \pi/A}{b - a}\, \varepsilon + \frac{2M\pi}{A} < 2\varepsilon,$$

and, thus, also

$$\left| \int_{a}^{b} \phi(t) \sin At\, dt \right| \leqq 3\varepsilon.$$

Since ε is arbitrary, the relation (67) follows.

It is clear that formula (67) holds more generally, namely when, by removing a finite number of exceptional points, the interval $a < t < b$ can be broken up into open intervals in each of which $\phi(t)$ is continuous and bounded.

Now let $f(t)$ be a function defined for all t that satisfies the assumptions 1–3 stated on p. 476–7. In order to prove our main theorem in the form (60), we first replace the infinite intervals of integration by finite ones so that we may reverse the order of integration. For positive A, B, (and a fixed x), we introduce the expression

$$I_A = \frac{1}{\pi} \int_0^A d\tau \int_{-\infty}^{\infty} f(t) \cos \tau(t - x)\, dt. \tag{68a}$$

By assumption 3,

$$\int_{-\infty}^{\infty} |f(t)|\, dt$$

converges. Consequently, given $\varepsilon > 0$, we have

$$\left|\int_{|t|>B} f(t)\cos\tau(t-x)\,dt\right| \leqq \int_{|t|>B} |f(t)|\,dt < \varepsilon$$

for all sufficiently large B. It follows that

$$\lim_{B\to\infty}\int_{-B}^{+B} f(t)\cos\tau(t-x)\,dt = \int_{-\infty}^{\infty} f(t)\cos\tau(t-x)\,dt \tag{68b}$$

converges uniformly in τ.

Formula (60), which we want to prove, states that

$$f(x) = \lim_{A\to\infty} I_A. \tag{69}$$

In the integral (68a) defining I_A, we can interchange the integrations [see (54b), p. 466] since the integral (68b) converges uniformly.[1] Thus,

$$\begin{aligned} I_A &= \frac{1}{\pi}\int_{-\infty}^{\infty} dt \int_0^A f(t)\cos\tau(t-x)\,d\tau \\ &= \frac{1}{\pi}\int_{-\infty}^{\infty} f(t)\frac{\sin A(t-x)}{t-x}\,dt = \frac{1}{\pi}\int_{-\infty}^{+\infty} f(t+x)\frac{\sin At}{t}\,dt. \end{aligned}$$

Using the identity

$$\int_0^{\infty}\frac{\sin At}{t}\,dt = \frac{\pi}{2} \qquad \text{for} \qquad A > 0$$

[see (57), p. 472], we can write this result in the form

$$\begin{aligned} I_A &= \frac{1}{\pi}\int_0^{\infty}[f(x+t)+f(x-t)]\frac{\sin At}{t}\,dt \\ &= \frac{f(x+0)+f(x-0)}{2} + \frac{1}{\pi}\int_0^{\infty}\phi(t)\sin At\,dt \\ &= \frac{f(x+0)+f(x-0)}{2} + \frac{1}{\pi}\int_0^{C}\phi(t)\sin At\,dt + \frac{1}{\pi}\int_C^{\infty}\phi(t)\sin At\,dt, \end{aligned}$$

[1]We apply the theorem on p. 466 separately to

$$\int_0^{\infty} f(t)\cos\tau(t-x)\,dt \qquad \text{and} \qquad \int_{-\infty}^{0} f(t)\cos\tau(t-x)\,dt.$$

The function f may have a finite number of jump-discontinuities in any finite interval without changing the proof of (54b).

where C is any positive constant and

$$\phi(t) = \frac{f(x+t) - f(x+0)}{t} + \frac{f(x-t) - f(x-0)}{t}.$$

The function $\phi(t)$ satisfies all the assumptions of the Riemann-Lebesgue lemma (67): It obviously is continuous except possibly at a finite number of points, since this is true for f. At a point of discontinuity $t \neq 0$ the function $\phi(t)$ stays bounded, since f has jump-discontinuities only. The boundedness of $\phi(t)$ near $t = 0$ follows from the differentiability of f and the boundedness of f', since by the mean value theorem of differential calculus,

$$\phi(t) = f'(x + \theta t) - f'(x - \eta t),$$

where θ and η are certain values intermediate between 0 and 1.[1] Applying (67), we conclude that for any $c > 0$

$$\lim_{A\to\infty} \frac{1}{\pi} \int_0^C \phi(t) \sin At\, dt = 0.$$

Moreover,

$$\frac{1}{\pi} \int_C^\infty \phi(t) \sin At\, dt = \frac{1}{\pi} \int_C^\infty \frac{f(x+t) + f(x-t)}{t} \sin At\, dt$$
$$- \frac{f(x+0) + f(x-0)}{\pi} \int_{AC}^\infty \frac{\sin t}{t}\, dt.$$

Here the second integral tends to 0 for $A \to \infty$ and any C, whereas by choosing C sufficiently large, the first one can be made arbitrarily small *uniformly for all* $A > 0$. It follows that

$$\lim_{A\to\infty} I_A = \frac{f(x+0) + f(x-0)}{2}.$$

This is equivalent to (69), since we assumed that

$$f(x) = \frac{f(x+0) + f(x-0)}{2}.$$

[1]Notice that to apply the mean value theorem we only require existence of the derivative in the interior of the interval and continuity in the closed interval (see Volume I, p. 174). These assumptions are satisfied by the function defined by $f(x + t)$ for small positive t and by $f(x + 0)$ for $t = 0$, as well as for the function defined by $f(x - t)$ for small positive t and by $f(x - 0)$ for $t = 0$.

d. Rate of Convergence in Fourier's Integral Theorem

The reciprocal formulas (61a, b) have been established under the assumptions 1–3 on the function $f(x)$ stated on p. 476–7. A consequence of the requirement

$$\int_{-\infty}^{\infty} |f(x)|\, dx = C < \infty$$

is that the Fourier transform $g(\tau)$ given by (61a) is absolutely and uniformly convergent. Indeed, if we put

$$g_B(\tau) = \frac{1}{\sqrt{2\pi}} \int_{-B}^{B} f(t) e^{-i\tau t}\, dt, \tag{70a}$$

then

$$|g(\tau) - g_B(\tau)| = \left| \frac{1}{\sqrt{2\pi}} \int_{|t|>B} f(t) e^{-i\tau t}\, dt \right| \leqq \frac{1}{\sqrt{2\pi}} \int_{|t|>B} |f(t)|\, dt.$$

Hence, given $\varepsilon > 0$, it is possible to find a B so large that

$$|g(\tau) - g_B(\tau)| < \varepsilon \qquad \text{for all } \tau.$$

It follows that g, as uniform limit of continuous functions g_B, is itself continuous.

We cannot be sure in general of the uniform convergence of the integral in the reciprocal formula (61b). The approximating functions

$$f_A(x) = \frac{1}{\sqrt{2\pi}} \int_{-A}^{A} g(\tau)\, e^{ix\tau}\, d\tau \tag{70b}$$

certainly are continuous and converge to $f(x)$ for each x. However, the convergence cannot be *uniform* if f has discontinuities, as in our Example 1 on p. 479. Sufficient for uniform convergence of the $f_A(x)$ toward $f(x)$ is again the existence of the improper integral

$$\int_{-\infty}^{\infty} |g(\tau)|\, d\tau.$$

This condition clearly is violated in the example mentioned, where $g(\tau) = 2 \sin \tau/\sqrt{2\pi}\, \tau$.

For many applications, it is convenient to work only with integrals that are uniformly and absolutely convergent. Interchanges of limit operations are usually much harder to justify for integrals that converge only conditionally. It is easy to impose additional restrictions on f that guarantee the integrability of $|g|$ over the whole axis, and, hence, the uniform convergence of the $f_A(x)$. *It is sufficient to require that $f(x)$ have continuous first and second derivatives $f'(x)$ and $f''(x)$ and that all three integrals*

$$\int_{-\infty}^{\infty} |f(x)|\,dx, \quad \int_{-\infty}^{\infty} |f'(x)|\,dx, \quad \int_{-\infty}^{\infty} |f''(x)|\,dx$$

are convergent.

First, the convergence of

$$\int_{-\infty}^{\infty} |f'(x)|\,dx$$

implies that

$$\lim_{x\to\infty} f(x) = \lim_{x\to\infty} \left[f(0) + \int_0^x f'(t)dt\right] = f(0) + \int_0^\infty f'(t)dt$$

exists. Obviously,

$$\lim_{x\to\infty} f(x)$$

can only have the value 0, since otherwise

$$\int_{-\infty}^{\infty} |f(x)|\,dx$$

could not converge. Thus, $\lim_{x\to\infty} f(x) = 0$ and, by the same argument, $\lim_{x\to-\infty} f(x) = 0$. Similarly, the convergence of

$$\int_{-\infty}^{\infty} |f''(x)|\,dx$$

implies that

$$\lim_{x\to\pm\infty} f'(x) = 0,$$

also. Integration by parts applied twice to formula (70a) yields

$$\text{(71a)} \qquad g_B(\tau) = \frac{1}{i\sqrt{2\pi}\,\tau}\left[-f(B)e^{-iB\tau} + f(-B)e^{iB\tau} + \int_{-B}^{B} f'(t)e^{-i\tau t}\,dt\right]$$

$$= \frac{e^{-iB\tau}[f'(B) + i\tau f(B)] - e^{iB\tau}[f'(-B) + i\tau f(-B)]}{\sqrt{2\pi}\,\tau^2}$$

$$- \frac{1}{\sqrt{2\pi}\,\tau^2}\int_{-B}^{B} f''(t)e^{-i\tau t}\,dt.$$

Hence, for $B \to \infty$

$$\text{(71b)} \qquad g(\tau) = \frac{1}{i\tau\sqrt{2\pi}}\int_{-\infty}^{+\infty} f'(t)e^{-i\tau t}\,dt = -\frac{1}{\sqrt{2\pi}\,\tau^2}\int_{-\infty}^{\infty} f''(t)e^{-i\tau t}\,dt$$

and thus,

$$\text{(71c)} \qquad |g(\tau)| \leqq \frac{1}{\sqrt{2\pi}\,\tau^2}\int_{-\infty}^{+\infty} |f''(t)|\,dt = 0\left(\frac{1}{\tau^2}\right).$$

This estimate for $g(\tau)$ clearly implies that

$$\int_{-\infty}^{\infty} |g(\tau)|\,d\tau$$

converges (see Volume I, p. 307) and, hence, that

$$f(x) = \lim_{A\to\infty} f_A(x) = \lim_{A\to\infty} \frac{1}{\sqrt{2\pi}}\int_{-A}^{A} g(\tau)e^{ix\tau}\,d\tau$$

uniformly for all x. In fact, under the assumptions made on f, it does not matter how the upper and lower limit in the integral tend to $\pm\infty$; in general,

$$f(x) = \lim_{\substack{A\to\infty \\ B\to-\infty}} \frac{1}{\sqrt{2\pi}}\int_{B}^{A} g(\tau)e^{ix\tau}\,d\tau.$$

Equation (71b) can be interpreted as stating that the function $f'(t)$ has the Fourier transform $i\tau g(\tau)$ and $f''(t)$, the Fourier transform $-\tau^2 g(\tau)$, where g is the Fourier transform of f. Thus, under suitable regularity assumptions *differentiation of f corresponds to multiplication of the Fourier transform of f by the factor $i\tau$*. This fact is crucial for many applications of the Fourier transformation.

e. *Parseval's Identity for Fourier Transforms*

For Fourier series, we proved (Volume I, p. 614) the Parseval identity connecting the integral of the square of a periodic function with the sum of squares of the Fourier coefficients. A remarkable analogous identity exists for Fourier integrals; it is even more symmetric in form because of the reciprocity between a function f and its Fourier transform g. Since, even for real f, the Fourier transform g will generally be complex-valued, one has to use the square of the absolute value rather than the square of the function. The Parseval identity then states that the integral of the square of the absolute value extended over the whole axis is the same for the function f and its Fourier transform g:

$$\int_{-\infty}^{\infty} |f(x)|^2\, dx = \int_{-\infty}^{\infty} |g(\tau)|^2\, d\tau. \tag{72}$$

We shall not prove this identity under the most general assumptions for which it holds, but merely for f restricted in the same way as at the end of the last section, namely, when the three functions f, f', f'' are all continuous and absolutely integrable over the whole x-axis.[1]

As before, we define the approximations $g_B(\tau)$ to g and $f_A(x)$ to f by the equations (70a) and (70b). Then we form the expression

$$\begin{aligned}
J_{A,B} &= \int_{-B}^{B} |f(x) - f_A(x)|^2\, dx \\
&= \int_{-B}^{B} [f(x) - f_A(x)][\overline{f(x)} - \overline{f_A(x)}]\, dx \\
&= \int_{-B}^{B} [f(x)\overline{f(x)} - f(x)\overline{f_A(x)} - f_A(x)\overline{f(x)} + f_A(x)\overline{f_A(x)}]\, dx,
\end{aligned}$$

where the bar above an expression indicates the complex conjugate value. Now, interchanging integrations, we find that

$$\begin{aligned}
\int_{-B}^{B} f(x)\overline{f_A(x)}dx &= \frac{1}{\sqrt{2\pi}} \int_{-B}^{B} f(x)dx \int_{-A}^{A} \overline{g(\tau)}e^{-ix\tau}\, d\tau \\
&= \frac{1}{\sqrt{2\pi}} \int_{-A}^{A} \overline{g(\tau)}\, d\tau \int_{-B}^{B} f(x)e^{-ix\tau}\, dx \\
&= \int_{-A}^{A} \overline{g(\tau)}g_B(\tau)\, d\tau,
\end{aligned}$$

[1]The identity can be extended to more general f by suitably approximating f by functions of the restricted class used here.

whence, taking the complex conjugate, we find

$$\int_{-B}^{B} f_A(x)\,\overline{f(x)}dx = \int_{-A}^{A} g(\tau)\,\overline{g_B(\tau)}\,d\tau.$$

Hence,

$$J_{A,B} = \int_{-B}^{B} (|f(x)|^2 + |f_A(x)|^2)dx - \int_{-A}^{A} [\overline{g(\tau)}\,g_B(\tau) + g(\tau)\,\overline{g_B(\tau)}]d\tau. \tag{73}$$

Since our assumptions about $f(x)$ guarantee that

$$\lim_{A\to\infty} f_A(x) = f(x)$$

uniformly in x (see p. 487), we also have

$$\lim_{A\to\infty} |f(x) - f_A(x)|^2 = 0$$

uniformly in x. Consequently,

$$\lim_{A\to\infty} J_{A,B} = \lim_{A\to\infty} \int_{-B}^{B} |f(x) - f_A(x)|^2\,dx = 0.$$

Thus, identity (73) yields for $A \to \infty$

$$0 = 2\int_{-B}^{B} |f(x)|^2\,dx - \int_{-\infty}^{\infty} [\overline{g(\tau)}\,g_B(\tau) + g(\tau)\,\overline{g_B(\tau)}]d\tau. \tag{74}$$

Since

$$\lim_{B\to\infty} g_B(\tau) = g(\tau)$$

uniformly in τ and since $g_B(\tau)$ is bounded uniformly, and

$$g(\tau) = 0\left(\frac{1}{\tau^2}\right),$$

we can let B tend to ∞ in identity (74) to obtain in the limit the Parseval relation (72).

f. The Fourier Transformation for Functions of Several Variables

In one dimension the Fourier integral identity yields a representation of a function $f(x)$ as a linear combination of exponential functions $e^{ix\xi}$ that depend on a parameter ξ. For each value ξ of the parameter, we multiply the function $e^{ix\xi}$ with a suitable "weight factor" $g(\xi)/\sqrt{2\pi}$ and integrate with respect to ξ. The appropriate factor $g(\xi)$ is the Fourier transform of f.

Similar formulae exist for decomposition of functions of several variables into exponential functions. Functions $f(x, y)$ of two independent variables x, y are represented as combinations of exponential functions of the form $e^{i(x\xi+y\eta)}$ that depend on the parameters ξ, η. Similarly, functions $f(x, y, z)$ of three independent variables are built up from exponentials $e^{i(x\xi+y\eta+z\zeta)}$ depending on the parameters, ξ, η, ζ. Such decompositions of general functions into exponentials constitute one of the most powerful tools of mathematical analysis. For a given set of parameters ξ, η, ζ the function $e^{i(x\xi+y\eta+z\zeta)}$ depends on the single combination $s = x\xi + y\eta + z\zeta$, which is constant along each plane with direction numbers ξ, η, ζ in x, y, z-space. If we introduce a new rectangular coordinate system in which one of these planes is a coordinate plane, then $e^{i(x\xi+y\eta+z\zeta)}$ becomes a function of a single coordinate. In this way, Fourier's formulae yield a decomposition of $f(x, y, z)$ into functions that depend only on a single coordinate (where, however, the direction of the corresponding coordinate axis depends on the parameters ξ, η, ζ).

Such exponential expressions are intimately connected with the *plane waves* encountered in physics. Multiplying the exponential function $e^{i(x\xi+y\eta+z\zeta)}$ by a time dependent exponential factor $e^{-i\omega t}$, we obtain the expression

$$u(x, y, z, t) = e^{i(x\xi+y\eta+z\zeta)}\, e^{-i\omega t} = e^{i(\xi x+\eta y+\zeta z-\omega t)}. \tag{75a}$$

Here u has a fixed value e^{is} for all times t at all locations (x, y, z) with the same "phase" value

$$s = x\xi + y\eta + z\zeta - \omega t.$$

For fixed s, this represents at each time t a plane ("wave front") in x, y, z-space with direction numbers ξ, η, ζ for its normal. As t varies, this plane moves parallel to itself. Since (see p. 135) the quantity

$$p = \frac{s + \omega t}{\sqrt{\xi^2 + \eta^2 + \zeta^2}}$$

is the distance of the plane from the origin at time t, the plane moves with speed

$$c = \frac{dp}{dt} = \frac{\omega}{\sqrt{\xi^2 + \eta^2 + \zeta^2}}. \tag{75b}$$

This is the *speed of propagation* of the wave fronts, corresponding to a "frequency" ω of the wave.

We shall state and prove the Fourier integral theorem for a function $f(x, y)$ of two independent variables under conditions on f that are sufficient for the validity of the theorem (although far from necessary) and are convenient for applications.

Let $f(x, y)$ be defined and have continuous derivatives of first, second, and third orders for all values x, y. The absolute values of f and its derivatives of order $\leqq 3$ shall be absolutely integrable over the whole plane; that is, for any nonnegative integers i, k with $i + k \leqq 3$ the improper integrals

$$\iint \left| \frac{\partial^{i+k} f(x, y)}{\partial x^i\, \partial y^k} \right| dx\, dy, \tag{76}$$

extended over the whole x, y-plane, shall converge. The Fourier transform $g(\xi, \eta)$ of f is defined by the formula

$$g(\xi, \eta) = \frac{1}{2\pi} \iint e^{-i(x\xi + y\eta)} f(x, y)\, dx\, dy. \tag{77a}$$

The function f is then expressed in terms of its Fourier transform by the reciprocal formula

$$f(x, y) = \frac{1}{2\pi} \iint e^{i(x\xi + y\eta)} g(\xi, \eta)\, d\xi\, d\eta. \tag{77b}$$

Here, all integrals are extended over the whole plane and converge absolutely.

An analogous statement holds for functions $f(x_1, \ldots, x_n)$ of n independent variables. We only have to assume that f and its derivatives of order $\leqq n + 1$ exist and are absolutely integrable over the whole space. The *Fourier transform* $g(\xi_1, \xi_2, \ldots, \xi_n)$ is then defined by

$$g = (2\pi)^{-n/2} \int \cdots \int e^{-i(x_1\xi_1 + \cdots + x_n\xi_n)} f(x_1, \ldots, x_n)\, dx_1 \cdots dx_n. \tag{77a}$$

The reciprocal formula for $f(x_1, \ldots, x_n)$ here becomes

$$f = (2\pi)^{-n/2} \int \cdots \int e^{i(x_1\xi_1 + \cdots + x_n\xi_n)} g(\xi_1, \ldots, \xi_n)\, d\xi_1 \cdots d\xi_n. \tag{77b}$$

The proof for n dimensions is exactly the same as the proof for the two-dimensional case that will be given now.

We shall first prove the Fourier integral theorem for a function $f(x, y)$ of class C^3 and of *compact support*, meaning that f has continuous derivatives of order ≤ 3 and vanishes outside some bounded set. For this situation the Fourier formula for f follows immediately from the formula for functions of a single variable, as we now show.

The Fourier transform

$$g(\xi, \eta) = \frac{1}{2\pi} \iint e^{-i(x\xi + y\eta)} f(x, y)\, dx\, dy$$

is given by a proper integral, since f vanishes outside a bounded region. Introducing the "intermediate" Fourier transform with respect to y alone, namely,

$$\gamma(x, \eta) = \frac{1}{\sqrt{2\pi}} \int e^{-iy\eta} f(x, y)\, dy, \tag{77c}$$

we can write g in the form

$$g(\xi, \eta) = \frac{1}{\sqrt{2\pi}} \int e^{-ix\xi} \gamma(x, \eta)\, dx.$$

Obviously, for each value of η, we have in $\gamma(x, \eta)$ a function of the single variable x of class C^3 and of bounded support. Its Fourier transform is $g(\xi, \eta)$. The theorem of p. 477 applies and yields

$$\gamma(x, \eta) = \frac{1}{\sqrt{2\pi}} \int e^{ix\xi} g(\xi, \eta)\, d\xi. \tag{78}$$

On the other hand, $\gamma(x, \eta)$ for fixed x is the Fourier transform of $f(x, y)$ considered as a function of y alone. Hence, the reciprocal formula

$$f(x, y) = \frac{1}{\sqrt{2\pi}} \int e^{iy\eta} \gamma(x, \eta)\, d\eta$$

holds. Substituting here for γ its expression from (78) yields

$$f(x, y) = \frac{1}{\sqrt{2\pi}} \int d\eta \int e^{i(x\xi + y\eta)} g(\xi, \eta)\, d\xi.$$

In this formula, the repeated integral (first with respect to ξ and then with respect to η) can be replaced by a double integral over the whole ξ, η-plane, which leads to formula (77b). This step is valid (see p. 466), since the single integral

$$\int_{-\infty}^{+\infty} |g(\xi, \eta)| \, d\xi \tag{79a}$$

converges uniformly in η for all η and, in addition, the double integral

$$\iint |g(\xi, \eta)| \, d\xi \, d\eta \tag{79b}$$

converges. Both convergence results follow if we can show that an estimate of the form

$$|g(\xi, \eta)| \leq \frac{M}{(1 + \xi^2 + \eta^2)^{3/2}} \tag{79c}$$

holds for g with a suitable constant M. The convergence of the double integral (79b) is a consequence of (79c). The uniform convergence of the single integral (79a) follows from (79c) since for $A > 1$

$$\begin{aligned}\int_{|\xi|>A} |g(\xi, \eta)| \, d\xi &\leq M \int_{|\xi|>A} \frac{d\xi}{(1 + \xi^2 + \eta^2)^{3/2}} \\ &\leq M \int_{|\xi|>A} \frac{2|\xi|}{(1 + \xi^2)^2} \, d\xi = \frac{M}{1 + A^2};\end{aligned}$$

the right side tends to 0 for $A \to \infty$ independently of η.

Inequality (79c) is established from (77a) by repeated integration by parts. Since f has compact support, we find that

$$\iint e^{-i(x\xi+y\eta)} \frac{\partial^3 f(x, y)}{\partial x^3} \, dx \, dy = 2\pi(i\xi)^3 g(\xi, \eta)$$

$$\iint e^{-i(x\xi+y\eta)} \frac{\partial^3 f(x, y)}{\partial y^3} \, dx \, dy = 2\pi(i\eta)^3 g(\xi, \eta)$$

and, hence, that

$$\begin{aligned}&2\pi(1 + |\xi|^3 + \eta|^3) \, |g(\xi, \eta)| \\ &\quad = 2\pi |g(\xi, \eta)| + |2\pi(i\xi)^3 g(\xi, \eta)| + |2\pi(i\eta)^3 g(\xi, \eta)| \\ &\quad \leq \iint \left(|f(x, y)| + \left| \frac{\partial^3 f(x, y)}{\partial x^3} \right| + \left| \frac{\partial^3 f(x, y)}{\partial y^3} \right| \right) dx \, dy.\end{aligned}$$

For any ξ, η let the largest of the three quantities 1, $|\xi|$, $|\eta|$ be denoted by ζ. Then

$$(1 + \xi^2 + \eta^2)^{3/2} \leqq (\zeta^2 + \zeta^2 + \zeta^2)^{3/2} = 3\sqrt{3}\,\zeta^3 \leqq 3\sqrt{3}(1 + |\xi|^3 + |\eta|^3).$$

This yields the inequality (79c) with the value

$$M = \frac{3\sqrt{3}}{2\pi}\iint\left(|f(x, y)| + \left|\frac{\partial^3 f(x, y)}{\partial x^3}\right| + \left|\frac{\partial^3 f(x, y)}{\partial y^3}\right|\right)dx\,dy \tag{79b}$$

for the constant and completes the proof of the Fourier theorem for functions $f(x, y)$ of class C^3 and of compact support.

The proof of the theorem for the most general f of class C^3 for which the integrals (76) converge follows by approximating such f by functions $f_n(x, y)$ of compact support. For this purpose we multiply $f(x, y)$ with a suitable "cut-off" function $\phi_n(x, y)$ so that the product $f_n = \phi_n f$ has compact support, but agrees with f in the disk $x^2 + y^2 \leqq n^2$. Here we only require an auxiliary function $\phi_n(x, y)$ with these properties:

1. $\phi_n(x, y)$ has compact support and belongs to C^3;
2. $\phi_n(x, y) = 1$ for $x^2 + y^2 \leqq n^2$;
3. The absolute values of $\phi_n(x, y)$ and of all its derivatives of orders $\leqq 3$ do not exceed a fixed quantity N independently of x, y and n.

Suitable functions ϕ_n can be constructed easily in a variety of ways.[1]

Denote by $g_n(\xi, \eta)$ the Fourier transform of $f_n = \phi_n f$:

$$g_n(\xi, \eta) = \frac{1}{2\pi}\iint e^{-i(x\xi + y\eta)}\,\phi_n(x, y) f(x, y)\,dx\,dy. \tag{80a}$$

Then

$$|g(\xi, \eta) - g_n(\xi, \eta)| = \left|\frac{1}{2\pi}\iint e^{-i(x\xi + y\eta)}(1 - \phi_n) f\,dx\,dy\right|$$

[1]For example, define the function $h(s)$ by

$$h(s) = \begin{cases} 1 & \text{for } s \leqq 0 \\ (1 - s^4)^4 & \text{for } 0 \leqq s \leqq 1 \\ 0 & \text{for } 1 \leqq s. \end{cases}$$

Then

$$\phi_n(x, y) = h(x - n)h(-n - x)h(y - n)h(-n - y)$$

has all the desired properties.

$$\leqq \frac{1}{2\pi} \iint_{x^2+y^2>n^2} |(1-\phi_n)f| \, dx\, dy$$

$$\leqq (N+1) \iint_{x^2+y^2>n^2} |f| \, dx\, dy.$$

From the assumed convergence of the integral of $|f|$ over the whole plane it follows that

(80b) $$\lim_{n\to\infty} g_n(\xi,\eta) = g(\xi,\eta)$$

uniformly for all (ξ, η). In order to see that $g(\xi,\eta)$ again satisfies an inequality of the form (79c), we observe that by Leibnitz's rule

$$\left|\frac{\partial^3 f_n}{\partial x^3}\right| = \left|\frac{\partial^3}{\partial x^3}\phi_n f\right|$$

$$\leqq N\left(\left|\frac{\partial^3 f}{\partial x^3}\right| + 3\left|\frac{\partial^2 f}{\partial x^2}\right| + 3\left|\frac{\partial f}{\partial x}\right| + |f|\right).$$

A similar estimate holds for the third y-derivative of f_n. Let I be the largest of the integrals taken over the whole plane, of the absolute values of f and its derivatives of orders $\leqq 3$. Then

$$\iint \left(|f_n| + \left|\frac{\partial^3}{\partial x^3} f_n\right| + \left|\frac{\partial^3}{\partial y^3} f_n\right|\right) dx\, dy \leqq (1+8+8)\, NI = 17NI.$$

Applying the inequality (79c, d) to the function f_n, we find that for any n and all ξ, η, the inequality

(80c) $$|g_n(\xi,\eta)| \leqq \frac{M}{(1+\xi^2+\eta^2)^{3/2}}$$

holds with

$$M = \frac{51\sqrt{3}}{2\pi} NI.$$

It follows from (80b) that

$$|g(\xi,\eta)| \leqq \frac{M}{(1+\xi^2+\eta^2)^{3/2}}$$

for all (ξ, η), with the same constant M.

Since f_n has compact support, the reciprocal formula

$$f_n(x, y) = \frac{1}{2\pi} \iint e^{i(x\xi+y\eta)} g_n(\xi, \eta)\, d\xi\, d\eta \tag{80d}$$

is known already to be valid. For a given (x, y) we have $f_n(x, y) = f(x, y)$, once n is so large that $n^2 > x^2 + y^2$. For $n \to \infty$ we obtain then from (80d), using (80b) and (80c), the reciprocity law (77b) for f itself.

Parseval's identity for multiple Fourier integrals takes the form

$$\iint |f(x, y)|^2\, dx\, dy = \iint |g(\xi, \eta)|^2\, d\xi\, d\eta \tag{81}$$

where the integrations are extended over the whole plane. The proof can be carried out by exactly the same arguments as those used in Section e, p. 488, for the Parseval identity for functions of a single variable, provided we make the same assumptions about $f(x, y)$ as for the derivation of the Fourier integral formula. Modifying the expressions used on pp. 488 appropriately, we consider the integral

$$J_{A,B} = \iint_{x^2+y^2<B^2} |f(x, y) - f_A(x, y)|^2\, dx\, dy,$$

where

$$f_A(x, y) = \frac{1}{2\pi} \iint_{\xi^2+\eta^2<A^2} e^{i(x\xi+y\eta)} g(\xi, \eta)\, d\xi\, d\eta$$

$$g_B(\xi, \eta) = \frac{1}{2\pi} \iint_{x^2+y^2<B^2} e^{-i(x\xi+y\eta)} f(x, y)\, dx\, dy.$$

Here, instead of (73) we obtain the identity

$$J_{A,B} = \iint_{x^2+y^2<B^2} (|f(x, y)|^2 + |f_A(x, y)|^2)\, dx\, dy$$

$$- \iint_{\xi^2+\eta^2<A^2} [\overline{g(\xi, \eta)}\, g_B(\xi, \eta) + g(\xi, \eta)\, \overline{g_B(\xi, \eta)}]\, d\xi\, d\eta.$$

For $A \to \infty$ and $B \to \infty$ the identity (81) follows in the same manner as before.

Exercises 4.13

1. Find the Fourier transforms of the following functions:

(a) $f(x) = \begin{cases} c, \text{ for } 0 < x < a \\ 0, \text{ for } x < 0 \text{ or } x > a. \end{cases}$

(b) $f(x) = \begin{cases} e^{-ax}, \text{ for } x > 0, \ (a > 0) \\ 0, \text{ for } x < 0 \end{cases}$

(c) $J_n(x)/x^n$ (with J_n defined as in 4.12, Exercise 8).

4.14 The Eulerian Integrals (Gamma Function)[1]

One of the most important examples of a function defined by an improper integral involving a parameter is the gamma function $\Gamma(x)$, which we shall discuss in some detail.

a. Definition and Functional Equation

In volume I (p. 308) we defined $\Gamma(x)$ for every $x > 0$ by the improper integral

$$\Gamma(x) = \int_0^\infty e^{-t}\, t^{x-1}\, dt. \tag{82a}$$

We can split up the integral into one extended over the unbounded portion of the t-axis from $t = 1$ to $t = \infty$ with a continuous integrand and one extended over the finite interval from $t = 0$ to $t = 1$, where—at least for values of x between 0 and 1—the integrand is singular. The tests developed on p. 000 show at once that the integral (82a) converges for any $x > 0$, the convergence being uniform in every closed interval of the positive x-axis that does not include the point $x = 0$. *The function $\Gamma(x)$ is therefore continuous for $x > 0$.*

The integrals obtained by formal differentiation of formula (82a) also converge uniformly in any interval $0 < a \leqq x \leqq b$. Consequently (see p. 465), $\Gamma(x)$ has continuous first and second derivatives given by

$$\Gamma'(x) = \int_0^\infty e^{-t} t^{x-1} \log t\, dt \tag{82b}$$

$$\Gamma''(x) = \int_0^\infty e^{-t} t^{x-1} \log^2 t\, dt. \tag{82c}$$

[1] A discussion related to the present one is given by E. Artin, *The Gamma Function* (English translation by Michael Butler), Holt, Rinehart and Winston: New York, 1964.

By simple substitution the integral (82a) for $\Gamma(x)$ can be transformed into other forms that are frequently used. Here we only mention the substitution $t = u^2$, which transforms the gamma function into the form

$$\Gamma(x) = 2\int_0^\infty e^{-u^2} u^{2x-1}\, du.$$

Thus, for $\alpha = 2x - 1$,

$$\int_0^\infty e^{-u^2} u^\alpha\, du = \frac{1}{2}\,\Gamma\left(\frac{1+\alpha}{2}\right) \qquad (\alpha > -1) \tag{82d}$$

[cf. formula (48d), p. 458].

As in Volume I (p. 308), integration by parts in formula (82a) yields the relation

$$\Gamma(x+1) = x\Gamma(x) \tag{83a}$$

for any $x > 0$. This equation is called the *functional equation of the gamma function.*

Clearly, $\Gamma(x)$ is not uniquely defined by the property of being a solution of this functional equation since we obtain other solutions merely by multiplying $\Gamma(x)$ by an arbitrary function $p(x)$ with period unity. The expression

$$u(x) = \Gamma(x)\, p(x) \tag{83b}$$

where

$$p(x+1) = p(x) \tag{83c}$$

represents the most general solution of equation (83a), for if $u(x)$ is any solution, the quotient

$$p(x) = \frac{u(x)}{\Gamma(x)}$$

[which can always be formed since $\Gamma(x) \neq 0$] satisfies equation (83c).

Instead of $\Gamma(x)$ it is frequently more convenient to consider the function $u(x) = \log \Gamma(x)$; this is defined for all positive x, since $\Gamma(x) > 0$ for $x > 0$. The function satisfies the functional equation (a "difference equation")

$$u(x+1) - u(x) = \log x. \tag{83d}$$

We obtain other solutions of (83d) by adding to $\log \Gamma(x)$ an arbitrary function with period unity. In order to characterize the function $\log \Gamma(x)$ uniquely, we must supplement the functional equation (83d) by other conditions. One very simple condition of this type is given by the following theorem of H. Bohr and H. Mollerup:

Every convex solution of the difference equation

$$u(x+1) - u(x) = \log x \tag{84a}$$

for $x > 0$ is identical with the function $\log \Gamma(x)$, except perhaps for an additive constant.

b. Convex Functions. Proof of Bohr and Mollerup's Theorem

A function $f(x)$ with continuous second derivative is called convex (see Volume I, p. 357) if $f'' \geqq 0$. A more general definition, applicable even to functions that are not twice differentiable, is the following:

The function $f(x)$ defined in an interval (posssibly extending to infinity) is called convex if for any values x_1, x_2 of its domain and any positive numbers α, β with $\alpha + \beta = 1$ the inequality

$$f(\alpha x_1 + \beta x_2) \leqq \alpha f(x_1) + \beta f(x_2) \tag{84b}$$

holds. Geometrically (84b) means that for any two points of the curve $y = f(x)$ with abscissa x_1, x_2, the chord joining them never lies beneath the curve (cf. Fig. 4.20).

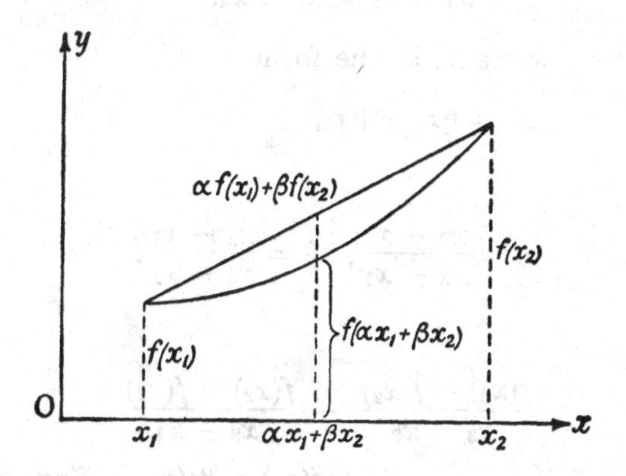

Figure 4.20 A convex function.

For a twice continuously differentiable function f, we find, using the mean value theorem of differential calculus and the fact that α and β are positive numbers with sum 1,

$$\begin{aligned} \text{(84c)} \qquad & \alpha f(x_1) + \beta f(x_2) - f(\alpha x_1 + \beta x_2) \\ &= \beta[f(x_2) - f(\alpha x_1 + \beta x_2)] - \alpha[f(\alpha x_1 + \beta x_2) - f(x_1)] \\ &= \alpha\beta(x_2 - x_1)f'(\xi_2) - \alpha\beta(x_2 - x_1)f'(\xi_1) \\ &= \alpha\beta(x_2 - x_1)(\xi_2 - \xi_1)f''(\eta), \end{aligned}$$

where ξ_1, ξ_2, η are suitable intermediate values with

$$\text{(84d)} \qquad x_1 < \xi_1 < \alpha x_1 + \beta x_2 < \xi_2 < x_2, \qquad \xi_1 < \eta < \xi_2.$$

It follows immediately from (84c) that (84b) is satisfied if $f''(\eta) \geqq 0$ for all η in the domain of f. Conversely, we find from (84b), (84c), using (84d), that $f''(\eta) \geqq 0$; for fixed α, β and $x_2 \to x_1$ it follows from the continuity of f'' that $f''(x_1) \geqq 0$ for any x_1 in the domain. Hence, *a twice continuously differentiable function f is convex in the sense of (84b) if and only if $f'' \geqq 0$.*

To be convex, a function need not be twice, or even once, differentiable. An example is furnished by $f(x) = |x|$. However, *a convex function necessarily is continuous at interior* points of its domain. This follows from the inequality

$$\text{(84e)} \qquad \frac{f(x_2) - f(x_1)}{x_2 - x_1} \leqq \frac{f(x_4) - f(x_3)}{x_4 - x_3}$$

satisfied by a convex function for any x_i in its domain for which

$$x_1 < x_2 < x_3 < x_4.$$

To prove (84e) we write x_2 in the form

$$x_2 = \alpha x_1 + \beta x_3,$$

where

$$\alpha = \frac{x_3 - x_2}{x_3 - x_1}, \quad \beta = \frac{x_2 - x_1}{x_3 - x_1}.$$

Then

$$\begin{aligned} &\frac{f(x_3) - f(x_2)}{x_3 - x_2} - \frac{f(x_2) - f(x_1)}{x_2 - x_1} \\ &\qquad = \frac{\alpha f(x_1) + \beta f(x_3) - f(\alpha x_1 + \beta x_3)}{\alpha\beta(x_3 - x_1)} \geqq 0, \end{aligned}$$

and, similarly,

$$\frac{f(x_4) - f(x_3)}{x_4 - x_3} - \frac{f(x_3) - f(x_2)}{x_3 - x_2} \geqq 0,$$

which implies (84e). In words, (84e) states *that the difference quotients of the convex function f formed for disjoint intervals are increasing.* It follows that

$$\frac{f(x_2) - f(x_1)}{x_2 - x_1} \leqq \frac{f(\xi_2) - f(\xi_1)}{\xi_2 - \xi_1} \leqq \frac{f(x_4) - f(x_3)}{x_4 - x_3}$$

for any values ξ_1, ξ_2 between x_2 and x_3. Thus, f satisfies a *Lipschitz condition* in the interval $x_2 < x < x_3$ and, hence, is continuous in that interval. For any x in the interior of the domain of f we can always find suitable x_1, x_2, x_3, x_4, showing that f is continuous at x.

In order to prove that the function $\log \Gamma(x)$ is convex, it is sufficient to show that

$$\text{(84f)} \qquad \frac{d^2 \log \Gamma}{dx^2} = \frac{\Gamma''\Gamma - \Gamma'^2}{\Gamma^2} \geqq 0.$$

The relation (84f) follows from the Cauchy-Schwarz inequality[1] for integrals, since, here by (82a, b, c),

$$\begin{aligned}\Gamma'^2 &= \left(\int_0^\infty e^{-t}t^{x-1} \log t \, dt\right)^2 \\ &= \left(\int_0^\infty (e^{-t/2}\sqrt{t^{x-1}})(e^{-t/2}\sqrt{t^{x-1}} \log t) \, dt\right)^2 \\ &\leqq \int_0^\infty e^{-t}t^{x-1} \, dt \int_0^\infty e^{-t}t^{x-1} \log^2 t \, dt = \Gamma\Gamma''.\end{aligned}$$

[1]From the Cauchy-Schwarz inequality for sums (Volume I, p. 15) we find for any continuous functions $f(x)$, $g(x)$ and any subdivision of their domain by points x_i into intervals of length Δx_i that

$$\left(\sum_i f(x_i)g(x_i)\Delta x_i\right)^2 \leqq \left(\sum_i f^2(x_i)\Delta x_i\right)\left(\sum_i g^2(x_i)\Delta x_i\right).$$

Refining the subdivisions we find in the limit the *Cauchy-Schwarz inequality for integrals:*

$$\left(\int_a^b f(x)g(x) \, dx\right)^2 \leqq \left(\int_a^b f^2(x) \, dx\right)\left(\int_a^b g^2(x) \, dx\right).$$

This inequality is extended immediately from proper Riemann integrals of continuous functions to improper integrals by passage to the limit with respect to the domain of integration.

Now let $u(x)$ be an arbitrary convex solution of the functional equation (84a) for $x > 0$. We form the expression

$$v_h(x) = u(x + h) - 2u(x) + u(x - h)$$

for $0 < h < x$. Applying relation (84e) which is valid for convex u, we find for $0 < h < k < x$ that

$$\begin{aligned} v_k(x) - v_h(x) &= [u(x + k) - u(x + h)] - [u(x - h) - u(x - k)] \\ &= (k - h)\left[\frac{u(x + k) - u(x + h)}{k - h} - \frac{u(x - h) - u(x - k)}{-h + k}\right] \geqq 0. \end{aligned}$$

For fixed x, therefore, $v_h(x)$ is a continuous nondecreasing function of h. Now, the functional equation for u yields

$$\begin{aligned} v_1(x) &= u(x + 1) - 2u(x) + u(x - 1) \\ &= [u(x + 1) - u(x)] - [u(x) - u(x - 1)] \\ &= \log x - \log(x - 1). \end{aligned}$$

Hence, for $0 < h < 1 < x$,

$$\begin{aligned} 0 = v_0(x) &\leqq v_h(x) \\ &= u(x + h) - 2u(x) + u(x - h) \\ &\leqq v_1(x) = \log \frac{x}{x - 1}. \end{aligned} \tag{84g}$$

Since

$$\lim_{x\to\infty} \log \frac{x}{x - 1} = \log 1 = 0,$$

we find from (84g) that for every convex solution of (84a)

$$\lim_{x\to\infty} [u(x + h) - 2u(x) + u(x - h)] = 0 \qquad (0 < h < 1).$$

If then $p(x)$ is the difference of two convex solutions of (84a), we find that also

$$\lim_{x\to\infty} [p(x + h) - 2p(x) + p(x - h)] = 0.$$

Since $p(x)$ is periodic with period 1, so also is the function

$$p(x + h) - 2p(x) + p(x - h)$$

and it approaches 0 as a limit for $x \to \infty$. Obviously, such a function must vanish identically. Hence,

$$p(x+h) - 2p(x) + p(x-h) = 0 \qquad (0 \leqq h < 1). \tag{84h}$$

Let $M = p(\xi)$ be the largest value of the continuous function $p(x)$ in the interval $1 \leqq x \leqq 2$. Then $p(x) \leqq M$ for all $x > 0$ and by (84h)

$$2M = 2p(\xi) = p(\xi + h) + p(\xi - h) \leqq 2M \qquad (0 \leqq h < 1).$$

Hence,

$$p(\xi - h) = p(\xi + h) = M \qquad (0 \leqq h < 1),$$

and since p has period 1,

$$p(x) = M = \text{constant} \qquad (\text{all } x > 0).$$

This shows that any two convex solutions of (84a) differ at most by a constant and completes the proof of Bohr and Mollerup's theorem.

c. *The Infinite Product for the Gamma Function*

Bohr and Mollerup's theorem can be used to derive the infinite products representations for the gamma function found by Gauss and Weierstrass.

For any given function $g(x)$ we can easily verify that a special solution $w(x)$ of the difference equation

$$w(x+1) - w(x) = g(x)$$

is given by the infinite series

$$\begin{aligned} w(x) &= -\sum_{j=0}^{\infty} g(x+j) \\ &= -g(x) - g(x+1) - g(x+2) - \cdots, \end{aligned}$$

provided that series converges. We cannot apply this observation directly to equation (84a) with $g(x) = \log x$, since the resulting series diverges. However, the difference equation for $w = u''$ obtained by differentiating (84a) twice can be solved in this way. A special solution of the equation

$$w(x+1) - w(x) = -\frac{1}{x^2} \qquad (x > 0) \tag{85a}$$

is given by

$$w(x) = \frac{1}{x^2} + \sum_{j=1}^{\infty} \frac{1}{(x+j)^2} \qquad (x > 0). \tag{85b}$$

Here, the infinite series converges uniformly in every finite interval $0 \leqq x \leqq b$ (see Volume I, p. 535) since

$$\frac{1}{(x+j)^2} \leqq \frac{1}{j^2} \qquad (x \geqq 0).$$

Consequently, w is continuous for $x > 0$. Moreover, term-by-term integration of the series is permitted (see Volume I, p. 537) and leads to a function

$$\begin{aligned} v(x) &= -\frac{1}{x} + \sum_{j=1}^{\infty} \int_0^x \frac{d\xi}{(\xi + j)^2} \\ &= -\frac{1}{x} - \sum_{j=1}^{\infty} \left(\frac{1}{x+j} - \frac{1}{j}\right), \end{aligned} \tag{85c}$$

where the series occuring in this formula again converges uniformly in any interval $0 \leqq x \leqq b$. Thus $v(x) + 1/x$ is a continuous function of x for $x \geqq 0$ that vanishes for $x = 0$. By the foregoing construction

$$v'(x) = w(x) \qquad (x > 0). \tag{85d}$$

Since, by (85a, d),

$$\frac{d}{dx}[v(x+1) - v(x)] = -\frac{1}{x^2} \qquad (x > 0),$$

it follows that

$$v(x+1) - v(x) = \frac{1}{x} + c \qquad (x > 0), \tag{85e}$$

where c is a constant. In order to determine the value of c, we observe that by (85e)

$$\begin{aligned} -c &= \lim_{x \to 0} \left[v(x) + \frac{1}{x}\right] - \lim_{x \to 0} v(x+1) = -v(1) \\ &= 1 + \sum_{j=1}^{\infty} \left(\frac{1}{1+j} - \frac{1}{j}\right) \\ &= 1 + \left(\frac{1}{2} - 1\right) + \left(\frac{1}{3} - \frac{1}{2}\right) + \left(\frac{1}{4} - \frac{1}{3}\right) + \cdots = 0. \end{aligned}$$

Integration of (85c) leads to a function

$$U(x) = -\log x - \sum_{j=1}^{\infty} \int_0^x \left(\frac{1}{\xi + j} - \frac{1}{j}\right) d\xi \tag{85f}$$

$$= -\log x - \sum_{j=1}^{\infty} \left[\log(x + j) - \log j - \frac{x}{j}\right],$$

where the infinite series again converges uniformly in any interval $0 \leq x \leq b$. As before we conclude that $U(x)$ is a continuous function of x for $x > 0$ satisfying

$$U'(x) = v(x), \lim_{x \to 0} (U(x) + \log x) = 0$$

$$U(x + 1) - U(x) - \log x = \text{constant} = C. \tag{85g}$$

Here,

$$C = \lim_{x \to 0} U(x + 1) - \lim_{x \to 0} [U(x) + \log x] = U(1)$$

$$= -\sum_{j=1}^{\infty} \left[\log(1 + j) - \log j - \frac{1}{j}\right]$$

$$= -\lim_{n \to \infty} \sum_{j=1}^{n-1} \left[\log(1 + j) - \log j - \frac{1}{j}\right]$$

$$= \lim_{n \to \infty} \left(1 + \frac{1}{2} + \cdots + \frac{1}{n-1} - \log n\right).$$

It follows that C is identical with *Euler's constant*

$$C = \lim_{n \to \infty} \left(1 + \frac{1}{2} + \frac{1}{3} + \cdots + \frac{1}{n} - \log n\right) \tag{85h}$$

introduced in Volume I (p. 526).

By (85g) the function

$$u(x) = U(x) - Cx$$

satisfies the difference equation

$$u(x + 1) - u(x) = \log x.$$

Moreover, by (85b)

$$u''(x) = w(x) > 0 \qquad (x > 0),$$

so that $u(x)$ is *convex*. Since, in addition,

$$u(1) = U(1) - C = 0 = \log \Gamma(1),$$

it follows from Bohr's theorem that $u(x)$ and $\log \Gamma(x)$ are identical:

$$\log \Gamma(x) = -Cx - \log x - \sum_{j=1}^{\infty} \left(\log \frac{x+j}{j} - \frac{x}{j}\right). \tag{86a}$$

Our derivation also shows that

$$\frac{\Gamma'(x)}{\Gamma(x)} = -C + v(x) = -C - \frac{1}{x} - \sum_{j=1}^{\infty} \left(\frac{1}{x+j} - \frac{1}{j}\right), \tag{86b}$$

$$\frac{d^2 \log \Gamma(x)}{dx^2} = w(x) = \frac{1}{x^2} + \sum_{j=1}^{\infty} \frac{1}{(x+j)^2}. \tag{86c}$$

Forming the exponential function of both sides of equation (86a), we arrive at the *Weierstrass infinite product* for $1/\Gamma(x)$:

$$\frac{1}{\Gamma(x)} = xe^{Cx} \prod_{j=1}^{\infty} \left(1 + \frac{x}{j}\right) e^{-x/j} \qquad (x > 0). \tag{86d}$$

We can write (86d) in a slightly different form not involving the Euler constant C. From (86a), (85h),

$$\begin{aligned}
\log \Gamma(x) &= -\log x + \lim_{n\to\infty} \sum_{j=1}^{n} \left(\frac{x}{j} - \log \frac{x+j}{j}\right) - Cx \\
&= -\log x + \lim_{n\to\infty} \left[x \left(\sum_{j=1}^{n} \frac{1}{j} - C - \log n\right)\right. \\
&\qquad \left. + x \log n - \sum_{j=1}^{n} \log \frac{x+j}{j}\right] \\
&= -\log x + \lim_{n\to\infty} \left[x \log n + \sum_{j=1}^{n-1} \log j - \sum_{j=1}^{n-1} \log (x+j)\right].
\end{aligned}$$

Consequently, we obtain the formula

$$\Gamma(x) = \lim_{n\to\infty} \frac{1 \cdot 2 \cdot 3 \cdots (n-1)}{x(x+1)(x+2)(x+3) \cdots (x+n-1)} n^x \qquad (x > 0), \tag{86e}$$

which is *Gauss's infinite product for the gamma function.*

The limit on the right-hand side of (86e) exists not only for positive values of x but all $x \neq 0, -1, -2, \ldots$: for a given x let the positive integer m be chosen so large that $x + m > 0$. Then, replacing n by $n + m$ under the limit sign, we obtain

$$\lim_{n\to\infty} \frac{1\cdot 2\cdots(n-1)}{x(x+1)(x+2)\cdots(x+n-1)} n^x$$

$$= \lim_{n\to\infty} \frac{1\cdot 2\cdots(n+m-1)}{x(x+1)(x+2)\cdots(x+n+m-1)}(n+m)^x$$

$$= \lim_{n\to\infty} \left[\frac{n(n+1)\cdots(n+m-1)(n+m)^x}{x(x+1)\cdots(x+m-1)n^{x+m}}\right] \left[\frac{1\cdot 2\cdots(n-1)n^{x+m}}{(x+m)(x+m+1)\cdots(x+m+n-1)}\right]$$

$$= \frac{\Gamma(x+m)}{x(x+1)\cdots(x+m-1)}.$$

Thus, we can use Gauss's formula (86e) to define $\Gamma(x)$ for all values of x other than zero or negative integers. When x approaches one of these exceptional values, $\Gamma(x)$ becomes infinite. The extended function $\Gamma(x)$ obviously still satisfies the functional equation

$$\Gamma(x+1) = x\Gamma(x). \tag{86f}$$

d. The Extension Theorem

The values of the gamma function for negative values of x can also easily be obtained from the values for positive values of x by means of the so-called extension theorem. We form the product $\Gamma(x)\Gamma(-x)$, which is

$$\lim_{n\to\infty} \frac{1\cdot 2\cdots(n-1)}{x(x+1)\cdots(x+n-1)} n^x \lim_{n\to\infty} \frac{1\cdot 2\cdots(n-1)}{-x(1-x)(2-x)\cdots(n-1-x)} n^{-x}$$

and combine the two limiting processes into one, to obtain

$$\Gamma(x)\Gamma(-x) = -\frac{1}{x^2}\lim_{n\to\infty} \frac{1}{\{1-(x/1)^2\}\{1-(x/2)^2\}\cdots\{1-[x/(n-1)]^2\}},$$

provided x is not an integer. But, by employing the infinite product for the sine,

$$\frac{\sin \pi x}{\pi x} = \prod_{\nu=1}^{\infty}\left(1-\left(\frac{x}{\nu}\right)^2\right),$$

from Volume I (p. 603), we obtain

$$\Gamma(x)\Gamma(-x) = -\frac{\pi}{x\sin \pi x}.$$

Hence,

$$\Gamma(-x) = -\frac{\pi}{x \sin \pi x} \frac{1}{\Gamma(x)}.$$

We can put this relation in a somewhat different form by calculating the product $\Gamma(x)\Gamma(1-x)$. Since by (86f)

$$\Gamma(1-x) = -x\Gamma(-x),$$

we obtain the *extension theorem*

(97a) $$\Gamma(x)\Gamma(1-x) = \frac{\pi}{\sin \pi x}.$$

Thus, if we put $x = \frac{1}{2}$, we have $\Gamma(\frac{1}{2}) = \sqrt{\pi}$. Since

$$\Gamma\left(\frac{1}{2}\right) = 2\int_0^\infty e^{-u^2}\, du,$$

we have here a new proof for the fact that the integral

$$\int_0^\infty e^{-u^2}\, du$$

has the value $\frac{1}{2}\sqrt{\pi}$ (see p. 415). In addition, we can calculate the gamma function for the arguments $x = n + \frac{1}{2}$, where n is any positive integer:

(97b) $$\Gamma\left(n+\frac{1}{2}\right) = \left(n-\frac{1}{2}\right)\left(n-\frac{3}{2}\right)\cdots\frac{3}{2}\frac{1}{2}\,\Gamma\left(\frac{1}{2}\right)$$
$$= \frac{(2n-1)(2n-3)\cdots 3\cdot 1}{2^n}\sqrt{\pi}.$$

e. The Beta Function

Another important function defined by an improper integral involving parameters is *Euler's beta function.* The beta function is defined by

(98a) $$B(x, y) = \int_0^1 t^{x-1}(1-t)^{y-1}\, dt.$$

If either x or y is less than unity, the integral is improper. By the criterion of p. 465, however, it converges uniformly in x and y, provided we restrict ourselves to intervals $x \geqq \varepsilon$, $y \geqq \eta$, where ε and η are arbitrary positive numbers. It therefore represents a continuous function for all positive values of x and y.

We obtain a somewhat different expression for $B(x, y)$ by using the substitution $t = \tau + \frac{1}{2}$:

$$B(x, y) = \int_{-1/2}^{1/2} \left(\frac{1}{2} + \tau\right)^{x-1} \left(\frac{1}{2} - \tau\right)^{y-1} d\tau. \tag{98b}$$

If we now put $\tau = t/2s$, where $s > 0$, we obtain

$$(2s)^{x+y-1} B(x, y) = \int_{-s}^{s} (s + t)^{x-1}(s - t)^{y-1}\, dt. \tag{98c}$$

If, finally, we put $t = \sin^2\phi$ in formula (98a), we obtain

$$B(x, y) = 2 \int_0^{\pi/2} \sin^{2x-1}\phi \cos^{2y-1}\phi\, d\phi. \tag{98d}$$

We shall now show how the beta function can be expressed in terms of the gamma function, by using a few transformations which, at first sight, may seem strange.

If we multiply both sides of the equation (98c) by e^{-2s} and integrate with respect to s from 0 to A, we have

$$B(x, y) \int_0^A e^{-2s}(2s)^{x+y-1}\, ds = \int_0^A e^{-2s}\, ds \int_{-s}^{s} (s + t)^{x-1}(s - t)^{y-1}\, dt.$$

The double integral on the right may be regarded as an integral of the function

$$e^{-2s}(s + t)^{x-1}(s - t)^{y-1}$$

over the isosceles triangle in the s, t-plane bounded by the lines $s \pm t = 0$ and $s = A$. If we apply the transformation

$$\sigma = s + t,$$

$$\tau = s - t,$$

this integral becomes

$$\frac{1}{2} \iint_R e^{-\sigma-\tau} \sigma^{x-1} \tau^{y-1}\, d\sigma\, d\tau.$$

The region of integration R is now the triangle in the σ, τ-plane bounded by the lines $\sigma = 0$, $\tau = 0$, and $\sigma + \tau = 2A$.

If we let A increase beyond all bounds, the left-hand side, by (82a), tends to the function

$$\frac{1}{2}B(x, y)\Gamma(x + y).$$

Therefore, the right side must also converge and its limit is the double integral over the whole first quadrant of the σ, τ-plane, the quadrant being approximated to by means of isosceles triangles. Since the integrand is positive in this region and the integral converges for a monotonic sequence of regions (by Chapter 4, p. 414) this limit is independent of the mode of approximation to the quadrant. In particular, we can use squares of side A and accordingly write

$$B(x, y)\Gamma(x + y) = \lim_{A\to\infty} \int_0^A \int_0^A e^{-\sigma-\tau}\sigma^{x-1}\tau^{y-1}\, d\sigma\, d\tau$$
$$= \int_0^\infty e^{-\sigma}\sigma^{x-1}\, d\sigma \int_0^\infty e^{-\tau}\tau^{y-1}\, d\tau.$$

We therefore obtain the important relation[1]

(99a)
$$B(x, y) = \frac{\Gamma(x)\Gamma(y)}{\Gamma(x + y)}.$$

From this relation we see that the beta function is related to the binomial coefficients

$$\binom{n+m}{n} = \frac{(n + m)!}{n!m!}$$

[1]This equation can also be obtained from Bohr's theorem. We first show that $B(x, y)$ satisfies the functional equation

$$B(x + 1, y) = \frac{x}{x + y} B(x, y),$$

so that the function

$$u(x, y) = \Gamma(x + y)\, B(x, y),$$

considered as a function of x, satisfies the functional equation of the gamma function,

$$u(x + 1) = xu(x).$$

The convexity of $\log B(x, y)$ and, hence, that of $\log u(x)$ follows from the Cauchy-Schwarz inequality in the same way as that of $\log \Gamma(x)$ on p. 501. Thus, we have

$$\Gamma(x + y)\, B(x, y) = \Gamma(x) \cdot \alpha(y),$$

and finally, if we put $x = 1$, $\alpha(y) = \Gamma(1 + y)\, B(1, y) = \Gamma(y)$.

in roughly the same way as the gamma function is related to the numbers $n!$ For integers n, m in fact,

$$\binom{n+m}{m} = \frac{1}{(n+m+1)B(n+1, m+1)}. \tag{99b}$$

Finally, we mention that the definite integrals

$$\int_0^{\pi/2} \sin^\alpha t \, dt \quad \text{and} \quad \int_0^{\pi/2} \cos^\alpha t \, dt,$$

which by (98d) are identical with the functions

$$\frac{1}{2} B\left(\frac{\alpha+1}{2}, \frac{1}{2}\right) = \frac{1}{2} B\left(\frac{1}{2}, \frac{\alpha+1}{2}\right),$$

can be simply expressed in terms of the gamma function:

$$\int_0^{\pi/2} \sin^\alpha t \, dt = \int_0^{\pi/2} \cos^\alpha t \, dt = \frac{\sqrt{\pi}}{\alpha} \frac{\Gamma(1+\alpha/2)}{\Gamma(\alpha/2)}. \tag{99c}$$

f. Differentiation and Integration to Fractional Order. Abel's Integral Equation

Using our knowledge of the gamma function, we now carry out a simple process of generalization of the concepts of differentiation and integration. We have already seen (p. 78) that the formula

$$F(x) = \int_0^x \frac{(x-t)^{n-1}}{(n-1)!} f(t) dt = \frac{1}{\Gamma(n)} \int_0^x (x-t)^{n-1} f(t) dt \tag{100a}$$

gives the n-times-repeated integral of the function $f(x)$ between the limits 0 and x. If D symbolically denotes the operator of differentiation and if D^{-1} denotes the operator

$$\int_0^x \cdots dx,$$

which is an inverse of differentiation, we may write

$$F(x) = D^{-n} f(x). \tag{100b}$$

The mathematical statement conveyed by this formula is that the function $F(x)$ and its first $(n-1)$ derivatives vanish at $x = 0$ and that the nth derivative of $F(x)$ is $f(x)$. But it is now very natural to construct a definition for the operator $D^{-\lambda}$ even when the positive number λ is not necessarily an integer. *The integral of order* λ *of the function* $f(x)$ *between the limits* 0 *and* x is defined by the expression

$$D^{-\lambda}f(x) = \frac{1}{\Gamma(\lambda)}\int_0^x (x-t)^{\lambda-1}f(t)dt. \tag{100c}$$

This definition may now be used to generalize nth-order differentiation, symbolized by the operator D^n or d^n/dx^n, to μth-order differentiation, where μ is an arbitrary nonnegative number. Let m be the least integer greater than μ, so that $\mu = m - \rho$, where $0 < \rho \leqq 1$. Then our definition is

$$D^{\mu}f(x) = D^m D^{-\rho}f(x) = \frac{d^m}{dx^m}\frac{1}{\Gamma(\rho)}\int_0^x (x-t)^{\rho-1}f(t)\,dt. \tag{101a}$$

A reversal of the order of the two processes would give the definition

$$D^{\mu}f(x) = D^{-\rho}D^m f(x) = \frac{1}{\Gamma(\rho)}\int_0^x (x-t)^{\rho-1}f^{(m)}(t)\,dt.$$

It is left to the reader (see Exercise 12) to employ the formulas for the gamma function to prove that

$$D^{\alpha}D^{\beta}f(x) = D^{\beta}D^{\alpha}f(x), \tag{101b}$$

where α and β are arbitrary real numbers. He should show that these relations and the generalized process of differentiation have a meaning whenever the function $f(x)$ is differentiable in the ordinary way to a sufficiently high order for all x and vanishes for $x \leqq 0$. In general $D^{\mu}f(x)$ exists if $f(x)$ has continuous derivatives up to, and including, the mth order.

In connection with these ideas, we mention *Abel's integral equation*, which has important applications. Since

$$\Gamma\left(\frac{1}{2}\right) = \sqrt{\pi},$$

the integral of a function $f(x)$ to the order $\frac{1}{2}$ is given by the formula

$$D^{-1/2}f(x) = \frac{1}{\sqrt{\pi}}\int_0^x \frac{f(t)}{\sqrt{x-t}}\,dt = \psi(x). \tag{102}$$

Formula (102) is called Abel's integral equation when it is to be solved for an unknown function $f(x)$, the function $\psi(x)$ on the right side being given. If the function $\psi(x)$ is continuously differentiable and vanishes at $x = 0$, the solution of the equation is given by the formula

$$f(x) = D^{1/2}\psi(x), \tag{103a}$$

or

$$f(x) = \frac{1}{\sqrt{\pi}}\frac{d}{dx}\int_0^x \frac{\psi(t)}{\sqrt{x-t}}\,dt. \tag{104}$$

Exercises 4.14

1. Verify that for nonnegative integral n,

$$\Gamma\left(n+\frac{1}{2}\right) = \frac{(2n)!\,\sqrt{\pi}}{n!4^n}.$$

2. Find $\Gamma(\frac{1}{2} - n)$ where n is a positive integer.
3. Show that

$$B(x, x) = 2^{1-2x}B\left(x, \frac{1}{2}\right).$$

4. Prove

$$I = \int_0^1 \frac{dt}{\sqrt{1-t^x}} = \frac{\sqrt{\pi}}{x}\,\frac{\Gamma\left(\frac{1}{x}\right)}{\Gamma\left(\frac{1}{x}+\frac{1}{2}\right)}.$$

5. Establish the following relations:

(a) $$\int_0^1 \frac{x^{2n+1}}{\sqrt{1-x^2}}\,dx = \frac{(n!)^2\,2^{2n}}{(2n+1)!},$$

(b) $$\int_0^1 \frac{x^{2n}}{\sqrt{1-x^2}}\,dx = \frac{(2n)!\,\pi}{2^{2n+1}(n!)^2}.$$

6. Prove that the volume of the positive octant bounded by the planes $x = 0$, $y = 0$, $z = h$ and the surface $x^m/a^m + y^m/b^m = z/c$, where $m > 0$, is

$$abh\left(\frac{h}{c}\right)^{2/m}\frac{\Gamma(1+1/m)^2}{\Gamma(2+2/m)}.$$

7. Prove that

$$\iiint f\left(\frac{x^2}{a^2}+\frac{y^2}{b^2}+\frac{z^2}{c^2}\right)x^{p-1}y^{q-1}z^{r-1}\,dx\,dy\,dz$$

taken throughout the positive octant of the ellipsoid $x^2/a^2 + y^2/b^2 + z^2/c^2 \leqq 1$ is equal to

$$\frac{a^p b^q c^r}{8} \frac{\Gamma\left(\frac{p}{2}\right)\Gamma\left(\frac{q}{2}\right)\Gamma\left(\frac{r}{2}\right)}{\Gamma\left(\frac{p+p+r}{2}\right)} \int_0^1 f(\xi)\xi^{(p+q+r-2)/2}\, d\xi.$$

(Hint: Introduce new variables ξ, η, ζ by writing

$$\frac{x^2}{a^2} + \frac{y^2}{b^2} + \frac{z^2}{c^2} = \xi \qquad \text{or} \qquad x = a\sqrt{\xi(1-\eta)}$$

$$\frac{y^2}{b^2} + \frac{z^2}{c^2} = \xi\eta \qquad \text{or} \qquad y = b\sqrt{\xi\eta(1-\zeta)}$$

$$\frac{z^2}{c^2} = \xi\eta\zeta \qquad \text{or} \qquad z = c\sqrt{\xi\eta\zeta}$$

and perform the integrations with respect to η and ζ.)

8. Find the x-coordinate of the center of mass of the solid

$$\left(\frac{x}{a}\right)^{1/n} + \left(\frac{y}{b}\right)^{1/n} + \left(\frac{z}{c}\right)^{1/n} \leqq 1, \qquad x \geqq 0,\ y \geqq 0,\ z \geqq 0.$$

9. Find the moment of inertia of the area enclosed by the astroid $x^{2/3} + y^{2/3} = R^{2/3}$ with respect to the x-axis.

10. Prove that the $(n+1)$-fold integral

$$\int \cdots \int f(x_0 + \cdots + x_n) x_0^{a_0 - 1} \cdots x_n^{a_n - 1} dx_0 \cdots dx_n$$

taken over the positive orthant $x_k \geqq 0$ for $k = 0, \ldots, n$ bounded by the hyperplane $x_0 + \cdots + x_n = 1$ is equal to

$$\frac{\Gamma(a_0) \cdots \Gamma(a_n)}{\Gamma(a_0 + \cdots + a_n)} \int_0^1 f(t)\, t^{a_0 + \cdots + a_n - 1}\, dt.$$

11. Prove that

$$2^{2x} \frac{\Gamma(x)\Gamma\left(x + \frac{1}{2}\right)}{\Gamma(2x)} = 2\sqrt{\pi}.$$

12. (a) Show that for any positive real numbers α and β

$$D^\alpha D^\beta f(x) = D^\beta D^\alpha f(x)$$

where the derivatives are defined by (101a) and f has ordinary derivatives up to $(p+q)$-th order that vanish at $x = 0$, p and q being the least integers greater than α and β, respectively.

(b) Under the foregoing conditions, is it always true that $D^\alpha D^\beta f(x) = D^{\alpha+\beta} f(x)$?

(c) Extend the foregoing result to the case in which α or β may be negative.

Appendix: Detailed Analysis of the Process of Integration[1]

A.1 Areas

The area of a set S can be defined rigorously along the lines suggested by intuition, as explained on pp. 368. Essentially one uses a subdivision of the plane into squares by lines parallel to the coordinate axes. One adds up the areas of the squares completely contained in S. This yields a lower bound for the area of S. Adding up the areas of all squares having points in common with S, we obtain an upper bound for the area of S. If these lower and upper bounds converge toward one and the same value as the subdivision of the plane is refined indefinitely, we identify this common value with the area of S. This construction for the area of a region incorporates the same ideas of approximating the region from inside and outside by regions composed of rectangles that led us to the notion of the Riemann integral of a function $f(x)$.

The concept of area, as defined here, is named the *Jordan measure* (after one of the initiators of modern precise analysis) or *content* of S. This is not the only way to introduce areas. (An extremely important definition that applies to more general sets yields the so-called *Lebesgue measure* of S.) The Jordan measure, which will occupy us here exclusively, has the advantage of greater intuitive immediacy and is quite adequate for those portions of analysis that lie within the scope of this book.

For simplicity, we shall work mainly in the plane. However, our treatment will apply to higher dimensions with only such changes of terminology as the replacement of the term *area* by *volume*, *square* by *cube* and so on.

a. *Subdivisions of the Plane and the Corresponding Inner and Outer Areas*

To define at the area of a set S in the x, y-plane, we use successive subdivisions of the plane into squares of side $1, \frac{1}{2}, \frac{1}{4}, \frac{1}{8}, \ldots$ by equidistant parallels to the coordinate axes.[2] The nth subdivision (where n is a positive integer) is produced by the lines

[1]Before reading this Appendix the reader would do well to review the arguments leading to the Riemann integral in Volume I (pp. 192–195).

[2]It is helpful at this stage to introduce area through a quite specific set of subdivisions of the plane into squares. Later, it will turn out that much more general subdivisions lead to the same area.

$$(1) \qquad x = \frac{i}{2^n}, \quad y = \frac{k}{2^n},$$

where i and k range over all integers. The plane is then divided into the closed squares R_{ik}^n given by

$$(2) \qquad R_{ik}^n: \frac{i}{2^n} \leqq x \leqq \frac{i+1}{2^n}, \quad \frac{k}{2^n} \leqq y \leqq \frac{k+1}{2^n}.$$

Let now S be any *bounded* set of points in the plane.[1] We form approximations from below and from above to the prospective area A of S by forming the sum A_n^- of the areas of all squares R_{ik}^n that are completely contained in S, and the sum A_n^+ of the area of all squares R_{ik}^n that have points in common with S. Here the area of a square R_{ik}^n that has side 2^{-n} is defined to be 2^{-2n}. Using the symbolic notation for relation between sets explained on p. 114, we have, accordingly,[2]

$$(3) \qquad A_n^- = \sum_{\substack{i,k \\ R_{ik}^n \subset S}} 2^{-2n}, \qquad A_n^+ = \sum_{\substack{i,k \\ R_{ik}^n \cap S \neq \emptyset}} 2^{-2n}$$

(see Fig. 4–1).

It is clear from the definition that

$$(4) \qquad 0 \leqq A_n^- \leqq A_n^+.$$

As we pass from the nth to the $(n+1)$-st subdivision, each square R_{ik}^n is broken up into four squares R_{rs}^{n+1}. If R_{ik}^n is contained in S, somust be its parts R_{rs}^{n+1}. If, on the other hand, a part R_{rs}^{n+1} contains a point of S, then the same holds for the whole square R_{ik}^n.
It follows[3] that successive sums satisfy the inequalities

$$(5) \qquad A_n^- \leqq A_{n+1}^- \leqq A_{n+1}^+ \leqq A_n^+.$$

We see from (5) that the sums A_n^- form a nondecreasing sequence with the upper bound A_1^+, hence, they converge to a limit,

$$A^- = \lim_{n\to\infty} A_n^-.$$

[1]Areas, properly speaking, will only be defined for bounded sets, although an "improper" area is defined for some unbounded sets as limit of "proper" areas.

[2]If no square R_{ik}^n is contained completely in S, we put $A_n^- = 0$.

[3]We have used here that the sum of the areas of the four squares R_{rs}^{n+1} making up R_{ik}^n equals the area of R_{ik}^n, which, in this context, follows from the arithmetical identity

$$4 \cdot 2^{-2(n+1)} = 2^{-2n}.$$

Similarly, the sums A_n^+ form a nonincreasing sequence with lower bound A_1^- and converge:

$$A^+ = \lim_{n\to\infty} A_n^+.$$

By (5), we have for all n

$$0 \leq A_n^- \leq A^- \leq A^+ \leq A_n^+. \tag{6}$$

We call A^- the *inner area* and A^+ the *outer area*[1] of S. Every bounded set S has an inner and an outer area, which we denote by $A^-(S)$ and $A^+(S)$.

The inner area $A^-(S)$ *has the value* 0 *if and only if* S *has no interior points,* for a set with no interior points contains no square R_{ik}^n, so that $A_n^- = 0$ for all n, and thus, $A^- = 0$. A set with interior points contains some square R_{ik}^n for sufficiently large n, so that $A_n^- > 0$ for large n, and hence, $A^- > 0$.

b. Jordan-Measurable Sets and Their Areas

We call a bounded set S Jordan-measurable if the inner area A^- and the outer area A^+ of S coincide.[2] We denote the common value by A and call it the *area* or the *Jordan measure* of S:

$$A^-(S) = A^+(S) = A(S).$$

Note that for the squares R_{ik}^n used in our definitions, the original notion of "area" and the new one, the Jordan measure, coincide. Each square R_{ik}^n has the Jordan measure 2^{-2n} in the sense of the general definition, since for $S = R_{ik}^n$ and $m > n$

$$A_m^-(S) = (2^{m-n})^2 2^{-2m} = 2^{-2n},$$

$$A_m^+ = [(2^{m-n})^2 + 4(2^{m-n}) + 4]2^{-2m} = 2^{-2n} + 2^{2-m-n} + 2^{2-2m}.$$

More generally, any rectangle S with sides parallel to the coordinate axes:

[1]The terms *interior Jordan measure* or *interior content*, or, respectively, *exterior Jordan measure* or *exterior content*, are also commonly used.

[2]Instead of using the phrase "the set S is Jordan-measurable," we shall simply say, "S has an area." The term *measure* has the advantage of being independent of dimension and can be used equally well for *length* in one dimension, as for *area* in two dimensions, and for *volume* in higher dimensions.

$$S:\quad a \leqq x \leqq b,\quad c \leqq y \leqq d$$

has the area $(b - a)(d - c)$, as expected from elementary geometry; for, given a positive integer n, we can find integers $\alpha, \beta, \gamma, \delta$ such that

$$\alpha 2^{-n} < a \leqq (\alpha + 1)2^{-n},\quad \beta 2^{-n} \leqq b < (\beta + 1)2^{-n}$$
$$\gamma 2^{-n} < c \leqq (\gamma + 1)2^{-n},\quad \delta 2^{-n} \leqq d < (\delta + 1)2^{-n}.$$

Then,

$$A_n^-(S) = (\beta - \alpha - 1)(\delta - \gamma - 1)2^{-2n} \geqq (b - a - 2^{1-n})(d - c - 2^{1-n}),$$
$$A_n^+(S) = (\beta - \alpha + 1)(\delta - \gamma + 1)2^{-2n} \leqq (b - a + 2^{1-n})(d - c + 2^{1-n}),$$

so that for $n \to \infty$,

$$\lim_{n\to\infty} A_n^-(S) = \lim_{n\to\infty} A_n^+(S) = (b - a)(d - c).$$

Our next task is to find criteria for measurability of a set S. We shall prove quite generally that *necessary and sufficient for a bounded set S to have an area is that its boundary ∂S have area zero.*

In proof, consider a subdivision of the plane into squares R_{ik}^n and form the corresponding sums $A_n^-(S)$ and $A_n^+(S)$ as in (3). Obviously, $A_n^+ - A_n^-$ represents the sum of the areas of the squares R_{ik}^n that contain points in S as well as points not in S. Let σ_n be the set of those squares. Each square of σ_n contains a boundary point of S, for on the line segment joining a point P of R_{ik}^n in S to a point Q not in S but in the same square R_{ik}^n there certainly lies a boundary point of S. Hence, each square of σ_n has points in common with ∂S, and consequently,

$$A_n^+(S) - A_n^-(S) \leqq A_n^+(\partial S).$$

If ∂S has area 0 (or, what is the same, *outer area* 0) the right-hand side tends to 0 for $n \to \infty$, and we find that $A^+(S) - A^-(S) = 0$, or that S has an area.

Conversely, let S have an area, so that

$$\lim_{n\to\infty} [A_n^+(S) - A_n^-(S)] = 0. \tag{7}$$

A point P in the plane that for a fixed n belongs only to squares R_{ik}^n contained in S must be an interior point of S.[1] Similarly, a point be-

[1]Remember that our squares R_{ik}^n are closed. Hence, P could belong to as many as four squares.

longing only to squares free of points of S must be an exterior point of S. Let P be a boundary point of S. If P did not lie in any square of σ_n, it would have to belong to a square contained in S as well as to a square free of points of S. But this is impossible since two such squares cannot have a common point. Hence, every P in ∂S is contained in a squre R_{ik}^n of the set σ_n. The total area of those squares is $A_n^+(S) - A_n^-(S)$. Any square R_{ik}^n having a point in common with ∂S either is then a square in σ_n or one of the eight neighbors of such a square, having a point in common with it. Hence, the total area of the squares R_{ik}^n having points in common with ∂S cannot exceed nine times the total area of the squares in σ_n:

$$A_n^+(\partial S) \leqq 9[A_n^+(S) - A_n^-(S)].$$

Hence, (7) implies that $A^+(\partial S) = 0$ and, thus, that ∂S has area 0.

An example of a set that does not have an area A in our sense is furnished by the set of rational points in the unit square, that is, the set S consisting of the points (x, y), where x and y are rational numbers between 0 and 1. Here the boundary ∂S is the set of all (x, y) with $0 \leqq x \leqq 1, 0 \leqq y \leqq 1$ and, hence, has area 1. It follows from our theorem that S is not Jordan-measurable.

c. Basic Properties of Area

Let S and T be two bounded sets with S contained in T. A square R_{ik}^n that contains a point of S necessarily contains a point of T, so that

$$A_n^+(S) \leqq A_n^+(T).$$

For $n \to \infty$ we find that generally

$$A^+(S) \leqq A^+(T) \quad \text{for} \quad S \subset T. \tag{8}$$

In the particular case that $A^+(T) = 0$, we conclude that also $A^+(S) = 0$. Hence:

Any subset of a set of area 0 has area 0.

For any two bounded sets S, T the totality of squares R_{ik}^n covering S and T also covers their union $S \cup T$. Hence

$$A_n^+(S \cup T) \leqq A_n^+(S) + A_n^+(T).$$

For $n \to \infty$ we find that

$$A^+(S \cup T) \leqq A^+(S) + A^+(T). \tag{9}$$

More generally, for any finite number of sets $S_1, S_2, \ldots, S_N$ we have the *finite subadditivity of outer areas* expressed by the formula

$$A^+\left(\bigcup_{i=1}^{N} S_i\right) \leqq \sum_{i=1}^{N} A^+(S_i). \tag{10}$$

If in (10) all the S_i have area 0 the same follows for the union:

The union of any finite number of sets of area 0 has area 0. In particular, any finite set of points has area 0.

By definition, a set of area 0 can be covered by a finite number of squares R_{ik}^n of arbitrarily small total area A_n^+. More generally, *a set S has area 0 if for each $\varepsilon > 0$ we can find a finite number of sets $S_1, \ldots, S_N$ covering S, the sum of whose outer areas is less than* ε, for then by (8) and (9) the outer area of S is less than ε, and hence, since ε is an arbitrary positive number, $A^+(S) = 0$.

For example, a continuous arc C in the plane given nonparametrically by an equation

$$y = f(x) \qquad (a \leqq x \leqq b)$$

has area 0. For the proof we only have to use the fact that a continuous function defined in a closed and bounded interval is uniformly continuous. For, given $\varepsilon > 0$, we can find an n so large that f differs by less than ε for any two arguments in its domain that have distance $< 2^{-n}$. We can find integers α, β such that

$$\alpha 2^{-n} \leqq a < (\alpha + 1)2^{-n}, \qquad \beta 2^{-n} < b \leqq (\beta + 1)2^{-n}.$$

The portion of the graph of $f(x)$ corresponding to values x with $i2^{-n} < x < (i + 1)2^{-n}$ is contained in a rectangle with sides that are parallel to the coordinate axes and have the lengths 2^{-n} and 2ε. Hence, C is contained in the union of these rectangles with sides parallel to the axes of total area

$$(\beta + 1 - \alpha)2^{-n}(2\varepsilon) \leqq (b - a + 2^{1-n})2\varepsilon.$$

For $n \to \infty$ it follows that

$$A^+(C) \leqq 2(b - a)\varepsilon,$$

and thus, since ε is an arbitrary positive number, that the arc C has area 0.

Most of the regions of practical interest have boundaries consisting of a finite number of continuous arcs of the form $y = f(x)$ or $x = g(y)$. Since the union of a finite number of sets of area 0 has itself area 0, we conclude that such regions have a boundary of area 0 and, hence, are Jordan-measurable:

Let the boundary of a set S be contained in the union of a finite number of arcs, each of which is given either by an equation $y = f(x)$ or by an equation $x = g(y)$ with the respective function f or g defined and continuous in a finite closed interval. Then S has an area.[1]

We now consider the *union* and *intersection* of S and T, where S and T are any two Jordan-measurable sets. A point that is interior to S or to T is interior to $S \cup T$; a point exterior to S and to T is exterior to $S \cup T$. Hence, a boundary point of $S \cup T$ must be boundary point of either S or T. Similarly, boundary points of $S \cap T$ must be boundary points of either S or of T. Hence, the boundaries of $S \cup T$ and $S \cap T$ lie in the union of ∂S and ∂T and have area 0, since the boundaries ∂S and ∂T have area 0. This proves the fundamental fact:

The union and intersection of two Jordan-measurable sets are again Jordan measurable.

Applying (9), we conclude:

If the sets S and T have an area, their union $S \cup T$ also has an area and

$$A(S \cup T) \leqq A(S) + A(T). \tag{11}$$

Furthermore, if S and T do not overlap (i.e., interior points of either one of the sets are exterior to the other), we can even conclude that

$$A(S \cup T) = A(S) + A(T). \tag{12}$$

For then a square R_{ik}^n cannot be contained in both S and T. Hence, for the nth subdivision

$$A_n^-(S \cup T) \geqq A_n^-(S) + A_n^-(T).$$

For $n \to \infty$ it follows that

$$A^-(S \cup T) \geqq A^-(S) + A^-(T).$$

[1]More generally, it follows in the same way that a set S in n dimensions is Jordan-measurable if its boundary is contained in the union of a finite number of surfaces, each given by an equation of the form

$$x_j = f(x_1, \cdots, x_{j-1}, x_{j+1}, \cdots, x_n)$$

with f continuous in a bounded closed set of $x_1 \cdots x_{j-1}\, x_{j+1} \cdots x_n$-space.

Since S, T and $S \cup T$ are Jordan-measurable this implies that

$$A(S \cup T) \geqq A(S) + A(T),$$

so that (12) follows from (11).

This result can be extended immediately to any finite number of Jordan-measurable sets and constitutes the *finite additivity of areas:*

If each of the finite number of sets $S_1, \ldots, S_N$ has an area and no two sets overlap, then the union S of $S_1, \ldots, S_N$ also has an area, and

$$A(S) = A(S_1) + A(S_2) + \cdots + A(S_N). \tag{13}$$

This addition theorem can be supplemented by a *subtraction theorem.* Given two sets S, T with $S \subset T$, we denote by $T - S$ the set of points of T that are not contained in S. We shall prove that when S and T have areas and $S \subset T$, then $T - S$ has an area and

$$A(T - S) = A(T) - A(S). \tag{14}$$

It is easily seen again that the boundary of $T - S$ is contained in the union of the boundaries of T and of S, so that $T - S$ has an area. Moreover, S and $T - S$ have no points in common hence do not overlap, and have union T, so that by the additivity rule (12)

$$A(T) = A(S) + A(T - S),$$

which is equivalent to (14).

A more symmetric combination of the addition and subtraction rules for areas consists in the identity

$$A(S \cap T) + A(S \cup T) = A(S) + A(T) \tag{15}$$

valid for any two Jordan-measurable sets S and T. Indeed, we have the identity

$$S \cup T - T = S - S \cap T$$

between the four sets S, T, $S \cap T$, $S \cup T$. Since all four sets have an area, we can apply (14), and (15) follows.

The preceding theorems permit us to free the notion of area from any reference to the special squares R_{ik}^n used in its definition. We shall see that area may be defined in terms of much more general methods of subdivision of the plane, including, for example, subdivisions of the plane into rectangles with sides parallel to the axes.

First, we observe that for a Jordan-measurable set S all points sufficiently close to the boundary ∂S of S can be enclosed in a set of arbitrarily small area, for, since ∂S has area 0, we can for a given $\varepsilon > 0$ find an $n = n(\varepsilon)$ such that the set σ_n of squars R_{ik}^n having points in common with ∂S has total area $< \varepsilon/9$. Let P be a point of the plane that has distance $< 2^{-n}$ from some point of ∂S. Then P either belongs to one of the squares in σ_n or to one of the eight neighbors of such a square. The union of the set of all squares in σ_n and of their neighbors is then a set of area $< \varepsilon$ that contains all points of distance $< 2^{-n}$ from the points of ∂S.

Now take a subdivision Σ of the whole plane into closed rectangles with sides parallel to the coordinate axes. The rectangles need not be congruent, but we require that the subdivision be so fine that all of the rectangles ρ have diameters[1] less than $2^{-n(\varepsilon)}$. We form the sum $A_\Sigma^-(S)$ of the areas of all rectangles ρ of our subdivision that are contained in S and also the sum $A_\Sigma^+(S)$ of all ρ that have points in common with S. Clearly,

$$A_\Sigma^-(S) \leqq A(S) \leqq A_\Sigma^+(S).$$

Moreover, $A_\Sigma^+(S) - A_\Sigma^-(S)$ represents the sum of the areas of all rectangles ρ that contain both points in S and points not in S. These rectangles necessarily contain boundary points of S. Since their diameter is less than 2^{-n}, each point of such a rectangle ρ will have a distance less than 2^{-n} from some point of ∂S. Hence, the total area of these rectangles will be less than ε. Thus,

$$A_\Sigma^+(S) - A_\Sigma^-(S) < \varepsilon,$$

and consequently,

$$A(S) - A_\Sigma^-(S) < \varepsilon, \quad A_\Sigma^+(S) - A(S) < \varepsilon.$$

Taking a sequence of subdivisions Σ_n of the plane into rectangles with the largest diameter of any rectangle in Σ_n tending to zero, we find that the corresponding sums $A_n^+(S)$ and $A_n^-(S)$ tend to the area $A(S)$ of our set.

The argument used applies equally well to sequences of much more general subdivisions Σ_n of the whole plane into sets ρ. We need require only that the individual sets ρ be Jordan-measurable, closed, and connected and that the maximum diameter of any set ρ in a subdivision tend to 0 as $n \to \infty$.

[1]The diameter of a set is defined generally as the least upper bound (or, in the case of a closed and bounded set, as the maximum) of the distances of any two points in the set. In the case of a rectangle ρ this is the length of the diagonals.

A.2 Integrals of Functions of Several Variables

a. *Definition of the Integral of a Function f(x, y)*

We first define the integral of a function $f(x, y)$ over the whole x, y-plane. Throughout this section we make the assumption that the function $f(x, y)$ is defined for all (x, y) but has the value 0 outside some bounded set, that is that $f(x, y) = 0$ for all (x, y) sufficiently far away from the origin (such functions are said to have *compact support*). Moreover, we assume that f is bounded.

In defining the integral of such a function f we make use of the same kind of subdivision of the plane into closed squares R_{ik}^n as in the case of areas. Let M_{ik}^n be the supremum and m_{ik}^n the infimum[1] of f in the square R_{ik}^n. We then associate with f and the nth subdivision of the plane the *upper sum*

$$F_n^+ = \sum_{i,k} M_{ik}^n 2^{-2n}$$

and the *lower sum*[2]

$$F_n^- = \sum_{i,k} m_{ik}^n 2^{-2n}.$$

Only a finite number of terms in these sums are different from 0, since $f = 0$ for distant points. Since $m_{ik}^n \leqq M_{ik}^n$, we have

$$F_n^- \leqq F_n^+. \tag{16}$$

In passing from the nth to the $(n + 1)$-st subdivision, each square R_{ik}^n is divided into four squares R_{js}^{n+1} of area 2^{-2n-2} for which, obviously,

$$m_{ik}^n \leqq m_{js}^{n+1} \leqq M_{js}^{n+1} \leqq M_{ik}^n.$$

It follows that

$$F_n^- \leqq F_{n+1}^- \leqq F_{n+1}^+ \leqq F_n^+. \tag{17}$$

Since bounded monotone sequences converge (see Volume I, p. 96), the upper and lower sums have limits

---

[1]See the definitions in Volume I, p. 97

[2]The factor 2^{-2n} represents the area of the squares R_{ik}^n produced in the nth subdivision. In three dimensions, where we subdivide space into cubes of side 2^{-n}, the factor becomes 2^{-3n} and, similarly, in k dimensions, 2^{-kn}.

(18) $$F^- = \lim_{n\to\infty} F_n^-, \quad F^+ = \lim_{n\to\infty} F_n^+,$$

where, of course,

(19) $$F^- \leqq F^+.$$

We call F^+ the *upper integral* and F^- the *lower integral* of the function $f(x, y)$.

DEFINITION. *The function $f(x, y)$ is called integrable*[1] *if its upper integral F^+ and its lower integral F^- have the same value, which is then called the integral of f and is denoted by*

$$\iint f \, dx \, dy.$$

Since

$$F^+ - F^- = \lim_{n\to\infty} (F_n^+ - F_n^-),$$

we immediately have the following integrability condition: *Necessary and sufficient for the integrability of f is that*

(20) $$\lim_{n\to\infty} (F_n^+ - F_n^-) = \lim_{n\to\infty} \sum_{i,k} (M_{ik}^n - m_{ik}^n) 2^{-2n} = 0.$$

We can associate with the nth subdivision a *Riemann sum*

$$F_n = \sum_{i,k} f(\xi_{ik}^n, \eta_{ik}^n) 2^{-2n},$$

where $(\xi_{ik}^n, \eta_{ik}^n)$ is an arbitrary point of the square R_{ik}^n. Clearly,

(21) $$F_n^- \leqq F_n \leqq F_n^+.$$

We conclude from (18):

If f is integrable, the Riemann sums F_n converge to the value of $\iint f \, dx \, dy$ irrespective of the choice of the intermediate points $(\xi_{ik}^n, \eta_{ik}^n)$ in R_{ik}^n.

[1]More precisely, "Riemann-integrable." The definition given here differs from the common one in so far as only the restricted class of subdivisions into squares R_{ik}^n is considered, but is equivalent to it.

b. Integrability of Continuous Functions and Integrals over Sets

For applications of the notion of integral the following theorem is basic:

A continuous function f vanishing outside some bounded set S is integrable.

For the proof we can assume that S is a square

$$|x| \leqq N, \quad |y| \leqq N,$$

where N is a positive integer. Then in the nth subdivision $M_{ik}^n = m_{ik}^n = 0$ for R_{ik}^n not contained in S. In the closed bounded set S the continuous function f is uniformly continuous. Consequently, given $\varepsilon > 0$, there exists a $\delta > 0$ such that the values of f differ by less than ε for any two points in S having distance less than δ. Hence,

$$M_{ik}^n - m_{ik}^n \leqq \varepsilon,$$

provided n is so large that

$$\sqrt{2}\, 2^{-n} < \delta.$$

Thus,

$$F_n^+ - F_n^- \leqq \sum \varepsilon 2^{-2n},$$

where the summation is extended over all i, k for which the square R_{ik}^n is contained in S. Since the sum of the areas of those squares equals the area $4N^2$ of S, it follows that

$$F_n^+ - F_n^- \leqq 4N^2\varepsilon$$

for all sufficiently large n and, hence, that f satisfies the integrability condition (20).

The continuous functions are not the only integrable ones. We shall not try to determine the most general integrable functions. However, we do consider one important class of discontinuous functions that are integrable, namely, the characteristic functions of bounded Jordan-measurable sets. With any set S in the plane we associate the *characteristic function* ϕ_S defined by

$$\phi_S(x, y) = \begin{cases} 1 & \text{for} \quad (x, y) \in S \\ 0 & \text{for} \quad (x, y) \notin S. \end{cases}$$

The points where ϕ_S is discontinuous are exactly the boundary points of S.

We take now a bounded set S and investigate the integrability of the function $\phi_S(x, y)$. The boundedness of S implies that ϕ_S vanishes outside some bounded set. Obviously, for this function $M_{ik}^n = 1$ for all squares R_{ik}^n having points in common with S, and $M_{ik}^n = 0$ for the others. Hence, the upper sum F_n^+ is just the sum $A_n^+(S)$ of the areas of all squares R_{ik}^n that have points in common with S. Thus, for the function ϕ_S the upper integral $F^+ = \lim_{n\to\infty} F_n^+$ is identical with the outer area $A^+(S)$. Similarly, F_n^- equals the total area $A_n^-(S)$ of the squares R_{ik}^n contained in S, so that the lower integral F^- is the inner area $A^-(S)$. Hence, integrability of ϕ_S is equivalent with $A^+(S) = A^-(S)$, that is, with Jordan-measurability of S. When ϕ_S is integrable, the value F of its integral is, of course, the area $A(S)$. We have proved:

The sets S whose characteristic function ϕ_S is integrable are exactly those that have an area. The integral of ϕ_S is the area of S:

$$\iint \phi_S \, dx \, dy = A(S).$$

From continuous functions and characteristic functions of Jordan-measurable sets, we can construct other integrable functions by applying the rule:

The product of two integrable functions is integrable.

Let f and g be integrable, which for us implies that they are bounded and vanish outside some bounded set. Let M_{ik}^n, M'^n_{ik}, M''^n_{ik} denote the supremum and m_{ik}^n, m'^n_{ik}, m''^n_{ik} the infimum of the three functions fg, f, g in the square R_{ik}^n. For any two points (ξ', η'), (ξ'', η''), we have

$$\begin{aligned} &f(\xi', \eta')g(\xi', \eta') - f(\xi'', \eta'')g(\xi'', \eta'') \\ &\quad = f(\xi', \eta')[g(\xi', \eta') - g(\xi'', \eta'')] + g(\xi'', \eta'')[f(\xi', \eta') - f(\xi'', \eta'')]. \end{aligned}$$

Hence, denoting by N an upper bound for $|f|$ and $|g|$:

$$M_{ik}^n - m_{ik}^n \leqq N(M''^n_{ik} - m''^n_{ik}) + N(M'^n_{ik} - m'^n_{ik}).$$

It follows immediately that fg satisfies the integrability condition (20) if it is satisfied by f and by g.

Given a function $f(x, y)$ and a set S in the y, z-plane, we say that f *is integrable over the set S* if the function $f\phi_S$ is integrable in the sense used before; we then define *the integral of f over S* by

$$\iint_S f \, dx \, dy = \iint f\phi_S \, dx \, dy. \tag{22}$$

We have from our product theorem:

An integrable function f is integrable over every Jordan-measurable set S. In particular, every continuous function of compact support is integrable over Jordan-measurable sets.

If f is integrable over the set S, the value of the integral

$$\iint_S f\,dx\,dy$$

does not depend on the values of f at points not in S, since the function $f\phi_S$ is determined by the values of f in the points of S. It is not even necessary to have f defined everywhere. As long as S belongs to the domain of a function f, we can define $f\phi_S$ to be equal to f at the points of S and 0 everywhere else.

For any integrable $f(x, y)$, we can always interpret

$$\iint f\,dx\,dy$$

as

$$\iint_S f\,dx\,dy,$$

where S is some sufficiently large square outside of which f vanishes.

c. Basic Rules for Multiple Integrals

We saw already that the product of two integrable functions f and g is again integrable. Even more trivial is the fact that $f + g$ also is integrable; this follows from the integrability condition (20) and the observation that for any set

$$\sup(f + g) - \inf(f + g) \leq (\sup f - \inf f) + (\sup g - \inf g).$$

The representation of integrals as limits of Riemann sums then shows that

$$\iint (f + g)dx\,dy = \iint f\,dx\,dy + \iint g\,dx\,dy. \tag{23}$$

An estimate analogous to the *mean value theorem of integral calculus* for functions of a single variable is basic for all work with integrals. Let S be a Jordan measurable set and f an integrable function. Let M be an upper bound and m a lower bound for f in S. We can approximate the integral of $f\phi_S$ by Riemann sums

$$F_n = \sum_{i,k} f(\xi_{ik}^n, \eta_{ik}^n)\phi_S(\xi_{ik}^n, \eta_{ik}^n)2^{-2n},$$

where we take care to choose for $(\xi_{ik}^n, \eta_{ik}^n)$ a point of S if the square R_{ik}^n contains such a point. Thus,

$$F_n = \sum f(\xi_{ik}^n, \eta_{ik}^n)2^{-2n}$$

where the sum is extended over all i, k for which R_{ik}^n has points in common with S. Since $m \leqq f \leqq M$ in S, we find that

$$mA_n^+(S) \leqq F_n \leqq MA_n^+(S).$$

For $n \to \infty$ it follows that

$$mA^+(S) \leqq F \leqq MA^+(S);$$

since, by assumption, S has an area, we conclude that the inequality

$$mA(S) \leqq \iint_S f\,dx\,dy \leqq MA(S) \tag{24}$$

holds.

Let S' and S'' be Jordan-measurable sets that do not overlap (that is, interior points of one are exterior to the other); let S be their union and s their intersection. The characteristic functions of these sets satisfy the relation

$$\phi_S + \phi_s = \phi_{S'} + \phi_{S''}.$$

Hence, for any integrable function f we find, on applying (23), the relation

$$\iint f\phi_S\,dx\,dy + \iint f\phi_s\,dx\,dy = \iint f\phi_{S'}\,dx\,dy + \iint f\phi_{S''}\,dx\,dy;$$

that is,

$$\iint_S f\,dx\,dy + \iint_s f\,dx\,dy = \iint_{S'} f\,dx\,dy + \iint_{S''} f\,dx\,dy.$$

Here, by assumption, s contains only boundary points of S' and of S''. Thus, $A(s) = 0$, and, hence, by (24), also

$$\iint_s f\,dx\,dy = 0.$$

This proves the *law of additivity for integrals:*

If the sets S' and S'' have areas and do not overlap and if f is integrable, the relation

$$\iint_{S'\cup S''} f\,dx\,dy = \iint_{S'} f\,dx\,dy + \iint_{S''} f\,dx\,dy \tag{25}$$

holds.

More generally, if S is the union of the Jordan-measurable sets $S_1, \ldots, S_N$, no two of which overlap, and if f is integrable, we have

$$\iint_S f\,dx\,dy = \sum_{i=1}^{N} \iint_{S_i} f\,dx\,dy. \tag{26}$$

This rule opens up the possibility of approximating integrals over a set S by Riemann sums based on much more general subdivisions than the ones we have considered so far. Assume, for simplicity, that S is a closed Jordan-measurable set and f a function continuous in S. A "general subdivision" Σ of S shall mean a representation of S as the union of the Jordan-measurable sets $S_1, \ldots, S_N$, no two of which overlap. In each S_i we pick an arbitrary point (ξ_i, η_i) and form the generalized Riemann sum

$$F_\Sigma = \sum_{i=1}^{N} f(\xi_i, \eta_i)A(S_i). \tag{27}$$

We shall prove that F tends to the integral of f over the set S as the subdivision is refined indefinitely. The continuous function f is uniformly continuous in the bounded closed set S. Given an $\varepsilon > 0$, we can find a $\delta > 0$ such that f varies by less than ε between any two points of S having distance less than δ. Assume that the subdivision Σ is so fine that all the S_i have diameter $< \delta$, that is, that any two points in the same S_i have distance less than δ. Then,

$$f(\xi_i, \eta_i) - \varepsilon \leqq f(\xi, \eta) \leqq f(\xi_i, \eta_i) + \varepsilon$$

for all (ξ, η) in S_i. It follows from (24) that

$$[f(\xi_i, \eta_i) - \varepsilon]A(S_i) \leqq \iint_{S_i} f(\xi, \eta)dx\,dy \leqq [f(\xi_i, \eta_i) + \varepsilon]A(S_i).$$

Hence, by (26), (27), (13),

$$F_\Sigma - \varepsilon A(S) \leqq \iint_S f\,dx\,dy \leqq F_\Sigma + \varepsilon A(S).$$

It follows that the generalized Riemann sums F_Σ differ arbitrarily little from the value of the integral of f over S, for all sufficiently fine subdivisions Σ.

d. Reduction of Multiple Integrals to Repeated Single Integrals

The computation of the value of a triple integral can usually be reduced to the evaluation of single and double integrals—and, similarly, that of double integrals to single integrals and generally that of an integral in n-space to integrals in $(n - 1)$-space—by use of the following theorem:

Let $f(x, y, z)$ be an integrable function defined in x, y, z-space. Assume that for any fixed values of x, y we have in $f(x, y, z)$ a function of the single variable z that is integrable,[1] *and let*

$$\int f(x, y, z)dz = h(x, y). \tag{28}$$

Then $h(x, y)$ as function of x, y is integrable and

$$\iiint f(x, y, z)\, dx\, dy\, dz = \iint h(x, y)\, dx\, dy. \tag{29}$$

For the proof we consider the nth subdivision of x, y, z-space into cubes C^n_{ijk} given by

$$C^n_{ijk}: \frac{i}{2^n} \leqq x \leqq \frac{i+1}{2^n}, \quad \frac{j}{2^n} \leqq y \leqq \frac{j+1}{2^n}, \quad \frac{k}{2^n} \leqq z \leqq \frac{k+1}{2^n}.$$

We form the upper sum for the triple integral of f:

$$F^+_n = \sum_{i,j,k} M^n_{ijk}\, 2^{-3n},$$

where M^n_{ijk} is the supremum of $f(x, y, z)$ in C^n_{ijk}, and, similarly, form the lower sum F^-_n. We now take any fixed point (x, y) in the square R^n_{ij}

$$R^n_{ij}: \frac{i}{2^n} \leqq x \leqq \frac{i+1}{2^n}, \quad \frac{j}{2^n} \leqq y \leqq \frac{j+1}{2^n}.$$

Then M^n_{ijk} is an upper bound for $f(x, y, z)$ as a function of z in the interval

$$I^n_k: \frac{k}{2^n} \leqq z \leqq \frac{k+1}{2^n}.$$

[1]Here, of course, single integrals are taken in the same sense as double integrals; they are defined with the help of the special subdivisions on the line into intervals $i2^{-n} \leqq z \leqq (i + 1)2^{-n}$, taking lower and upper sums, and so on.

It follows from (24) and (26) that for $x, y, \in R_{ij}^n$

$$h(x, y) = \int f(x, y, z)\, dz$$

$$= \sum_k \int_{I_k^n} f(x, y, z)\, dz \leqq \sum_k M_{ijk}^n 2^{-n}.$$ [1]

Denote by H_n^+ and H_n^- the upper and lower sums for the integral of $h(x, y)$ in the nth subdivision. It follows that

$$H_n^+ \leqq \sum_{i,j} (\sum_k M_{ijk}^n\, 2^{-n}) 2^{-2n} = F_n^+,$$

and similarly,

$$H_n^- \geqq F_n^-.$$

Since

$$\lim_{n\to\infty} F_n^+ = \lim_{n\to\infty} F_n^- = \iiint f(x, y, z)\, dx\, dy\, dz,$$

it follows that $h(x, y)$ is integrable and that (29) holds.

Under appropriate assumptions we can further reduce the double integral

$$\iint h(x, y)\, dx\, dy$$

to a repeated single integral

$$\int g(x)\, dx,$$

where for each fixed x the function $g(x)$ is defined by

$$g(x) = \int h(x, y)\, dy$$

To apply this reduction we only have to know that for each fixed x we have in $h(x, y)$ an integrable function of y. This follows, however, from the two-dimensional analogue of formula (29) if we make *the*

[1]Implicit in our assumptions is, of course, that f vanishes outside some bounded region, so that only a finite number of the intervals I_k^n are involved.

additional assumption that $f(x, y, z)$ for any fixed x is an integrable function in the y,z-plane, so that

$$\iint f(x, y, z)\, dx\, dy = \int h(x, y)\, dy = g(x).$$

Hence, we can evaluate the original triple integral by repeated single integrations:

$$\iiint f(x, y, z)\, dx\, dy\, dz = \int\left\{\int\left[\int f(x, y, z)\, dz\right]dy\right\}dx. \tag{30}$$

A simple application, familiar from elementary calculus, is provided by the formula for the reduction of a volume integral over a cylindrical region to a double integral.

Assume that S, a closed set in the x, y-plane, has an area and that $\alpha(x, y)$, $\beta(x, y)$ are continuous functions defined in S with $\alpha(x, y) \leqq \beta(x, y)$. Let C denote the *cylindrical* region

$$C\colon (x, y) \in S, \qquad \alpha(x, y) \leqq z \leqq \beta(x, y).$$

The boundary of C consists of the surfaces $z = \alpha(x,y)$, and $z = \beta(x, y)$, which, by p. 521, have volume 0, and of the points in C for which (x, y) lies on the boundary S_b of S. Since S_b has area 0, this latter set also has volume 0. This shows that C is Jordan-measurable. Now let $f(x, y, z)$ be a continuous function defined in C. Then $f(x, y, z)\phi_C(x, y, z)$ is integrable and

$$\iiint_C f\, dx\, dy\, dz = \iiint f(x, y, z)\phi_C(x, y, z)\, dx\, dy\, dz$$

exists. Now for any fixed $(x, y) \in S$ the expression $f(x, y, z)\phi_C(x, y, z)$ vanishes outside the interval

$$\alpha(x, y) \leqq z \leqq \beta(x, y)$$

(which might shrink to a point) and is continuous in the interval. Hence, $f(x, y, z)\phi_C(x, y, z)$ is integrable and has the integral

$$h(x, y) = \int f(x, y, z)\phi_C(x, y, z)\, dz = \int_{\alpha(x,y)}^{\beta(x,y)} f(x, y, z)\, dz,$$

where we have made use of the ordinary notation for definite integrals over intervals. For $(x, y) \notin S$ we have $f(x, y, z)\phi_C(x, y, z) = 0$ for all z. Hence, for any (x, y)

$$h(x, y) = \phi_S(x, y) \int_{\alpha(x,y)}^{\beta(x,y)} f(x, y, z)\, dx\, dy.$$

Consequently, in this case, the identity (29) yields

$$\iiint_C f(x, y, z)\, dx\, dy\, dz = \iint_S \left[\int_{\alpha(x,y)}^{\beta(x,y)} f(x, y, z)\, dz \right] dx\, dy. \tag{31}$$

A.3 Transformation of Areas and Integrals

a. *Mappings of Sets*

Our aim will be to derive the rule by which a multiple integral is transformed when we change the variables of integration. Such a change of the independent variables x, y in the plane is a *mapping* T of the form

$$\xi = f(x, y), \quad \eta = g(x, y), \tag{32}$$

where f and g are defined in a set Ω, the *domain* of the mapping. (Similar mappings define a change of variable in higher dimensions.) Each point (x, y) in Ω has a unique image (ξ, η). The images form the *range* $\omega = T(\Omega)$ of the mapping T (see p. 242). More generally, for any subset S of Ω we denote by $T(S)$ the set consisting of the images of all the points of S.

For the mappings T considered here, we make the following assumptions:

1. The domain Ω of T is an open bounded set in the x, y-plane.
2. The mapping functions f, g are continuous and have continuous first derivatives: f_x, f_y, g_x, g_y in Ω.
3. The Jacobian Δ of the mapping does not vanish in Ω:

$$\Delta = \frac{d(\xi, \eta)}{d(x, y)} = \begin{vmatrix} f_x & f_y \\ g_x & g_y \end{vmatrix} = f_x g_y - f_y g_x \neq 0. \tag{33}$$

4. The mapping is 1–1; that is, each point (ξ, η) in ω is the image of a *single* point (x, y) of Ω.

Formula (33) has the important consequence (see p. 261) that for every ε-neighborhood N_ε of a point (x_0, y_0) of Ω there exists a δ-neighborhood of the image point (ξ_0, η_0) contained in $T(N_\varepsilon)$. This implies that for any subset S of Ω an interior point of S is mapped into an interior point of $T(S)$. Thus, *open sets S are mapped onto open sets $T(S)$.*[1] In particular, the range ω of our mapping is open.

Condition 4 states that there exists an *inverse mapping* T^{-1}, which associates with every (ξ, η) in ω the unique (x, y) in Ω that is mapped by T onto (ξ, η). The inverse mapping is given by functions

$$x = \alpha(\xi, \eta), \quad y = \beta(\xi, \eta)$$

defined in the open set ω, which are continuous and have continuous first derivatives

$$\alpha_\xi = g_y/\Delta, \quad \alpha_\eta = -f_y/\Delta, \quad \beta_\xi = -g_x/\Delta, \quad \beta_\eta = f_x/\Delta$$

(see p. 261). The Jacobian of the inverse mapping is

$$\frac{d(x, y)}{d(\xi, \eta)} = \begin{vmatrix} \alpha_\xi & \alpha_\eta \\ \beta_\xi & \beta_\eta \end{vmatrix} = \alpha_\xi\beta_\eta - \alpha_\eta\beta_\xi = \frac{1}{\Delta}$$

and, of course, is also different from zero.

Hence, in short, the inverse mapping T^{-1} has all the properties we postulated for T.

In order to arrive at the area of the image of a set S, we first consider a closed square R^n_{ik} contained in Ω and estimate the area of $T(R^n_{ik})$. We assume that we are given an upper bound μ for $|f_x|, |f_y|, |g_x|, |g_y|$ and an upper bound M for $|\Delta|$ in R^n_{ik}. We assume also that we have an upper bound ε for the amount by which any of the quantities f_x, f_y, g_x, g_y varies in R^n_{ik}. Introducing the abbreviations $x_i = i2^{-n}, y_k = k2^{-n}$ for the coordinates of the lower left-hand corner of R^n_{ik}, we can approximate f and g in R^n_{ik} by the *linear* functions

$$f^n_{ik}(x, y) = f(x_i, y_k) + f_x(x_i, y_k)(x - x_i) + f_y(x_i, y_k)(y - y_k)$$
$$g^n_{ik}(x, y) = g(x_i, y_k) + g_x(x_i, y_k)(x - x_i) + g_y(x_i, y_k)(y - y_k).$$

By the mean value theorem of differential calculus (see p. 67), we have for every (x, y) in R^n_{ik}

$$f(x, y) = f(x_i, y_k) + f_x(x', y')(x - x_i) + f_y(x', y')(y - y_k)$$
$$g(x, y) = g(x_i, y_k) + g_x(x'', y'')(x - x_i) + g_y(x'', y'')(y - y_k),$$

where (x', y') and (x'', y'') are suitable intermediate points on the line joining (x, y) and (x_i, y_k). It follows that for any (x, y) in R^n_{ik},

[1]We say that T is an *open mapping.*

$$|f(x, y) - f_{ik}^n(x, y)|$$
$$= |[f_x(x', y') - f_x(x_i, y_k)](x - x_i) + [f_y(x', y') - f_y(x_i, y_k)](y - y_k)| \leqq 2\varepsilon 2^{-n},$$

and similarly,

$$|g(x, y) - g_{ik}^n(x, y)| \leqq 2\varepsilon 2^{-n}.$$

Now, the *linear mapping*

$$\xi = f_{ik}^n(x, y), \quad \eta = g_{ik}^n(x, y) \tag{34}$$

takes the square R_{ik}^n into the parallelogram π_{ik}^n with vertices

$$(f, g), \quad (f + 2^{-n}f_x, g + 2^{-n}g_x), \quad (f + 2^{-n}f_y, g + 2^{-n}g_y),$$
$$(f + 2^{-n}f_x + 2^{-n}f_y, g + 2^{-n}g_x + 2^{-n}g_y),$$

where f, g, f_x, f_y, g_x, g_y are to be taken at the point (x_i, y_k). The area of this parallelogram is the absolute value of the determinant (p.195)

$$\begin{vmatrix} 2^{-n}f_x & 2^{-n}f_y \\ 2^{-n}g_x & 2^{-n}g_y \end{vmatrix} = 2^{-2n}\Delta.$$

The coordinates (ξ, η) of any point of $T(R_{ik}^n)$ differ at most by $2\varepsilon 2^{-n}$ from the corresponding coordinates of a point in π_{ik}^n obtained by the linear mapping. Hence, every point in $T(R_{ik}^n)$ either lies in π_{ik}^n or at a distance at most $2^{3/2}\varepsilon 2^{-n}$ from one of the sides of π_{ik}^n. Each side of π_{ik}^n has length at most $\sqrt{2}\, 2^{-n}\mu$. The set of points lying within the distance $2^{3/2}\varepsilon 2^{-n}$ from one side has an area at most

$$(4\sqrt{2}\, 2^{-n}\varepsilon)(\sqrt{2}\, 2^{-n}\mu) + \pi(2\sqrt{2}\, 2^{-n}\varepsilon)^2 = 8\varepsilon(\pi\varepsilon + \mu)2^{-2n}.$$

Since the area of π_{ik}^n does not exceed $M2^{-2n}$, we find that $T(R_{ik}^n)$ is contained in a set whose area is at most

$$(M + 32\pi\varepsilon^2 + 32\mu\varepsilon)2^{-2n}. \tag{35}$$

Take now any square R_{jr}^N arising in the Nth subdivision contained in Ω. In the closed set R_{jr}^N the quantities $|f_x|, |f_y|, |g_x|, |g_y|$ have a common upper bound μ. Since f_x, g_x, f_y, g_y are uniformly continuous in R_{jr}^N, we can find a finer subdivision into squares R_{ik}^n such that these functions vary by less than ε in each square $R_{ik}^n \subset R_{jr}^N$. If M_{ik}^n denotes the supremum of $|\Delta|$ in R_{ik}^n, we find from (35) that $T(R_{jr}^N)$ is covered by sets of total area at most

$$\sum_{R_{ik}^n \subset R_{jr}^N} (M_{ik}^n + 32\pi\varepsilon^2 + 32\mu\varepsilon)2^{-2n} = F_n^+ + (32\pi\varepsilon^2 + 32\mu\varepsilon)2^{-2N},$$

where F_n^+ is the upper sum corresponding to the nth subdivision for the integral

$$\iint_{R_{jr}^N} |\Delta|\, dx\, dy.$$

For $n \to \infty$ the upper sums F_n^+ tend to the value of the integral, since the function $|\Delta|$ is continuous and, thus, integrable over R_{jr}^N. Since ε is an arbitrary positive number we find [see (8), (10), p. 519,520] that the outer area of the image of the square R_{jr}^N satisfies the inequality

$$A^+[T(R_{jr}^N)] \leqq \iint_{R_{jr}^N} |\Delta|\, dx\, dy, \tag{36}$$

which represents the first step in our computation of the area of image sets.

Now take any Jordan-measurable set S, which together with its boundary ∂S lies in the open set Ω. We can find a closed set $S' \subset \Omega$ and an N such that for $n > N$ any square R_{ik}^n of side 2^{-n} that has points in common with S lies completely in S'.[1]
For $n > N$, let the union of the squares R_{ik}^n having points in common with S be denoted by S_n. The image of S_n is covered by the images of those squares. Hence, (36) yields the estimate for the outer area of $T(S)$

$$A^+[T(S)] \leqq A^+[T(S_n)] \leqq \sum_{R_{ik}^n \subset S_n} A^+[T(R_{ik}^n)]$$

$$\leqq \sum_{R_{ik}^n \subset S_n} \iint_{R_{ik}^n} |\Delta|\, dx\, dy = \iint_{S_n} |\Delta|\, dx\, dy.$$

For $n \to \infty$ the integral of $|\Delta|$ over S_n tends to the integral over S, since $|\Delta|$ is bounded in S' and the total area of the R_{ik}^n that have points in common with S without lying completely in S tends to 0 for the Jordan-measurable set S. Thus, we have proved that

$$A^+[T(S)] \leqq \iint_S |\Delta|\, dx\, dy \tag{37}$$

[1]We only have to choose for S' the union of all R_{jr}^N having points in common with S, where we take N sufficiently large.

for any Jordan-measurable set whose closure lies in Ω.

Under the same assumptions on S, we can also apply (37) to the boundary ∂S of S which is a closed subset of Ω of area 0. Then, by (37),

$$A^+[T(\partial S)] \leqq \iint_{\partial S} |\Delta|\,dx\,dy \leqq (\operatorname*{Max}_{\partial S}|\Delta|)A(\partial S) = 0.$$

Hence $T(\partial S)$ has area 0. Let (ξ, η) be a boundary point of $T(S)$ and consider a sequence of points (ξ_n, η_n) in $T(S)$ with the limit (ξ, η). The (ξ_n, η_n) are images of points (x_n, y_n) in S. A subsequence of the (x_n, y_n) converges to a point (x, y) in the closure of S and, hence, in Ω. The continuity of the mapping T implies that (ξ, η) is the image of (x, y). Here (x, y) cannot be an interior point of S, since then (ξ, η) would have to be an interior point of $T(S)$ and not a boundary point. Hence, (x, y) is a boundary point of S. Thus, the boundary of $T(S)$ consists of images of boundary points of S, and, hence, is a subset of the set $T(\partial S)$ that has been shown to have area 0. Thus, *the boundary of $T(S)$ also has area* 0, and we have proved that $T(S)$ *is Jordan-measurable.* We can then replace $A^+[T(S)]$ in (37) by the area $A[T(S)]$ and find that $A[T(S)]$ *exists and satisfies*

$$A[T(S)] \leqq \iint_S |\Delta|\;dx\,dy = \iint_S \left|\frac{d(\xi, \eta)}{d(x, y)}\right| dx\,dy \tag{38}$$

for any Jordan-measurable set S whose closure lies in Ω.

We saw that the boundary of $T(S)$ is contained in $T(\partial S)$ and, hence, in ω. Thus, $T(S)$ is a Jordan-measurable set whose closure lies in $\omega = T(\Omega)$. Since T and T^{-1} have the same properties we can apply formula (38) to the inverse mapping and find that also

$$A(S) \leqq \iint_{T(S)} \left|\frac{d(x, y)}{d(\xi, \eta)}\right| d\xi\,d\eta = \iint_{T(S)} \left|\frac{1}{\Delta}\right| d\xi\,d\eta. \tag{39}$$

If we apply this last formula to a square R_{ik}^n contained in Ω, we find that

$$2^{-2n} = A(R_{ik}^n) \leqq \iint_{T(R_{ik}^n)} \left|\frac{1}{\Delta}\right| d\xi\,d\eta \leqq \frac{1}{m_{ik}^n} A[T(R_{ik}^n)],$$

where m_{ik}^n is the greatest lower bound of $|\Delta|$ in R_{ik}^n. Thus,

$$A[T(R_{ik}^n)] \geqq m_{ik}^n 2^{-2n}.$$

For any Jordan-measurable set S with closure in Ω, let the union of the $R_{ik}^n \subset S$ be denoted by S_n. Then

$$A[T(S)] \geqq A[T(S_n)] = \sum_{R_{ik}^n \subset S} A[T(R_{ik}^n)] \geqq \sum_{R_{ik}^n \subset S} m_{ik}^n 2^{-2n} = F_n^-,$$

where F_n^- is the lower sum for the integral of $|\Delta|$ over the set S. For $n \to \infty$ we conclude that

$$A[T(S)] \geqq \iint_S |\Delta| \, dx \, dy.$$

Combined with (38) we have thus proved the fundamental fact:

Let S be a Jordan-measurable set whose closure lies in the domain Ω of the mapping T. Then the image $T(S)$ also has an area and this area is given by the formula

$$A[T(S)] = \iint_{T(S)} d\xi \, d\eta = \iint_S \left| \frac{d(\xi, \eta)}{d(x, y)} \right| dx \, dy. \tag{40}$$

b. Transformation of Multiple Integrals

It is easy to pass from formula (40), which represents the law of transformation of areas, to the more general formula for transformation of integrals. We make the same assumptions on the mapping T as before. Now let S be a *closed* Jordan-measurable set contained in Ω and let $F(x, y)$ be a function that is defined and continuous for (x, y) in S. Since the inverse mapping $x = \alpha(\xi, \eta)$, $y = \beta(\xi, \eta)$ is continuous in Ω, the function $F(\alpha(\xi, \eta), \beta(\xi, \eta))$ is defined and continuous in the set $T(S)$. We again denote this function of ξ and η by the letter F. The law of transformation for integrals then takes the form

$$\iint_{T(S)} F \, d\xi \, d\eta = \iint_S F \left| \frac{d(\xi, \eta)}{d(x, y)} \right| dx \, dy. \tag{41}$$

For the proof, we use the representation of integrals of continuous functions by generalized Riemann sums (see p. 530). We consider a general subdivision of S:

$$S = \bigcup_{i=1}^{n} S_i,$$

where the S_i are closed Jordan-measurable subsets of S that do not overlap. The image sets $T(S_i)$ furnish a corresponding subdivision of the set $T(S)$. Since the mapping T is uniformly continuous in the closed set S, the diameters of the image sets $T(S_i)$ tend to 0 when those of the S_i do. Take a subdivision so fine that f varies by less than ε in each S_i. Let (x_i, y_i) be a point in S_i. Then $F(x_i, y_i)$ is also one of the values taken by the function $F(\alpha(\xi, \eta), \beta(\xi, \eta))$ in the set $T(S_i)$. We form the Riemann sum corresponding to the left-hand integral in (41):

$$\begin{aligned}\sum_i F(x_i, y_i)A[T(S_i)] &= \sum_i \iint_{S_i} F(x_i, y_i)\,|\Delta(x, y)|\,dx\,dy\\ &= \sum_i \iint_{S_i} F(x, y)\,|\Delta(x, y)|\,dx\,dy + r\\ &= \iint_S F(x, y)\,|\Delta(x, y)|\,dx\,dy + r,\end{aligned}$$

where

$$\begin{aligned}|r| &= \Big|\sum_i \iint_{S_i} [F(x_i, y_i) - F(x, y)]\,|\Delta(x, y)|\,dx\,dy\Big|\\ &\leqq \varepsilon \sum_i \iint_{S_i} |\Delta(x, y)|\,dx\,dy = \varepsilon A[T(S)].\end{aligned}$$

As the subdivision becomes finer, the Riemann sum tends to the integral of F over the set $T(S)$. For $\varepsilon \to 0$ we obtain the identity (41).

A.4 Note on the Definition of the Area of a Curved Surface

In Section 4.8 (p. 423) we defined the area of a curved surface in a way somewhat dissimilar to that in which we defined the length of arc in Volume I (p. 348). In the definition of length, we started with inscribed polygons, while in the definition of area we used tangent planes instead of inscribed polyhedra.

In order to see why we cannot use *inscribed* polyhedra, we consider that part of the cylinder with the equation $x^2 + y^2 = 1$ in x, y, z-space, which lies between the planes $z = 0$ and $z = 1$. The area of this cylindrical surface is 2π. In it we now inscribe a polyhedral surface, all of whose faces are identical triangles, as follows: We first subdivide the circumference of the unit circle into n equal parts, and on the cylinder we consider the m equidistant horizontal circles $z = 0, z = h, z = 2h, \ldots, z = (m-1)h$, where $h = 1/m$. We subdivide each of these circles into n equal parts in such a way that the points of division of each circle lie above the centers of the arcs of the preceding circle. We now consider a polyhedron inscribed in the cylinder whose edges consist of the chords of the circles and of the lines joining neighboring points of division of neighboring circles. The faces of this polyhedron are congruent isosceles triangles, and if n and m are chosen sufficiently large, this polyhedron will lie as close as we please to the cylindrical surface. If we now keep n fixed, we can choose m so large that each of the triangles is as nearly parallel as we please to the x, y-plane and therefore makes an arbitrarily steep angle with the surface of the cylinder. Then we can no longer expect that the sum of the areas of the triangles will be an approximation to the area of the cylinder. In fact, the bases of the individual triangles have the length $2 \sin \pi/n$, and the altitude, by the Pythagorean theorem, the length

$$\sqrt{\frac{1}{m^2} + \left(1 - \cos\frac{\pi}{n}\right)^2} = \sqrt{\frac{1}{m^2} + 4\sin^4\frac{\pi}{2n}}.$$

Since the number of triangles is obviously $2mn$, the surface area of the polyhedron is

$$F_{n,m} = 2mn \sin\frac{\pi}{n}\sqrt{\frac{1}{m^2} + 4\sin^4\frac{\pi}{2n}} = 2n\sin\frac{\pi}{n}\sqrt{1 + 4m^2\sin^4\frac{\pi}{2n}}.$$

The limit of this expression is not independent of the way in which m and n tend to infinity. If, for example we keep n fixed and let $m \to \infty$, the expression increases beyond all bounds. If, however, we make m and n tend to ∞ together putting $m = n$, the expression tends to 2π. If we put $m = n^2$, we obtain the limit

$$2\pi\sqrt{1 + \pi^4/4},$$

and so on. From the above expression $F_{n,m}$ for the area of the polyhedron we see that the lower limit (lower point of accumulation) of the set of numbers $F_{n,m}$ is 2π, where m tends to infinity with n in any manner whatsoever.[1] This follows at once from $F_{n,m} \geqq 2n \sin \pi/n$ and $\lim_{n\to\infty} 2n \sin \pi/n = 2\pi$.

---

[1]The *lower limit* L of a bounded sequence F_n (denoted by $L = \liminf_{n\to\infty} F_n$) can be defined in several equivalent ways:

a) L is the greatest lower bound of the limits of all convergent subsequences of the F_n.

b) L is the limit for $N \to \infty$ of the greatest lower bounds of the sets obtained from the F_n by omitting the first N terms.

In conclusion we mention—without proof—a theoretically interesting fact of which the example just given is a particular instance. If we have any arbitrary sequence of polyhedra tending to a given surface, we have seen that the areas of the polyhedra need not tend to the area of the surface. But the limit of the areas of the polyhedra (if it exists) or, more generally, any point of accumulation of the values of these areas is always greater than, or at least equal to, the area of the curved surface. If for every sequence of such polyhedral surfaces we find the lower limit of the area, these numbers form a definite set of numbers associated with the curved surface. *The area of the surface can be defined as the greatest lower bound of this set of numbers.*[1]

---

c) L is the *lower point of accumulation* (see Volume I. p. 95) of the F_n that is L is the smallest number with the property that every neighborhood of L contains points F_n for infinitly many n.

d) For every positive ε we have $F_n < L - \varepsilon$ for at most a finite number of n, and $F_n < L + \varepsilon$ for infinitly many n.

The *upper limit* $M = \limsup_{n\to\infty} F_n$ of the sequence F_n is defined analogously. The sequence converges if and only if $L = M$.

[1]This remarkable property of the area is called *semicontinuity* or, more precisely, *lower semicontinuity.*

List of Biographical Dates

Abel, Niels Henrik (1802-1829)
Amsler, Jakob (1823-1912)
Archimedes (287?-212 B.C.)
Bernoulli, Jakob (1654-1705)
Bernoulli, John (1667-1748)
Bessel, Friedrich Wilhelm (1784-1846)
Birkhoff, George David (1884-1944)
Bohr, Harald (1887-1951)
Bolzano, Bernhard (1781-1848)
Borel, Felix Édouard Émile (1871-1956)
Brouwer, Luitzen Egbertus Jan (1881-1966)
Cauchy, Augustin (1789-1857)
Cavalieri, Francesco Bonaventura (1598-1647)
Chebyshev, Pafnuti Lvovich (1821-1894)
Clairaut, Alexis Claude (1713-1765)
Cramer, Gabriel (1704-1752)
Coulomb, Charles Augustin de (1736-1806)
De Moivre, Abraham (1667-1754)
Descartes, (Cartesius) René (1596-1650)
Dirac, Paul Adrien Maurice (1902-)
Dirichlet, Gustav Lejeune (1805-1859)
Du Bois-Reymond, Paul (1831-1889)
Euler, Leonhard (1707-1783)
Fermat, Pierre de (1601-1665)
Fourier, Joseph (1768-1830)
Frenet, Fréderic-Jean (1816-1900)
Fréchet, Maurice René (1878-)
Fresnel, Augustin Jean (1788-1827)
Gauss, Carl Friedrich (1777-1855)
Gram, Jörgen Pederson (1850-1916)
Green, George (1793-1841)
Guldin, Paul (1577-1643)
Hamilton, Sir William Roan (1805-1865)
Heine, Heinrich Eduard (1821-1881)
Helmholtz, Hermann Ludwig Ferdinand von (1821-1894)
Hermite, Charles (1822-1901)
Heron (of Alexandria) (third century A.D.)
Hölder, Otto (1860-1937)
Holditch, Hamnet (1800-1867)

Huygens, Christian (1629-1695)
Jacobi, Carl Gustav Jacob (1804-1851)
Kepler, Johannes (1571-1630)
Lagrange, Joseph Louis (1736-1813)
Laplace, Pierre Simon (1749-1827)
Lebesgue, Henri (1875-1941)
Legendre, Adrien-Marie (1752-1833)
Leibnitz, Gottfried Wilhelm von (1646-1716)
Lipschitz, Rudolf Otto (1832-1903)
Lissajous, Jules Antoine (1822-1880)
Maxwell, James Clerk (1831-1879)
Möbius, August Ferdinand (1790-1868)
Mollerup, Peter Johannes (1872-1937)
Morera, Giacinto (1856-1909)
Morse, Marston Harold (1892-)
Newton, Isaac (1642-1727)
Parseval-Deschênes, Marc Antoine (B? -1836)
Plateau, Joseph Antoine Ferdinand (1801-1883)
Poincaré, Henri (1854-1912)
Poisson, Siméon Denis (1781-1840)
Riccati, Jacopo Francesco (1676-1754)
Riemann, Bernhard (1826-1866)
Schuler, Maximilian Joseph Johannes Eduard (1882-)
Schwarz, Hermann Amandus (1843-1921)
Steiner, Jacob (1796-1863)
Stokes, George Gabriel (1819-1903)
Taylor, Brook (1685-1731)
Wallis, John (1616-1703)
Weierstrass, Karl (1815-1897)
Wronski, (Hoene), Jozef Maria (1778-1853)

Index

Abel's integral equation, 512
Absolute value, 769
Absolutely convergent, 771
Acceleration, normal-, 214
  tangential-, 214
  -vector, 214
Active interpretation of transformation, 148
Additivity for, -areas, 372
  -integrals, 93
  -masses, 387
Admissibility for variational problem, 740
Affine, -coordinates, 144
  -mapping, 148, 242
  -transformation, 179, 276
Algebraic functions, 13, 229
Alternating, -differential, forms, 307, 324
  -functions, 167, 170, 175
Amplitude of complex number, 769
Analytic, -extension, 814–818
  -function, 780, 791
Anchor ring, 285
Angle, -between curves, 234
  -between curves on surface, 285
  -between directions, 127–131
  -between surfaces, 239
  solid-, 619, 720
Angular magnitude, 721
Anticommutative law of multiplication, 181
Apparent magnitude, 721
Approximation, linear-, 50
  polynomial-, 64
  successive-, 267
  Weierstrass theorem on, 81
Arc tangent, power series, 777
  principal branch, 12
Archimedes'-principle, 52, 607
Area, 367–374, 515
  additivity for-, 372, 522
  basic properties, 519–523
  -derivative, 566
  inner-, 369, 517
  -law, 667
    of curved surface, 424, 428, 540
  -of hypersurface, 453, 460
  -of n-dimensional sphere, 455–458
    of polygon, 203
  -of spherical surface, 426
    outer-, 369, 517, 520
  -swept out by moving curves, 448–453
  -vector, 621
Argument of complex number, 769
Associative law, 132, 152
Astroid, 298
Averaging of function, 82

Ball, 9
Base of vectors, 143
Beam, loaded, 675–678
Bernoulli's, -differential equation, 683, 690
  -numbers, 802
Bessel function, 475
Beta function, 508–511
Binomial, coefficients, 510
  series, 801–802
Binormal vector, 216
Bohr-Mollerup theorem, 499
Bolzano-Weierstrass principle of the point of accumulation, 107
Boundary, -of oriented region, 580
  -of set, 6, 8, 10
  -value problem, 719, 724
Bounded sequence, 2
Brachistochrone problem, 737, 751, 756
Buoyancy, 607

Cable, loaded, 672–675
Calculus, -of errors, 52–53
  of variations, 737

Cardiod, 302
Cartesian, coordinate system, 127, 146, 156
  product of sets, 117
Catenary, 751, 768
Catenoid, 287
Cauchy-Riemann equations, 58, 288, 780, 786
Cauchy-Schwarz inequality, 129, 182, 343
  for integrals, 501
Cauchy's, -convergence test, 3, 108
  -formula, 799
  -symbol, 28
  -theorem, 789, 803
Caustic, 302
Cell, 10
Center of mass, 432
Centroid, 432
Chain rule of differentiation, 55
Characteristic function of set, 526
Circle of convergence, 773
Circular disk, 5, 6
Circulation, 572, 615
Clairaut equation, 296, 708
Closed, -set, 8
  -differential form, 314
Closure of set, 9, 10, 11, 118
Columns of matrix, 147
Commutative law, 132
Compact, -set, 86, 109
  -support, 492
Comparison test, 772
Complement of a set, 116, 118, 119
Complementary minor, 189
Components, -of set, 102
  -of vector, 122, 131, 143
Compound, -functions, 53–55, 62–63
  pendulum, 436–438
Cone, 59
Confocal, -conics, 256
  -parabolas, 234, 701
  -quadrics, 287
Conformal transformation, 256, 288, 785, 786
Conjugate, -functions, 803, 805
  number, 767, 777
Connected, -region, 102
  simply-, 103
  -surface, 579
Connectivity, 358
Conservation, -of energy, 656–658, 759
  of mass, 567, 571, 603
Conservative field, 616, 657
Constraint, 340
Content, 369, 515–517
Continuity, -and partial derivatives, 34
  -equation, 571, 603
  modulus of-, 67
  -of integral with respect to a parameter, 74, 464
  uniform-, 112
Continuous, -deformation, 103
  -function, 17–22, 112–113
Continuously differentiable, 42
Contour integration, 807–814
Convergence, absolute-, 771
  Cauchy's intrinsic test for-, 3
  circle of-, 773
  -of improper integrals, 411
  of sequence, 2
  radius of-, 773, 802
  uniform-, 771
Convex, set, 102, 103
  functions, 499–500
  hull, 739
Coordinate(s), affine-, 144
  Cartesian-, 127, 146, 156
  -curves, 247
  curvilinear-, 246, 251
  cylindrical-, 250
  focal-, 256, 257
  general-, 249
  -lines on surface, 282
  -net, 243, 247
  parabolic-, 248
  polar-, 248
  right-handed-, 184
  spherical-, 249
  -surfaces, 250
  -transformation of, 246
  -vector, 129, 133, 143
Cosines, law of, 71, 127
Coulomb's law, 445, 714
Cramer's rule, 163, 177
Critical points, 326, 352
Cross product of vectors, 181, 182
Curl of a vector, 209, 313
Curvature, center of-, 213, 214, 232
  -of curve, 213, 230, 232
  radius of-, 213, 232
  -vector, 213

Curve(s), coordinate-, 247
curvature of-, 213, 230, 232
discriminant-, 293
double points of-, 360
envelope of-, 293
evolute of-, 301
family of-, 291–302
-in implicit form, 230–237
isolated point of-, 361
length of-, 283
multiple point of-, 236
normal of-, 231
parallel-, 365
pedal-, 303
polygonal-, 112
sectionally smooth-, 88
singular point of-, 236, 360
space-, 282
tangent of-, 212, 231
tangential representation of-, 365
torsion of-, 216
Curvilinear coordinates, 246–251
Cusp, 299, 361
Cut-off function, 494

Deformation, 244
Degenerate transformation, 274
Degree, -of freedom, 757
-of mapping, 562
-of polynomial, 13, 119
Density, 386, 566
Dependent, -functions, 272, 273, 684
linearly-, 137, 684
-variables, 11
vectors, 137
Derivative, -at boundary points, 27
directional-, 43, 45, 206
exterior-, 312
Fréchet-, 268
normal-, 557
-of an implicit function, 223
-of function of complex variable, 779
-of mapping, 268
-of vector, 212
partial-, 27
radial-, 45, 62
Determinants, 160–202
definition of-, 166–170
expansion of-, 170, 187
functional-, 253
geometrical interpretation of-, 180–187
Gram-, 193
Jacobian-, 253
n^{th} order-, 171
-of matrix, 170
matrix, 175
of product, 172
second order-, 161
third order-, 161
Diagonal, -rule, 162
-matrix, 177
Diameter of set, 376, 523
Difference, of function, 66
of points, 125
Differentiability, 40–42
complex variable, 779
Differential, exact-, 314
-of function, 49–51
-of higher order, 50
-operator, 209, 684
total-, 49, 50, 314, 322
Differential equations, 654–734
constant of integration for-, 699
existence and uniqueness of solution of-, 702–706
fundamental theorem on linear-, 687
homogeneous-, 688
integral curves of-, 697
integration of-, 656
linear-, 680, 696
non-homogeneous-, 691
-of family of curves, 699–702
-of first order, 678–682
-of higher order, 683–690
-of second order, 688
ordinary-, 654–712
partial-, 713–735
-systems of, 709–710
-with constant coefficients, 696, 699, 812–814
Differential form, alternating-, 307–324
closed-, 314
exterior-, 316
integral of-, 589–601, 647–653
linear-, 84
non-alternating-, 308
quadratic-, 283
Differentiation area-, 565
change of order of-, 36–39
-for inverse functions, 252

-to fractional order, 511–512
under the integral sign, 74–80, 466–468
Dipole, 717
Dirac function, 674
Direction, -cosines, 129
-numbers, 130
Directional derivative, 44
Dirichlet's discontinuous factor, 479
Disconnected, 102
Discontinuous, 18
Discriminant, 304, 347
Disjoint sets, 116
Disk, 5, 6
Distance, -from hyperplane, 135
-from surface, 343
-of points, 127, 146
Distributive law, 132, 152, 165
Div, 208
Divergence, -of a vector, 208–210
theorem, 549, 554, 637–642, 651
Domain of a function, 11, 12
Double, -integral, 80, 374–386
-integral over oriented region, 589–592
-layer, 717, 719, 720
Doublet, 717

Element of matrix, 147
-of area, 425, 628
Elementary surface, 624–627, 645–647
Ellipsoid, 240
greatest axis of-, 345
moment of inertia of-, 443
momental-, 443
volume of, 417, 462
Elliptic integral, 78
Energy, conservation of-, 656, 657, 759
kinetic-, 656, 758
potential-, 657
Envelopes, 292–295, 303–306, 735
Epicycloid, 302
ϵ-neighborhood, 1, 9
Equilibrium, 659–663
Equipotential surfaces, 715
Errors, 52–53
Eulerian integrals, 497–511
Euler's, -Beta function, 508
-constant, 505
-differential equation, 743, 748, 755, 761, 766
-partial differential equation for homogeneous functions, 120, 761
-representations of motion, 363
Even permutation, 170
Evolute, 301–302
Exp, 457
Exact differential form, 84
Exponential function, 782–785, 792, 793
Extension of function, 20
Exterior, -content, 517
differential forms, 312–313, 321–324
-Jordan measure, 517
-normal, 580, 633
-point, 7, 9, 118
Extremals, 755
Extreme values, 325, 326, 333, 334, 336, 345

Families, of curves, 290, 291
of surfaces, 291
Fermat's principle of least time, 740
Field, direction-, 697
gradient-, 352
vector-, 204
Final point of vector, 125
Fixed point of mapping, 270, 359, 787
Fluid flow, 602–605
Flux, 597, 732
Focal coordinates, 256, 611
Folium of Descartes, 224, 238
Force, electric-, 733
field of-, 204
flux of-, 597
gravitational-, 207, 655
magnetic-, 733
surface-, 606
Form(s), 13, 83, 84
alternating-, 168, 169, 175
bilinear-, 164, 165, 167, 168, 179
differential-, 84, 283, 307–324
linear-, 83, 163, 164
multilinear-, 166, 169, 175
quadratic-, 165, 347
trilinear-, 165, 168
Fourier, -integral, 476–496
-integral theorem, 477, 481, 485, 491
-transform, 478, 491
Fréchet derivative, 268
Free surface, 606
Freely falling particle, 658
Frenet's formulae, 216

Fresnel's integrals, 473
Function(s), 11, 19
algebraic-, 13, 229
alternating-, 167–170
analytic-, 780, 791
characteristic-, 526
compound-, 54, 55, 62
continuous-, 17, 18, 19, 20, 112
conjugate-, 803, 805
convex-, 499
cut-off-, 494
dependent-, 273–275, 684
differentiable-, 41, 42, 45
domain of-, 11, 12, 16, 17
extreme values of-, 333
geometric representation of-, 13–15
harmonic-, 719
Hölder-continuous-, 19
implicit-, 218–230
independent-, 274
inverse-, 252
limit of-, 19
Lipschitz-continuous-, 19
many valued-, 814
-of class C^n, 42
-of compact support, 492
-of functions, 53
potential-, 719, 803, 805
rational-, 18
rational integral-, 12
support-, 365
transcendental-, 229
uniformly continuous-, 18
variation of-, 742
Functional, 740
Functional equation of gamma function, 498
Fundamental quantities of surface, 283
Fundamental system of solutions, 688
Fundamental theorem, -of algebra, 806
-on integrability of linear differential forms, 95, 104, 616
-on linear dependence, 138, 158

Gamma function, 497–508, 818
Gauss, divergence theorem, 544, 597–610, 637–642, 651
-infinite product, 506
Gaussian fundamental quantities of surface, 283
Geodesics, 739, 757, 765
Geometric series, 771
Global, 222
Grad, 206
Gradient, -field, 352
-vector, 206, 207, 210, 231
Gram determinant, 193, 194
Gravitational, -constant, 207, 655
-field of force, 207, 655
-potential, 439
-vector field, 622
Green's, 543
-integral theorems, 556–558, 607–608
Guldin's rule, 429, 452

Half-spaces, 135
Hamilton's principle, 757, 758
Heine-Borel covering theorem, 109–110, 119
Helix, 92, 767
Hemisphere, 14, 279
Hermite polynomials, 71
Heron's formula, 341
Higher order of vanishing, 22
Hölder, -condition, 19
-continuous, 19
-inequality, 343
Holomorphic, 780
Homogeneous, -differential equations, 684, 688
-fluid, 604
-functions, 119–121, 124
-linear system of equations, 138–140
-medium, 571
-polynomials, 13, 119
positively-, 120
Homotopic, 103
Huyghens' theorem, 435
Hyperbolic paraboloid, 14
Hyperboloid, 280, 287
Hyperplanes, 133–135, 201
Hypersurface, 453, 460

Identities, 252
Identity, mapping, 126, 153
transformation, 63
Imaginary part, 769
Implicit, -function theorem, 221, 228, 265
-functions, 218–230, 261, 265
-representation, 231, 238

Improper integrals, 407–416, 462–468
  differentiation of-, 467
  integration of-, 467
Inclination, 249, 353
Incompressible fluid, 571, 604, 617
Increment, 83
Indefinite quadratic form, 346
Independent, 139
  -functions, 274
  -variables, 11, 60
  -vectors, 137
Index of closed curve, 352, 355
Inflection point, 231, 232
Initial point of vectors, 125
Inner area, 517
Integrability conditions for differential, 84, 98, 314
Integrability of continuous functions, 526
Integrable, 407, 525–528
Integral(s), -curves, 699
  double-, 374–385
  -estimates, 383–385
  Eulerian-, 497
  Fourier-, 476
  Fresnel's, 473
  -identities in higher dimensions, 622
  improper-, 406–416, 462–468
  law of additivity for-, 383, 529
  Lebesgue-, 407
  line-, 82–106
  multiple-, 367, 388, 531
  -of analytic function, 788
  -of continuous functions, 526
  -of differential forms, 589–597, 634, 647–653
  -of functions of several variables, 524–525
  -over an elementary surface, 627
  -over regions in more dimensions, 385
  -over sets, 526
  -over simple surfaces, 594–597
  over unbounded regions, 414–416
  reduction of double-, 392
  repeated-, 78
  Riemann-, 89, 407
  transformation of multiple-, 539, 562
Integration, 78, 80, 515, 656
  -constant, 699
  -of analytic functions, 787–789
  -of rational functions, 809
  -of total differentials, 95
  -to fractional order, 511
Interchange of, -differentiations, 36–39
  -integrations, 80
Interior, -content, 517
  -normal, 580
  -of set, 8
  -points, 6, 7, 8, 9, 118
Interval, 10
Intrinsic convergence test, 3
Invariant, 317
Inverse, -functions, 252, 786
  -image, 242
  -mapping, 154, 242, 266
  -transformation, 261
Inversion, 243, 244, 256, 277, 787
Irrotational motion, 572, 616
Isoperimetric, -inequality, 365–366
  problem, 739, 767
  -subsidiary conditions, 765
Iteration, 267, 703

Jacobian, -determinant, 253, 254
  -matrix, 268, 272
  -of product of two transformations, 258, 276
Jordan, -measure, 367–370, 515, 517
  -measurable set, 517, 628

Kepler's, -equation, 671
  -laws, 665, 667, 669, 671
Kinetic energy, 656, 758
  -of rotating body, 435

Lagrange's, -equations, 759
  -multiplier, 332, 762–768
  -representation of motion, 363
Laplace, -equation, 58, 62, 573, 617, 713, 724, 762
  -operator, 211, 608
  -operator in polar coorindates, 62
  -operator in spherical coordinates, 610
Laplacian, 62, 211
Latitude, 249
Lebesgue, -area, 371
  -integral, 407
  -measure, 515
Left-handed screws, 185
Legendre's condition, 747, 768
Lemniscate, 223, 236, 238

Length, -of arc on surface, 283
-of vector, 146, 157
Level line, 14, 207, 233
Limit, 9, 19, 21
-for complex variable, 770, 774
of function, 19, 21
-of sequence, 2, 9, 21
Line, contour-, 14, 233
element, 283
level-, 14, 207, 233
parametric representation of-, 131
vector representation for-, 130
Line integrals, 85–91
additivity of-, 93
-independent of the path, 96, 104
Linear, -approximation, 50
-dependence, 137, 684
-equations, 137, 138, 175–177
-homogeneous function, 124
-differential form, 84, 93, 95
manifolds, 134, 144–146
mappings, 150
operations, 123
transformations, 202, 778
Lines of force, 597
Lipschitz, -condition, 19
-constant, 19
-continuous, 19, 35, 67
Lissajous figures, 665
Local, 222
Logarithm, 792–794
Longitude, 249
Lower, integral, 525
-limit, 541
-point of accumulation, 542

Main diagonal of matrix, 157
Manifold, 317, 543
abstract-, 653
linear-, 134, 144–146, 195, 198–200
vector-, 204
Mapping(s), 11, 242
affine-, 148, 242
-by reciprocal radii, 243
degree of-, 561–565
fixed point of-, 270, 359, 787
identity-, 126, 153
inverse-, 242, 266
linear-, 150
-of directions, 259
-of sets, 11, 534
-of vectors, 148
open-, 535
primitive-, 264
resultant-, 257
symbolic product of -, 152, 257
Mass, center of-, 432
conservation of-, 571, 603
moment of-, 431
total-, 387
Matrices, 147
addition of-, 151
columns of-, 147
determinants of-, 170
diagonal-, 177
elements of-, 147
Jacobian-, 268, 272
main diagonal of-, 151
minor of-, 189
multiplication of-, 151
nonsingular-, 150, 155, 175
operations with-, 150, 153
orthogonal-, 156, 175
product of-, 151–153, 172
reciprocal-, 153, 154, 155
rectangular-, 150, 153
rows of-, 147
singular-, 150, 155, 175
square-, 150, 153
transpose-, 157, 173
unit-, 153, 154, 177
upper triangular-, 178
zero-, 153
Maximum, absolute-, 325
-of continuous function, 112
relative-, 325, 347, 349
strict-, 325
value-, 327
-with subsidiary conditions, 330–334
Maxwell's equations, 731–734
Mean, arithmetic-, 341
-density, 387
geometric-, 341
Mean value theorem, -for functions, 67
-for potential functions, 722
Minimal surfaces, 762
Minimum, -of continuous function, 112
relative-, 325, 347–349
strict-, 325
-with subsidiary conditions, 330–334

Minor of a matrix, 189
Möbius band, 582, 589
Modulus, -of complex number, 769
of continuity, 18, 19, 67
-of elasticity, 675
Moment, -of dipole, 717
-of inertia, 433–435
-of inertia of ellipsoid, 443
-of mass distribution, 431–432
of momentum, 666
-of velocity, 666
Momental ellipsoid, 443
Momentum, 602, 655
Monomial, 13
Morera's theorem, 803
Motion, equations of-, 654–656
planetary-, 665–671
Multiplier, 334–340, 762–768

N-dimensional, ball, 459
-Euclidean space RN, 10, 124
sphere, 455
-surface, 645, 648
-vector space, 143
Negative definite quadratic form, 346
Neighborhood, 1, 9
Newton's, -law of attraction, 204, 665
-second law, 654
Non-homogeneous differential equation, 684
Non-overlapping sets, 368
Non-singular matrix, 150, 155, 175
Non-trivial solution, 138, 140
Normal, -acceleration, 214
-derivative, 557
-distance, 448
exterior-, 580
hyperplane, 135
outward-drawn-, 599
positive-, 593
-to curve, 230–231
-to hyperplane, 134–135
-to surface, 238, 283, 284
-velocity, 448

Odd permutation, 170
One sided surface, 582
Open, -mapping, 535
-set, 8
Orders of magnitude, 22
Orientability, 583
Orientation, continuously varying-, 578, 586
-of curves on surfaces, 587
-of hyperplanes, 200, 201
of parallel-epiped, 186, 195, 198, 199
-of parallel-ogram, 180
-of planes, 200, 201
opposite-, 86, 185, 196
standard-, 196
-transformed, 260
Oriented, area, 91
-boundary, 580
-hyperplanes, 201
-linear manifold, 200
-parallellepiped, 194, 195
-simple closed curve, 86, 91
-surface, 578, 580, 629, 633
-tangent plane, 577
Orthogonal, -curves, 234
-matrices, 156, 158, 175
-trajectories, 701, 707
-transformations, 157
-vectors, 133
Orthogonality relations, 145, 146
Orthonormal, -base, 145
-system of vectors, 145, 156, 158
Oscillations, 661–665
Osculating plane, 215
Outer area, 517, 520
Overlapping, 368

Parabolas, coaxial-, 244
confocal-, 234, 244, 248
Parabolic coordinates, 248
Paraboloid, hyperbolic-, 14
-of revolution, 14
Parallel curves, 365
Parallel displacements, 124
Parallelepiped, orientation of, 186, 195, 198, 199
rectangular-, 10, 12
-spanned by vectors, 186, 191
volume of-, 187, 191, 193, 194, 195, 197
Parallelogram, area of-, 182, 184, 190, 191
orientation of-, 180
Parametric representation, -of arc, 86
-of line, 131
-of surface, 278, 576
Parseval's identity for Fourier transforms, 488, 496
Partial, 27, 29, 34

-derivative, 26–30
-differential equation, 713–736
-sums, 771
Partition of unity, 635, 636
Passive interpretation of transformation, 148
Paths, 102
family of-, 103, 105
homotopic-, 103
-of rays of light, 740
support of-, 111
Pathwise simply connected, 102
Pendulum, 436–438
Permutation, 170
even-, 170
odd-, 170
Perpendicular, -distance, 192
-vectors, 133
Plane, osculating-, 215, 216
perpendicular distance from-, 192
tangent-, 239
-waves, 490, 729
Planetary motion, 665–671
Planimeter, 453
Plateau's problem, 762
Poincaré, -identity, 358
-index, 353
-lemma, 313
Point, boundary-, 6, 7
critical-, 326, 352
double-, 360
exterior-, 6, 7, 8, 118
fixed, 787
-in n-dimensional space, 10
interior-, 6, 7, 8, 118
isolated-, 361
-of inflection, 231, 232
rational-, 370
saddle-, 327, 347
sequences of-, 2
singular-, 360, 362
stationary-, 326
Poisson's integral formula, 724–726
Polar, -coordinates, 61
-planimeter, 453
-reciprocal, 303
Pole of analytic function, 805
Polygonal curve, 112
Polygonally connected, 68
Polynomial(s), 13, 18
Hermite-, 71
Taylor-, 64
trigonometric-, 124
Position vector, 126
Positive, -definite quadratic form, 346
-normal of surface, 579, 593
-side of oriented surface, 579
-side of plane, 201
Postiively homogeneous, 120
Potential, -due to a spherical surface, 441, 716
-energy, 439, 657, 758
-equation, 62, 211, 718–726
-functions, 719, 722, 802, 805
-of attracting charges, 714
-of ellipsoid of revolution, 444
-of forces, 657, 661
-of solid sphere, 716
-of straight line, 716–719
-of uniform double layer, 720
Power series, 772–777, 799–802
Pressure, 605
Primitive, -mappings, 264
-nth root, 11, 821
-transformation, 264
Principal, -branch of arc tangent, 12
-normal, 213, 265
-value of logarithm, 794–802
Product, cross-, 181
of differential forms, 311–312, 321
-of mappings, 257
-of matrices, 152
scalar-, 131–133
symbolic-, 152, 257
vector-, 181, 182, 187

Quadratures, 679
Quadratic form, discriminant of-, 347
indefinite-, 346
negative definite-, 346
positive definite-, 346
Quadratic, 179

Radius of convergence, 773, 802
Rational, -functions, 809
-integral function, 12
-points, 370
Reaction forces, 215, 659
Real part, 769
Reciprocal matrix, 153, 154, 155
Reflection with respect to unit circle, 243

Region, connected-, 4, 102
rectangular-, 7, 10
simply connected-, 4, 102–104
Relative, -boundary, 648
-closure, 648
-error, 53
-extremum, 326, 349
-maximum, 325, 347–349
-minimum, 325, 347–349
Relatively open, 648
Remainder in Taylor expansion, 69
Repeated integration, 78
Residue, -at point, 805
-theorem, 805
Restriction of function, 12
Resultant, -mapping, 257
-transformation, 257
Riccati's differential equation, 690, 691
Riemann, -integrable, 407, 525
-integral, 89, 407
-sum, 89, 525, 530
-zeta function, 797, 820
Riemann-Lebesgue lemma, 481
Right handed screws, 185
Rigid motions, 157, 202
Rolle's theorem, 352
Rotation, clockwise-, 200
counterclockwise-, 200
-of axes, 61, 202
sense of-, 200
Rows of matrix, 147

Saddle point, 347
Saddle-shaped, 15
Sag, -of beam, 675
-of cable, 672
Scalar, 123, 205, 318
gradient of a-, 205–208, 210
-multiplication of matrices, 151
-products of vectors, 131–133, 157
Sectionally smooth, 5, 88
Semi-continuity, 542
Sense, -of curves, 357
-of rotation, 200
of vectors, 185
Sequence, bounded-, 2
convergence of-, 2
limit of-, 2, 9, 21
lower limit of-, 541
-of complex numbers, 770
-of points, 2
Sequentially compact, 109
Separation of variables, 678
Series, 770
Set, boundary of-, 10, 118
closed-, 8, 109
closure of-, 10, 118
compact-, 109
complement of-, 116, 118, 119
connected-, 102
diameter of-, 376, 523
disjoint-, 116
empty-, 114
null-, 114
open, 8, 109
simply connected-, 102, 103
Sets, Cartesian product of-, 115
disjoint-, 116
family of-, 113
intersection of-, 115–117
Jordan-measurable-, 517
non-overlapping-, 368
Shell, spherical, 580
Shortest line joining two points, 764
Simple, -arc, 86
-surface, 631–634, 648
Simplex, 462
Simply connected sets, 102–103
Singular, -matrix, 150, 155, 175
-points of curves, 236, 360–362
surfaces, 362–363
-solutions, 701
Singularity of analytic function, 804
Sink, 574
Slope of surface, 27
Smoothing of function, 81
Solid angle, 619
Solutions, nontrivial-, 138
trivial-, 138, 140
-system of fundamental, 687, 688
Solvability of system of linear equations, 150
Source of mass, 574
Space differentiation, 387
Spanned by vectors, 144
Speed of propagation, 491
Spherical, -coordinates, 404
-law of cosines, 71
-pendulum, 663
-shell, 580

Square matrices, 150
Stability of equilibrium, 653–659
Statics, principles of-, 618
Stationary, -character, 737
-point, 345, 351, 742
-values, 331, 349, 754
Steady flow, 573
Stereographic projection, 280, 290
Stokes', -integral theorem, 554, 555, 572, 611–617, 642, 643
-formula in higher dimensions, 624, 651–653
Straight line, parametric representation of-, 131
vector representation of-, 131
String, plucked-, 735
vibrations of-, 727
Strophoids, 300
Subadditivity of outer areas, 520
Subset, 114
Subsidiary conditions, 330–336, 762–767
Successive approximation, 266, 703
Sum(s), lower-, 376, 524
-of vectors, 125
Riemann-, 89, 525, 530
upper-, 376, 524
Superposition, principle of-, 683–684
Support, compact-, 492
-function, 365
-of path, 111
Surface, -areas in any number of dimensions, 453–455
area of-, 424, 428
area of spherical-, 426, 458
connected-, 579
coordinate lines on-, 282
elementary-, 624–625, 632, 645–647
equipotential-, 715
-forces, 606
free-, 606
geodesics on-, 739, 757, 765
implicit representation of-, 238–240
in parametric representation, 278, 576
-integrals, 624, 645–653, 594–597
isobaric-, 606
m-dimensional-, 645, 648
minimal-, 762
-normal, 239, 283, 284
of revolution, 50, 429
one sided-, 582
orientation of-, 575–588
oriented-, 578, 580, 629, 633
simple-, 631–634, 648
tangent plane to-, 282
Symbolic product, -of mappings, 125, 152, 257
-of operators, 29
System, -of functions, 241
-of linear equations, 137, 138, 175–177
-of mappings, 241
-of transformations, 241
orthonormal-, 145, 156, 158

Tangent, -line, 231
-plane, 47, 239, 282
Tangential representation of curve, 365
Taylor's, -expansion, 65, 64–66
-series, 68–70, 776, 801
-theorem, 68–70
Tetrahedron, 141, 142
Torus, 102, 285, 286, 589
Total differentials, integration of-, 95–98
-of functions, 49–51, 97, 104
Transcendental functions, 229
Transformations, affine-, 179, 276
conformal-, 256, 288, 785
degenerate-, 274
inversion of, 261
-of coordinates, 246
primitive-, 264
product of two-, 257
resultant-, 257
Translations, 124
Transpose of matrix, 157
Trigonometric polynomial, 124
Triangle inequality, 769, 770
Trivial solution, 138, 140
Tube surface, 306
Twisted curve, 282

Undetermined, -coefficients, 711, 712
-multipliers, 334–340, 762–768
Uniform, -convergence, 464–771
-approximations, 81
Uniformly continuous, 18, 112
Unit matrix, 153, 154, 177
Unstable equilibrium, 663
Upper integral, 525
Upper-triangular matrix, 178

Variation, first-, 741–743
-of function, 742, 754
-of parameters, 681, 691–694
Vectors, acceleration-, 214
as differences of points, 125
base of-, 143
binormal-, 216
component of-, 122, 131
coordinate-, 123, 129, 133, 143
cross product of-, 180, 181, 182
curl of-, 209, 313
curvature-, 213
definitions of-, 122, 123
divergence of-, 208, 210
electric-, 731
families of-, 211, 212
fields of-, 204, 208, 211
geometric representation of-, 124–127
gradient-, 206, 207, 210, 231
inclination of-, 353
length of-, 127, 146, 157
linear dependence of-, 136, 141
linear forms of-, 163
magnetic-, 731
-manifold, 204
mapping of-, 148, 153
multilinear forms of-, 163–170
opposite-, 126
orthogonal-, 133
orthonormal-, 145, 156
perpendicular-, 133
position-, 126, 127, 212
principal normal-, 213
-product, 180, 188
-representation for lines, 130
scalar products of-, 131–133, 146, 157
spaces of-, 123, 142, 143
spanned by-, 144, 182
sum of-, 122, 125
triple product of-, 181
unit-, 130
vector product of-, 181, 182, 187, 188, 311
Zero-, 123, 129
Velocity, -of light, 741
-potential, 617
-vector, 214
Vibrations, -forced, 695
-of a string, 727
Volume, 146, 374, 419
-in any number of dimensions, 453
-of ellipsoid, 417, 418, 462
of n-dimensional ball, 459
-of parallelepipeds, 190–195, 201, 202
-of pyramid, 418
-of region bounded by surface, 600
Vortex, 575
Vorticity, 572, 616

Wallis's product, 469
Wave, -equation in one dimension, 727–728
-equation in three dimensions, 728, 729, 733, 735, 736
-fronts, 448, 490, 491
plane-, 490
spherical-, 730
traveling-, 728
Weierstrass', -approximation theorem, 81
-infinite product, 506
-principle of the point of accumulation, 107
Winding number, 100, 564
Work, 616, 657
Wronskian, 686
Wronski's condition, 688

Zero, -matrix, 153
-vector, 123, 129
Zeros, number of-, 806
-of analytic function, 803
Zeta function, 797, 820

Printing: Saladruck, Berlin
Binding: Buchbinderei Lüderitz & Bauer, Berlin

∞ CIM

M. Aigner Combinatorial Theory ISBN 978-3-540-61787-7
A. L. Besse Einstein Manifolds ISBN 978-3-540-74120-6
N. P. Bhatia, G. P. Szegő Stability Theory of Dynamical Systems ISBN 978-3-540-42748-3
J. W. S. Cassels An Introduction to the Geometry of Numbers ISBN 978-3-540-61788-4
R. Courant, F. John Introduction to Calculus and Analysis I ISBN 978-3-540-65058-4
R. Courant, F. John Introduction to Calculus and Analysis II/1 ISBN 978-3-540-66569-4
R. Courant, F. John Introduction to Calculus and Analysis II/2 ISBN 978-3-540-66570-0
P. Dembowski Finite Geometries ISBN 978-3-540-61786-0
A. Dold Lectures on Algebraic Topology ISBN 978-3-540-58660-9
J. L. Doob Classical Potential Theory and Its Probabilistic Counterpart ISBN 978-3-540-41206-9
R. S. Ellis Entropy, Large Deviations, and Statistical Mechanics ISBN 978-3-540-29059-9
H. Federer Geometric Measure Theory ISBN 978-3-540-60656-7
S. Flügge Practical Quantum Mechanics ISBN 978-3-540-65035-5
L. D. Faddeev, L. A. Takhtajan Hamiltonian Methods in the Theory of Solitons ISBN 978-3-540-69843-2
I. I. Gikhman, A. V. Skorokhod The Theory of Stochastic Processes I ISBN 978-3-540-20284-4
I. I. Gikhman, A. V. Skorokhod The Theory of Stochastic Processes II ISBN 978-3-540-20285-1
I. I. Gikhman, A. V. Skorokhod The Theory of Stochastic Processes III ISBN 978-3-540-49940-4
D. Gilbarg, N. S. Trudinger Elliptic Partial Differential Equations of Second Order ISBN 978-3-540-41160-4
H. Grauert, R. Remmert Theory of Stein Spaces ISBN 978-3-540-00373-1
H. Hasse Number Theory ISBN 978-3-540-42749-0
F. Hirzebruch Topological Methods in Algebraic Geometry ISBN 978-3-540-58663-0
L. Hörmander The Analysis of Linear Partial Differential Operators I – Distribution Theory and Fourier Analysis ISBN 978-3-540-00662-6
L. Hörmander The Analysis of Linear Partial Differential Operators II – Differential Operators with Constant Coefficients ISBN 978-3-540-22516-4
L. Hörmander The Analysis of Linear Partial Differential Operators III – Pseudo-Differential Operators ISBN 978-3-540-49937-4
L. Hörmander The Analysis of Linear Partial Differential Operators IV – Fourier Integral Operators ISBN 978-3-642-00117-8
K. Itô, H. P. McKean, Jr. Diffusion Processes and Their Sample Paths ISBN 978-3-540-60629-1
T. Kato Perturbation Theory for Linear Operators ISBN 978-3-540-58661-6
S. Kobayashi Transformation Groups in Differential Geometry ISBN 978-3-540-58659-3
K. Kodaira Complex Manifolds and Deformation of Complex Structures ISBN 978-3-540-22614-7
Th. M. Liggett Interacting Particle Systems ISBN 978-3-540-22617-8
J. Lindenstrauss, L. Tzafriri Classical Banach Spaces I and II ISBN 978-3-540-60628-4
R. C. Lyndon, P. E Schupp Combinatorial Group Theory ISBN 978-3-540-41158-1
S. Mac Lane Homology ISBN 978-3-540-58662-3
C. B. Morrey Jr. Multiple Integrals in the Calculus of Variations ISBN 978-3-540-69915-6
D. Mumford Algebraic Geometry I – Complex Projective Varieties ISBN 978-3-540-58657-9
O. T. O'Meara Introduction to Quadratic Forms ISBN 978-3-540-66564-9
G. Pólya, G. Szegő Problems and Theorems in Analysis I – Series. Integral Calculus. Theory of Functions ISBN 978-3-540-63640-3
G. Pólya, G. Szegő Problems and Theorems in Analysis II – Theory of Functions. Zeros. Polynomials. Determinants. Number Theory. Geometry ISBN 978-3-540-63686-1
W. Rudin Function Theory in the Unit Ball of $\mathbb{C}^n$ ISBN 978-3-540-68272-1
S. Sakai C*-Algebras and W*-Algebras ISBN 978-3-540-63633-5
C. L. Siegel, J. K. Moser Lectures on Celestial Mechanics ISBN 978-3-540-58656-2
T. A. Springer Jordan Algebras and Algebraic Groups ISBN 978-3-540-63632-8
D. W. Stroock, S. R. S. Varadhan Multidimensional Diffusion Processes ISBN 978-3-540-28998-2
R. R. Switzer Algebraic Topology: Homology and Homotopy ISBN 978-3-540-42750-6
A. Weil Basic Number Theory ISBN 978-3-540-58655-5
A. Weil Elliptic Functions According to Eisenstein and Kronecker ISBN 978-3-540-65036-2
K. Yosida Functional Analysis ISBN 978-3-540-58654-8
O. Zariski Algebraic Surfaces ISBN 978-3-540-58658-6